Primes and Knots

CONTEMPORARY MATHEMATICS

416

Primes and Knots

Toshitake Kohno
Masanori Morishita
Editors

American Mathematical Society
Providence, Rhode Island

This volume contains the proceedings of an AMS Special Session on Primes and Knots held January 15–16, 2003, in Baltimore, Maryland, and the 15th JAMI (Japan-U.S. Mathematics Institute) Conference held March 7–16, 2003, at the Johns Hopkins University, Baltimore, Maryland.

2000 *Mathematics Subject Classification.* Primary 11Gxx, 11Mxx, 11Rxx, 57Mxx.

Library of Congress Cataloging-in-Publication Data

Primes and knots : JAMI conference, knots and primes, March 7–16, 2003, Johns Hopkins University / Toshitake Kohno, Masanori Morishita, editors.
p. cm. — (Contemporary mathematics, ISSN 0271-4132 ; v. 416)
Includes bibliographical recferences.
ISBN-13: 978-0-8218-3456-5 (acid-free paper)
1. Algebraic number theory—Congresses. 2. Low-dimensional topology—Congresses. I. Kohno, Toshitake. II. Morishita, Masanori, 1961– III. Japan-U.S. Mathematics Institute. IV. Series: Contemporary mathematics (American Mathematical Society) ; v. 416.

QA247.P75 2006
512.7′4—dc22 2006040609

---

10 9 8 7 6 5 4 3 2 1   11 10 09 08 07 06

Contents

Preface

The fifteenth year program of the Japan-U.S. Mathematics Institute was entitled "Primes and Knots" and devoted to the connection between algebraic number theory and low-dimensional topology. The program culminated in two conferences, one held as a special session at the 109th AMS annual meeting in Baltimore on January 15 and 16, 2003, and another held as a week-long workshop at the Johns Hopkins University from March 7 to 16, 2003. The conferences were organized by Stavros Garoufalidis, Toshitake Kohno, Jack Morava, Masanori Morishita, and Steve Zucker. The theme of the conferences was the interaction between number theory and topology.

The conferences focused on the following topics:

1. Analogies between primes and knots, number rings and 3-manifolds
2. Avant-garde Galois theory—Galois representations, Grothendieck-Teichmüller groups
3. Classical and quantum topology of knots
4. Arithmetic non-commutative geometry and related mathematical physics.

The papers assembled for this volume include speakers' contributions which present extended or modified versions of the lectures delivered at the meetings.

The organizers and editors are grateful to the Department of Mathematics of the Johns Hopkins University and the National Science Foundation for their financial support. Thanks are also due to Professor Kenji Fukaya for his support by Grants-in-Aid for Scientific Research 13852001, Japan Society for the Promotion of Science.

We are pleased to celebrate that in March 2006 the Japan-U.S. Mathematics Institute was awarded the Mathematical Society of Japan's Seki Takakazu Prize for promoting cooperation between the two countries in mathematical research.

March 2006
Toshitake Kohno
Masanori Morishita

Contemporary Mathematics
Volume **416**, 2006

Categorification of the skein module of tangles

Marta M. Asaeda, Józef H. Przytycki, and Adam S. Sikora

Abstract. We generalize our previous work on categorification of Kauffman bracket skein module of surfaces, by extending our homology to tangles in cylinders over surfaces, $F \times [0,1]$. Our homology of 0-tangles and 1-tangles in D^3 coincides (up to normalization) with Khovanov link homology and the reduced Khovanov link homology.

We prove the basic properties of our homology. In particular, the short exact sequence of homologies of skein related tangles and the Künneth formula for the tensor product of tangles.

1. Introduction

In this paper we apply the ideas of [**APS**] to define "Khovanov like" homology of tangles in cylinders over surfaces.

A *marked surface* is an oriented compact surface, F, together with a distinguished finite set of "marked" points B in the boundary of F. *A framed tangle* in the cylinder over a marked surface (F, B) is a finite disjoint union of embedded bands, $b : [0,1]\times[0,1] \hookrightarrow F\times(-1,1)$, and of annuli, $a : S^1\times[0,1] \hookrightarrow F\times(-1,1)$, such that for any band b, $b([0,1]\times[0,1]) \cap \partial F \times (-1,1) = b([0,1]\times\{0\}) \cup b([0,1]\times\{1\})$. Neither the bands nor annuli are oriented. However, we assume that all bands in a tangle have an integral framing. The midpoints of the arcs $b([0,1] \times \{0\})$, $b([0,1] \times \{0\})$ are called the endpoints of T and we assume that the set of endpoints of T coincides with the set of marked points of F. Each tangle T is represented by a tangle diagram composed of closed loops and arcs in F which represent the cores of annuli and of bands of T with blackboard framing. Therefore, each tangle diagram is a compact 1-manifold D properly embedded into F and such that $D \cap \partial F = B$. Since tangles are considered up to an ambient isotopy fixing their endpoints, two

---

1991 *Mathematics Subject Classification.* Primary 57M27.

Key words and phrases. tangle, Kauffman bracket, Khovanov homology, skein module.

The first author was supported in part by NSF grant #DMS-0202613.
The second author was supported in part by grant #TMP-25051958.
The third author was sponsored in part by NSF grant #DMS-0307078.

tangle diagrams are equivalent if they differ by second, third, and balanced first Reidemeister moves.

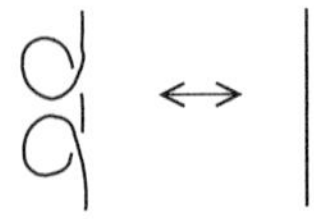

Fig. 1: Balanced first Reidemeister move

We denote the set of all tangles in the cylinder over marked surface (F, B) by $\mathcal{T}(F, B)$. Additionally, we denote the set of all diagrams of tangles in $\mathcal{T}(F, B)$ with no crossings and no contractible closed loops by $\mathcal{B}(F, B)$. Since different diagrams in $\mathcal{B}(F, B)$ represent non-isomorphic tangles, $\mathcal{B}(F, B) \subset \mathcal{T}(F, B)$. Note that $\mathcal{B}(F, \emptyset)$ is the set of all finite collections of non-intersecting nontrivial simple closed loops in F considered up to homotopy (including the empty collection). Observe also that if $B \subset \partial D^2$ is a set of $2n$ points then $\mathcal{B}(D^2, B)$ has the n-th Catalan number of elements, $\frac{1}{n+1}\binom{2n}{n}$.

For any tangle T in $\mathcal{T}(F, B)$ we define homology groups $H_{i,j,s}(T)$, for $i, j \in \mathbb{Z}$ and $s \in \mathcal{B}(F, B)$, which are invariant under isotopies of T and which generalize the homology groups of [**APS**]. Its properties are discussed in Sections 3 and 4.

2. Definition of Homology

Let $\mathcal{B}'(F, B) \subset \mathcal{B}(F, B)$ be the set of all tangle diagrams in F with no closed loops. Consequently, diagrams in $\mathcal{B}'(F, B)$ are composed of non-intersecting arcs only. Let $\mathcal{C}(F)$ be the set of all unoriented non-contractible simple closed curves in F considered up to homotopy. By separating arcs from closed loops in a tangle, we embed $\mathcal{B}(F, B)$ into $\mathbb{N}\mathcal{C}(F) \times \mathcal{B}'(F, B)$, where $\mathbb{N} = \{0, 1, 2, ...\}$. We are about to define homology groups of tangles T in $\mathcal{T}(F, B)$, $H_{i,j,s,b}(T)$, indexed by $i, j \in \mathbb{Z}$, $s \in \mathbb{Z}\mathcal{C}(F)$ and $b \in \mathcal{B}'(F, B)$. The following proposition is proved at the end of Section 3:

PROPOSITION 1. *(1)* $H_{i,j,s,b}(T)$ *is isomorphic to* $H_{i,j,|s|,b}(T)$, *where* $|s| = \sum_\gamma |s_\gamma| \cdot \gamma$ *for* $s = \sum_\gamma s_\gamma \cdot \gamma$ *in* $\mathbb{Z}\mathcal{C}(F)$.
(2) $H_{i,j,|s|,b}(T) = 0$ *if* $(|s|, b) \in \mathbb{N}\mathcal{C}(F) \times \mathcal{B}'(F, B)$ *does not lie in the image of* $\mathcal{B}(F, B)$.

COROLLARY 2. *Alternatively, our homology groups can be indexed by* $i, j \in \mathbb{Z}$ *and* $s \in \mathcal{B}(F, B)$ *(as promised in the introduction).*

Nonetheless, indexing by $s \in \mathbb{Z}\mathcal{C}(F)$ and $b \in \mathcal{B}'(F, B)$ is useful for technical reasons.

Consider a tangle diagram D in F whose crossings are ordered by consecutive integers $1, 2, ...$ Following [**Vi1, APS**], a *state* S of D is an assignment of $+$ or $-$ sign to each of the crossings of D and an additional assignment of $+$ or $-$ sign to each closed loop in the diagram D_S obtained by smoothing the crossings of D according to the following convention:

+1 marker −1 marker

(Note that arcs of D_S do not carry any labels.) Our construction of the chain complex associated with D is analogous to that of [**APS**]:

For any state S of D let

$$\begin{aligned} I(S) &= \sharp\{\textit{positive crossing markers}\} - \sharp\{\textit{negative crossing markers}\}, \\ J(S) &= I(S) + 2\sharp\{\textit{positive contr. circles}\} - 2\sharp\{\textit{negative contr. circles}\}, \end{aligned}$$

where "contr" stands for "contractible".

Let $\Phi(S)$ be an element of $\mathcal{B}(F, B)$ obtained from D_S by removing all closed loops. Denote the non-contractible loops in D_S by $\gamma_1, ..., \gamma_n$. If these loops are marked by $\varepsilon_1, ..., \varepsilon_n \in \{+1, -1\}$, then let

$$\Psi(S) = \sum_i \varepsilon_i \gamma_i \in \mathbb{Z}\mathcal{C}(F),$$

cf. [**APS**].

Let $\mathcal{S}_{i,j,s,b}(D)$ be the set of all states S of D with $I(S) = i$, $J(S) = j$, $\Psi(S) = s$, and $\Phi(S) = b$. Let $\mathcal{C}_{i,j,s,b}(D)$ be the free abelian group generated by states in $\mathcal{S}_{i,j,s,b}(D)$. We define the incidence number between states following [**APS**, Definition 3.1]:

$[S : S']_v = 1$ if the following four conditions are satisfied:

(1) the crossing v is marked by $+$ in S and by $-$ in S',
(2) S and S' assign the same markers to all the other crossings,
(3) the labels of the common circles in S and S' are unchanged,
(4) $J(S) = J(S')$, $\Psi(S) = \Psi(S')$, $\Phi(S) = \Phi(S')$.

Otherwise $[S : S']_v$ is equal to 0.

The types of incident states at a crossing v which involve loops only are listed in [**APS**, Table 2.1]. The only two other types of incident states are as follows:

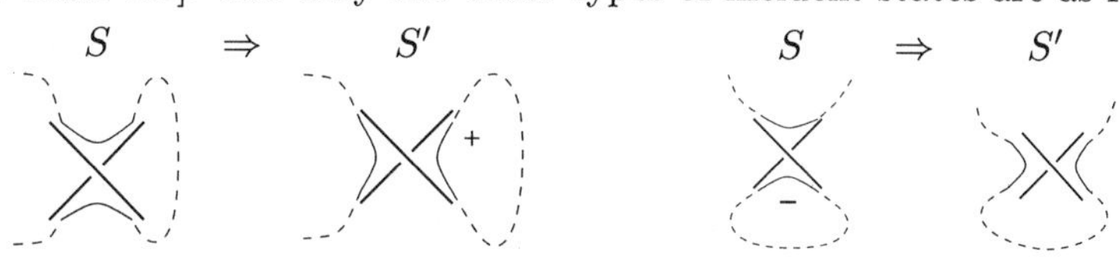

Fig 2. Incident states involving arcs

The labeled loops appearing in this diagram are contractible.

Let $t(S, v)$ denote the number of negative markers assigned to crossings bigger than v. Then

$$d_{i,j,s,b} : \mathcal{C}_{i,j,s,b}(D) \to \mathcal{C}_{i-2,j,s,b}(D)$$

is defined by

$$d_{i,j,s,b}(S) = \sum_v (-1)^{t(S,v)} d_{i,j,s,b,v}(S),$$

where

$$d_{i,j,s,b,v}(S) = \sum [S : S']_v S',$$

and the sum is over all states of D. Clearly, it is enough to consider states in $\mathcal{S}_{i-2,j,s,b}(D)$ only. As in [**APS**], d has degree -2 with respect to the first index. Note also that $\mathcal{C}_{i,*,*,*}(D) = 0$ for i not equal to the number of crossings of D mod 2.

Remark 3. *For any link diagram D in a surface F, $(\mathcal{S}_{*,*,*,\emptyset}(D), d)$ is the chain complex introduced in* [**APS**].

In the next section we will prove:

PROPOSITION 4. *For any marked surface* (F, B) *and* $T \in \mathcal{T}(F, B)$, $d^2 = 0$.

Therefore for any abelian group G, $(C_{*,j,s,b}(D) \otimes G, d_{*,j,s,b})$ is a chain complex whose homology groups we denote by $H_{*,j,s,b}(D; G)$, as usual abbreviating $H_{i,j,s,b}(D; \mathbb{Z})$ to $H_{i,j,s,b}(D)$. By the argument of [**APS**, Sec. 10.1], $H_{i,j,s,b}(D; G)$ does not depend (up to isomorphism) on the ordering of crossings of D.

3. Embeddings into surfaces, Reidemeister moves

Let (F, B) and (F', B') be marked surfaces with equal numbers of boundary components, such that there is a preserving orientation homeomorphism $\phi : \partial F \to \partial F'$ mapping B onto B'. Then for any $b \in \mathcal{B}'(F, B)$ and $b' \in \mathcal{B}'(F', B')$, $b \cup_\phi b'$ is a disjoint union of simple closed curves in the closed surface $F \cup_\phi F'$. We leave the proof of the following easy statement to the reader.

LEMMA 5. *For any marked surface* (F, B) *and any* $b \in \mathcal{B}'(F, B)$ *there is a marked surface* (F', B'), $b' \in \mathcal{B}'(F', B')$ *and a homeomorphism* $\phi : \partial F \to \partial F'$ *as above such that the components of* $b \cup_\phi b'$ *are not homotopic to any loops in* F *and are not homotopic one to another.*

Let (F, B), (F', B'), $b \in \mathcal{B}'(F, B)$, $b' \in \mathcal{B}'(F, B)$ satisfy the assumptions of Lemma 5 and let c be a state of $b \cup_\phi b'$. For any $T \in \mathcal{T}(F, B)$ and its state $S \in \mathcal{S}_{i,j,s,b}(T)$, let $\Lambda(S) \in \mathcal{S}_{i,j,s+c,\emptyset}(T \cup_\phi b')$ be the state $S \cup_\phi b'$ obtained by labeling the loops of $b \cup_\phi b'$ as in c.

LEMMA 6. Λ *extends to an isomorphism*

$$\Lambda : \mathcal{C}_{i,j,s,b}(T) \to \mathcal{C}_{i,j,s+c,\emptyset}(T \cup_\phi b')$$

commuting with the differentials.

PROOF. Λ is $1-1$ since no loops in $b \cup_\phi b'$ are homotopic to each other nor are homotopic to loops in s. Obviously, Λ is onto as well. Since states $S, S' \in \mathcal{C}_{*,j,s,b}(T)$ are incident if and only if $S \cup_\phi b'$ and $S' \cup_\phi b'$ are incident, the statement follows. □

Since by Remark 3, $\mathcal{C}_{*,j,s+c,\emptyset}(T \cup_\phi b')$ is a chain complex, Lemma 6 implies Proposition 4 and, together with (7) of [**APS**], it implies Proposition 1.

Now, Lemma 6 and [**APS**, Theorem 2] imply

THEOREM 7. *Let* $i, j \in \mathbb{Z}$, $s \in \mathbb{Z}\mathcal{C}(F)$, D *be a diagram of a tangle in* $\mathcal{T}(F, B)$, $b \in \mathcal{B}'(F, B)$,
(1) If D' *is related to* D *by the first Reidemeister move consisting of adding a negative kink to* D *then* $H_{i,j,s,b}(D') = H_{i-1,j-3,s,b}(D)$.
(2) $H_{i,j,s,b}(D)$ *is invariant (up to an isomorphism) under the second and third Reidemeister moves.*

Additionally, by Lemma 6 for any diagram D of a tangle in $\mathcal{T}(F, B)$ the groups $H_{i,j,s,b}(D)$ coincide with stratified homology groups $H_{i,j,s+c}(\bar{D})$ defined in [**APS**] for the link diagram $\bar{D} = D \cup_\phi b'$ constructed above.

Further properties of our homology groups are described below. From now on we index our homology groups by $i, j \in \mathbb{Z}$ and $b \in \mathcal{B}(F, B)$, as proposed in Corollary 2.

4. Properties

4.1. Twisting the endpoints. For any tangle $T \in \mathcal{T}(F,B)$ and any component $C \subset \partial F$ let $\lambda_C(T)$ denote a tangle obtained from T by shifting its endpoints in C counterclockwise by one:

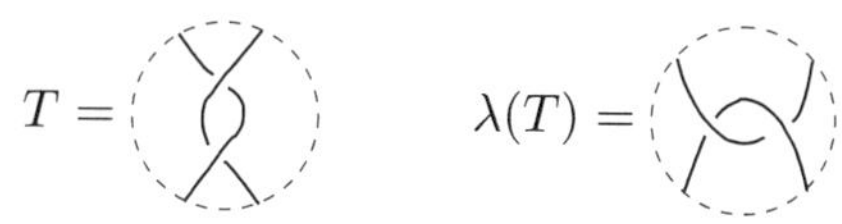

Note that $\lambda^n(T) = T$ for $T \in \mathcal{T}(F,B)$ if $|B| = n$ and $F = D^2$, but not necessarily for other surfaces. In any case, we have $H_{i,j,s,b}(T) = H_{i,j,s,\lambda_c(b)}(\lambda_c(T))$.

4.2. Short exact sequence. Any three skein related tangle diagrams in F

define a short exact sequence

$$0 \to C_{*jsb}(D_\infty) \xrightarrow{\alpha} C_{*,j-1,s,b}(D_p) \xrightarrow{\beta} C_{*,j-2,s,b}(D_0) \to 0 \tag{4.1}$$

where the maps $\alpha : C_{i,j,s,b}(D_\infty) \to C_{i-1,j-1,s,b}(D_p)$ and $\beta : C_{i,j-1,s,b}(D_p) \to C_{i-1,j-1,s,b}(D_0)$ are defined as in [**APS**, Sec 7]. This sequence leads to the long exact sequence
(4.2)

$$\ldots \to H_{i,j,s,b}(D_\infty) \xrightarrow{\alpha_*} H_{i-1,j-1,s,b}(D_p) \xrightarrow{\beta_*} H_{i-2,j-2,s,b}(D_0) \xrightarrow{\partial} H_{i-2,j,s,b}(D_\infty) \to \ldots$$

4.3. Categorification of the skein modules of tangles. Formal $\mathbb{Z}[A^{\pm 1}]$-linear combinations tangles in $\mathcal{T}(F,B)$ quotiented by the skein relations of the Kauffman bracket

 $= A$  $+ A^{-1}$ 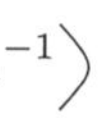, $\quad L \cup \bigcirc = -(A^2 + A^{-2})L,$

form the relative Kauffman bracket skein module $\mathcal{S}(F,B)$, cf. [**Pr**]. The set $\mathcal{B}(F,B)$ forms a natural basis of $\mathcal{S}(F,B)$. For any tangle $T \in \mathcal{T}(F,B)$, we denote its coordinates in this basis by $\langle T\rangle_b$, $b \in \mathcal{B}(F,B)$. Hence

$$T = \sum_{b \in \mathcal{B}(F,B)} \langle T\rangle_b b \in \mathcal{S}(F,B).$$

Let $\chi_A(H_{**})$ denote the polynomial Euler characteristic of a bigraded group H_{**} defined as in [**APS**, Sec 1]:

$$\chi_A(H_{**}) = \sum_{i,j} A^j(-1)^{\frac{j-i}{2}} rkH_{i,j}. \tag{4.3}$$

In the following proposition we index our tangle homology groups by $i,j \in \mathbb{Z}$ and $b \in \mathcal{B}(F,B)$ as in Corollary 2:

PROPOSITION 8. *For any* $T \in \mathcal{T}(F,B)$ *and* $b \in \mathcal{B}(F,B)$, $\chi_A(H_{**b}(T)) = \langle T\rangle_b$

PROOF. By definition

$\langle$  $\rangle_b = A\langle$  $\rangle_b + A^{-1}\langle$ 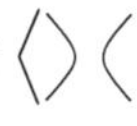 $\rangle_b$, $\quad \langle L \cup \bigcirc\rangle_b = -(A^2 + A^{-2})\langle L\rangle_b,$

and, by (4.2), analogous identifies hold for $\chi_A(H_{**b}(T))$. Therefore, it is enough to assume that the tangle diagram T has no crossings and no trivial components. Under these assumptions $T \in \mathcal{B}'(F, B)$ and

$$\chi_A(H_{**b}(T)) = \chi_A(C_{**b}(T)) = rk\, C_{0,0,b}(T) = \delta_{b,T} = \langle T \rangle_b,$$

where $\delta_{b,T} = \begin{cases} 1 & \text{if } b = T \\ 0 & \text{otherwise.} \end{cases}$ □

4.4. Reduced Link Homology. For any link diagram L with a specified one of its components, Khovanov defined its *reduced homology*, $\tilde{H}_{i,j}(L)$, [**K3, Sh**]. His construction has the following interpretation in our setting: If L' is a 1-tangle obtained by cutting L at an arbitrary point of its specified component and its end-points are p_1 and p_2, then $\tilde{H}_{i,j}(L)$ is isomorphic (up to normalization of indices) to $H_{i,j,\alpha}(L')$, where α is the unique element of $\mathcal{B}(D^2, \{p_1, p_2\})$. Since different cutting points on a given component of a link give isomorphic 1-tangles, we get another proof that $\tilde{H}_{i,j}(L)$ does not depend on the choice of the cutting point on the distinguished component L.

4.5. Tensor product of tangles. Consider a decomposition of a disk D^2 into two disks D_1, D_2 by a properly embedded arc whose endpoints are disjoint from a finite set $B \subset \partial D^2$. Let $B_1 = B \cap \partial D_1$ and $B_2 = B \cap \partial D_2$. In this situation, a union of tangles $T_i \in \mathcal{T}(D_i, B_i)$, $i = 1, 2$, is a tangle in $\mathcal{T}(D^2, B)$ which we denote by $T_1 \otimes T_2$.

T_1 T_2 $T_1 \otimes T_2$

Fig. 3: A tensor product of tangles.

Note that $T_1 \otimes T_2$ is not well defined for tangles T_1, T_2 in disjoint disks, since several ways of gluing of these disks are possible. This issue is irrelevant however in the calculation of homology of $T_1 \otimes T_2$. To see that, observe first that if $T_i \in \mathcal{T}(D_i, B_i)$, $i = 1, 2$, then $H_{**s}(T_1 \otimes T_2) = 0$ unless $s = s_1 \otimes s_2$ for some $s_1 \in \mathcal{B}(D_1, B_1)$, $s_2 \in \mathcal{B}(D_2, B_2)$. If $s = s_1 \otimes s_2$ then s_1 and s_2 are unique and $C_{**s}(T_1 \otimes T_2)$ is the tensor product of (filtered) chain complexes, $C_{**s_1}(T_1)$ and $C_{**s_2}(T_2)$. By the Künneth formula we have a short exact sequence

$$0 \to \bigoplus_{\substack{i_1+i_2=i,\\ j_1+j_2=j}} H_{i_1,j_1,s_1}(T_1) \otimes H_{i_2,j_2,s_2}(T_2) \to H_{i,j,s}(T_1 \otimes T_2) \to$$

$$\to \bigoplus_{\substack{i_1+i_2=i-1,\\ j_1+j_2=j}} Tor(H_{i_1,j_1,s_1}(T_1), H_{i_2,j_2,s_2}(T_2)) \to 0 \tag{4.4}$$

which is non-canonically split. In particular the homology groups of $T_1 \otimes T_2$ are determined by the homology groups of T_1 and T_2.

4.6. Reduced tensor products of tangles. Again, consider a decomposition of a disk D^2 into two disks D_1, D_2 by a properly embedded arc whose endpoints are disjoint from a finite set $B \subset \partial D^2$. Let p be a selected point inside that arc and let $B_1 = B \cap \partial D_1 \cup \{p\}$ and $B_2 = B \cap \partial D_2 \cup \{p\}$. In this situation, union of tangles $T_i \in \mathcal{T}(D_i, B_i)$, $i = 1, 2$, is a tangle in $\mathcal{T}(D^2, B)$ which we call a reduced tensor product of T_1 and T_2 and denote by $T_1 \otimes_1 T_2$.

T_1 T_2 $T_1 \otimes_1 T_2$

Fig. 4: A reduced tensor product of tangles

As for tensor products, the homology groups of reduced tensor products $T_1 \otimes_1 T_2$ are related to homology groups of T_1 and T_2 by Künneth formula. As before, if $H_{**s}(T_1 \otimes_1 T_2) \neq 0$ then $s = s_1 \otimes_1 s_2$ for unique $s_i \in \mathcal{B}(D_i, B_i)$, $i = 1, 2$. In this situation $C_{**s}(T_1 \otimes_1 T_2)$ is the tensor product of (filtered) chain complexes $C_{**s_1}(T_1)$ and $C_{**s_2}(T_2)$ inducing a non-canonically split short exact sequence

$$0 \to \bigoplus_{\substack{i_1+i_2=i,\\ j_1+j_2=j}} H_{i_1,j_1,s_1}(T_1) \otimes H_{i_2,j_2,s_2}(T_2) \to H_{i,j,s}(T_1 \otimes_1 T_2) \to$$

$$\to \bigoplus_{\substack{i_1+i_2=i-1,\\ j_1+j_2=j}} Tor(H_{i_1,j_1,s_1}(T_1), H_{i_2,j_2,s_2}(T_2)) \to 0 \tag{4.5}$$

4.7. Open questions and final comments. If T is union of tangles T_1, T_2 with more than a single pair of their endpoints identified then the dependence of homologies of T_1, T_2 and T is far more complicated. In the simplest case, when T_2 is a 1-tangle composed of a single arc whose endpoints are identified with two consecutive endpoints of T_1, one can construct a spectral sequence converging to $H_{***}(T)$ whose E^2 term is composed of homology groups of T_1.

Given a tangle T in $[0,1] \times [0,1]$ with specified top $2n$ and bottom $2m$ points, Khovanov constructed homology groups $\mathcal{H}_{i,j}(T)$, which are invariant under isotopies of T. Additionally each such group is an (H^n, H^m)-bimodule for certain rings H^n, H^m defined in [**K2, Ja**]. As groups, $\mathcal{H}_{i,j}(T)$ are sums of Khovanov homology groups $H_{i,j}(L)$ over all links L which are plat closures of T. (A plat closure of T is a link obtained by closing its $2n$ top points by n non-intersecting arcs and, analogously, its $2m$ bottom points by m non-intersecting arcs.) We expect that for any tangle T there is a spectral sequence converging to $\mathcal{H}_{*j}(T)$ whose summands of E^2 are sums of certain groups $H_{ijs}(T)$. We do not know if there is a more explicit relation between Khovanov's and our homology groups. Nor we can interpret the bimodule structure of Khovanov's groups in our setting.

There is a parallel theory of tangle homologies due to D. Bar-Natan. At present time we do not know the relation between his and our tangle homologies.

References

[APS] M. M. Asaeda, J. H. Przytycki, A. S. Sikora, *Categorification of the Kauffman bracket skein module of I-bundles over surfaces*, *Algebr. Geom. Topol.* **4** 2004, 1177–1210, arXiv: math.QA/0403527

[BN] D. Bar-Natan, *On Khovanov's categorification of the Jones polynomial,* Algebraic and Geometric Topology **2** 2002, 337–370, arXiv: math.QA/0201043

[K1] M. Khovanov, *A categorification of the Jones polynomial,* Duke Math. J. **101**, no. 3 2000, 359–426, arXiv: math.QA/9908171

[K2] M. Khovanov, *A functor-valued invariant of tangles,* Algebr. Geom. Topol. **2** 2002, 665–741, arXiv: math.QA/0103190

[K3] M. Khovanov, *Patterns in knot homology I,* Experiment. Math. **12**(3) 2003, 365–374, arXiv: math.QA/0201306

[Ja] M. Jacobsson, *Chewing the Khovanov homology of tangles,* Proceedings of the Banach Center Conference "Knots in Poland", Fund. Math. **184** 2004, 103–112.

[Pr] J. H. Przytycki, *Fundamentals of Kauffman bracket skein modules,* Kobe Math. J. **16**(1) 1999, 45–66, arXiv:math.GT/9809113

[PS] J. H. Przytycki, A. S. Sikora, *On Skein Algebras and $Sl_2(C)$-Character Varieties,* Topology, **39**(1) 2000, 115–148. arXiv:q-alg/9705011

[Sh] A. Shumakovitch, *Torsion of the Khovanov Homology,* Geom. and Topol. to appear, arXiv: math.GT/0405474

[Vi1] O. Viro, *Remarks on definition of Khovanov homology,* arxiv:math.GT/0202199

[Vi2] O. Viro, *Khovanov homology, its definitions and ramifications,* Proceedings of Knots in Poland, July 2003, Fund. Math. **184** 2004, 317–342.

Current address: Dept. of Mathematics, UCR, 900 Big Springs Drive, Riverside, CA 92521
E-mail address: marta@math.ucr.edu

Current address: Dept. of Mathematics, Old Main Bldg., 1922 F St. NW, The George Washington University, Washington, DC 20052
E-mail address: przytyck@gwu.edu

Current address: Dept. of Mathematics, 244 Math. Bldg., SUNY at Buffalo, Buffalo, NY 14260
E-mail address: asikora@buffalo.edu

Contemporary Mathematics
Volume **416**, 2006

The Double Shuffle Relations for p-adic Multiple Zeta Values

Amnon Besser and Hidekazu Furusho

ABSTRACT. We give a proof of double shuffle relations for p-adic multiple zeta values by developing higher dimensional version of tangential base points and discussing a relationship with two (and one) variable p-adic multiple polylogarithms.

0. Introduction

In this paper we will prove a set of formulas, known as double shuffle relations, relating the p-adic multiple zeta values defined by the second named author in [**F1**]. These formulas are analogues of formulas for the usual (complex) multiple zeta values. These have a very simple proof which unfortunately does not translate directly to the p-adic world.

Recall that the (complex) multiple zeta value $\zeta(\mathbf{k})$, where $\mathbf{k}$ stands for the multi-index $\mathbf{k} = (k_1, \ldots, k_m)$, is defined by the formula

$$\zeta(\mathbf{k}) = \sum_{\substack{0<n_1<\cdots<n_m \\ n_i \in \mathbf{N}}} \frac{1}{n_1^{k_1} \cdots n_m^{k_m}}, \tag{0.1}$$

The series is easily seen to be convergent assuming that $k_m > 1$.

Multiple zeta values satisfy two types of so called shuffle product formulas, expressing a product of multiple zeta values as a linear combination of other such values. The first type of formulas are called series shuffle product formulas (sometimes called by harmonic product formulas). The simplest example is the relation

$$\zeta(k_1) \cdot \zeta(k_2) = \zeta(k_1, k_2) + \zeta(k_2, k_1) + \zeta(k_1 + k_2), \tag{0.2}$$

which is easily obtained from the expression (0.1) by noting that the left hand side is a summation over an infinite square of pairs (n_1, n_2) of the summand in (0.1), and that summing over the lower triangle (respectively the upper triangle, respectively the diagonal) gives the three terms on the right hand side. Every series shuffle product formula has this type of proof.

The second type of shuffle product formulas, known as iterated integral shuffle product formulas, is somewhat harder to establish and follows from the description

---

2000 *Mathematics Subject Classification.* Primary 11G55; Secondary 11S80.

The first author was supported in part by an Israel Science Foundation grant.

of multiple zeta values in terms of multiple polylogarithms. More precisely. The one variable multiple polylogarithm is defined by the formula

$$\mathrm{Li}_{\mathbf{k}}(z) = \sum_{\substack{0<n_1<\cdots<n_m \\ n_i\in\mathbf{N}}} \frac{z^{n_m}}{n_1^{k_1}\cdots n_m^{k_m}}, \tag{0.3}$$

near $z=0$. It can then be extended as a multi-valued function to $\mathbf{P}^1(\mathbf{C})-\{0,1,\infty\}$. We clearly have the relation $\lim_{z\to 1}\mathrm{Li}_{\mathbf{k}}(z)=\zeta(\mathbf{k})$.

Multiple polylogarithms can be written using the theory of iterated integrals due to Chen [**Ch**]. In other words, they satisfy a system of unipotent differential equations. This gives an integral expression for multiple polylogarithms. By substituting $z=1$ and splitting the domain of integration in the right way we obtain the iterated integral shuffle product formulas, a simple example of which is the formula (0.4)

$$\zeta(k_1)\cdot\zeta(k_2) = \sum_{i=0}^{k_1-1}\binom{k_2-1+i}{i}\zeta(k_1-i,k_2+i)+\sum_{j=0}^{k_2-1}\binom{k_1-1+j}{j}\zeta(k_2-j,k_1+j).$$

In [**F1**] the second named author defined the p-adic version of multiple zeta values and studied some of their properties. The defining formula (0.1) can not be directly used p-adically because the defining series does not converge. Instead, one must use an indirect approach based on the theory of Coleman integration [**Co, Bes1**]. Coleman's theory defines p-adic analytic continuation for solutions of unipotent differential equations "along Frobenius". Coleman used his theory initially to define p-adic polylogarithms. In [**F1**] Coleman integration was used to define one variable p-adic multiple polylogarithms. Taking the limit at 1 in the right way one obtains the definition of p-adic multiple zeta values. It is by no means trivial that the limit even exists or is independent of choices, and this is the main result of [**F1**].

Given their definition, it is not surprising that for p-adic multiple zeta values it is the iterated integral shuffle product formulas that are easier to obtain p-adically. In [**F1**] the series shuffle product formulas were not obtained. The purpose of this work is to prove (Theorem 7.1) these formulas, and as a consequence the double shuffle relations (Corollary 7.2) for p-adic multiple zeta values.

To prove the main theorem it is necessary to use the theory of Coleman integration in several variables developed by the first named author in [**Bes1**]. The reason for this is quite simple - If one tries to replace multiple zeta values by multiple polylogarithms in the proof of (0.2) sketched above one easily establishes the formula

$$\mathrm{Li}_{k_1}(z)\,\mathrm{Li}_{k_2}(w) = \mathrm{Li}_{k_1,k_2}(z,w)+\mathrm{Li}_{k_2,k_1}(w,z)+\mathrm{Li}_{k_1+k_2}(zw)\,, \tag{0.5}$$

which is a two variable formula. It seems impossible to obtain a one variable version of the same formula. The proof of the main theorem thus consists roughly speaking of showing that (0.5) extends to Coleman functions of several variables and then taking the limit at $(1,1)$.

Since taking the limit turned out to be rather involved in [**F1**], we opted for an alternative approach, which was motivated by a letter of Deligne to the second named author [**D2**]. Deligne observes that taking the limit at 1 for the multiple polylogarithm can be interpreted as doing analytic continuation from tangent vectors at 0 and 1, using the theory of the tangential basepoint at infinity introduced in [**D1**]. To analytically continue (0.5) and obtain the series shuffle product formula we analyze a more general notion of tangential basepoint sketched in loc. cit. and examine among other things its relation with Coleman integration.

To give a precise meaning of the limit value to $(1,1)$, we work over the moduli space $\mathcal{M}_{0,5}$ of curves of $(0,5)$-type and the normal bundles for the divisors at infinity $\overline{\mathcal{M}_{0,5}} - \mathcal{M}_{0,5}$ ($\overline{\mathcal{M}_{0,5}}$: a compactification of $\mathcal{M}_{0,5}$). Two variable p-adic multiple polylogarithms are introduced. They are Coleman functions over $\mathcal{M}_{0,5}$. In §6 their analytic continuation to the normal bundle will be discussed. In particular, we will relate the behavior of the analytic continuation of two variable multiple polylogarithm to a normal bundle with one variable multiple polylogarithm and then we get p-adic multiple zeta values as "special values" of two variable multiple polylogarithms.

It is a pleasure to thank P. Deligne and A. Goncharov. The first named author was supported by a grant from the Israel Science foundation. The second named author is supported in part by JSPS Research Fellowships for Young Scientists. He would like to thank the Center for Advanced Studies in Mathematics at Ben-Gurion University for supporting a visit to Beer-Sheva in February 2003 when a large part of the research described in this work was carried out and he would also like to thank the Johns Hopkins University for their support and hospitality during the JAMI conference.

1. Tannakian duality and fundamental groups

This section is introductory, and contains some basics of the theory of Tannakian categories (for which the standard reference is [**DM**]) and some ways it can be used to construct fundamental groups and path spaces in algebraic geometry.

The theory of Tannakian categories has its roots in Tannaka-Krein duality, itself a generalization of Pontrjagin duality, which shows how to reconstruct a compact topological group out of the category of its representations. Subsequently, it was realized by Grothendieck that the reconstruction can be done, in the somewhat different context of affine group schemes, based on some formal properties of this category of representations. Consequently, any category with some additional structures and properties, formalized in the notion of a Tannakian category, gives rise to an affine group scheme.

A Tannakian category (or more precisely a neutral Tannakian category) over a field F is a rigid abelian F-linear tensor category possessing a fiber functor [**DM**, Definition 2.19]. An abelian F-linear category $\mathcal{T}$ becomes a tensor category with the addition of a bifunctor $\otimes : \mathcal{T} \times \mathcal{T} \to \mathcal{T}$ satisfying certain axioms mimicking those of the tensor product of vector spaces. It is rigid if it satisfies [**DM**, Definition 1.7]. A tensor functor $G : \mathcal{T} \to \mathcal{T}'$ is a functor together with a functorial isomorphism $G(X) \otimes G(Y) \to G(X \otimes Y)$ satisfying a few obvious properties [**DM**, Definition 1.8]

The category Vec_F of finite dimensional vector spaces over F is a rigid abelian F-linear tensor category. The final data in the definition of a Tannakian category is a fiber functor, which by definition is an exact F-linear faithful tensor functor $\omega : \mathcal{T} \to \mathrm{Vec}_F$.

Suppose now that G is an affine group scheme over F. One than has the category $\mathrm{Rep}_F(G)$ of finite dimensional F-rational representations of G. This is clearly a Tannakian category with fiber functor sending a representation to its underlying K-vector space. The main result of Tannakian duality [**DM**, Theorem 2.11] says that $\mathrm{Rep}_F(G)$ determines G uniquely, and conversely, any Tannakian category arises as $\mathrm{Rep}_F(G)$ for some G.

Recall that an affine group scheme over F can be defined as a group valued functor on the category of F-algebras which is represented by an F-algebra. The way that $\mathrm{Rep}_F(G)$ determines G is as follows: G is isomorphic to the group valued functor $\mathrm{Aut}^{\otimes}(\omega)$ whose value on an F-algebra R is

$$\mathrm{Aut}^{\otimes}(\omega)(R) := \mathrm{Hom}(\omega \otimes R, \omega \otimes R) ,$$

where $\omega \otimes R$ is the composition of ω with the change of ring functor and Hom is in the category of tensor functors. We will often write $\pi_1(\mathcal{T}, \omega)$ instead of $\mathrm{Aut}^{\otimes}(\omega)$ to indicate the dependency on the category $\mathcal{T}$.

Tannakian duality can be used to define all kinds of "fundamental groups" in algebraic geometry and algebraic topology. The basic example to keep in mind is that of the category Loc_X of local system of finite dimensional $\mathbf{Q}$-vector spaces on a connected topological space X. Given a choice of a point $x \in X$, this is equivalent to the category of finite dimensional $\mathbf{Q}$-representations of the fundamental group $\pi_1(X, x)$. It is clearly a Tannakian category over $\mathbf{Q}$. Each point $x \in X$ determines a fiber functor ω_x where $\omega_x(V) =$ fiber of V at x. The obvious map $\pi_1(X, x) \to \pi_1(\mathrm{Loc}_X, \omega_x)(\mathbf{Q})$ is explicitly given by

$$\gamma \mapsto (V \mapsto \text{monodromy along the loop } \gamma \in \mathrm{Aut}(\omega_x(V))) . \tag{1.1}$$

We may also use Tannakian duality to define generalized path spaces. Namely, if $\mathcal{T}$ is a Tannakian category and ω_1 and ω_2 are two fiber functors on $\mathcal{T}$, then there is an affine scheme $\pi_1(\mathcal{T}, \omega_1, \omega_2) = \mathrm{Hom}^{\otimes}(\omega_1, \omega_2)$ whose points in an F-algebra R are

$$\mathrm{Hom}^{\otimes}(\omega_1, \omega_2)(R) := \mathrm{Hom}(\omega_1 \otimes R, \omega_2 \otimes R) ,$$

Again Hom taken in the category of tensor functors. There is an obvious composition map,

$$\circ : \pi_1(\mathcal{T}, \omega_2, \omega_3) \times \pi_1(\mathcal{T}, \omega_1, \omega_2) \to \pi_1(\mathcal{T}, \omega_1, \omega_3) , \tag{1.2}$$

and in particular $\pi_1(\mathcal{T}, \omega_1, \omega_2)$ is a homogeneous space for an action from the left of $\pi_1(\mathcal{T}, \omega_1)$ as well as for an action from the right of $\pi_1(\mathcal{T}, \omega_2)$ and is principal for both actions. For $\mathcal{T} = \mathrm{Loc}_X$ we have a map $\pi_1(X, x, y) \to \pi_1(\mathrm{Loc}_X, \omega_x, \omega_y)(\mathbf{Q})$, where $\pi_1(X, x, y)$ is the space of paths between x and y on X up to homotopy, given by

$$\gamma \mapsto (V \mapsto \text{parallel transport along the path } \gamma \in \mathrm{Hom}(\omega_x(V), \omega_y(V))) . \tag{1.3}$$

A tensor functor $F : \mathcal{S} \to \mathcal{T}$ clearly induces a map of path spaces

$$F^* : \pi_1(\mathcal{T}, \omega_1, \omega_2) \to \pi_1(\mathcal{S}, \omega_1 \circ F, \omega_2 \circ F) \tag{1.4}$$

which is explicitly given by sending $\gamma \in \mathrm{Hom}^{\otimes}(\omega_1 \otimes R, \omega_2 \otimes R)$ to $F^*(\gamma)$ given by $F^*(\gamma)_S = \gamma_{F(S)}$. This is clearly compatible with composition. In the case of categories of local systems, when F is induced by the continuous map $f : X \to Y$ it is easy to see that this map is compatible, via (1.3) with the usual map induced by f on path spaces.

2. Coleman's p-adic integration

In this section we recall the theory of Coleman's p-adic integration in several variables introduced in [**Bes1**] and make some additional constructions which are required for the current work. In some respects, the similar theory of Vologodsky [**V**] is more suitable for the current setting. However, because it is much less explicit it seems harder to adapt it to the tangential constructions of §3.

Recall from [**Bes1**] that the basic setup for Coleman integration theory is as follows: We have a field of characteristic p, κ, which we assume for simplicity to be algebraically closed and a discrete valuation ring $\mathcal{V}$ with fraction field K and residue field κ. A rigid triple is a triple $T = (X, Y, P)$ consisting of P, a formal p-adic $\mathcal{V}$-scheme, Y a closed κ-subscheme of P which is proper over $\operatorname{Spec}(\kappa)$ and X an open κ-subscheme of Y such that P is smooth in a neighborhood of X.

A simple case where rigid triples arise (see [**Bes1**, Definition 5.1]) is that of tight rigid triples. Let $\mathcal{X} \subset \mathcal{Y}$ be an open immersion of $\mathcal{V}$-schemes such that $\mathcal{X}$ is smooth and $\mathcal{Y}$ is complete. The associated triple $T_{(\mathcal{X},\mathcal{Y})} := (\mathcal{X} \otimes_{\mathcal{V}} \kappa, \mathcal{Y} \otimes_{\mathcal{V}} \kappa, \hat{\mathcal{Y}})$, where $\hat{\mathcal{Y}}$ is the p-adic completion of $\mathcal{Y}$ is called a tight rigid triple. An affine rigid triple is a tight rigid triple $T_{(\mathcal{X},\mathcal{Y})}$ with $\mathcal{X}$ affine.

Given a rigid triple as above we have the category of unipotent overconvergent isocrystals on T, denoted $\mathcal{U}n(T)$. This is the category of sheaves on the tube of X in P, overconverging into the tube of Y in P in the sense of Berthelot [**Ber**], with an integrable connection. This category depends, up to a unique isomorphism, only on X, so we will often denote it simply by $\mathcal{U}n(X)$.

A Frobenius endomorphism of X is a κ-linear endomorphism of X, which is some power of the geometric Frobenius. A Frobenius endomorphism ϕ induces an auto-functor ϕ^* on $\mathcal{U}n(X)$. Suppose that $T = T_{(\mathcal{X},\mathcal{Y})}$ is a tight rigid triple and that ϕ is the reduction of an endomorphism φ of the pair $(\mathcal{X}, \mathcal{Y})$. Then the functor ϕ^* is given by the obvious functor induced by φ on $\mathcal{U}n(T)$.

DEFINITION 2.1. A Frobenius fiber functor on $\mathcal{U}n(X)$ is a fiber functor ω together with an isomorphism of fiber functors (which we write as an equality to simplify matters) $\omega \circ \phi^* = \omega$

For two Frobenius fiber functors ω_1 and ω_2 there is an obvious action of ϕ^* on the path space $\pi_1(\mathcal{U}n(X), \omega_1, \omega_2)$ coming from the action (1.4). The main result of [**Bes1**] (Corollary 3.2) can be stated, in a slightly generalized form, as the following:

THEOREM 2.2. *Given any two Frobenius fiber functors on* $\mathcal{U}n(X)$*,* ω_1 *and* ω_2*, there exists a canonical invariant path, i.e., an isomorphism of fiber functors,* $a_{\omega_1,\omega_2} : \omega_1 \to \omega_2$*, which is fixed by* ϕ^**. We will call the isomorphism* $\omega_1 \to \omega_2$ *analytic continuation along Frobenius. It has the basic compatibility property that* $a_{\omega_2,\omega_3} \circ a_{\omega_1,\omega_2} = a_{\omega_1,\omega_3}$*.*

PROOF. We only sketch the main idea, which is rather simple, and refer to reader to loc. cit. for the full story. The key point is that for any Frobenius fiber functor ω there is an action ϕ^* on the fundamental group $\pi_1(\mathcal{U}n(X))$, which is a pro-unipotent group, and its Lie algebra is controlled by negative tensor powers of $H^1_{\mathrm{rig}}(X/K)$. The action of ϕ^* on $H^1_{\mathrm{rig}}(X/K)$ has strictly positive weights, and consequently one can show that the map

$$g \mapsto \phi^*(g)g^{-1}\ ,\ \pi_1(\mathcal{U}n(X), \omega) \to \pi_1(\mathcal{U}n(X), \omega)\ ,$$

is a bijection. Now, the space $\pi_1(\mathcal{U}n(X),\omega_1,\omega_2)$ is a $\pi_1(\mathcal{U}n(X),\omega_1)$-principal space and its ϕ^* action is compatible with the group action. From this it is very easy to deduce that there exists a unique ϕ^*-invariant element. The uniqueness immediately implies the compatibility. □

Usually, Frobenius fiber functors as above are obtained from geometric points $x : \operatorname{Spec}\kappa \to X$ by pullback, $\omega_x = x^*$, assuming that x is fixed by ϕ. Concretely, for an overconvergent isocrystal (M,∇) the fiber functor $\omega_x(M,\nabla)$ is realized as the vector space of horizontal sections of ∇ on the residue disc

$$U_x =]x[_P \text{ (=points reducing to } x)$$

which is the tube in the sense of Berthelot. Since each point of X will be fixed by some Frobenius automorphism it is not hard to see that the invariant path is independent of the choice of the Frobenius endomorphism.

An abstract Coleman function on T is defined to be a triple (M,s,h) where

- $M = (M,\nabla)$ is a unipotent isocrystal on T.
- $s \in \operatorname{Hom}(M,\mathcal{O})$.
- h is a collection of sections, $\{h_x \in M(U_x),\ x \in X\}$, with $\nabla(h_x) = 0$, which correspond to each other via analytic continuation along Frobenius.

One can evaluate an abstract Coleman function on each residue disc U_x by taking $s(h_x)$. Coleman functions are then equivalence classes of abstract Coleman functions under a relation which essentially guarantees that their evaluations will be identical. In fact, a Coleman function is a connected component of the category of abstract Coleman functions, where a morphism $(M_1,s_1,h_1) \to (M_2,s_2,h_2)$ is a horizontal map $u : M_1 \to M_2$ such that $s_2 \circ u = s_1$ and $u(h_1) = h_2$. It is shown in [**Bes1**, Corollary 4.13] that two Coleman functions which coincide on an open subset of the tubular neighborhood $]X[_P$ of X in P coincide everywhere.

REMARK 2.3. It will be important to note how the theory can be used to define iterated integrals. For a similar and more detailed discussion see the proof of Theorem 4.15 in [**Bes1**]. Suppose that we have a collection of Coleman functions F_i, arising from the abstract Coleman functions (M_i,s_i,h_i), and one-forms ω_i for $i = 1,\dots,k$. Suppose now that the locally analytic differential form $\omega = \sum F_i\omega_i$ is closed, and that we wish to find a Coleman function F such that $dF = \omega$. We may begin by constructing the connection $M = (\oplus M_i) \oplus \mathcal{O}_T$ with the connection given by the formula

$$\nabla(m_1,\cdots,m_k,f) = \left(\nabla_1 m_1,\cdots,\nabla_k m_k, df - \sum \omega_i s_i(m_i)\right).$$

This is clearly a unipotent connection. If it is integrable we immediately obtain our required Coleman function represented by the abstract Coleman function (M,s,y), where s is the projection on the last factor and h is obtained by choosing one residue disc U, extending the sum of the h_i to a horizontal section of M on U, and then analytically continuing along Frobenius. In general, ∇ will not be integrable. However, there is a maximal integrable subconnection (M^{int},∇) of (M,∇) [**Bes1**, (2.3)]. The fact that $\sum F_i\omega_i$ is closed guarantees that we may choose h to be in $M^{\text{int}}(U)$ hence we obtain our Coleman function from (M^{int},s,h).

One can apply the theory to other fiber functors. A simple case was already discussed in [**Bes1**, Section 5]. Here we will need an easy generalization (to the two dimensional case for simplicity) as follows: Suppose that $T = (X,Y,P) = T_{(\mathcal{X},\mathcal{Y})}$ is a rigid triple and that $y \in D = Y - X$ is a closed point and that D is locally given near y by the reduction of two parameters t_1 and t_2 on Y. Let A be the ring of Laurent series with coefficients in K,

$$A := \left\{ f(t_1, t_2) = \sum_{(i,j) \in \mathbf{Z}^2} a_{ij} t_1^i t_2^j \;\middle|\; f \text{ converges on } r < |t_1|, |t_2| < 1 \text{ for some } r \right\}$$

It follows from [**Ber**] that overconvergent isocrystals on T give rise to a connection on a (free) A-module.

PROPOSITION 2.4. *An overconvergent unipotent connection on T has a full set of solutions in $B := A[\log(t_1), \log(t_2)]$.*

PROOF. Overconvergent unipotent crystals are solved by iterated integration and the ring B is easily seen to be closed under partial integration with respect to t_1 and t_2. □

Note that the ring B is independent of choice of parameters t_1 and t_2. The following operation will be required later.

DEFINITION 2.5. The constant term with respect to the parameters t_1 and t_2 of $f \in B$ is defined as follows: Let $f = \sum f_{ij} \log^i(t_1) \log^j(t_2)$. Let $f_{00} = \sum a_{ij} t_1^i t_2^j$. Then the constant term is a_{00}

Note that the constant term does depend on the choice of parameters.

DEFINITION 2.6. The fiber functor ω_y of $\mathcal{U}n(X)$ associates to a unipotent overconvergent isocrystal the vector space of solutions in B.

Suppose that we have an endomorphism φ of the pair $(\mathcal{X}, \mathcal{Y})$ reducing to a Frobenius endomorphism fixing y (a sufficiently high power of a Frobenius endomorphism will fix y). Then, pulling back by φ gives an isomorphism $\omega_y \circ \varphi \cong \omega_y$ making it a Frobenius fiber functor in the sens of Definition 2.1. It is therefore possible to analytically continue along Frobenius to or from ω_y.

To end this section we note that by the naturality of the analytic continuation along Frobenius, it clearly extends to pro-unipotent isocrystals. We also note that as long as the underlying connections are defined over a discretely valued subfield it is possible to work over $\mathbf{C}_p$.

3. Tangential basepoints

In this section we first recall the theory of the tangential basepoint in the de Rham setting due to Deligne [**D1**]. In its simplest form this theory allows one to define fiber functors for the category of integrable connections on a curve associated with a tangent point "at infinity". We explain Deligne's interpretation of constant terms in terms of these tangential basepoints. In loc. cit. a higher dimensional theory is also sketched with no details, but these can easily be filled in, as we will do. We then discuss the possibility of analytically continuing solutions of a unipotent integrable connection along Frobenius a la Coleman to such fiber functors. We finally explain how one can iterate the construction of tangential basepoints in two different ways and get the same result and we close with some applications to Coleman integration theory.

We first recall the construction of Deligne [**D1**, 15. Theorie algébrique]. Let C be a curve over a field K of characteristic 0, smooth at a point P, and let t be a parameter at P (Deligne immediately passes to the completion at P but we will not do the same). Suppose (M, ∇) is a connection on $C - \{P\}$ with logarithmic singularities along P. This means that locally near P the connection ∇ can be written as $\nabla = d + \Gamma$, where Γ is a section of $\mathrm{End}(M) \otimes \Omega^1_C(\log P)$ with at most a simple pole at P. The parameter t induces naturally a parameter $\bar{t}$ on the tangent space $T_P(C)$ (the linear parameter taking the value 1 at the derivation d/dt). We associate with this data the connection on $T_P(C) - 0$ on the trivial bundle with fiber M_P (fiber of M at P), given by $\mathrm{Res}_P(\nabla) := d + (\mathrm{Res}_P \Gamma)\,\mathrm{dlog}(\bar{t})$. This clearly defines a functor Res_P.

An easy computation shows that the functor Res_P does not depend on the parameter t. Deligne gives an alternative, coordinate free description of the same construction, making this fact evident. The valuation v_P on the fraction field $K(C)$ gives an algebra filtration F_P on this field and there is an obvious canonical isomorphism between the associated graded algebra $\mathrm{Gr}_P K(C)$ and the coordinate ring of $T_P(C) - 0$, given by sending a cotangent vector $d_P f$, thought of as a linear function on $T_P(C)$, to the image of f in $\mathrm{Gr}^1_P K(C)$.

Let $\Omega^1_{\mathcal{O}_C/K}(\log P)$ be the sheaf of differential forms with log singularities along P. We give $\Omega^1_{K(C)/K} = K(C) \otimes_{\mathcal{O}_C} \Omega^1_{\mathcal{O}_C/K}(\log P)$ the filtration induced from the filtration on the first term. The filtrations on $K(C)$ and on $\Omega^1_{K(C)/K}$ induces a filtration on $M \otimes_{\mathcal{O}_C} K(C)$ and on $M \otimes_{\mathcal{O}_C} \Omega^1_{K(C)/K}$ and the assumption that ∇ has log singularities implies that it preserves the filtration. It is easy to see that the induced connection on the associated graded is exactly $\mathrm{Res}_P(\nabla)$

The construction of Res_P gives us the option of producing more fiber functors for the category of connections on C by taking the fiber at a point of $T_P(C) - 0$. A particular case of this construction gives the notion of a constant term for a horizontal section of this connection as we now explain. Suppose that the connection ∇ is unipotent. Then it is very easy to see that one can find a basis of formal horizontal sections to ∇ near P with coefficients in the ring $K[[t]][\log(t)]$, where $\log(t)$ is treated as a formal variable whose derivative is dt/t. Let v be such a horizontal section. There is a sense in which we can specialize v to the fiber M_P, namely, taking the constant term.

Definition 3.1. The constant term of v is obtained by formally setting $t = \log(t) = 0$.

The justification for this definition is that over the complex numbers one has $\lim_{t\to 0} t\log(t) = 0$. Thus in situations where the solutions has coefficients in $K[[t]] + tK[[t]][\log(t)]$ this is indeed the constant term. The same can be argued p-adically, provided one takes the appropriate notion of limit. It is important to note that the situation described above is indeed what happened for p-adic multiple polylogarithms near 1, by the main result of [**F1**].

Similarly, there is a basis of (this time global) solutions to $\mathrm{Res}_P \nabla$ with coefficients in $K[\log(\bar{t})]$. In fact, all the solutions are of the form $\exp(\mathrm{Res}_P \Gamma \cdot \log \bar{t}) \cdot v$ with $v \in M_P$ and the exponential is a finite sum as $\mathrm{Res}_P \Gamma$ is nilpotent. In this case we can again take the constant term. This can now be interpreted as simply evaluating at 1 with the convention that $\log(1) = 0$. Thus, we may formally interpret taking the constant term as continuing to the tangential vector 1 at P. To make this more than a mere heuristic, though, one needs to introduce a topology.

It can be made precise in the complex case [**D1**] and we will show this also in the p-adic case (see below Proposition 4.5).

The higher dimensional generalization is now fairly clear. Suppose that a smooth variety X is given and in it a divisor $D = \sum_{i\in I} D_i$ with normal crossings. We assume that all the components D_i are smooth. For a subset $J \subset I$ we set $D_J = \cap_{j\in J} D_j$. Let N_J be the normal bundle to D_J and let N_J^0 be the complement in N_J of $N_{J'}|_{D_J}$ for $J' \subset J$. Note that one only need to take J' smaller by one index and that $N_\emptyset$ is considered as the zero section. Thus, for example, N_j^0 is the (one-dimensional) normal space to D_j minus the zero section. Finally, we denote by N_J^{00} the restriction of N_J^0 to $D_J^0 := D_J - \cup_{j\notin J} D_j$. Following Deligne we construct, given a connection on X with logarithmic singularities along D, a connection on every N_J^{00} with logarithmic singularities "at infinity". Infinity here means the union of the hyperplane at infinity for the normal bundle N_J, the hyperplanes $N_J - N_J^0$, and $N_J|_{D_J - D_J^0}$.

For each $j \in J$ consider the valuation v_j on $K(X)$ associated with the divisor D_j. Let $\mathcal{O}_X(D^{-1})$ be the localization of $\mathcal{O}_X$ at D. There exists a multi-filtration F_J on $\mathcal{O}_X(D^{-1})$, indexed by tuples $\chi = (\chi_j)_{j\in J}$, such that F_J^χ is the $\mathcal{O}_X$-module generated by $\{f \in \mathcal{O}_X(D^{-1}),\ v_j(f) \geq \chi_j \text{ for all } j \in J\}$. It is easy to see that $\operatorname{Spec}(\operatorname{Gr}_J \mathcal{O}_X(D^{-1}))$ is precisely N_J^{00}. Again, a natural map sends a differential form df, viewed as a function on N_J^{00}, linear in the bundle direction, to the image of f in $\operatorname{Gr}_J^1 \mathcal{O}_X(D^{-1})$.

REMARK 3.2. Let $\operatorname{Spec}(A)$ be an affine patch on X with coordinates $t_i \in A$, $i \in I$, such that $D_i \cap \operatorname{Spec}(A)$ is defined by $t_i = 0$. Let $B = A/(t_j, j \in J)$, such that $D_J \cap \operatorname{Spec}(A) = \operatorname{Spec}(B)$ Let $t^\chi := \prod_{j\in J} t_j^{\chi_j}$. Then $F_J^\chi = At^\chi$ and it follows easily that $\operatorname{Gr}_J \cong B[t_j^\pm, j \notin J]$

Now suppose we have a connection with logarithmic singularities along D, $\nabla : M \to M \otimes_{\mathcal{O}_X} \Omega_X^1(\log D)$. We give $\Omega_X^1(D^{-1}) = \Omega_X^1(\log D) \otimes \mathcal{O}_X(D^{-1})$ the induced filtrations from the filtration on $\mathcal{O}_X(D^{-1})$. It is easy to see that the differential d preserves the filtration. Now let $M(D^{-1}) = M \otimes \mathcal{O}_X(D^{-1})$ and $M \otimes \Omega_X^1(D^{-1})$ have the induced filtrations. It follows that the extended connection $\nabla : M(D^{-1}) \to M \otimes \Omega_X^1(D^{-1})$ respects the filtration. The connection we are looking for is the *gr* of this connection. It is evident from this construction that if ∇ is flat, so will be the resulting connection.

DEFINITION 3.3. We call the resulting connection the residue connection on N_J^{00} along D_J and denote it by $\operatorname{Res}_{D,J} \nabla$.

Note that the construction really depends on both D and J and not just on D_J. It is evident that $\operatorname{Res}_{D,J}$ is a tensor functor.

As a consequence of the construction above we obtain more fiber functors for connections.

DEFINITION 3.4. If $x \in N_J^{00}$ we let $\omega_{x,J}$ be the fiber functor $\omega_{x,J} := \omega_x \circ \operatorname{Res}_{D,J}$.

Suppose now that $f : (X', D') \to (X, D)$ is a morphism (i.e., takes D' to D) and that $f^* D_j = n_j D'_j$. We get an induced map

$$N_f : N_{J'}^{00} \to N_J^{00} ,$$

by looking at the map on the graded sheaf of rings, obtained from f, sending Gr_J^χ to $\mathrm{Gr}_{J'}^{\overline{n}\cdot\chi}$, where $\overline{n}\cdot\chi$ is the vector with components $n_j\chi_j$. The following Lemma is trivial.

LEMMA 3.5. *Suppose that* Fr *is the absolute Frobenius. Then* N_{Fr} *is also the absolute Frobenius.*

The constructions above can be iterated, and they commute in a sense we will now explain. For simplicity, we now assume that the divisor D has only two components, D_1 and D_2. Consider N_1. It has on it a divisor D' with normal crossings consisting of D_1' = the zero section and D_2' = restriction of N_1 to $D_{12} = D_1 \cap D_2$. The complement of this divisor is exactly N_1^{00}. We have the following result.

PROPOSITION 3.6. *Write* $N = N^{00}_{D_1',D_2'}$ *be the construction above performed on* N_1 *with respect to* D'. *Then there exists a natural isomorphism* $\psi : N \xrightarrow{\sim} N_{12}^{00}$ *in such a way that for any connection* M *on* X *with logarithmic singularities along* D *we have a natural isomorphism* $\psi^* \operatorname{Res}_{D,12} M \to \operatorname{Res}_{D',12} \operatorname{Res}_{D,1} M$.

PROOF. The isomorphism ψ is essentially the isomorphism, existing on any object with two filtrations, $\mathrm{Gr}_{12} \cong \mathrm{Gr}_2 \mathrm{Gr}_1$, applied to the filtrations we have. The isomorphism on the Res's is then a completely formal consequence of this. □

REMARK 3.7. To make very concrete the situation of the last Proposition, and to make it clear how it is going to be used, consider the simplest situation, X is the affine plane $\mathbf{A}^2$ with coordinates x and y, $D = D_1 \cup D_2 = \{x = 0\} \cup \{y = 0\}$. Clearly, N_1, the tangent space to $\{x = 0\}$, is again isomorphic to $\mathbf{A}^2$, and we may consider on it the curve C which is the section $x = 1$. Proposition 3.6, together with the functoriality of the Res construction, allows us to get the following: Consider the fiber functor on connections on X with logarithmic singularities along D: First apply Res_1, then restrict to C, finally consider the fiber functor at the tangential point 1 to C at $y = 0$. Then this fiber functor is naturally isomorphic to the fiber functor at the normal vector $(1, 1)$ at the point $(0, 0)$.

4. Analytic continuation to tangential points

We now turn to the relation between the tangential theory described in the previous section and the analytic continuation along Frobenius explained in §2. Suppose then that K is a p-adic field with ring of integers $\mathcal{O}_K$ and residue field κ. Let X be a smooth scheme over $\mathcal{O}_K$ and let $D = \sum D_i$ be a divisor with relative normal crossings on X. We assume as before that the components D_i are smooth. The schemes N_J^{00} are then also smooth over $\mathcal{O}_K$. Define the category $Un(X, D)$ of unipotent connections on X_K with logarithmic singularities along D_K. This is a Tannakian category over K. Suppose that $x \in (X - D)(\kappa)$ and $y \in N_J^{00}(\kappa)$ are two points on the special fibers. Restricting $(M, \nabla) \in Un(X, D)$ to U_x and taking horizontal sections we obtain a fiber functor ω_x on $Un(X, D)$ as in §2. Similarly, restricting $\operatorname{Res}_{D_K,J} \nabla$ to the residue disk U_y of y in N_J^{00} and taking horizontal sections we obtain a fiber functor ω_y. Then we have the following result.

THEOREM 4.1. *For any two points* x *and* y, *which are either in* $(X - D)(\kappa)$ *or in* $N_J^{00}(\kappa)$ *for some* J, *there exists a canonical isomorphism (analytic continuation* *along a Frobenius invariant path), $a_{x,y} : \omega_x \xrightarrow{\sim} \omega_y$, such that the following properties are satisfied.*

(1) *These isomorphisms are compatible in the sense that $a_{y,z} \circ a_{x,y} = a_{x,z}$.*
(2) *If $x, y \in (X - D)(\kappa)$ then $a_{x,y}$ is the isomorphism of Theorem 2.2 pulled back via $Un(X, D) \to \mathcal{U}n((X - D)_\kappa)$*
(3) *If $x, y \in N_J^{00}(\kappa)$ (for the same J) then $a_{x,y}$ is the isomorphism of Theorem 2.2 pulled back via $\mathrm{Res}_{D_K,J} : Un(X, D) \to \mathcal{U}n((N_J^{00})_\kappa)$*

The Frobenius invariant path is functorial with respect to morphisms $(X', D') \to (X, D)$ in the obvious sense.

PROOF. Since the nerve of an affine covering is contractible, we are easily reduced to the following case: $X = \mathrm{Spec}(A)$ is affine and D is defined on it by the vanishing of $t_1 \cdot t_2 \cdots t_k$, where t_i are parameters. Let $A^\dagger$ be the weak completion of A in the sense of [**Put**] and let $B^\dagger$ be the weak completion of $A[(t_1 \cdots t_k)^{-1}]$. We clearly have $A^\dagger \subset B^\dagger$. We may further assume that there exists a map, whose reduction is a Frobenius endomorphism, $\varphi : B^\dagger \to B^\dagger$ preserving $A^\dagger$ and sending t_i to t_i^q. We consider the de Rham complex of $A^\dagger$ with logarithmic singularities along D_K,

$$\mathcal{DR}(A_K^\dagger, D) : \ A_K^\dagger \xrightarrow{d} \sum \Omega^1(A_K^\dagger / K)\, \mathrm{dlog}(t_i) \xrightarrow{d} \cdots$$

and the de Rham complex $\mathcal{DR}(B_K^\dagger)$ of $B_K^\dagger$. Note that the cohomology of this last complex is the Monsky-Washnitzer = rigid cohomology $H_{\mathrm{rig}}((X - D)_\kappa)$. There is an obvious map

$$\mathcal{DR}(A_K^\dagger, D) \to \mathcal{DR}(B_K^\dagger) \tag{4.1}$$

and an action of φ^* on both complexes compatible with this map.

PROPOSITION 4.2. *The map (4.1) is a quasi-isomorphism.*

PROOF. This follows essentially from the arguments in [**BC**, Section 6] □

We only need the fact that the induced map on H^1 is an injection. This immediately implies the following.

COROLLARY 4.3. *The map $\varphi^* : H^1(\mathcal{DR}(A^\dagger, D)) \to H^1(\mathcal{DR}(A^\dagger, D))$ is independent of the choice of φ and has strictly positive weights,*

PROOF. The same is true for $H^1_{\mathrm{rig}}((X - D)_\kappa)$ as in the proof of Theorem 2.2. □

Define now the category $Un(A^\dagger, D)$ of unipotent integrable $A^\dagger$-connections with logarithmic singularities along D. There is an action of φ^* on this category, hence on its fundamental group and torsors. By the same argument as in the proof of Theorem 2.2 we have using Corollary 4.3, that there exists a unique invariant path between any two fiber functors on $Un(A^\dagger, D)$ and that this is independent of the choice of φ. We have such a fiber functor ω_x for $x \in (X - D)_\kappa$, and since there is an obvious residual functor $Un(A^\dagger, D) \to \mathcal{U}n((N_J^{00})_\kappa)$, defined in the same way as in Definition 3.3, we also have fiber functors ω_y for $y \in (N_J^{00})_\kappa)$. The fiber functors discussed before the theorem are the compositions of the ones we just defined with the restriction functor $Un(X, D) \to Un(A^\dagger, D)$ and we obtain our required path by pulling back along this restriction. With this definition, (1) and (2) are essentially clear, while (3) follows easily because by Lemma 3.5 we have that φ induces residually a morphism on N_J^{00} whose reduction is a Frobenius automorphism, and consequently the pullback via the residue functor of a Frobenius invariant path is φ-invariant. □

We can use the extension of Coleman functions described here to make a theory of Coleman functions "of algebraic origin" on the pair (X, D). The abstract Coleman functions are then triples (M, s, h) with $M \in Un(X, D)$ and $s : M \to \mathcal{O}_{X_K}$, while h is a compatible system of local horizontal sections as before. Theorem 4.1 allows us to define a map $\mathrm{Res}_{D,J}$ from these algebraic Coleman functions to Coleman functions on the normal bundle N_J^{00}. It is defined by

$$\mathrm{Res}_{D,J}(M, s, h) = (\mathrm{Res}_{D,J} M, \mathrm{Gr}\, s, h)$$

where h now refers to the induced compatible system of horizontal sections whose existence is guaranteed by Theorem 4.1.

In particular, we obtain from a Coleman function f as above a Coleman function $f^{(D)}$ on the normal bundle to D. The following propositions suggests how to compute it.

Proposition 4.4. *Suppose that we have the relation* $df = \sum \omega_i g_i$ *with* f *and* g_i *Coleman functions of algebraic origin. Then we have the relation* $df^{(D)} = \sum (\mathrm{Res}_D\, \omega_i) g_i^{(D)}$, *where, if* ω *is locally written as* $\omega' + h\, \mathrm{dlog}(t)$, *with* t *the defining parameter for* D, *then* $\mathrm{Res}_D(\omega) = \omega'|_D + h|_D\, \mathrm{dlog}(\bar{t})$.

Proof. We consider the way f can be written as a Coleman function following Remark 2.3. In that description suppose that the connection M constructed there was indeed integrable. Then the result is easily obtained by computing the residual connection for M. The result remains true if M is not integrable. To see this we first note that the M^{int} construction is algebraic by Lemma 2.4 in [**Bes2**] and the discussion following it. One further sees that it is a connection with logarithmic singularities along D_K. It follows that we may obtain f from the abstract Coleman function (M^{int}, s, h) as in Remark 2.3. The residual construction is not limited to integrable connections and is functorial. Thus there is a horizontal map of connections $\mathrm{Res}_{D,i}(M^{\mathrm{int}}) \to \mathrm{Res}_{D,i}(M)$. Consequently, $f^{(D)}$, which is defined on the normal bundle to D_i by $\mathrm{Res}_{D,i}(M^{\mathrm{int}}, \mathrm{Gr}\, s, h)$, still satisfies the differential equation given by $\mathrm{Res}_{D,i}(M)$ □

We can now give, in the p-adic case, a tangential basepoint interpretation of the constant term.

Proposition 4.5. *In the situation of Definition 2.5 the constant term with respect to the parameters* t_1 *and* t_2 *of a horizontal section of a unipotent overconvergent isocrystal* ∇ *on* T *is the evaluation at the normal vector* $\bar{t}_1 = 1, \bar{t}_2 = 1$ *of the analytic continuation along Frobenius of this horizontal section.*

Proof. We only prove the 1-dimensional analogue as the proof is similar. In fact, we prove something stronger. Suppose (using the notation of this section) that the connection is locally given by $d + \Gamma$. Let f be a horizontal section for the connection ∇ and let v be its constant term. We claim that the analytic continuation of f is precisely the solution $c(f) := \exp(\mathrm{Res}_P \Gamma \cdot \log \bar{t}) \cdot v$ (the result is then proved by specializing to $\bar{t} = 1$). We already know that this is indeed a solution of $\mathrm{Res}_P(\nabla)$. The association $\nabla \mapsto (f \mapsto c(f))$ is clearly compatible with direct sums and tensor products hence defines a path. To show that this is a Frobenius invariant path we may assume that φ is such that $\varphi^*(t) = t^p$ The constant term of $\varphi^*(f)$ remains v while $\varphi^*\nabla$ is associated with $\varphi^*\Gamma$. It follows that

$$c(\varphi^* f) = \exp(\mathrm{Res}_P(\varphi^*\Gamma) \cdot \log \bar{t}) \cdot v = \exp(p\,\mathrm{Res}_P\,\Gamma \cdot \log \bar{t}) \cdot v = c(f)(\bar{t}^p)\ ,$$

which proves our claim. □

5. The set up of the moduli space $\mathcal{M}_{0,5}$

This section is devoted to giving a quick review on basic materials of the moduli space $\mathcal{M}_{0,5}$ of genus 0 curves with 5 distinct marked points and its Deligne-Mumford compactification $\overline{\mathcal{M}_{0,5}}$. The two (and one) variable p-adic multiple polylogarithms are introduced as Coleman functions on $\mathcal{M}_{0,5}$.

The moduli space $\mathcal{M}_{0,5} = \{(P_i)_{i=1}^5 \in (\mathbf{P}^1)^5 | P_i \neq P_j \ (i \neq j)\}/PGL(2)$ is identified with

$$\{(x, y) \in \mathbf{A}^2\} \backslash \{x = 0\} \cup \{y = 0\} \cup \{x = 1\} \cup \{y = 1\} \cup \{xy = 1\}. \tag{5.1}$$

This identification is given by sending (x, y) to 5 marked points in $\mathbf{P}^1$ given by $(0, x, 1, \frac{1}{y}, \infty)$. The symmetric group S_5 acts on $\mathcal{M}_{0,5}$ by $\sigma(P_i) = P_{\sigma^{-1}(i)}$ $(1 \leqslant i \leqslant 5)$ for $\sigma \in S_5$. Especially for $c = (1, 3, 5, 2, 4) \in S_5$ its action is described by $x \mapsto \frac{1-y}{1-xy}$, $y \mapsto x$.

The Deligne-Mumford compactification of $\mathcal{M}_{0,5}$ is denoted by $\overline{\mathcal{M}_{0,5}}$. This space classifies stable curves of $(0, 5)$-type and the above S_5-action extends to the action on $\overline{\mathcal{M}_{0,5}}$. This space is the blow-up of $(\mathbf{P}^1)^2 (\supset \mathcal{M}_{0,5})$ at $(x, y) = (1, 1)$, $(0, \infty)$ and $(\infty, 0)$. The complement $\overline{\mathcal{M}_{0,5}} - \mathcal{M}_{0,5}$ is a divisor with 10 components: $\{x = 0\}$, $\{y = 0\}$, $\{x = 1\}$, $\{y = 1\}$, $\{xy = 1\}$, $\{x = \infty\}$, $\{y = \infty\}$ and 3 exceptional divisors obtained by blowing up at $(x, y) = (1, 1)$, $(\infty, 0)$ and $(0, \infty)$. In particular for our convenience we denote $\{y = 0\}$, $\{x = 1\}$, the exceptional divisor at $(1, 1)$, $\{y = 1\}$ and $\{x = 0\}$ by D_1, D_2, D_3, D_4 and D_5 (or sometimes D_0) respectively. It is because $c^i(D_0) = D_i$. These five divisors form a pentagon and we denote each vertex $D_i \cap D_{i-1}$ by P_i. Hence we have $c^i(P_0) = P_i$. The two dimensional affine space $U_1 = Spec\mathbf{Q}[x, y]$ gives an open affine subset of $\overline{\mathcal{M}_{0,5}}$. The S_5-action gives other open subsets $U_i = c^{i-1}(U_1) = Spec\mathbf{Q}[z_i, w_i]$ $(1 \leqslant i \leqslant 5)$ in $\overline{\mathcal{M}_{0,5}}$ where $(z_1, w_1) = (x, y)$, $(z_2, w_2) = (y, \frac{1-x}{1-xy})$, $(z_3, w_3) = (\frac{1-x}{1-xy}, 1 - xy)$, $(z_4, w_4) = (1 - xy, \frac{1-y}{1-xy})$ and $(z_5, w_5) = (\frac{1-y}{1-xy}, x)$.

For $\mathbf{a} = (a_1, \cdots, a_k) \in \mathbf{Z}_{>0}^k$, $\mathbf{b} = (b_1, \cdots, b_l) \in \mathbf{Z}_{>0}^l$, and x, $y \in \mathbf{Q}_p$ with $|x|_p < 1$ and $|y|_p < 1$ we define **two variable p-adic multiple polylogarithm** by

$$\mathrm{Li}_{\mathbf{a},\mathbf{b}}(x, y) := \sum_{\substack{0<m_1<\cdots<m_k \\ <n_1<\cdots<n_l}} \frac{x^{m_k} \qquad y^{n_l}}{m_1^{a_1} \cdots m_k^{a_k} n_1^{b_1} \cdots n_l^{b_l}} \in \mathbf{Q}_p[[x, y]],$$

and for $\mathbf{c} = (c_1, \cdots, c_h) \in \mathbf{Z}_{>0}^h$ **one variable p-adic multiple polylogarithm** by

$$\mathrm{Li}_{\mathbf{c}}(y) := \sum_{0<m_1<\cdots<m_h} \frac{y^{m_h}}{m_1^{c_1} \cdots m_h^{c_h}} \in \mathbf{Q}_p[[y]] \subset \mathbf{Q}_p[[x, y]].$$

It is easy to see that these functions satisfy the following differential equations.

(5.2)

$$\frac{d}{dx}\mathrm{Li}_{\mathbf{a},\mathbf{b}}(x,y)=$$
$$\begin{cases} \frac{1}{x}\mathrm{Li}_{(a_1,\cdots,a_{k-1},a_k-1),\mathbf{b}}(x,y) & \text{if } a_k\neq 1,\\ \frac{1}{1-x}\mathrm{Li}_{(a_1,\cdots,a_{k-1}),\mathbf{b}}(x,y)-\left(\frac{1}{x}+\frac{1}{1-x}\right)\mathrm{Li}_{(a_1,\cdots,a_{k-1},b_1),(b_2,\cdots,b_l)}(x,y) & \\ & \text{if } a_k=1,k\neq 1,l\neq 1,\\ \frac{1}{1-x}\mathrm{Li}_{\mathbf{b}}(y)-\left(\frac{1}{x}+\frac{1}{1-x}\right)\mathrm{Li}_{(b_1),(b_2,\cdots,b_l)}(x,y) & \text{if } a_k=1,k=1,l\neq 1,\\ \frac{1}{1-x}\mathrm{Li}_{(a_1,\cdots,a_{k-1}),\mathbf{b}}(x,y)-\left(\frac{1}{x}+\frac{1}{1-x}\right)\mathrm{Li}_{(a_1,\cdots,a_{k-1},b_1)}(xy) & \\ & \text{if } a_k=1,k\neq 1,l=1,\\ \frac{1}{1-x}\mathrm{Li}_{\mathbf{b}}(y)-\left(\frac{1}{x}+\frac{1}{1-x}\right)\mathrm{Li}_{\mathbf{b}}(xy) & \text{if } a_k=1,k=1,l=1, \end{cases}$$

$$\frac{d}{dy}\mathrm{Li}_{\mathbf{a},\mathbf{b}}(x,y)=\begin{cases} \frac{1}{y}\mathrm{Li}_{\mathbf{a},(b_1,\cdots,b_{l-1},b_l-1)}(x,y) & \text{if } b_l\neq 1,\\ \frac{1}{1-y}\mathrm{Li}_{\mathbf{a},(b_1,\cdots,b_{l-1})}(x,y) & \text{if } b_l=1,l\neq 1,\\ \frac{1}{1-y}\mathrm{Li}_{\mathbf{a}}(xy) & \text{if } b_l=1,l=1, \end{cases}$$

$$\frac{d}{dx}\mathrm{Li}_{\mathbf{c}}(xy)=\begin{cases} \frac{1}{x}\mathrm{Li}_{(c_1,\cdots,c_{h-1},c_h-1)}(xy) & \text{if } c_h\neq 1,\\ \frac{y}{1-xy}\mathrm{Li}_{(c_1,\cdots,c_{h-1})}(xy) & \text{if } c_h=1,h\neq 1,\\ \frac{y}{1-xy} & \text{if } c_h=1,h=1, \end{cases} \tag{5.3}$$

$$\frac{d}{dy}\mathrm{Li}_{\mathbf{c}}(xy)=\begin{cases} \frac{1}{y}\mathrm{Li}_{(c_1,\cdots,c_{h-1},c_h-1)}(xy) & \text{if } c_h\neq 1,\\ \frac{x}{1-xy}\mathrm{Li}_{(c_1,\cdots,c_{h-1})}(xy) & \text{if } c_h=1,h\neq 1,\\ \frac{x}{1-xy} & \text{if } c_h=1,h=1, \end{cases}$$

$$\frac{d}{dx}\mathrm{Li}_{\mathbf{c}}(y)=0 \tag{5.4}$$

$$\frac{d}{dy}\mathrm{Li}_{\mathbf{c}}(y)=\begin{cases} \frac{1}{y}\mathrm{Li}_{(c_1,\cdots,c_{h-1},c_h-1)}(y) & \text{if } c_h\neq 1,\\ \frac{1}{1-y}\mathrm{Li}_{(c_1,\cdots,c_{h-1})}(y) & \text{if } c_h=1,h\neq 1,\\ \frac{1}{1-y} & \text{if } c_h=1,h=1, \end{cases}$$

By the above differential equations, $Li_{\mathbf{a},\mathbf{b}}(x,y)$, $Li_{\mathbf{c}}(xy)$ and $Li_{\mathbf{c}}(y)$ are all iterated integrals of $\frac{dx}{x}$, $\frac{dx}{1-x}$, $\frac{dy}{y}$, $\frac{dy}{1-y}$ and $\frac{xdy+ydx}{1-xy}$, differential forms over $\mathcal{M}_{0,5}$. Whence they are obtained from some triple (M,s,y) over $\mathcal{M}_{0,5}$. We interpret them as Coleman functions over the rigid triple $(\mathcal{M}_{0,5},\overline{\mathcal{M}_{0,5}})$. This means that they are analytically continued to $\mathcal{M}_{0,5}(\mathbf{Q}_p)$ as Coleman functions by the methods of analytically continuation along Frobenius in §2.

6. Analytic continuation of two variable p-adic multiple polylogarithms

In this section two (and one) variable p-adic multiple polylogarithms are analytically continued into $N_{D_i}^{00}$ ($i \in \mathbf{Z}/5$), which is a Zariski open subset of the normal bundle N_{D_i} of the divisor D_i introduced in the previous section.

NOTATION 6.1. For a Coleman function f over $\mathcal{M}_{0,5}$, $f^{(D_i)}$ means the analytic continuation of f to $N_{D_i}^{00}$ ($i \in \mathbf{Z}/5$). For $\mathbf{a} = (a_1, \cdots, a_k) \in \mathbf{Z}_{>0}^k$ and $\mathbf{b} = (b_1, \cdots, b_l) \in \mathbf{Z}_{>0}^l$, $F_{\mathbf{a},\mathbf{b}}$ stands for the Coleman function $Li_{\mathbf{a},\mathbf{b}}(x,y) - Li_{\mathbf{ab}}(xy)$ and for $\mathbf{c} = (c_1, \cdots, c_h) \in \mathbf{Z}_{>0}^h$, $G_{\mathbf{c}}$ stands for the Coleman function $Li_{\mathbf{c}}(xy) - Li_{\mathbf{c}}(y)$ over $\mathcal{M}_{0,5}$.

LEMMA 6.2. *$F_{\mathbf{a},\mathbf{b}}^{(D_1)} = 0$ and $G_{\mathbf{c}}^{(D_1)} = 0$ for any index* $\mathbf{a}$, $\mathbf{b}$ *and* $\mathbf{c}$.

PROOF . The constant terms of $Li_{\mathbf{a},\mathbf{b}}(x,y)$, $Li_{\mathbf{c}}(xy)$ and $Li_{\mathbf{c}}(y)$ at the origin P_5 are zero because there are no constant terms in their power series expansions. We take their differentials and take their residues at $y = 0$. It gives 0 by induction because each term will be a multiple polylogarithm with one lower weight than the original one. Whence $Li_{\mathbf{a},\mathbf{b}}(x,y)$, $Li_{\mathbf{c}}(xy)$ and $Li_{\mathbf{c}}(y)$ are identically zero. It gives our claim. □

LEMMA 6.3. *$F_{\mathbf{a},\mathbf{b}}^{(D_2)} \equiv 0$ if* $\mathbf{a}$ *is admissible* [1] *and $G_{\mathbf{c}}^{(D_2)} \equiv 0$ for any index* $\mathbf{c}$.

PROOF . On the affine coordinate (z_2, w_2) for U_2, the divisor D_2 is defined by $w_2 = 0$. By using $dx = \frac{w_2(1-w_2)}{(z_2w_2-1)^2}dz_2 + \frac{z_2-1}{(z_2w_2-1)^2}dw_2$ and $dy = dz_2$, we obtain the following from the differential equations in (5.2) $\sim$ (5.4).

$$\frac{d}{dz_2}Li_{\mathbf{a},\mathbf{b}}(x,y) = \begin{cases} \frac{w_2}{1-z_2w_2}Li_{(a_1,\cdots,a_{k-1},a_k-1),\mathbf{b}}(x,y) + \frac{1}{z_2}Li_{\mathbf{a},(b_1,\cdots,b_{l-1},b_l-1)}(x,y) \\ \qquad \text{if } a_k > 1, b_l \neq 1, \\ \frac{w_2}{1-z_2w_2}Li_{(a_1,\cdots,a_{k-1},a_k-1),\mathbf{b}}(x,y) + \frac{1}{1-z_2}Li_{\mathbf{a},(b_1,\cdots,b_{l-1})}(x,y) \\ \qquad \text{if } a_k > 1, b_l = 1, l \neq 1, \\ \frac{w_2}{1-z_2w_2}Li_{(a_1,\cdots,a_{k-1},a_k-1),\mathbf{b}}(x,y) + \frac{1}{1-z_2}Li_{\mathbf{a}}(xy) \\ \qquad \text{if } a_k > 1, b_l = 1, l = 1, \end{cases}$$

$$\frac{d}{dz_2}Li_{\mathbf{c}}(xy) = \begin{cases} \left(\frac{w_2}{1-z_2w_2} + \frac{1}{z_2}\right) \cdot Li_{(c_1,\cdots,c_{h-1},c_h-1)}(xy) & \text{if } c_h \neq 1, \\ \left(\frac{z_2w_2(1-w_2)}{(z_2-1)(z_2w_2-1)} + \frac{w_2-1}{z_2-1}\right) \cdot Li_{(c_1,\cdots,c_{h-1})}(xy) & \text{if } c_h = 1, h \neq 1, \\ \frac{z_2w_2(1-w_2)}{(z_2-1)(z_2w_2-1)} + \frac{w_2-1}{z_2-1} & \text{if } c_h = 1, h = 1, \end{cases}$$

$$\frac{d}{dz_2}Li_{\mathbf{c}}(y) = \begin{cases} \frac{1}{z_2}Li_{(c_1,\cdots,c_{h-1},c_h-1)}(y) & \text{if } c_h \neq 1, \\ \frac{1}{1-z_2}Li_{(c_1,\cdots,c_{h-1})}(y) & \text{if } c_h = 1, h \neq 1, \\ \frac{1}{1-z_2} & \text{if } c_h = 1, h = 1. \end{cases}$$

[1]An index $\mathbf{a} = (a_1, \cdots, a_k)$ ($a_i \in \mathbf{N}$) is called *admissible* if $a_k > 1$.

Following Proposition 4.4, we compute the residue around D_2

$$\frac{d}{d\bar{z}_2}Li^{(D_2)}_{\mathbf{a},\mathbf{b}}(x,y)=\begin{cases}\frac{1}{\bar{z}_2}Li^{(D_2)}_{\mathbf{a},(b_1,\cdots,b_{l-1},b_l-1)}(x,y) & \text{if } a_k>1, b_l\neq 1,\\ \frac{1}{1-\bar{z}_2}Li^{(D_2)}_{\mathbf{a},(b_1,\cdots,b_{l-1})}(x,y) & \text{if } a_k>1, b_l=1, l\neq 1,\\ \frac{1}{1-\bar{z}_2}Li^{(D_2)}_{\mathbf{a}}(xy) & \text{if } a_k>1, b_l=1, l=1,\end{cases}$$

$$\frac{d}{d\bar{z}_2}Li^{(D_2)}_{\mathbf{c}}(xy)=\begin{cases}\frac{1}{\bar{z}_2}Li^{(D_2)}_{(c_1,\cdots,c_{h-1},c_h-1)}(xy) & \text{if } c_h\neq 1,\\ \frac{1}{1-\bar{z}_2}Li^{(D_2)}_{(c_1,\cdots,c_{h-1})}(xy) & \text{if } c_h=1, h\neq 1,\\ \frac{1}{1-\bar{z}_2} & \text{if } c_h=1, h=1,\end{cases}$$

$$\frac{d}{d\bar{z}_2}Li^{(D_2)}_{\mathbf{c}}(y)=\begin{cases}\frac{1}{\bar{z}_2}Li^{(D_2)}_{(c_1,\cdots,c_{h-1},c_h-1)}(y) & \text{if } c_h\neq 1,\\ \frac{1}{1-\bar{z}_2}Li^{(D_2)}_{(c_1,\cdots,c_{h-1})}(y) & \text{if } c_h=1, h\neq 1,\\ \frac{1}{1-\bar{z}_2} & \text{if } c_h=1, h=1.\end{cases}$$

Similarly we compute

$$\frac{d}{d\bar{w}_2}Li^{(D_2)}_{\mathbf{a},\mathbf{b}}(x,y)=0 \qquad \text{if } a_k>1,$$
$$\frac{d}{d\bar{w}_2}Li^{(D_2)}_{\mathbf{c}}(xy)=0,$$
$$\frac{d}{d\bar{w}_2}Li^{(D_2)}_{\mathbf{c}}(y)=0.$$

These computations imply that differentials of $F^{(D_2)}_{\mathbf{ab}}$ and $G^{(D_2)}_{\mathbf{c}}$ with respect to $\bar{z}_2$ and $\bar{w}_2$ are zero by induction. Therefore they must be constant. By Lemma 6.2 their constant terms at P_1 is zero. So they are identically zero. □

LEMMA 6.4. *$F^{(D_3)}_{\mathbf{a},\mathbf{b}}=0$ if $\mathbf{a}$ and $\mathbf{b}$ are admissible and $G^{(D_3)}_{\mathbf{c}}=0$ if $\mathbf{c}$ is admissible.*

PROOF . On the affine coordinate (z_3,w_3) for U_3, the divisor D_3 is defined by $w_3=0$. By using $dx=-w_3dz_3-z_3dw_3$ and $dy=\frac{w_3(1-w_3)}{(z_3w_3-1)^2}dz_3+\frac{z_3-1}{(z_3w_3-1)^2}dw_3$, we obtain the following from the differential equations in (5.2) $\sim$ (5.4).

$$\frac{d}{dz_3}Li_{\mathbf{a},\mathbf{b}}(x,y)=\frac{w_3}{z_3w_3-1}Li_{(a_1,\cdots,a_{k-1},a_k-1),\mathbf{b}}(x,y)$$
$$+\frac{w_3}{1-z_3w_3}Li_{\mathbf{a},(b_1,\cdots,b_{l-1},b_l-1)}(x,y) \qquad \text{if } a_k>1 \text{ and } b_l>1,$$
$$\frac{d}{dz_3}Li_{\mathbf{c}}(xy)=0 \qquad \text{if } c_h>1,$$
$$\frac{d}{dz_3}Li_{\mathbf{c}}(y)=\frac{w_3}{1-z_3w_3}Li_{(c_1,\cdots,c_{h-1},c_h-1)}(y) \qquad \text{if } c_h>1.$$

Following Proposition 4.4, we compute the residue around D_3

$$\frac{d}{d\bar{z}_3} Li^{(D_3)}_{\mathbf{a},\mathbf{b}}(x,y) = 0 \qquad \text{if } a_k > 1 \text{ and } b_l > 1,$$
$$\frac{d}{d\bar{z}_3} Li^{(D_3)}_{\mathbf{c}}(xy) = 0 \qquad \text{if } c_h > 1,$$
$$\frac{d}{d\bar{z}_3} Li^{(D_3)}_{\mathbf{c}}(y) = 0 \qquad \text{if } c_h > 1.$$

Similarly we also compute

$$\frac{d}{d\bar{w}_3} Li^{(D_3)}_{\mathbf{a},\mathbf{b}}(x,y) = 0 \qquad \text{if } a_k > 1 \text{ and } b_l > 1,$$
$$\frac{d}{d\bar{w}_3} Li^{(D_3)}_{\mathbf{c}}(xy) = 0 \qquad \text{if } c_h > 1,$$
$$\frac{d}{d\bar{w}_3} Li^{(D_3)}_{\mathbf{c}}(y) = 0 \qquad \text{if } c_h > 1.$$

These computations imply that differentials of $F^{(D_3)}_{\mathbf{a},\mathbf{b}}$ and $G^{(D_3)}_{\mathbf{c}}$ with respect to $\bar{z}_3$ and $\bar{w}_3$ are zero by induction. Therefore they must be constant. By Lemma 6.3 their constant term at P_2 is zero. So they are identically zero. □

In [**F1**] it was shown that the limit (in a certain way) to $z = 1$ of $\mathrm{Li}_{k_1,\cdots,k_m}(z)$, which is a Coleman function over $\mathbf{P}^1\backslash\{0,1,\infty\}$, exists when $k_m > 1$ (loc. cit. Theorem 2.18) and p-adic multiple zeta value $\zeta_p(k_1,\cdots,k_m)$ is defined to be this limit value (loc. cit. Definition 2.17), but by using the terminologies in §3 we reformulate its definition as follows.

DEFINITION 6.5. For $k_m > 1$, the **p-adic multiple zeta value** $\zeta_p(k_1,\cdots,k_m)$ is the constant term of $\mathrm{Li}_{k_1,\cdots,k_m}(z)$ at $z = 1$.

In the case for $k_m = 1$, the constant term of $\mathrm{Li}_{k_1,\cdots,k_m}(z)$ at $z = 1$ is actually equal to the (canonical) regularization $(-1)^m I_p(BA^{k_m-1-1}B\cdots A^{k_1-1}B)$ of p-adic multiple zeta values by loc. cit. Theorem 2.22 (for this notation, see loc. cit. Theorem 3.30).

The following is important to prove double shuffle relations for p-adic multiple zeta values.

PROPOSITION 6.6. (1) *The analytic continuation* $\mathrm{Li}^{(D_3)}_{\mathbf{a},\mathbf{b}}(x,y)$ *is constant and equal to* $\zeta_p(\mathbf{a},\mathbf{b})$ *if* $\mathbf{a}$ *and* $\mathbf{b}$ *are admissible.*

(2) *The analytic continuation* $\mathrm{Li}^{(D_3)}_{\mathbf{c}}(xy)$ *and* $\mathrm{Li}^{(D_3)}_{\mathbf{c}}(y)$ *are constant and take value* $\zeta_p(\mathbf{c})$ *if* $\mathbf{c}$ *is admissible.*

PROOF . By Lemma 6.4 it is enough to prove this for $Li^{(D_3)}_{\mathbf{c}}(y)$. By the argument in Lemma 6.2 $Li^{(D_1)}_{\mathbf{c}}(y) = 0$. By the computation in Lemma 6.3 $Li^{(D_2)}_{\mathbf{c}}(y) = Li^{(D_2)}_{\mathbf{c}}(\bar{z}_2)$. So the constant term of $Li^{(D_2)}_{\mathbf{c}}(y)$ at P_2 is equal to the constant term of $Li_{\mathbf{c}}(\bar{z}_2)$ at $\bar{z}_2 = 1$, which is $\zeta_p(\mathbf{c})$. By the computation in Lemma 6.4 $Li^{(D_3)}_{\mathbf{c}}(y)$ must be constant if $c_h > 1$. Since this constant term must be the constant term of $Li^{(D_2)}_{\mathbf{c}}(y)$, $Li^{(D_3)}_{\mathbf{c}}(y) \equiv \zeta_p(\mathbf{c})$ for $c_h > 1$. □

By discussing on the opposite divisors D_5, D_4 and D_3, we also obtain the following.

PROPOSITION 6.7. (1) *The analytic continuation* $\mathrm{Li}^{(D_3)}_{\mathbf{a},\mathbf{b}}(y,x)$ *is constant and equal to* $\zeta_p(\mathbf{a},\mathbf{b})$ *if* $\mathbf{a}$ *and* $\mathbf{b}$ *are admissible.*
(2) *The analytic continuation* $\mathrm{Li}^{(D_3)}_{\mathbf{c}}(xy)$ *and* $\mathrm{Li}^{(D_3)}_{\mathbf{c}}(x)$ *are constant and equal to* $\zeta_p(\mathbf{c})$ *if* $\mathbf{c}$ *is admissible.*

7. The double shuffle relations

In this section, we prove double shuffle relations for p-adic multiple zeta values (Definition 6.5). Firstly we recall double shuffle relations for complex multiple zeta values. Let $\mathbf{a}=(a_1,\cdots,a_k)$ and $\mathbf{b}=(b_1,\cdots,b_l)$ be admissible indices (i.e. $a_k>1$ and $b_l>1$). The **series shuffle product formulas** (called by harmonic product formulas in [**F1**] and first shuffle relations in [**G1**]) are relations

$$\zeta(\mathbf{a})\cdot\zeta(\mathbf{b})=\sum_{\sigma\in Sh^{\leqslant}(k,l)}\zeta(\sigma(\mathbf{a},\mathbf{b})) \tag{7.1}$$

which is obtained by expanding the summation on the left hand side into the summation which give multiple zeta values. Here

$$Sh^{\leqslant}(k,l):=\bigcup_N\Big\{\sigma:\{1,\cdots,k+l\}\to\{1,\cdots,N\}\Big|\ \sigma \text{ is onto},$$
$$\sigma(1)<\cdots<\sigma(k),\sigma(k+1)<\cdots<\sigma(k+l)\Big\}$$

and $\sigma(\mathbf{a},\mathbf{b})=(c_1,\cdots,c_N)$ where N is the cardinality of the image of σ and

$$c_i=\begin{cases} a_s+b_{t-k} & \text{if } \sigma^{-1}(i)=\{s,t\} \text{ with } s<t,\\ a_s & \text{if } \sigma^{-1}(i)=\{s\} \quad \text{with } s\leqslant k.\\ b_{s-k} & \text{if } \sigma^{-1}(i)=\{s\} \quad \text{with } s>k.\end{cases}$$

One of the easiest example of (7.1) is (0.2).

On the other hand, multiple zeta values admit an iterated integral expression (cf. [**G1**], [**IKZ**] see also [**F0**])

$$\zeta(\mathbf{a})=\int_0^1\underbrace{\frac{du}{u}\circ\cdots\circ\frac{du}{u}\circ\frac{du}{1-u}}_{a_k}\circ\frac{du}{u}\circ\cdots\cdots\circ\frac{du}{1-u}$$
$$\circ\underbrace{\frac{du}{u}\circ\cdots\circ\frac{du}{u}\circ\frac{du}{1-u}}_{a_1}.$$

Here for differential 1-forms $\omega_1,\omega_2,\ldots,\omega_n$ ($n\geqslant 1$) on $\mathbf{C}$ an iterated integral $\int_0^1\omega_1\circ\omega_2\circ\cdots\circ\omega_n$ is defined inductively as $\int_0^1\omega_1(t_1)\int_0^{t_1}\omega_2\circ\cdots\circ\omega_n$. There are the well-known shuffle product formulas (for example see loc. cit.) of iterated integration

$$\int_0^1\omega_1\circ\cdots\circ\omega_k\cdot\int_0^1\omega_{k+1}\circ\cdots\circ\omega_{k+l}=\sum_{\tau\in Sh(k,l)}\int_0^1\omega_{\tau(1)}\circ\cdots\circ\omega_{\tau(k+l)},$$

where $Sh(k,l)$ is the set of shuffles defined by

$$Sh(k,l) := \Big\{\tau : \{1,\cdots,k+l\} \to \{1,\cdots,k+l\} \Big| \ \tau \text{ is bijective},$$
$$\tau(1) < \cdots < \tau(k), \tau(k+1) < \cdots < \tau(k+l)\Big\}.$$

They induce the **iterated integral shuffle produce formulas** (called by shuffle product formulas simply in [**F1**] and second shuffle relations in [**G1**]) for multiple zeta values

$$\zeta(\mathbf{a}) \cdot \zeta(\mathbf{b}) = \sum_{\tau \in Sh(N_{\mathbf{a}}, N_{\mathbf{b}})} \zeta(I_{\tau(W_{\mathbf{a}}, W_{\mathbf{b}})}) \tag{7.2}$$

where $N_{\mathbf{a}} = a_1 + \cdots + a_k$, $N_{\mathbf{b}} = b_1 + \cdots + b_l$. For $\mathbf{c} = (c_1, \cdots, c_h)$ with $h, c_1, \ldots, c_h \geqslant 1$ the symbol $W_{\mathbf{c}}$ means a word $A^{c_h - 1}BA^{c_{h-1}-1}B \cdots A^{c_1 - 1}B$ and conversely for given such W we denote its corresponding index by I_W. For words, $W = X_1 \cdots X_k$ and $W' = X_{k+1} \cdots X_{k+l}$ with $X_i \in \{A, B\}$, and $\tau \in Sh(k,l)$ the symbol $\tau(W, W')$ stands for the word $Z_1 \cdots Z_{k+l}$ with $Z_i = X_{\tau^{-1}(i)}$. One of the easiest example of (7.2) is (0.4).

The **double shuffle relations** for multiple zeta values are linear relations which are obtained by combining two shuffle relations (7.3), i.e. series shuffle product formulas (7.1) and iterated integral shuffle produce formulas (7.2)

$$\sum_{\sigma \in Sh^{\leqslant}(k,l)} \zeta(\sigma(\mathbf{a}, \mathbf{b})) = \sum_{\tau \in Sh(N_{\mathbf{a}}, N_{\mathbf{b}})} \zeta(I_{\tau(W_{\mathbf{a}}, W_{\mathbf{b}})}). \tag{7.3}$$

The following is the easiest example of the double shuffle relations obtained from (0.2) and (0.4):

$$\zeta(k_1, k_2) + \zeta(k_2, k_1) + \zeta(k_1 + k_2)$$
$$= \sum_{i=0}^{k_1 - 1} \binom{k_2 - 1 + i}{i} \zeta(k_1 - i, k_2 + i) + \sum_{j=0}^{k_2 - 1} \binom{k_1 - 1 + j}{j} \zeta(k_2 - j, k_1 + j)$$

for $k_1, k_2 > 1$.

THEOREM 7.1. *p-adic multiple zeta values in convergent case (i.e. for admissible indices) satisfy the series shuffle product formulas, i.e.*

$$\zeta_p(\mathbf{a}) \cdot \zeta_p(\mathbf{b}) = \sum_{\sigma \in Sh^{\leqslant}(k,l)} \zeta_p(\sigma(\mathbf{a}, \mathbf{b})) \tag{7.4}$$

for admissible indices **a** *and* **b**.

PROOF . Put $\mathbf{a} = (a_1, \cdots, a_k)$ and $\mathbf{b} = (b_1, \cdots, b_l)$. By the power series expansion of $\mathrm{Li}_{\mathbf{a},\mathbf{b}}(x, y)$ and $\mathrm{Li}_{\mathbf{a}}(x)$ in §5, we obtain the following formula

$$\mathrm{Li}_{\mathbf{a}}(x) \cdot \mathrm{Li}_{\mathbf{b}}(y) = \sum_{\sigma \in Sh^{\leqslant}(k,l)} \mathrm{Li}^{\sigma}_{\mathbf{a},\mathbf{b}}(x, y). \tag{7.5}$$

Here

$$\mathrm{Li}^{\sigma}_{\mathbf{a},\mathbf{b}}(x, y) := \sum_{(m_1, \cdots, m_k, n_1, \cdots, n_l) \in Z^{\sigma}_{++}} \frac{x^{m_k} \qquad y^{n_l}}{m_1^{a_1} \cdots m_k^{a_k} n_1^{b_1} \cdots n_l^{b_l}}$$

with

$$Z^{\sigma}_{++} = \left\{(c_1, \cdots, c_{k+l}) \in \mathbf{Z}^{k+l}_{>0} \mid c_i < c_j \text{ if } \sigma(i) < \sigma(j), c_i = c_j \text{ if } \sigma(i) = \sigma(j)\right\}.$$

Then for each $\sigma \in Sh^{\leqslant}(k,l)$, $\mathrm{Li}^{\sigma}_{\mathbf{a},\mathbf{b}}(x,y)$ can be written $\mathrm{Li}_{\mathbf{a}',\mathbf{b}'}(x,y)$, $\mathrm{Li}_{\mathbf{a}',\mathbf{b}'}(y,x)$ or $\mathrm{Li}_{\mathbf{a}',\mathbf{b}'}(xy)$ for some indices $\mathbf{a}'$ and $\mathbf{b}'$. We note that, if $\mathbf{a}$ and $\mathbf{b}$ are admissible, then these $\mathbf{a}'$ and $\mathbf{b}'$ are also admissible. By Proposition 6.6 and Proposition 6.7, we know that analytic continuations $\mathrm{Li}^{(D_3)}_{\mathbf{a},\mathbf{b}}(x,y)$, $\mathrm{Li}^{(D_3)}_{\mathbf{b},\mathbf{a}}(y,x)$, $\mathrm{Li}^{(D_3)}_{\mathbf{a},\mathbf{b}}(xy)$, $\mathrm{Li}^{(D_3)}_{\mathbf{a}}(x)$ and $\mathrm{Li}^{(D_3)}_{\mathbf{b}}(y)$ are all constant and take values $\zeta_p(\mathbf{a},\mathbf{b})$, $\zeta_p(\mathbf{b},\mathbf{a})$, $\zeta_p(\mathbf{a},\mathbf{b})$, $\zeta_p(\mathbf{a})$ and $\zeta_p(\mathbf{b})$ respectively when $\mathbf{a}$ and $\mathbf{b}$ are admissible. Therefore by taking an analytic continuation along Frobenius of both hands sides of (7.5) into $N^{00}_{D_3}(\mathbf{Q}_p)$, we obtain the series shuffle product formulas (7.4) for p-adic multiple zeta value in convergent case. □

By this theorem we say for example

$$\zeta_p(k_1)\cdot\zeta_p(k_2) = \zeta_p(k_1,k_2) + \zeta_p(k_2,k_1) + \zeta_p(k_1+k_2)$$

for $k_1, k_2 > 1$ which is a p-adic analogue of (0.2).

COROLLARY 7.2. *p-adic multiple zeta values in convergent case satisfy double shuffle relations. Namely*

$$\sum_{\sigma\in Sh^{\leqslant}(k,l)} \zeta_p(\sigma(\mathbf{a},\mathbf{b})) = \sum_{\tau\in Sh(N_{\mathbf{a}},N_{\mathbf{b}})} \zeta_p(I_{\tau(W_{\mathbf{a}},W_{\mathbf{b}})}).$$

holds for $a_k > 1$ and $b_l > 1$.

PROOF . It was shown in [**F1**] Corollary 3.46 that p-adic multiple zeta values satisfy iterated integral shuffle product formulas

$$\zeta_p(\mathbf{a})\cdot\zeta_p(\mathbf{b}) = \sum_{\tau\in Sh(N_{\mathbf{a}},N_{\mathbf{b}})} \zeta_p(I_{\tau(W_{\mathbf{a}},W_{\mathbf{b}})}). \tag{7.6}$$

By combining it with Theorem 7.1, we obtain double shuffle relations for p-adic multiple zeta values. □

Therefore we say for example

$$\begin{aligned}&\zeta_p(k_1,k_2) + \zeta_p(k_2,k_1) + \zeta_p(k_1+k_2)\\ &\quad= \sum_{i=0}^{k_1-1}\binom{k_2-1+i}{i}\zeta_p(k_1-i,k_2+i) + \sum_{j=0}^{k_2-1}\binom{k_1-1+j}{j}\zeta_p(k_2-j,k_1+j)\end{aligned}$$

for $k_1, k_2 > 1$ which is a p-adic analogue of (0.4).

In complex case there are two regularizations of multiple zeta values in divergent case, integral regularization and power series regularization (see [**IKZ**], [**G1**]§2.9 and §2.10). The first ones satisfy iterated integral shuffle product formulas, the second ones satisfy series shuffle product formulas and these two regularizations are related by regularization relations. Actually these provide new type of relations among multiple zeta values. In the case of p-adic multiple zeta values, p-adic analogue of integral regularization appear on coefficients of p-adic Drinfel'd associator (see [**F1**]) and they satisfy iterated integral shuffle product formulas like (7.6). On the other hand, it is not clear at all to say that p-adic analogue of power series regularization satisfy series shuffle product formulas and regularization relation. It is because that in the complex case the definition of this regularization and the proof of their series shuffle product formulas and regularization relation essentially based on the asymptotic behaviors of power series summations of multiple zeta values (see [**G1**] Proposition 2.19) however in the p-adic case our p-adic multiple zeta values do not have power series sum expression like (0.1). Recently the validity of these type of relations among p-adic multiple zeta values were achieved in [**FJ**] by using several variable p-adic multiple polylogarithm and a stratification of the moduli $\mathcal{M}_{0,N+3}$ $(N \geqslant 3)$.

References

[BC] Baldassarri, F., and Chiarellotto, B.; Algebraic versus rigid cohomology with logarithmic coefficients. Barsotti Symposium in Algebraic Geometry (Abano Terme, 1991), Perspect. Math. 15 11–50. Academic Press, San Diego, CA, 1994.

[Ber] Berthelot, P.; Cohomologie rigide et cohomologie rigide a supports propres, premièr partie. Preprint 96-03 of the university of Rennes, available at `http://www.maths.univ-rennes1.fr/~berthelo/`, 1996.

[Bes1] Besser, A.; Coleman integration using the Tannakian formalism, Mathematische Annalen 322 (2002) 1, 19-48.

[Bes2] ——— p-adic Arakelov theory. J. Number Theory 111(2) 2005 318–371.

[Ch] Chen, K.T.; Algebras of iterated path integrals and fundamental groups. Trans. Amer. Math. Soc. 156 1971 359–379.

[Co] Coleman, R.; Dilogarithms, regulators and p-adic L-functions, Invent. Math. 69 (1982), no. 2, 171–208.

[D1] Deligne, P.; Le groupe fondamental de la droite projective moins trois points, Galois groups over Q (Berkeley, CA, 1987), 79–297, Math. Sci. Res. Inst. Publ., 16, Springer, New York-Berlin, 1989.

[D2] ——— A letter to H.Furusho on 27th February, 2003.

[DM] Deligne, P. and Milne, J.; Tannakian categories, in Hodge cycles, motives, and Shimura varieties (P.Deligne, J.Milne, A.Ogus, K.-Y.Shih editors), Lecture Notes in Mathematics 900, Springer-Verlag, 1982.

[F0] Furusho, H.; The multiple zeta value algebra and the stable derivation algebra, Publ. Res. Inst. Math. Sci. Vol 39. no 4. (2003). 695-720.

[F1] ——— p-adic multiple zeta values I — p-adic multiple polylogarithms and the p-adic KZ equation, Inventiones Mathematicae, Volume 155, Number 2, 253-286(2004).

[F2] ——— p-adic multiple zeta values II – tannakian interpretations. Preprint, `arXive:math.NT/0506117`.

[FJ] ——— and Jafari,A.; Regularization and generalized double shuffle relations for p-adic multiple zeta values, preprint, `arXive:math.NT/0510681`.

[G1] Goncharov, A. B.; Multiple polylogarithms and mixed Tate motives, preprint, `arXive:math.AG/0103059`.

[G2] ——— Periods and mixed motives, preprint, `arXive:math.AG/0202154`.

[IKZ] Ihara, K., Kaneko, M. and Zagier, D.; Derivation and double shuffle relations for multiple zeta values, to appear in Compositio Math.

[Put] van der Put, M.; The cohomology of Monsky and Washnitzer. Mém. Soc. Math. France (N.S.) 23 1986 33–59. Introductions aux cohomologies p-adiques (Luminy, 1984).

[V] Vologodsky, V.; Hodge structure on the fundamental group and its application to p-adic integration, Moscow Mathematical Journal 3 no. 1, 2003.

Department of Mathematics, Ben-Gurion University of the Negev, P.O.B. 653, Be'er-Sheva 84105, Israel

E-mail address: `bessera@math.bgu.ac.il`

Graduate School of Mathematics, Nagoya University, Chikusa-ku, Furo-cho, Nagoya, 464-8602, Japan

E-mail address: `furusho@math.nagoya-u.ac.jp`

Contemporary Mathematics
Volume **416**, 2006

GALOIS p-GROUPS UNRAMIFIED AT p - A SURVEY

NIGEL BOSTON

ABSTRACT. We survey the theory of Galois groups of p-extensions of number fields. These are well understood when p is ramified, and related to the much applied theory of p-adic Galois representations. When p is unramified, the situation is completely different with very little known about the structure of the Galois groups. We can, however, present glimpses of a rich theory, culminating in the study of Galois actions on locally finite, rooted trees. Analogies with link groups are mentioned.

1. Introduction.

Throughout p is fixed to be a rational prime, K a number field, i.e. a finite extension of $\mathbf{Q}$, and S a finite set of primes of its ring of integers O_K. We shall investigate Galois extensions L of K that are ramified only at primes in S (allowing ramification at any infinite primes - equivalently, the discriminant of L/K is divisible only by primes in S) and whose degree is a power of p. Such an extension or a union of them will be called a p-extension. The union of all p-extensions of K unramified outside S will be denoted K_S and its Galois group over K by G_S. Alternatively this is the maximal pro-p quotient of the étale fundamental group of $\mathrm{Spec}(O_K) - S$. The big question then is to identify G_S, given p, K, and S. One case of particular interest is that where $S = \emptyset$, when K_S is called the Hilbert p-class field of K.

If S contains the primes S_p above p, then much is known about G_S and in many cases it can be explicitly presented [27]. On the other hand, assume for the rest of this introduction that $S \cap S_p = \emptyset$ (we say that L/K is unramified at p). Then we know only a few things about G_S. Class field theory tells us that all its subgroups of finite index have finite abelianization (a property we denote FIFA). It is a finitely presented pro-p group [37]. For certain K and S, we can actually compute what G_S is or at least a short list of candidates for G_S [9, 11, 17], although in all these cases G_S is finite. For many years [22], it has been known that there exist number fields K for which $G_\emptyset$ is infinite and the theorem of Golod-Shafarevich can be used to show that many other G_S are infinite. Not one example, however, is known of an explicit presentation for an infinite G_S. We obtain conjectural presentations in some simple cases.

On the other hand, there are fundamental conjectures that concern infinite G_S. In particular, the Fontaine-Mazur conjecture [19] says that G_S should have no

2000 *Mathematics Subject Classification.* 11R32, 11Y40, 12F10, 20D15.

Key words and phrases. Explicit Galois group, class tower, just-infinite, branch, Fontaine-Mazur.

The author was supported by NSF DMS.

infinite p-adic analytic quotients, meaning quotients that embed in $GL_n(\mathbf{Z}_p)$ for some n. This is equivalent [7] to claiming that the just-infinite quotients of G_S are not p-adic analytic. Classification of all just-infinite pro-p groups is a major activity of group theorists today [16] and has yielded various families and in particular Grigorchuk's dichotomy [23]. My extension [7] of the Fontaine-Mazur conjecture proposes that the just-infinite quotients of G_S always lie on one side of this dichotomy, namely are branch. Branch pro-p groups are certain groups of automorphisms of locally finite, rooted trees, and so this extension yields interesting actions of $\mathrm{Gal}(\overline{K}/K)$ on these trees.

We know much about actions of $\mathrm{Gal}(\overline{K}/K)$ on p-adic vector spaces such as étale cohomology groups of varieties over K, and the theory has found many applications, in particular in the proof of Fermat's Last Theorem [39, 41] and similar results. The Fontaine-Mazur conjecture however predicts that such actions are always ramified at p or are trivial (meaning they factor through a finite quotient). Thus the theory of p-adic Galois representations reveals little about p-extensions of K unramified at p. My extension of the Fontaine-Mazur conjecture implies that Galois actions on trees fill this gap. This theory, which may also be of use in 3-manifold theory, is introduced in the last section.

2. Some Group-Theoretical Preliminaries.

Whenever we talk about profinite groups, homomorphisms will always be continuous maps, subgroups closed, and generators will generate topologically. If G is a pro-p group, then it can be written as a quotient F/R of a free pro-p group F, i.e. given by a presentation. The minimal number of generators of G, denoted $d(G)$, is the smallest rank of such an F, and by Burnside's lemma equals $d(G/\Phi(G))$, where $\Phi(G)$ is the Frattini subgroup of G, i.e. the intersection of all open maximal subgroups of G. Since $G/\Phi(G)$ is the largest elementary abelian quotient of G, $G/\Phi(G) \cong C_p^{d(G)}$. Dualizing, $d(G) = \dim H^1(G, \mathbf{Z}/p)$.

Likewise, if F has rank $d(G)$, then the minimal number of generators of R as a normal subgroup of F will be denoted $r(G)$ and in fact $r(G) = \dim H^2(G, \mathbf{Z}/p)$. The most important result here is the inequality of Golod and Shafarevich [GS] that states that if $r(G) \leq d(G)^2/4$, then G is infinite, not p-adic analytic [30] (except for the cases of $G = \mathbf{Z}_p$ and $\mathbf{Z}_p \times \mathbf{Z}_p$), and not just-infinite [42] (see below).

We shall need certain filtrations of G. The derived series of subgroups of G is defined by $G^{(0)} = G, G^{(n)} = [G, G^{(n-1)}]$. The smallest n for which $G^{(n)} = \{1\}$, if it exists, is called the derived length of G. Groups of derived length 1 are abelian, length 2 metabelian. The p-central series of G is defined by $P_0(G) = G, P_n(G) = P_{n-1}(G)^p[G, P_{n-1}(G)]$. The smallest n for which $P_n(G) = \{1\}$, if it exists, is called the p-class of G. Note that $P_1(G) = \Phi(G)$ and that if G is finite, then the p-class and derived length both exist, the former being at least the latter.

If G is any infinite, finitely generated pro-p group, then it has minimally infinite quotients called *just-infinite* pro-p groups [26]. The attempted classification [16] of just-infinite pro-p groups has so far yielded four classes of just-infinite pro-p groups, namely

I. Solvable (and so linear over $\mathbf{Z}_p$).

II. Nonsolvable and linear over $\mathbf{Z}_p$.

III. Nonsolvable and linear over $\mathbf{F}_p[[T]]$.

IV. The rest!

At present, the rest consists of certain subgroups of the Nottingham groups R_k (k a finite field) (in particular open subgroups, the Fesenko groups [18], and the Barnea-Klopsch groups [3]), where R_k consists of the automorphisms $T \mapsto T + a_2T^2 + a_3T^3 + ...$ of $k[[T]]$, and Grigorchuk-type (or 'branch') groups [23], which are certain pro-p automorphism groups of the regular p-ary rooted tree. This tree has p^n vertices at distance n from its root with each vertex above the root having valency $p+1$. In fact, Grigorchuk [23] has proved that *every* just-infinite pro-p group either is branch or contains an open subgroup of the form $H \times ... \times H$ (finitely many factors), where H is hereditarily just-infinite (i.e. every open subgroup of H is just-infinite).

Note also that each of the types of just-infinite pro-p groups is attached to a kind of Galois representation. Linear representations over $\mathbf{Z}_p$ and $\mathbf{F}_p[[T]]$ are well-known and important. The field of norms construction yields representations of local Galois groups into R_k, something that Fesenko has attempted to globalize. In section 7, we explore possible Galois representations into branch groups.

3. The Ramified Case.

A good reference for this section is Koch's book [27]. We assume $p > 2$ in this section. Let K be a finite Galois extension of $\mathbf{Q}$ with Galois group H. Let r_2 be the number of complex places of K ($[K : \mathbf{Q}]/2$ if K is complex, else 0). If v is a place of K, then K_v will denote the completion of K at v, a local field. If F is a field, set $\delta(F) = 1$ if F contains a nontrivial pth root ζ_p of 1, $\delta(F) = 0$ otherwise.

Let Z_S be the set of nonzero elements x of K such that the fractional ideal (x) is a pth power and such that x is a pth power in each completion K_v for $v \in S$. Then $K^{*p} \subseteq Z_S$. Let B_S denote the quotient $\mathbf{F}_p[H]$-module Z_S/K^{*p}.

Proposition ([27, 34])

$$(a)\ d(G_S) = r_2 + 1 + r(G_S)$$

$$(b)\ r(G_S) = (\sum_{v \in S} \delta(K_v)) - \delta(K) + \dim_{\mathbf{F}_p} B_S$$

Note that $d(G_S) - r(G_S) = r_2 + 1 > 0$ confirming that G_S has an infinite abelian quotient (in fact coming from the cyclotomic extension $\mathrm{Gal}(\cup K(\zeta_{p^n})/K)$).

Let $\overline{E}$ and $\overline{E}_v$ denote the units modulo pth powers of the rings of integers of K and K_v respectively. If the class number $h(K)$ is prime to p, then global class field theory gives an exact sequence of $\mathbf{F}_p[H]$-modules:

$$(*)\ 0 \to B_S \to \overline{E} \to \oplus_{v \in S} \overline{E}_v \to G_S/\Phi(G_S) \to 0.$$

For each rational prime ℓ, let H_ℓ be a decomposition subgroup of H for ℓ and let H_∞ be the subgroup generated by a chosen complex conjugation. Let μ_p be the group of pth roots of 1 in K with the Galois H (and so H_ℓ) actions (note: μ_p can be trivial).

Proposition ([10]) *As $F_p[H]$-modules, if H has order prime to p,*

$$\oplus_{v\in S}\overline{E}_v \cong \mathbf{F}_p[H] \oplus (\oplus_{\ell\in S_0} Ind^H_{H_\ell}\mu_p),$$

$$\overline{E} \oplus \mathbf{F}_p \cong \mu_p \oplus Ind^H_{H_\infty}\mathbf{F}_p.$$

The relations coming from B_S are called the unknown relations, since they do not come from the relations of the local Galois groups, which are well-known to be tame or Demushkin. This situation is remedied either by noting that increasing S decreases B_S (and so auxiliary primes are routinely added to reduce B_S to 0) or by employing the following fact.

Proposition ([27]) If $\delta(K) = 1$, then B_S is a quotient $\mathbf{F}_p[H]$-module of the ideal class group modulo pth powers.

The maximal metabelian quotient of G_S is studied by means of Iwasawa theory [40]. In certain cases, G_S can be presented exactly - for example, if $K = \mathbf{Q}, p = 2$, and $S = \{2\}$, then Markscheitis [31] showed that $G_S = < x, y \mid x^2 >$.

4. The Analogy With 3-manifold Theory.

As noted elsewhere [33], there are analogues between these Galois groups and link groups. Assume that S contains the primes above p. Then G_S has an infinite abelian quotient isomorphic to $\mathbf{Z}_p$, corresponding to the union of cyclotomic fields. The action of this on $G'_S/G_S^{(2)}$ yields Iwasawa's characteristic polynomial [31], which Mazur and Wiles [32] proved coincides with the Leopoldt-Kubota p-adic L-function. The fundamental group of a knot complement has similar properties, leading to the Alexander polynomial. In the unramified case, there are still similarities with link groups, mentioned below.

5. The Unramified Case.

Suppose for simplicity that $K = \mathbf{Q}$. In the unramified case, we assume $p \notin S$. Suppose $S = \{p_1, ..., p_n\}$ and that p^{e_i} exactly divides $p_i - 1$. If $e_i = 0$, then in every p-extension of $\mathbf{Q}$, p_i is unramified and so can be omitted from S. We therefore assume, without loss of generality, that $e_i \geq 1$ for all i. Then it is known [20, 37] that $d(G_S) = r(G_S) = |S|$. Moreover the inertia subgroups I_i at p_i are procyclic subgroups $< \tau_i >$ of G_S, defined up to conjugacy, and the generators of G_S are the generators $\tau_i (1 \leq i \leq n)$.

As for the relations of G_S, these are the local tame relations $\sigma_i^{-1}\tau_i\sigma_i = \tau_i^{p_i}$, where the decomposition subgroup of G_S at p_i is $< \sigma_i, \tau_i >$. We do not know formulae for σ_i in terms of $\tau_1, ..., \tau_n$, although approximations to it modulo $G_S^{(2)}$ are given in [27]. These are related to linking numbers and Milnor invariants [33]. In general, we cannot give an explicit presentation for G_S.

What we can do is to define an NT-group of type $[m_1, ..., m_n]$, where $m_1, ..., m_n$ are powers of p, to be a pro-p group with presentation of the form $< \tau_1, ..., \tau_n \mid \tau_1^{\sigma_1} =$

$\tau_1^{1+m_1}, ..., \tau^{\sigma_n} = \tau_n^{1+m_n} >$ and satisfying FIFA (Finite-Index Finite-Abelianization), meaning that every subgroup of finite index has finite abelianization. The fact that G_S is an NT-group of type $[p^{e_1}, ..., p^{e_n}]$ follows from the previous paragraph together with the fact from class field theory that the abelianizations of finite index subgroups equal certain ray class groups, which are always finite in the unramified case. Thus, the purely group-theoretical problem of classifying NT-groups emerges, and in fact yields a group-theoretical strategy for tackling the Fontaine-Mazur conjecture [8].

Note that the abelianization of an NT-group of type $[m_1, ..., m_n]$ has elementary factors $m_1, ..., m_n$. Thus a classic result of Taussky implies that all $[2, 2]$ NT-groups are finite. The corresponding G_S are those with $p = n = 2$ and $p_1, p_2 \equiv 3 \pmod 4$ and were completely described in [11], together with some easier cases of $p_1 \equiv 3 \pmod 4), p_2 \equiv 5 \pmod 8$ (i.e. $[2, 4]$ NT-groups).

In [9], harder cases of this were studied by using O'Brien's p-group generation algorithm [35], which given a prime p and positive integer d constructs a tree whose vertices are the finite p-groups G with exactly d generators and whose edges connect G and $G/P_{n-1}(G)$ if G has p-class n. We pruned this tree by eliminating all groups for which the abelianizations of their low index subgroups were too large, given the ray class group information. This enabled us to hunt down G_S, usually obtaining a short list of very similar possible isomorphism types for it. In each of the $[2, 4]$ cases tried, this culminated in G_S being a finite group. Some conjectures from [9] on the family of groups so obtained have been partially proven by [17].

[9] also considered some cases with $p = 3$. The 3-groups obtained there fit into a family previously studied by Andoszkii and Cvetkov [2]. They are all related to the 3-adic analytic group $H = \ker(GL_2(\mathbf{Z}_3) \to GL_2(\mathbf{F}_3))$ in that if G_S has 3-class n, then $G_S/P_{n-1}(G_S) = H/P_{n-1}(H)$. We call H the governing group of the family. This seems to be a general phenomenon - for instance, Bush [14, 15] has recently studied maximal everywhere unramified p-extensions of imaginary quadratic fields K. In the case $p = 2$, he obtains results similar to [9] with consequences for the root-discriminant problem [24]; in the case $p = 3$, the Galois groups are Schur σ-groups [28], meaning that $r(G_S) = d(G_S)$ and that G_S has an automorphism of order 2 (complex conjugation in fact) that inverts every element of its abelianization. Bush obtains new families of Schur σ-groups, one of which has governing group a Sylow 3-subgroup of $PSL_2(\mathbf{Z}_3)$.

For general K, a theorem of Shafarevich [37] says that $r(G_S) \leq d(G_S) + [K : \mathbf{Q}] - 1$ and so by Golod-Shafarevich, G_S is infinite if $d(G_S) \geq 2 + 2\sqrt{[K : \mathbf{Q}]}$. Schur σ-groups are always infinite if $d(G_S) \geq 3$ by refinements of Golod-Shafarevich [28].

6. A New Approach To Fontaine-Mazur.

If L/K is an infinite p-extension unramified at the primes above p, then the structure of Gal(L/K) is much more of a mystery, although the conjecture of Fontaine-Mazur [19] says that it should not be p-adic analytic. In fact, no explicit presentation of such a group is known in the case of L/K ramified at finitely many primes. If we allow infinitely many primes to ramify, the recent work of Khare, Larsen, and Ramakrishna [25] exhibits a p-extension ramified at infinitely many primes but not at p, with p-adic analytic Galois group. We, however, shall focus on the finitely ramified case. In this section, we pick the simplest example where G_S is known to be infinite, list all the properties it must satisfy, and obtain a conjecture as to its structure. For more details see [8].

Let $p = 2, K = \mathbf{Q}$, and $S = \{q, r\}$ where q and r are distinct odd primes. Then G_S is known by the last section to be the NT-group with pro-2 presentation $< x, y | x^a = x^q, y^b = y^r >$. NT-groups of type $[2, 2]$ are known to be finite, and the evidence from section 5 suggests that NT-groups of type $[2, 4]$ are always finite. We therefore try the next simplest case of $[4, 4]$ NT-groups, i.e. $q, r \equiv 5 \pmod 8$. For many such q, r, the method of section 5 leads to G_S being finite. If, however, one of the primes q, r is a 4th power modulo the other but not vice versa, then a refinement of Golod-Shafarevich due to Kuhnt [29] shows that G_S is infinite.

We, therefore, conduct in this case a search using the computer algebra system MAGMA [6], for presentations of the given form that satisfy everything we know of G_S. Picking words a, b in x, y at random, we filter out those abstract groups $G =< x, y | x^a = x^q, y^b = y^r >$ that have low index subgroups (with core of 2-power index) with infinite abelianization (i.e. fail FIFA) and those for which the sequence $|G/P_n(G)|$ stabilizes (i.e. have finite pro-2 completion). Amazingly, the groups that survive these filters fit into a simply described family, leading to the following conjecture.

Conjecture. There exists a subset $\mathcal{F}$ of the free pro-2 group on x, y such that any $[4, 4]$ NT-group (such as G_S for $S = \{q, r\}, q, r \equiv 5 \pmod 8), q$ a 4th power modulo r but not vice versa) has presentation $< x, y | x^a = x^5, y^4 = 1 >$ with $a \in \mathcal{F}$.

The shortest elements in $\mathcal{F}$ have length 6 and there are 48 of them, for instance $y^2xyxy, y^2xyx^{-1}y^{-1}, \ldots$.

Moreover, the sequence $\log_2(|G/P_n(G)|)$ is always the same, namely

$(\Sigma) : 2, 5, 8, 11, 14, 16, 20, 24, 30, 36, 44, 52, 64, 76, 93, 110, 135, 160, 196, 232, 286, 340, 419, 498, 617, 736, 913, 1090, 1357, 1634, \ldots$

Putting sequence $\Delta_n := \log_2 |P_n(G)/P_{n+1}(G)|$ into Sloane's On-Line Encyclopedia of Integer Sequences http://www.research.att.com/~ njas/sequences/Seis.html yields A001461, arising in the paper [12] concerning knot theory and quantum field theory. If so, $\Delta_{2n-2} = \Delta_{2n-1} = \sum_{m=1}^{n} (1/m) \sum_{d|m} \mu(m/d)(F_{d-1} + F_{d+1})$, where μ is the usual Möbius function and F_n the nth Fibonacci number (so in fact $F_{d-1} + F_{d+1}$ is the dth Lucas number).

Letting $L(G) = \oplus P_n(G)/P_{n+1}(G)$, the $\mathbf{F}_p$-Lie algebra of G, it appears further that all the $[4, 4]$ NT-groups possess the same $\mathbf{F}_p$-Lie algebra. There are algebras arising in other areas of mathematics whose graded pieces have the same dimensions, namely (i) the free Lie algebra generated by one generator in degree 1 and one in degree 2 (arising in work on multi-zeta values and quantum field theory [12]) and (ii) Cameron's permutation group algebra [21] of $C_2 \wr A$, where A is the group of all order-preserving permutations of the rationals. This suggests the following amazing possibility.

Conjecture. If G is a $[4, 4]$ NT-group, then $L(G)$ is the $\mathbf{F}_p$-Lie algebra in (i) or (ii) above.

7. Tree Representations.

In the unramified case, the Fontaine-Mazur conjecture [19] says that the just-infinite quotients of G_S are never p-adic analytic. My extension [7] gives evidence that these just-infinite quotients always lie on the other side of Grigorchuk's dichotomy [23]:

Conjecture. Assume $S \cap S_p = \emptyset$. The just-infinite quotients of G_S are branch.

Corollary to Conjecture. The unramified Fontaine-Mazur conjecture [19] (and its generalization in [7]) holds.

This suggests a study of Galois representations into automorphisms of locally finite, rooted trees, where the image is branch. At first glance there are too many of these to be useful since every finitely generated pro-p group embeds in the automorphism group W of the regular p-ary rooted tree T [23]. Our conjecture, however, concerns representations with large image. This is a familiar idea from the theory of p-adic representations, although here 'large' cannot mean 'of finite index' since W is not finitely generated, whereas our Galois groups are. To define 'large', we need the notion of Hausdorff dimension [4].

Let W_n be the quotient of W given by its action on the subtree of vertices of distance $\leq n$ from the root of T. If G is a closed subgroup of W, its Hausdorff dimension is defined to be $\liminf_{n\to\infty} \frac{\log|G_n|}{\log|W_n|}$, where G_n is the image of G in W_n. So, for instance, the first branch group discovered (by Grigorchuk) [23] has Hausdorff dimension 5/8. If G is any pro-p group, then we define its Hausdorff dimension to be the supremum of its Hausdorff dimensions over all embeddings into W.

Conjecture. A just-infinite pro-p group is branch if and only if its Hausdorff dimension is nonzero.

This conjecture is purely group-theoretical. The evidence for it is that all hereditarily just-infinite pro-p groups have Hausdorff dimension zero, whereas, so far, every branch group whose Hausdorff dimension has been calculated, is of nonzero dimension. Abért and Virág [1] has made some progress on the conjecture.

Combining the two conjectures we see that when G_S has infinite quotients, it should have maps to W with 'large' image. These arise by mapping G_S onto a just-infinite quotient J and then embedding J in W with nonzero Hausdorff dimension.

We concentrate on the $p = 2$ case here. There are straightforward generalizations to other p, but this is computationally most convenient. Let $W_1 = C_2$ and $W_n = W_{n-1} \wr C_2$ $(n \geq 2)$. Then W_n is a finite 2-group of order 2^{2^n-1}, that naturally arises as the automorphism group of the binary tree T_n, which consists of a single root vertex with two edges above each vertex to the next level extending to level n (in fact, W_n is isomorphic to the Sylow 2-subgroup of the symmetric group on 2^n letters). Thus W_{n-1} is naturally a quotient of W_n and we set $W = \varprojlim W_n$, an infinitely generated pro-2 group, which is the automorphism group of the infinite binary tree T. We shall be interested in representations of $G_{\mathbf{Q}} := \mathrm{Gal}(\overline{\mathbf{Q}}/\mathbf{Q})$ into $W = \mathrm{Aut}(T)$.

The only place where representations of this kind have been studied so far is by Odoni [36], who in work completed by Stoll [38], showed that the Galois group of the nth iterate of $x^2 + 1$ is W_n, so that letting $n \to \infty$, we obtain a surjective representation $G_{\mathbf{Q}} \to W$. In showing surjectivity, their approach was to prove that new primes were ramified at each level. Since W is not finitely generated, any surjective representation must be ramified at infinitely many primes.

Pink has also studied Galois groups of iterates in the case of $\mathbf{Q}$ replaced by the function field of a finite field [5], so iterating morphisms $f : X \to X$. Here, unlike the characteristic 0 case, there are finitely many ramification points, leading to finitely generated subgroups of W, which can sometimes be computed. It is easy to see using the Riemann-Hurwitz genus formula that if X is an irreducible curve of genus g, then if there exists a non-constant morphism $f : X \to X$ of degree > 1, then $g = 0$ or 1. In this latter case, f is unramified and the image of the geometric fundamental group is abelian with ≤ 2 generators. Thus, the case of $X = \mathbf{P}^1$ is forced upon us.

An example of a representation $G_S \to W_4$ with large image and S not containing 2 is given in the last section of my work with Leedham-Green [9].

Some questions that arise from this are:

(i) Is there a "natural" way to construct finitely ramified tree representations in the number field case, perhaps by using a series of quadratic covering maps (the Odoni-Stoll case corresponding to the case of $\mathbf{P}^1 \to \mathbf{P}^1$ given by $x \mapsto x^2+1$)? Tools from algebraic geometry could then reenter.

(ii) Given a finitely ramified tree representation, do the images of Frobenius elements carry arithmetical information? Are there objects related to the representation in a way similar to modular forms being related to p-adic representations via the traces of the images of Frobenius elements? The conjugacy classes of the automorphism groups of p-ary trees have been studied - see e.g. [13]. In particular, there are uncountably many conjugacy classes. Is there a countable collection to which images of Frobenius elements are constrained? (Compare the situation of representations into $GL_n(\mathbf{Z}_p)$, which has uncountably many conjugacy classes. Representations of arithmetical interest map Frobenius elements into the countably many conjugacy classes that have characteristic polynomials with algebraic coefficients.) Pink has suggested that perhaps a finite automaton might describe the permissible conjugacy classes.

BIBLIOGRAPHY

[1] M.Abért and B.Virág, Dimension and randomness in groups acting on rooted trees, J. Amer. Math. Soc. **18** (2005), 157–192.

[2] Andoszkii and Cvetkov, A certain series of finite closed p-groups (Russian), Izvestiya Akad. Nauk SSSR Ser. Mat. **38** (1974), 278–290.

[3] Y.Barnea and B.Klopsch, Index-subgroups of the Nottingham group, Adv. Math. **180** (2003), 187–221.

[4] Y.Barnea and A.Shalev, Hausdorff dimension, pro-p groups, and Kac-Moody algebras, Trans. Amer. Math. Soc. **349** (1997), no. 12, 5073–5091.

[5] L.Bartholdi, Amenability of groups acting on trees, preprint on arXiv math.GR/0204076 (2003).

[6] W.Bosma and J.Cannon, Handbook of MAGMA Functions, Sydney: School of Mathematics and Statistics, University of Sydney (1993).

[7] N.Boston, Some Cases of the Fontaine-Mazur Conjecture II, J. Number Theory **75** (1999), 161–169.

[8] N.Boston, Reducing the Fontaine-Mazur conjecture to group theory, in "Progress in Galois Theory," Proceedings of Thompson's 70th Birthday Conference (2005), 39–50.

[9] N.Boston and C.R.Leedham-Green, Explicit computation of Galois p-groups unramified at p, J. Algebra **256** (2002), 402–413.

[10] N.Boston and B.Mazur, Explicit universal deformations of Galois representations, Advanced Studies in Pure Mathematics **17** (1989), 1–21

[11] N.Boston and D.Perry, Maximal 2-extensions with restricted ramification, J. Algebra **232** (2000), 664–672.

[12] D.J.Broadhurst, On the enumeration of irreducible k-fold Euler sums and their roles in knot theory and field theory, preprint.

[13] A.M.Brunner and S.Sidki, On the automorphism group of the one-rooted binary tree, J. Algebra **195** (1997), no. 2, 465–486.

[14] M.R.Bush, Computation of Galois groups associated to the 2-class towers of some quadratic fields, J. Number Theory **100** (2003), 313–325.

[15] M.R.Bush, Maximal unramified p-extensions of imaginary quadratic fields and p-groups of deficiency zero, preprint (2005).

[16] M.DuSautoy, D.Segal, and A.Shalev, New Horizons in pro-p Groups, Birkhauser, Boston 2000.

[17] B.Eick and H.Koch, On maximal 2-extensions of $\mathbf{Q}$ with given ramification, to appear in the Proc. St.Petersburg Math. Soc. (Russian), AMS Translations (English version).

[18] I.Fesenko, On just infinite pro-p-groups and arithmetically profinite extensions of local fields, J. Reine Angew. Math. **517** (1999), 61–80.

[19] J.-M.Fontaine and B.Mazur, Geometric Galois representations, *in* "Elliptic curves and modular forms, Proceedings of a conference held in Hong Kong, December 18-21, 1993," International Press, Cambridge, MA and Hong Kong.

[20] A.Fröhlich, Central Extensions, Galois groups, and ideal class groups of number fields, *in* "Contemporary Mathematics," Vol. **24**, AMS, 1983.

[21] J.Gilbey, Permutation groups, a related algebra, and a conjecture of Cameron, Journal of Algebraic Combinatorics **19** (2004), 25–45.

[22] E.S.Golod and I.R.Shafarevich, On class field towers (Russian), Izv. Akad. Nauk. SSSR **28** (1964), 261–272. English translation in AMS Trans. (2) **48**, 91–102.

[23] R.Grigorchuk, Just infinite branch groups, *in* "New Horizons in pro-p Groups," (eds. du Sautoy, Segal, Shalev), Birkhauser, Boston 2000.

[24] F.Hajir and C.Maire, Tamely ramified towers and discriminant bounds for number fields. II, J. Symbolic Comput. **33** (2002), 415–423.

[25] C.Khare, M.Larsen, and R.Ramakrishna, Constructing semisimple p-adic Galois representations with prescribed properties, American J. of Mathematics **127** (2005).

[26] G.Klaas, C.R.Leedham-Green, and W.Plesken, Linear pro-p groups of finite width, Springer Lecture Notes in Math 1674, 1997.

[27] H.Koch, Galois theory of p-extensions. With a foreword by I. R. Shafarevich. Translated from the 1970 German original by Franz Lemmermeyer. With a postscript by the author and Lemmermeyer. Springer Monographs in Mathematics. Springer-Verlag, Berlin, 2002.

[28] H.Koch and B.Venkov, The p-tower of class fields for an imaginary quadratic field (Russian), Zap. Nau. Sem. Leningrad Otdel. Mat. Inst. Steklov (LOMI) **46** (1974), 5–13.

[29] T.Kuhnt, UIUC Ph.D. thesis (2002).

[30] A.Lubotzky, Group presentation, p-adic analytic groups and lattices in $SL_2(\mathbf{C})$, Ann. of Math. **118** (1983), 115–130.

[31] H.Markscheitis, On p-extensions with one critical prime (Russian), Izvestiya Akad. Nauk SSSR Ser. Mat. **27** (1963), 463–466.

[32] B.Mazur and A.Wiles, Class fields of abelian extensions of $\mathbf{Q}$, Invent. Math. **76** (1986), 179–330.

[33] M. Morishita, On certain analogies between knots and primes, J. Reine Angew. Math. **550** (2002), 141–167.

[34] O.Neumann, On p-closed number fields and an analogue of Riemann's existence theorem, *in* Algebraic Number Fields Fröhlich ed., Academic Press, London 1977, 625–647.

[35] E.A.O'Brien, The p-group generation algorithm, J. Symbolic Comput. **9** (1990), 677–698.

[36] R.W.K.Odoni, Realising wreath products of cyclic groups as Galois groups, Mathematika **35** (1988), no. 1, 101–113.

[37] I.R.Shafarevich, Extensions with prescribed ramification points (Russian), IHES Publ. Math. 18, 71–95 (1964).

[38] M.Stoll, Galois groups over $\mathbf{Q}$ of some iterated polynomials, Arch. Math. (Basel) **59** (1992), no. 3, 239–244.

[39] R.Taylor and A.Wiles, Ring-theoretic properties of certain Hecke algebras, Ann. of Math. **141** (1995), 553–572.

[40] L.Washington, Introduction to cyclotomic fields, Graduate Texts in Math 83, Springer-Verlag, New York, 1997.

[41] A.Wiles, Modular elliptic curves and Fermat's last theorem, Ann. of Math. **141** (1995), 443–551.

[42] J.Wilson, On a theorem of Zelmanov, J. Algebra **220** (1999), 629–635.

Department of Mathematics, University of Wisconsin, Madison, WI 53706
E-mail address: boston@math.wisc.edu

Contemporary Mathematics
Volume **416**, 2006

On capitulation theorems for infinite groups

Kazuyuki Fuluta

Dedicated to Professor Masanori Morishita

ABSTRACT. We prove group-transfer theorems for infinite groups which generalize the previous results proven by Ph. Furtwängler, K. Miyake, H. Suzuki etc for finite groups. As their applications, we show topological capitulation theorems for the first homology groups of manifolds. We also construct a counter-example for a generalized problem on the group-transfer kernel.

Introduction

The capitulation problem is originally concerned with the structure of ideal classes which capitulate in an extension of number fields. Via the Artin reciprocity map, the problem is translated into that on the kernel of the group-transfer. Among many works on the subject, H. Suzuki [**Su1, Su2**] established a fundamental theorem on this group-theoretic version, which includes, as its consequences, three classical theorems: Hilbert's Theorem 94, Principal ideal theorem and Terada-Tannaka's theorem.

Recently, M. Morishita initiated a systematic study of the analogies between number theory and knot theory [**Mo3**], and he showed, among other things, a topological capitulation theorem for the first homology groups of manifolds which is again stated in terms of the group-transfer via the Hurewicz isomorphism [**Mo2**]. However his theorem deals essentially with the case where the involved groups have finite abelianizations (cf. Theorem 1.3 below). Since a homology group of a manifold is an infinite abelian group in general, it is desirable for a topological application to extend Morishita's theorem to the case where the involved groups have infinite abelianizations.

In this paper, we prove the following group-transfer theorems for infinite groups and give topological capitulation theorems which generalize Morishita's theorem.

Throughout this paper, we denote by G' and $G^{\mathrm{ab}} := G/G'$ the commutator subgroup and the abelianization of a group G respectively and by A_{tors} the maximal torsion subgroup of an abelian group A. For a group G and its subgroup H, we

2000 *Mathematics Subject Classification.* Primary 55R12 : Secondary 20E36 : 57M12.
Key words and phrases. Capitulation theorem; Group transfer; Homology group; Arithmetic Topology.

denote by $V_{G/H} : G^{\rm ab} \to H^{\rm ab}$ the group-transfer. For a set S, $\#S$ stands for the cardinality of S.

THEOREM 1. *Let G be a group such that $(G^{\rm ab})_{\rm tors}$ is finite and H a normal subgroup such that G/H is finite abelian. Then $\ker(V_{G/H})$ is finite and its order is divisible by $\gcd(\#(G^{\rm ab})_{\rm tors}, [G : H])$.*

Theorem 1 gives a generalization of Suzuki's theorem [**Su1**]. More generally, we show the following theorem.

THEOREM 2. *Let G be a group such that $(G^{\rm ab})_{\rm tors}$ is finite. Let H be a normal subgroup so that G/H is finite abelian. We assume that H consists of all $g \in G$ such that g mod G' has infinite order. Suppose that σ is an endomorphism of G such that $\sigma(H) \subset H$ and induces the identity map on G/H. Then the group*

$$\{\gamma G' \in \ker(V_{G/H})\,;\, \sigma(\gamma)\gamma^{-1} \in G'\}$$

is finite and its order is divisible by $[G : H]$.

As applications of Theorem 1 and 2 above, we generalize Morishita's theorem (Theorem 4.1 and 4.2 below).

We note that one can not replace $[G : H]$ in the conclusion of Theorem 2 by $\gcd(\#(G^{\rm ab})_{\rm tors}, [G : H])$ under the assumption of Theorem 1, namely, without the assumption $(*)$, as the following theorem shows (cf. Example 3.2 below).

THEOREM 3. *Let U be a finite abelian group and W a group so that $W^{\rm ab}$ is torsion free and not equal to $\{e\}$. Let G denote the free product $U * W$ and let H be a subgroup of G represented by $U * W_0$, where W_0 is a subgroup of W so that $W_0 \supset W'$ and $[W : W']$ is divisible by the order of U. Let σ be an endomorphism of G such that σ sends any element of U to its inverse element and $\sigma|_W$ is an endomorphism of W such that $\sigma(W_0) \subset W_0$ and induces the identity map on W/W_0. Then we have*

$$\{\gamma G' \in \ker(V_{G/H})\,;\, \sigma(\gamma)\gamma^{-1} \in G'\} \simeq (\mathbb{Z}/2\mathbb{Z})^{{\rm rank}_2 U}$$

where ${\rm rank}_2 U$ is the 2-rank of U.

1. Two capitulation theorems

In this section, we recall two capitulation theorems by Suzuki and Morishita in number theory and topology.

Let K/k be a finite extension of number fields and ${\rm Cl}_k$ denotes the ideal class group of k. Then the capitulation problem is concerned with the structure of the finite abelian group

$$C_{K/k} := \ker(t_{k/k} : {\rm Cl}_k \to {\rm Cl}_K).$$

Here $t_{K/k}$ is the natural homomorphism induced by an extension of an ideal under the natural inclusion $k \to K$. For a useful survey on this topic, we refer to [**Mi**]. The fundamental theorem, due to Suzuki, is the following theorem.

THEOREM 1.1 ([**Su2**]). *Let k/k_0 be a finite cyclic extension of number field with Galois group G_0 and K/k an unramified extension so that K is abelian over k_0. Then the number of G_0-invariant ideal classes in $C_{K/k}$ is divisible by $[K : k]$.*

We note that Theorem 1.1 includes three classical theorems: Hilbert's Theorem 94, Principal ideal theorem and Terada-Tannaka's theorem.

Now, we recall a part of the analogies between number theory and 3-dimensional topology, which have been discovered by B. Mazur, Y. Manin, M. Kapranov, A. Reznikov and M. Morishita.

number fields K	$\longleftrightarrow$	orientable 3-dim. closed manifolds M
(resp. un) ramified extension L/K	$\longleftrightarrow$	(resp. un) branched covering $N \to M$
prime ideal (prime ideals)	$\longleftrightarrow$	knot (link)
$\pi_1(\mathrm{Spec}(O_K))$	$\longleftrightarrow$	$\pi_1(M)$
Cl_K	$\longleftrightarrow$	$\mathrm{H}_1(M)$
Artin reciprocity map	$\longleftrightarrow$	Hurewicz isomorphism.

where O_K stands for the ring of integers of K and homology groups have the integral coefficient. For more analogies, the reader may consult [**Mo1, Mo2, Mo3, Mo4**],[**Ra**],[**Re1, Re2**],[**Si**].

Based on these analogies, Morishita [**Mo2**] considered a topological analogue of the capitulation problem, namely the estimation of the group $C_{N/M}$ defined by

$$C_{N/M} := \ker(t_{N/M} : \mathrm{H}_1(M) \to \mathrm{H}_1(N)), \quad t_{N/M}([c]) := [p^{-1}(c)],$$

for a covering map $p : N \to M$ of path-connected topological manifolds. Among other things, he showed a topological counterpart of Suzuki's theorem.

THEOREM 1.2 ([**Mo2**]). *Let $M \to M_0$ be a finite cyclic branched covering of path-connected compact manifolds with Galois group G_0 and $N \to M$ a finite covering so that the composite $N \to M_0$ is an abelian covering. Assume that $\mathrm{H}_1(M)$ is finite. Then the number of G_0-invariant first homology classes in $C_{N/M}$ is divisible by a covering index $[N : M]$.*

By the Hurewicz isomorphism, Theorem 1.2 follows immediately from the following theorem.

THEOREM 1.3 ([**Mo2**]). *Let G be a finitely generated group such that G^{ab} is finite and H a normal subgroup such that G/H is finite abelian. Let σ be an endomorphism*[1] *of G such that $\sigma(H) \subset H$ and induces the identity map on G/H. Then the order of the subgroup*

$$\{\gamma G' \in \ker(V_{G/H})\,;\, \sigma(\gamma)\gamma^{-1} \in G'\}$$

is divisible by $[G : H]$.

In the following, we shall generalize Theorem 1.3 by removing the assumption that G^{ab} is finite so that we have a general topological capitulation theorem without the assumption in Theorem 1.2 that $\mathrm{H}_1(M)$ is finite.

2. Transfer theorem

In this section, we prove Theorem 1 and 3 in Introduction. Firstly, we prepare some lemmas.

LEMMA 2.1. *Let G be a finitely generated group such that G^{ab} is finite and H a normal subgroup such that G/H is abelian and let m be an integer. We define the homomorphism $V^m_{G/H}$ by*

$$V^m_{G/H} : G^{\mathrm{ab}} \ni g \mapsto V_{G/H}(g)^m \in H^{\mathrm{ab}}.$$

[1]Though σ is assumed to be an automorphism in [**Mo2**], a similar proof works for any endomorphism σ.

Then the order of $\ker(V_{G/H})$ *is divisible by* $\gcd(\#(G^{\mathrm{ab}}), m \cdot [G:H])$.

PROOF. Let f_m be a m-fold map. Then we have

$$\#\ker V_{G/H}^m = \#\ker V_{G/H} \times \#\ker f_m|_{\mathrm{Im}V_{G/H}}.$$

Since $\mathrm{Im}(V_{G/H})$ is finite abelian, the order of $\ker f_m\,|_{\mathrm{Im}V_{G/H}}$ is divisible by $\gcd(\#\mathrm{Im}(V_{G/H})\,,\,m)$. Therefore using theorem 1.3, we have

$$\begin{aligned} \gcd(\#G^{\mathrm{ab}}\,,\,m \times [G:H]) \quad &\Big|\quad \gcd(\#G^{\mathrm{ab}}\,,\,m \cdot \#\ker V_{G/H}) \\ &= \quad \#\ker V_{G/H} \times \gcd(\#\mathrm{Im}V_{G/H}\,,\,m) \\ &\Big|\quad \#\ker V_{G/H} \times \#\ker f_m|_{\mathrm{Im}V_{G/H}} \\ &= \quad \#\ker V_{G/H}^m, \end{aligned}$$

where $A\,\big|\,B$ means that B is divisible by A. □

The following is a key lemma to prove our group-transfer theorem.

LEMMA 2.2. *Let* G *be a group represented by the amalgamated product* $A *_\Gamma B$ *of groups* A *and* B *over a subgroup* Γ *of* $A \cap B$. *Then* $G' \cap A$ *is equal to* A'.

PROOF. A similar argument to [**L**] and the definition of amalgamated product yields this lemma. □

LEMMA 2.3. *Let* G *be a group represented by the amalgamated product* $A *_\Gamma B$ *of a finitely generated group* A *and groups* B *over* Γ. *Furthermore we assume that* A^{ab} *is finite and* B^{ab} *is torsion free. Let* H *be a normal subgroup of* G *such that* G/H *is finite abelian. Then* $\ker(V_{G/H})$ *is finite and its order is divisible by* $\gcd(\#(G^{\mathrm{ab}})_{\mathrm{tors}}, [G:H])$.

PROOF. By Lemma 2.2, $G' \cap A = A'$ and $G' \cap B = B'$, and comparing G^{ab} with $A^{\mathrm{ab}} \times B^{\mathrm{ab}}$, we find that G^{ab} and $(G^{\mathrm{ab}})_{\mathrm{tors}}$ are isomorphic to $A^{\mathrm{ab}} \times B^{\mathrm{ab}}$ and A^{ab} respectively. By a property of the group-transfer, the composed map of $V_{G/H}$ and an embedding map $\imath : H^{\mathrm{ab}} \to G^{\mathrm{ab}}$ is $[G:H]$-fold map, from which we see $\ker V_{G/H}$ is finite, that is, $\ker V_{G/H} \subset A \pmod{G'}$. Since we have

$$G/H \hookleftarrow HA/H \simeq A/A \cap H,$$

$[G:H]$ is divisible by $[A : A \cap H]$.

So comparing the structure of $V_{G/H}$ with that of $V_{A/A\cap H}^{[G:H]/[A:A\cap H]}$, we find the following embedding map

$$V_{G/H}((G^{\mathrm{ab}})_{\mathrm{tors}}) \hookrightarrow \{V_{A/A\cap H}^{[G:H]/[A:A\cap H]}(A^{\mathrm{ab}})\}.$$

Namely $\#\ker V_{G/H}$ is divisible by $\#\ker V_{A/A\cap H}^{[G:H]/[A:A\cap H]}$.

Therefore we have the following estimation from Lemma 2.1

$$\begin{aligned} \gcd(\#(G^{\mathrm{ab}})_{\mathrm{tors}}\,,\,[G:H]) \quad &= \quad \gcd\left(\#U^{\mathrm{ab}}\,,\,\frac{[G:H]}{[A:A\cap H]} \times [A:A\cap H]\right) \\ &\Big|\quad \#\ker V_{A/A\cap H}^{[G:H]/[A:A\cap H]} \\ &\Big|\quad \#\ker V_{G/H}. \end{aligned}$$

□

We can deduce Theorem 1 from the above two lemmas.

PROOF OF THEOREM 1. Since $(G^{\rm ab})_{\rm tors}$ is finite, we can represent G by $\langle \{g_i\}_{i=1}^n \cup \{h_j\}_{j\in\Lambda} | R\rangle$ for a natural number n and relators R so that the order of g_iG' is finite and greater than 1, and the order of h_jG' is infinite or 1. Here U (resp. W) denotes $\langle \{g_i\}_{i=1}^n\rangle$ (resp. $\langle \{h_j\}_{j\in\Lambda}\rangle$). Now we obtain the isomorphism

$$f : G \to UR/R *_\Gamma WR/R.$$

where Γ is some group.

Thus UR/R is finitely generated so that $(UR/R)^{\rm ab}$ is finite, and $(WR/R)^{\rm ab}$ is torsion free. Furthermore by Lemma 2.2, $(UR/R)^{\rm ab}$ is isomorphic to $(G^{\rm ab})_{\rm tors}$. Therefore we are done by Lemma 2.3. □

REMARK 2.4. If the set $\{h_j\}_{j\in\Lambda}$ is equal to $\{e\}$, then $\gcd(\#(G^{\rm ab})_{\rm tors}\,, [G:H])$ is equal to $[G:H]$ by the above two proofs and the two statements in Theorem 1.3 and Theorem 1 are same.

EXAMPLE 2.5. Let n be a natural number greater than one and consider the case $G := \mathbb{Z}$ and $H := n\mathbb{Z}$. Then $V_{G/H}(x) = nx$ and so $\ker(V_{G/H})$ is trivial, hence $\#\ker(V_{G/H})$ is not divisible by $[G:H]$. Therefore, our estimation of $\#\ker V_{G/H}$ using $\gcd(\#(G^{\rm ab})_{\rm tors}, [G:H])$ is more precise for the case where $G^{\rm ab}$ may be infinite.

As a corollary of Theorem 1, we can show Theorem 3.

PROOF OF THEOREM 3. By Theorem 1, the order of $\ker V_{G/H}$ is divisible by $\gcd(\#(G^{\rm ab})_{\rm tors}\,, [G:H])$, that is, $[G:H]$. In fact $\ker(V_{G/H})$ is equal to $(G^{\rm ab})_{\rm tors}$. Hence an element of $\ker V_{G/H}$ is σ-invariant if and only if the element is the inverse element, that is, the order of the element is 1 or 2. We finished the proof. □

3. Endomorphism version

In this section, we give a proof of Theorem 2.

PROOF OF THEOREM 2. We use the same notations as in the proof of Theorem 1. We put $\tilde\sigma := f^{-1}w_0 f\sigma$, where w_0 is an endomorphism of $UR/R *_\Gamma WR/R$ which sends any element of WR/R in $f\sigma(G)$ to the unit element. Note that $\tilde\sigma$ is an endomorphism of $\langle U|R\rangle$. If a is an element of $\langle U|R\rangle \cap H$, then we see that

$$\begin{aligned}\tilde\sigma(a) \in f^{-1}w_0f\sigma(\langle U|R\rangle) \cap f^{-1}w_0f\sigma(H) &\subset f^{-1}w_0(f(\langle U|R\rangle)f(G')) \cap f^{-1}w_0f(H)\\ &\subset \langle U|R\rangle G' \cap H = \langle U|R\rangle \cap H.\end{aligned}$$

Next we show that $\tilde\sigma$ induces the identity map on $\langle U|R\rangle/(\langle U|R\rangle \cap H)$.

If a is a representative element of $\langle U|R\rangle/(\langle U|R\rangle \cap H)$ and γ is an element of $\langle U|R\rangle \cap H$, we have

$$\tilde\sigma(a\gamma) = f^{-1}w_0f\sigma(a\gamma) = f^{-1}w_0(f(a)f(\gamma')) = af^{-1}w_0f(\gamma')\ (\gamma' \in H),$$

since a is also a representative element of G/H. Hence $f^{-1}w_0(\gamma') \in \langle U|R\rangle \cap H$ and we see that $\tilde\sigma$ induces the identity map on $\langle U|R\rangle/(\langle U|R\rangle \cap H)$.

Finally we prove

$$\#\{\gamma \in \ker V_{\langle U|R\rangle/(\langle U|R\rangle\cap H)}\,;\, \tilde\sigma(\gamma)\gamma^{-1} \in \langle U|R\rangle'\} \mid \#\{\gamma \in \ker V_{G/H}\,;\, \sigma(\gamma)\gamma^{-1} \in G'\}.$$

As the proof of Theorem 1 shows, $\# \ker V_{G/H}$ is divisible by $\# \ker V_{\langle U|R\rangle/(\langle U|R\rangle \cap H)}$. Let γ be an element of $\langle U|R\rangle$ such that $\tilde{\sigma}(\gamma)\gamma^{-1}$ is the element of $\langle U|R\rangle'$. Noting that $\sigma(\langle U|R\rangle) \subset \langle U|R\rangle G'$, we have by the above assumption

$$\sigma(\gamma)\gamma^{-1} \equiv \gamma\gamma^{-1} \equiv e \pmod{G'}.$$

Next we show the converse. it is clear that $\ker V_{G/H}$ is included in $\langle U|R\rangle$ mod G' by the proof of Theorem 1. If γ is an element of $\langle U|R\rangle$ mod G' such that $\sigma(\gamma)\gamma^{-1} \in G'$, then we have

$$\tilde{\sigma}(\gamma)\gamma^{-1} \equiv f^{-1}w_0 f\sigma(\gamma)\gamma^{-1} \equiv f^{-1}w_0 f(\sigma(\gamma)\gamma^{-1}) \equiv e \pmod{G'}.$$

Obviously $\tilde{\sigma}(\gamma)\gamma^{-1} \in \langle U|R\rangle$. Therefore we get the following by Lemma 2.2

$$\tilde{\sigma}(\gamma)\gamma^{-1} \in G' \cap \langle U|R\rangle = \langle U|R\rangle'.$$

Note that G/H is isomorphic to $\langle U|R\rangle/(\langle U|R\rangle \cap H)$, in particular, $[G:H]$ is equal to $[\langle U|R\rangle : \langle U|R\rangle \cap H]$ by the assumption, we finish the proof by Theorem 1.3. □

REMARK 3.1. If the set $\{h_j\}_{j\in\Lambda}$ is equal to $\{e\}$, then two statements in Theorem 1.3 and Theorem 2 are same.

The following example shows that we can not replace $[G:H]$ in the conclusion of Theorem 2 by $\gcd(\#(G^{\mathrm{ab}})_{\mathrm{tors}}, [G:H])$, when we remove the assumption $(*)$.

EXAMPLE 3.2. Let G and H be $P * \mathbb{Z}$ and $P * l\mathbb{Z}$ respectively, where P is a finite abelian group with odd order, and l is a natural number which is divisible by $\#P$. Then by Theorem 3, we see that the group

$$\{\gamma G' \in \ker(V_{G/H})\,;\, \sigma(\gamma)\gamma^{-1} \in G'\}$$

is trivial. Therefore the order is not divisible by $\gcd(\#(G^{\mathrm{ab}})_{\mathrm{tors}}, [G:H])$.

4. Applications

Since homology groups of compact manifolds are finitely generated, Theorem 1, 2 yield immediately the following theorems:

THEOREM 4.1. *Let $N \to M$ a finite abelian covering of compact path-manifolds. Then $C_{N/M}$ is finite and the order is divisible by* $\gcd(\#\mathrm{H}_1(M)_{\mathrm{tors}}, [N:M])$.

THEOREM 4.2. *Let $M \to M_0$ be a finite cyclic branched covering of path-connected compact manifolds with Galois group G_0 and $N \to M$ a finite covering so that the composite $N \to M_0$ is an abelian covering. We suppose that for a system of generators g_i of G, N corresponds to the subgroup of $\pi_1(M)$ which contains all $g_i \in \pi_1(M)$ so that the order of g_i modulo $\pi_1(M)'$ is infinite.*

Then the number of G_0-invariant classes in $C_{N/M}$ is finite and divisible by $[N:M]$.

EXAMPLE 4.3. Let M be the 6-fold cyclic branched cover over knot 8_{10} in the 3-sphere with Galois group G_0. Then we have the following isomorphism by Hempel ([**H**], Example 5.6.):

$$\mathrm{H}_1(M) \simeq \mathbb{Z}^2 + (\mathbb{Z}/4\mathbb{Z})^2 + \mathbb{Z}/9\mathbb{Z}.$$

Now we can represent $\pi_1(M)$ by $\langle\{x_1, x_2, y, z_1, z_2\} \cup \{w_j\}_{j\in\Lambda}|R\rangle$ where R stands for the relators, and the orders of $x_i\pi_1(M)'$ $(i = 1,2)$, $y\pi_1(M)'$, $w_i\pi_1(M)'$ $(i \in \Lambda)$

and $z_i\pi_1(M)'$ $(i = 1, 2)$ are 4, 9, 1 and ∞ respectively. Let N be the covering space corresponding to the subgroup

$$G_N := \langle \{y, z_1, z_2\} \cup \pi_1(M)' | R \rangle.$$

Then we see $G_N \supset \pi_1(M)'$ and $[\pi_1(M) : G_N] = 16$. Let σ be a generator of G_0. Then from Theorem 2, we find that the subgroup

$$\{\gamma \in \ker V_{N/M} \,;\, \sigma(\gamma) = \gamma\}$$

is finite and the order is divisible by 16. Hence it follows

$$\{\gamma \in \ker V_{N/M} \,;\, \sigma(\gamma) = \gamma\}_{(4)} = \mathrm{H}_1(M)_{(4)} \simeq (\mathbb{Z}/4\mathbb{Z})^2.$$

Therefore any homology class of the maximal 4-torsion subgroup of $\mathrm{H}_1(M)$ is σ-invariant and capitulates in N.

ACKNOWLEDGEMENTS . I would like to thank Prof. M. Morishita for useful communications and pointing out some mistakes in this paper. Many thanks go to M. Morishita and J. Morava for giving me an opportunity to submit this paper to the Proceedings "Primes and Knots".

References

[B] K. S. Brown, *Cohomology of Groups*, GTM **87**, Springer-Verlag.

[H] J. Hempel, *Homology of coverings*, Pacific J. Math., **121**(1985), 83–112.

[L] F. W. Levi, *The commutatorgroup of a free product*, J. Indian Math. Soc. (N.S.), **4**(1940), 136–144.

[Mi] K. Miyake, *Algebraic investigation of Hilbert's Theorem 94, the principal theorem and the capitulation problem*, Expo. Math., **7**(1989), 289–346.

[Mo1] M. Morishita, *On certain analogies between knots and primes*, J. reine angew. Math., **550**(2002), 141–167

[Mo2] —, *On capitulation problem for 3-manifolds*, "Galois theory and Modular forms", 305–313. Dev. Math., **11**, Kluwer Acad. Publ., Boston, MA, 2004.

[Mo3] —, *Analogies between knots and primes, 3-manifolds and number rings*, to appear in Sugaku Expositions (Sugaku **58**, No.1 (2006), 40–63).

[Mo4] —, *Milnor invariants and Massey products for prime numbers*, Compositio Math., **140**(2004), 69–83.

[Ra] N. Ramachandran, *A note on arithmetic topology*, C. R. Math. Acad. Sci. Soc. R. Can., **23**(2001), 130–135.

[Re1] A. Reznikov, *Three-manifolds class field theory* (Homology of coverings for a nonvirtually b_1-positive manifold), Selecta Math. New Ser., **3**(1997), 361–399.

[Re2] —, *Embedded incompressible surfaces and homology of ramified coverings of three-manifolds*, Selecta Math. New Ser., **6**(2000), 1–39.

[Si] A. Sikora, *Analogies between group actions on the 3-manifolds and number fields*, Comm. Math. Helv., **78**(2003), 832–844.

[Su1] H. Suzuki, *A generalization of Hilbert's theorem 94*, Nagoya Math. J., **121**(1991), 161–169.

[Su2] —, *On the Capitulation Problem*, Class Field Theory—its centenary and prospect—(Tokyo 1998), Adv. Stud. Pure Math., **30**, Math. Soc. Japan, Tokyo, (2001), 483–507.

MATHEMATICAL INSTITUTE TOHOKU UNIVERSITY SENDAI 980-8578 JAPAN
E-mail address: `flt_link@mac.com`

Contemporary Mathematics
Volume **416**, 2006

Multiple zeta values and Grothendieck-Teichmüller groups

Hidekazu Furusho

Abstract. We get a canonical embedding from the spectrum of the algebra of multiple zeta values modulo π^2 into the graded version of the Grothendieck-Teichmüller group by using relations of the Drinfel'd associator. On the other hand, it is known that the rational pro-l Galois image algebraic group is embedded into the Grothendieck-Teichmüller pro-algebraic group for each prime l. Via these embeddings, we get two kinds of relationship between the spectrum of the algebra of multiple zeta values and the pro-l Galois image algebraic group.

0. Introduction

The aim of this article is to establish a certain relationship among Grothendieck-Teichmüller groups, multiple zeta values and certain Galois image implicitly built in Drinfel'd's great paper in 1991 [**Dr**]. In the pro-finite group case, on Galois Side, Y. Ihara showed explicitly that the *absolute Galois group* $G_{\mathbf{Q}} := Gal(\overline{\mathbf{Q}}/\mathbf{Q})$ can be embedded into the Grothendieck-Teichmüller (pro-finite) group $\widehat{GT}$ in [**Ih94**] following the lines suggested in [**Dr**] (another group theoretical proof can be found in [**Na**]). But here we work on the unipotent pro-algebraic group case, especially on Hodge Side. We concentrate on the construction of an embedding from the spectrum of the algebra of multiple zeta values modulo π^2 into one of Grothendieck-Teichmüller unipotent pro-algebraic groups (§3) following the lines of [**Dr**], from which we shall get two kinds of interesting relationship (§5 and §6) between Hodge Side (§3) and Galois Side (§4).

In connection with the study of Galois representations on the algebraic fundamental group of the projective line minus 3 points, the Grothendieck-Teichmüller (pro-finite) group $\widehat{GT}$ recently considered in consecutive papers [**HS**], [**Ih90**]~[**Ih00**], [**LNS**], [**NS**], [**S**] and [**SL**]. This pro-finite group $\widehat{GT}$ was constructed by V.G. Drinfel'd in [**Dr**] and it is said that it may (or may not?) coincide with a certain combinatorial pro-finite group predicted by A. Grothendieck in [**Gr**]Ch 2. But, in fact, Drinfel'd originally invented and studied the pro-algebraic group $\underline{GT}$ over $\mathbf{Q}$ instead of the pro-finite group $\widehat{GT}$ in his study of the deformation of quasi-triangular quasi-Hopf quantized universal enveloping algebras. In this paper, we shall call this pro-algebraic group $\underline{GT}$ the *Grothendieck-Teichmüller* (pro-algebraic)

2000 *Mathematics Subject Classification*: Primary 11M41.

group. There he also introduced another pro-algebraic group $\underline{GRT}$ over $\mathbf{Q}$, which we call the *graded Grothendieck-Teichmüller group* as a graded version of $\underline{GT}$. In this paper, their unipotent parts $\underline{GT}_1$ and $\underline{GRT}_1$ play a role in *Galois Side* and *Hodge Side* respectively.

In §4.2, from the pro-l Galois representation

$$p_1^{(l)} : Gal\left(\overline{\mathbf{Q}}/\mathbf{Q}(\mu_{l^\infty})\right) \to Aut\ \pi_1^{(l)}(\mathbf{P}^1_{\overline{\mathbf{Q}}} - \{0,1,\infty\}) \quad ,$$

we shall associate the embedding of pro-algebraic group

$$\Phi_{\mathbf{Q}}^{(l)} : \underline{Gal}_{\mathbf{Q}}^{(l)} \hookrightarrow \underline{GT}_1$$

for each prime l, where $\underline{Gal}_{\mathbf{Q}}^{(l)}$ is the associated pro-algebraic group over $\mathbf{Q}$ (for definition, see §4.2). On this embedding $\Phi_{\mathbf{Q}}^{(l)}$, we shall see in §4.2 that it is natural to conjecture that $\Phi_{\mathbf{Q}}^{(l)}$ gives an isomorphism. Namely

CONJECTURE B . $\underline{Gal}_{\mathbf{Q}}^{(l)} \cong \underline{GT}_1$ for all prime l.

In contrast, we get a Hodge counterpart of the embedding $\Phi_{\mathbf{Q}}^{(l)}$, which is our main result. Let $Z_{\bullet}$ be the graded $\mathbf{Q}$-algebra generated by all multiple zeta values and $Z_{\bullet}/(\pi^2)$ be its quotient algebra modulo the graded principal ideal generated by π^2 (see §3.1). In §3.1, we shall associate a scheme *Spec* $Z_{\bullet}/(\pi^2)$ and show that

Theorem 3.2.5. *There is a canonical embedding of schemes*

$$\Phi_{DR} : Spec\ Z_{\bullet}/(\pi^2) \hookrightarrow \underline{GRT}_1 \ .$$

In §3.4, we shall also see that it is natural to conjecture that Φ_{DR} gives an isomorphism. Namely

CONJECTURE A . *Spec* $Z_{\bullet}/(\pi^2) \cong \underline{GRT}_1$.

In §5 and §6, we will discuss two kinds of relationship between $\underline{GT}_1$ and $\underline{GRT}_1$. The first one is on the relationship between $\mathbf{C}$-structures (cf. *Notation* 4.2.1) of $\underline{GT}_1$ and $\underline{GRT}_1$.

Proposition 5.1.2. *There exists a natural isomorphism* $p : \underline{GT}_1 \times_{\mathbf{Q}} \mathbf{C} \overset{\sim}{\to} \underline{GRT}_1 \times_{\mathbf{Q}} \mathbf{C}$. *If we identify their groups of* $\mathbf{C}$*-valued points by this isomorphism, two groups of rational points* $\underline{GT}_1(\mathbf{Q})$ *and* $\underline{GRT}_1(\mathbf{Q})$ *become inner conjugate to each other in* $\underline{GRT}_1(\mathbf{C})$.

By combining this proposition with two embeddings $\Phi_{\mathbf{Q}}^{(l)}$ and Φ_{DR}, we get Figure 1. Here $Hom_{\mathbf{Q}\text{-alg}}\left(Z_{\bullet}/(\pi^2), \mathbf{Q}\right)$ stands for the set of $\mathbf{Q}$-algebra homomorphisms from $Z_{\bullet}/(\pi^2)$ to $\mathbf{Q}$.

The second one is

Theorem 6.3.2. *The pro-algebraic group* $\underline{GrGT}_1$ *(for definition, see* (6.3.2)*) is isomorphic to* $\underline{GRT}_1$, *i.e.*

$$\underline{GrGT}_1 \cong \underline{GRT}_1 \ .$$

By combining this theorem with two embeddings $\Phi_{\mathbf{Q}}^{(l)}$ and Φ_{DR}, we get Figure 2.

In *Remark* 6.5.1, we shall see that the main result of our previous article [**F**] can be deduced from Figure 2. As a corollary of *Theorem* 6.3.2, we get an interesting correspondence between the l-adic Ihara associator $\Phi_{Ih}^{(l)}$ (§6.6.1) and the Drinfel'd associator Φ_{KZ} (§6.6.1).

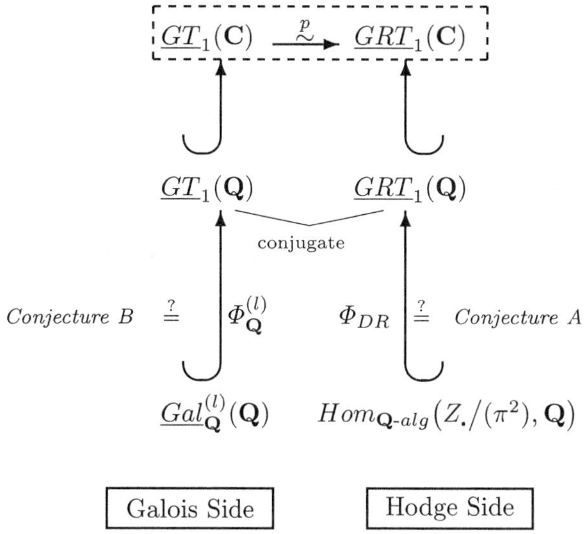

FIGURE 1

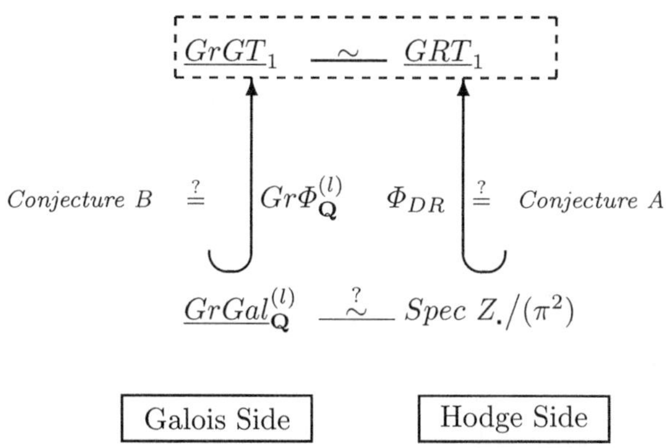

FIGURE 2

Proposition 6.6.1.

$$Gr\Phi^{(l)\sharp}_{\mathbf{Q}\langle\langle A,B\rangle\rangle}(\Phi_{GRT}) = Gr\Phi^{(l)}_{Ih} \quad , \quad \Phi^{\sharp}_{DR\langle\langle A,B\rangle\rangle}(\Phi_{GRT}) = \Phi_{KZ} \mod \pi^2,$$

from which we get Figure 3.

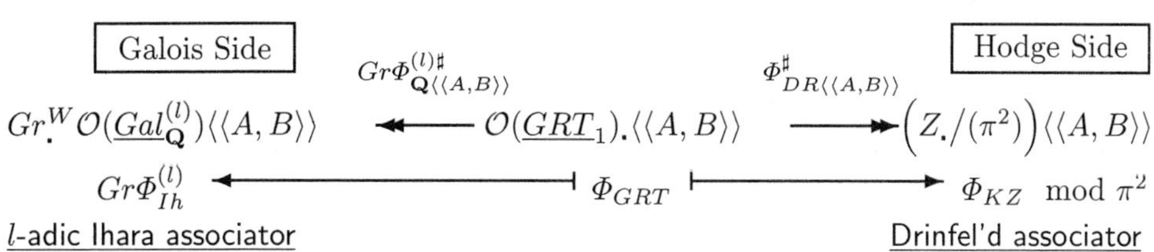

FIGURE 3

The pictures in *Corollary* 6.6.4 and *Corollary* 6.6.6 should be better called cases of meta-abelian quotient of Figure 3.

This paper is organized as follows. In §1, we shall make a brief review of the notion of weight filtrations of negatively weighted extensions [**HM**]. §2 is devoted to a (long) review and detailed description of $\underline{GT}$, $\underline{GRT}$ and $\underline{M}$ constructed by Drinfel'd [**Dr**]. In §3, we shall recall the definition of multiple zeta values and construct an embedding Φ_{DR} in Theorem 3.2.5, which is our main result. In §4, we shall discuss the embedding $\Phi_{\mathbf{Q}}^{(l)}$. Finally in §5 and §6, we shall give two kinds of interesting relationship between $\underline{GT}_1$ and $\underline{GRT}_1$ and then make an interesting comparison between Hodge Side (§3) and Galois Side (§4) in §6.6.

ACKNOWLEDGMENTS . The author expresses a special thanks to his previous advisor Professor Y. Ihara for his continuous encouragement and he is deeply grateful to Professor A. Tamagawa for carefully reading this manuscript.

1. On weight filtrations by Hain-Matsumoto

This section is devoted to a brief review of the notion of weight filtrations on modules of negatively weighted pro-algebraic groups, which is introduced in [**HM**]. In §1.3, we will introduce a helpful proposition (*Proposition* 1.3.2), which will be used later.

1.1. Negatively weighted extension. We recall the definition of negatively weighted extension in [**HM**]§3.

NOTATION 1.1.1. Let k be a field with characteristic 0. Let S be a reductive algebraic group over k. Let $\varpi : \mathbb{G}_m \to S$ be a *central cocharacter*, which means a homomorphism whose image is contained in the center of S. Let G be an algebraic group over k which is an extension of S by a unipotent algebraic group U ; $1 \to U \to G \to S \to 1$. By [**HM**]*Proposition 2.3*, the maximal abelian quotient $H_1(U)$ of this algebraic group U becomes an S-module. Therefore it becomes a $\mathbb{G}_m$-module via ϖ. Thus we can decompose *uniquely* as $H_1(U) = \bigoplus_{k\in\mathbf{Z}} W_k$, where W_k is the $\mathbb{G}_m$-module whose $\mathbb{G}_m$-action is given by the k-th power multiplication.

DEFINITION 1.1.2. A *pro-algebraic group* over a field k is a projective limit of algebraic groups over k. A pro-algebraic group $\mathcal{G} = \varprojlim_{\alpha} G_\alpha$ is called *unipotent* if each G_α is unipotent.

DEFINITION 1.1.3 ([**HM**]§3). The extension algebraic group G in *Notation* 1.1.1 is called negatively weighted with respect to ϖ if $W_k = 0$ for all $k \geqslant 0$. A *pro*-algebraic group $\mathcal{G}$ which is an extension of a reductive algebraic group S by a unipotent *pro*-algebraic group $\mathcal{U}$ is also called negatively weighted with respect to ϖ if it is a projective limit of algebraic groups which are negatively weighted extensions of S with respect to ϖ.

1.2. Weight filtration. We will review the fact shown in [**HM**]§3 that each representation of negatively weighted pro-algebraic group is equipped with a natural weight filtration.

NOTATION 1.2.1. Let $\mathcal{G}$ be a pro-algebraic group over a field k with characteristic 0 which is a negatively weighted extension of a reductive algebraic group S over k by a unipotent pro-algebraic group $\mathcal{U}$ over k with respect to a central cocharacter $\varpi : \mathbb{G}_m \to S$. By [**HM**] *Lemma 3.1*, there exists a homomorphism $\tilde{\varpi} : \mathbb{G}_m \to \mathcal{G}$ which is a lift of ϖ. Let V be a finite dimensional k-vector space equipped with $\mathcal{G}$-action. Then this $\mathcal{G}$-module V becomes a $\mathbb{G}_m$-module via $\tilde{\varpi}$. So we can decompose as $V = \bigoplus_{a \in \mathbf{Z}} V_a$, where V_a is the $\mathbb{G}_m$-module whose $\mathbb{G}_m$-action is given by the a-th power multiplication.

In this subsection, we assume *Notation* 1.2.1.

DEFINITION 1.2.2 ([**HM**]§3). The weight filtration of $\mathcal{G}$-module V is the ascending filtration $W = \{W_n V\}_{n \in \mathbf{Z}}$ of k-linear subspaces defined by $W_n V = \bigoplus_{a \leqslant n} V_a$ for each $n \in \mathbf{Z}$.

In [**HM**]*Proposition 3.8*, it was proved that this weight filtration is natural in the sense that it does not depend on the choice of lift $\tilde{\varpi}$ above and it was also shown that $W_n V$ is stable by the $\mathcal{G}$-action. Moreover,

PROPOSITION 1.2.3 ([**HM***]*Proposition 4.10*). *The weight filtration W of a finite dimensional $\mathcal{G}$-module V is the unique ascending filtration $W = \{W_n V\}_{n \in \mathbf{Z}}$ of $\mathcal{G}$-submodules that satisfies the following properties:*

(a) $\bigcap_{n \in \mathbf{Z}} W_n V = 0, \quad \bigcup_{n \in \mathbf{Z}} W_n V = V.$
(b) *The action of $\mathcal{U}$ on $Gr_n^W V$ $(:= W_n V / W_{n-1} V)$ is trivial for all $n \in \mathbf{Z}$.*
(c) *The action of $\mathbb{G}_m$ on $Gr_n^W V$ via ϖ is given by the n-th power multiplication for all $n \in \mathbf{Z}$.*

1.3. Filtered Hopf algebras.
Assume *Notation* 1.2.1.

NOTATION 1.3.1. Let $\mathcal{U} = \varprojlim_{\alpha} U_\alpha$ denote a projective limit of unipotent algebraic groups U_α. The regular function ring $\mathcal{O}(\mathcal{U})$ of $\mathcal{U}$ is defined to be the inductive limit $\mathcal{O}(\mathcal{U}) := \varinjlim_{\alpha} \mathcal{O}(U_\alpha)$ of those of U_α. This k-algebra $\mathcal{O}(\mathcal{U})$ is equipped with a structure of Hopf algebra over k. Let A be an arbitrary k-algebra. Take any element $g \in \mathcal{G}(A)$. Then the inner automorphism $\tau_g^{-1} : \mathcal{U} \times A \to \mathcal{U} \times A$ defined by $x \mapsto g^{-1} x g$ induces the isomorphism $(\tau_g^{-1})^\sharp : \mathcal{O}(\mathcal{U})_A \to \mathcal{O}(\mathcal{U})_A$ of Hopf algebras. By the correspondence $\tau : \mathcal{G} \to \underline{Aut}\mathcal{O}(\mathcal{U})$ defined by $g \mapsto (\tau_g^{-1})^\sharp$, we regard $\mathcal{O}(\mathcal{U})$ as a left module of $\mathcal{G}$.

Although $\mathcal{O}(\mathcal{U})$ is infinite-dimensional, it is equipped with the following weight filtration.

PROPOSITION 1.3.2. *The regular function ring $\mathcal{O}(\mathcal{U})$ is equipped with a weight filtration, which is an ascending filtration of finite dimensional k-linear sub-spaces:*

$$W : \cdots = W_{-2}\mathcal{O}(\mathcal{U}) = W_{-1}\mathcal{O}(\mathcal{U}) = 0 \subseteq W_0\mathcal{O}(\mathcal{U}) \subseteq W_1\mathcal{O}(\mathcal{U}) \subseteq \cdots\cdots\cdots \subseteq W_n\mathcal{O}(\mathcal{U}) \subseteq W_{n+1}\mathcal{O}(\mathcal{U}) \subseteq \cdots\cdots .$$

It satisfies the following properties as in Proposition *1.2.3:*

(a) $\bigcap_{n \in \mathbf{Z}} W_n\mathcal{O}(\mathcal{U}) = 0, \quad \bigcup_{n \in \mathbf{Z}} W_n\mathcal{O}(\mathcal{U}) = \mathcal{O}(\mathcal{U}).$
(b) *The action of $\mathcal{U}$ on $Gr_n^W \mathcal{O}(\mathcal{U})$ by τ is trivial for all $n \in \mathbf{Z}$.*
(c) *The action of $\mathbb{G}_m$ on $Gr_n^W \mathcal{O}(\mathcal{U})$ via τ and ϖ is the n-th power multiplication for all $n \in \mathbf{Z}$.*

Moreover this filtration is compatible with all structure morphisms of Hopf algebras, i.e. $(\mathcal{O}(\mathcal{U}), W)$ *becomes a filtered Hopf algebra.*

PROOF . Regard $Lie\mathcal{U}$ as a left $\mathcal{G}$-module by its adjoint representation. In [**HM**] *Proposition 4.5*, it is shown that $Lie\mathcal{U}$ is equipped with a weight filtration of finite dimensional k-linear subspaces

$$W : \cdots\cdots \subseteq W_{-n-1}Lie\mathcal{U} \subseteq W_{-n}Lie\mathcal{U} \subseteq \cdots\cdots$$
$$\cdots \subseteq W_{-1}Lie\mathcal{U} = W_0 Lie\mathcal{U} = W_1 Lie\mathcal{U} = \cdots\cdots = Lie\mathcal{U}$$

which satisfies properties (a)$\sim$(c) in *Proposition* 1.2.3. Since $\mathcal{U}$ is unipotent, $\mathcal{O}(\mathcal{U})$ is isomorphic to the dual of the universal enveloping algebra $\mathbb{U}Lie\mathcal{U}$ of $Lie\mathcal{U}$ as left $\mathcal{G}$-modules. Thus $\mathcal{O}(\mathcal{U})$ is equipped with the induced filtration from the above one on $Lie\mathcal{U}$, from which the first half of our statement can be deduced. The second half is immediate since the $\mathcal{G}$-action on $\mathcal{O}(\mathcal{U})$ by τ is consistent with all structure morphisms of Hopf algebras, i.e. the product map, the unit map, the co-product map, the co-unit map and the antipode map. □

2. A detailed analysis of Drinfel'd's GT's

This section is devoted to a long review and detailed analysis on $\underline{GT}$, $\underline{GRT}$ and $\underline{M}$ constructed by Drinfel'd [**Dr**] in terms of weight filtration (§1) by Hain-Matsumoto. In §2.1 (resp. §2.2), we shall recall the definition of $\underline{GRT}$ (resp. $\underline{GT}$) [**Dr**] and discuss the weight filtration of the regular function ring $\mathcal{O}(\underline{GRT}_1)$ (resp. $\mathcal{O}(\underline{GT}_1)$). §2.3 will be devoted to a review of the definition of the Drinfel'd associator φ_{KZ} and the pro-torsor $\underline{M}$ [**Dr**].

2.1. On $\underline{GRT}$. §2.1.1 is devoted to reviewing the definition of the pro-algebraic group $\underline{GRT}$ that appeared in [**Dr**]. In §2.1.2, we will endow the regular function ring $\mathcal{O}(\underline{GRT}_1)$ of the unipotent part $\underline{GRT}_1$ of $\underline{GRT}$ with a grading and show that it becomes a graded Hopf algebra.

2.1.1. *The graded Grothendieck-Teichmüller group.*

NOTATION 2.1.1. Let k be any $\mathbf{Q}$-algebra. Let $k\langle\langle A, B\rangle\rangle$ be the graded non-commutative formal power series ring over k with 2 variables A and B with degrees given by $degA = degB = 1$. Denote the subset of $k\langle\langle A, B\rangle\rangle$ consisting of formal Lie series in $k\langle\langle A, B\rangle\rangle$ by $\mathbb{L}_k^{\wedge}$. Let R be a completed (non-commutative) associative k-algebra and $\phi : k\langle\langle A, B\rangle\rangle \to R$ be a completed k-algebra homomorphism uniquely defined by $\phi(A) = a$, $\phi(B) = b$ for certain elements $a, b \in R$. For $g \in k\langle\langle A, B\rangle\rangle$, we denote the image $\phi(g) \in R$ by $g(a, b)$.

DEFINITION 2.1.2 ([**Dr**]§5). The graded [1] Grothendieck-Teichmüller (pro-algebraic) group $\underline{GRT}$ is the pro-linear algebraic group over $\mathbf{Q}$ whose set of k-valued points is defined as follows:

$$\underline{GRT}(k) = \{(c, g) \in k^{\times} \times k\langle\langle A, B\rangle\rangle \big|\ g \text{ satisfies } (0)\sim \text{(iii) below.}\}$$

[1] The word *graded* means the natural grading (see §2.1.2).

$$\begin{cases} (0)\ \ g \in exp[\mathbb{L}_k^\wedge, \mathbb{L}_k^\wedge] \\ (\mathrm{i})\ \ g(A,B)g(B,A) = 1 \\ (\mathrm{ii})\ \ g(C,A)g(B,C)g(A,B) = 1 \ \text{ for } \ A+B+C=0 \\ (\mathrm{iii})\ \ g(X_{1,2},X_{2,3})g(X_{3,4},X_{4,5})g(X_{5,1},X_{1,2})g(X_{2,3},X_{3,4})g(X_{4,5},X_{5,1}) = 1 \\ \qquad\qquad\qquad \text{in} \quad \underline{U\mathfrak{P}_\bullet^{(5)}}(k) \quad (\text{see } Note 2.1.3). \end{cases}$$

The multiplication map [2] m' of the graded Grothendieck-Teichmüller group is given as follows:

$$\begin{aligned} m' : \underline{GRT}(k) \times \underline{GRT}(k) &\longrightarrow \underline{GRT}(k) \\ (c_2, g_2) \times (c_1, g_1) &\longmapsto (c_2, g_2) \circ (c_1, g_1) \quad , \end{aligned}$$

where $(c_2, g_2) \circ (c_1, g_1) := \Big(c_1c_2 \ , g_2(A,B) \ g_1\big(\frac{A}{c_2} \ , g_2(A,B)^{-1}\frac{B}{c_2}g_2(A,B)\big)\Big)$.

NOTE 2.1.3. (1) The defining relation (i) (resp. (ii), (iii)) is sometimes called 2 (resp. 3, 5)-cycle relation.

(2) On (0), for any element h (in the topological commutator $[\mathbb{L}_k^\wedge, \mathbb{L}_k^\wedge]$) of $\mathbb{L}_k^\wedge$, we define $exp\ h := 1 + \frac{h}{1!} + \frac{h^2}{2!} + \frac{h^3}{3!} + \cdots \in k\langle\langle A, B\rangle\rangle$. This series converges since h has no constant (degree 0) term.

(3) On (iii), $\underline{U\mathfrak{P}_\bullet^{(5)}}(k)$ means the completion by degree (i.e. that is respected to the filtration induced by degree) of the universal enveloping algebra of the pure sphere 5-braid graded Lie algebra $\mathfrak{P}_\bullet^{(5)}$ tensored with k and $X_{i,j}$ $(1 \leqslant i,j \leqslant 5)$ stand for the standard generators of $\mathfrak{P}_\bullet^{(5)}$ (for definitions, see [**Ih90**]~[**Ih92**] and also [**F**] §2.1.).

(4) It can be checked easily that (iii) implies (i).

We remark that the same pro-algebraic group also appeared and was studied by Z. Wojtkowiak [**Wo**]. In this paper, especially we examine its unipotent part $\underline{GRT}_1$.

DEFINITION 2.1.4 ([**Dr**]§5). The unipotent graded Grothendieck-Teichmüller (pro-algebraic) group $\underline{GRT}_1$ is the unipotent sub-pro-algebraic group of $\underline{GRT}$ whose set of k-valued points is

$$\begin{aligned} \underline{GRT}_1(k) &= \{\ (1,g) \in \underline{GRT}(k)\ \} \\ &= \{\ g \in k\langle\langle A,B\rangle\rangle \ \big| \ g \text{ satisfies } (0)\sim \text{ (iii) in } Definition 2.1.2.\} \ . \end{aligned}$$

Note that we get the following exact sequence of pro-algebraic groups

$$(2.1.1) \qquad \begin{aligned} 1 \longrightarrow \underline{GRT}_1 \longrightarrow \underline{GRT} &\longrightarrow \mathbb{G}_m \longrightarrow 1 \\ (c,g) &\longmapsto c \quad . \end{aligned}$$

REMARK 2.1.5. The above exact sequence is equipped with a standard section

$$\begin{aligned} s_0 : \mathbb{G}_m &\to \underline{GRT} \\ c &\mapsto (c, 1) \quad . \end{aligned}$$

This property may distinguish $\underline{GRT}$ from $\underline{GT}$ (§2.2).

LEMMA 2.1.6. *The pro-algebraic group $\underline{GRT}$ is a negatively weighted extension (*Definition*1.1.3) of $\mathbb{G}_m$ by the unipotent pro-algebraic group $\underline{GRT}_1$ with respect to the central cocharacter $\varpi : \mathbb{G}_m \to \mathbb{G}_m$ defined by $x \mapsto x$.*

---

[2]For our convenience, we reverse the original definition of the multiplication of $\underline{GRT}$ in [**Dr**]

PROOF . It follows from

$$(c,1)\circ\big(1,g(A,B)\big)\circ(c^{-1},1)=\Big(1,g(\frac{A}{c},\frac{B}{c})\Big).$$

□

2.1.2. *The graded Hopf algebra $\mathcal{O}(\underline{GRT}_1)_\bullet$.* In this subsection, we will see that the regular function ring $\mathcal{O}(\underline{GRT}_1)_\bullet$ of $\underline{GRT}_1$ is naturally equipped with a structure of graded Hopf algebra.

NOTATION 2.1.7. By a *word* we mean a monic and monomial element W of $\mathbf{Q}\langle\langle A,B\rangle\rangle$ (§2.1.1) whose degree is greater than 0. Note that we *do not* include 1 among words. Suppose that R is an arbitrary $\mathbf{Q}$-algebra. Then each element $g\in R\langle\langle A,B\rangle\rangle$ can be expanded uniquely as $g=x_1(g)+\sum_{W:\text{words}} x_W(g)W$ where $x_1(g)$ and $x_W(g)\in R$. Let W be a word or 1. We define the map $x_W:\underline{GRT}_1(R)\to R$ which is determined by $g\mapsto x_W(g)$. Then these x_W's generate the algebra $\mathcal{O}(\underline{GRT}_1)$ i.e. $\mathcal{O}(\underline{GRT}_1)=\mathbf{Q}[x_1,x_W]_{W:\text{words}}$. We remark that $x_1(g)=1$ for all $g\in\underline{GRT}_1(R)$ because of *Definition* 2.1.2 (0).

By *Proposition* 1.3.2, $\mathcal{O}(\underline{GRT}_1)$ is equipped with a weight filtration $W=\{W_n\mathcal{O}(\underline{GRT}_1)\}_{n\in\mathbf{Z}}$ satisfying properties (a)∼(c) in *Proposition* 1.3.2 with respect to ϖ (*Lemma* 2.1.6) and the pair $(\mathcal{O}(\underline{GRT}_1),W)$ becomes a filtered Hopf algebra. Recall that s_0 (*Remark* 2.1.5) is the standard section of the exact sequence (2.1.1). By imitating the prescription described in §1.2, $\mathcal{O}(\underline{GRT}_1)$ and $W_n\mathcal{O}(\underline{GRT}_1)$ can be decomposed into $\mathcal{O}(\underline{GRT}_1)=\bigoplus_{0\leqslant a}V_a$ and $W_n\mathcal{O}(\underline{GRT}_1)=\bigoplus_{0\leqslant a\leqslant n}V_a$ as $\mathbb{G}_m$-modules with respect to s_0, where V_a is the $\mathbb{G}_m$-module whose $\mathbb{G}_m$-action (given by $\tau\circ s_0$) is the a-th power multiplication. Since this grading on $\mathcal{O}(\underline{GRT}_1)_\bullet:=\bigoplus_{0\leqslant a}V_a$ is natural, it provides a natural isomorphism

$$\tilde{s_0}:Gr^W_\bullet\mathcal{O}(\underline{GRT}_1)\to\mathcal{O}(\underline{GRT}_1)_\bullet \quad .\tag{2.1.2}$$

Therefore it follows that

PROPOSITION 2.1.8. (a) *The Hopf algebra $\mathcal{O}(\underline{GRT}_1)$ is equipped with a weight filtration W.*
(b) *The pair $(\mathcal{O}(\underline{GRT}_1),W)$ becomes a filtered Hopf algebra.*
(c) *By* (2.1.2), *$\mathcal{O}(\underline{GRT}_1)$ is equipped with a structure of graded Hopf algebra.*

NOTE 2.1.9. This grading on $\mathcal{O}(\underline{GRT}_1)_\bullet=\mathbf{Q}[x_W]_{W:\text{words}}$ is given by $deg\ x_W=deg\ W$.

2.1.3. *The stable derivation algebra.* We shall review the definition of the stable derivation algebra $\mathfrak{D}_\bullet$ which was constructed by Y. Ihara in his series of works on the Galois representation on the pro-l fundamental group $\pi_1^{(l)}(\mathbf{P}^1_{\overline{\mathbf{Q}}}-\{0,1,\infty\})$ (see [**Ih90**]∼[**Ih92**] and [**Ih02**]).

NOTATION 2.1.10. Let $\mathbb{L}_w$ ($w\geqslant 1$) denote the degree w-part of the free completed Lie algebra $\mathbb{L}^\wedge_{\mathbf{Q}}$ (see §2.1.1) of rank 2 and let $\mathbb{L}_\bullet=\bigoplus_{w\geqslant 1}\mathbb{L}_w$ be the free (non-completed) graded Lie algebra over $\mathbf{Q}$. For f in $\mathbb{L}_\bullet$, we define the *special derivation* $D_f:\mathbb{L}_\bullet\to\mathbb{L}_\bullet$ which is the derivation determined by $D_f(A)=0$ and $D_f(B)=[B,f]$. It can be checked easily that $[D_f,D_g]=D_h$ with $h=[f,g]+D_f(g)-D_g(f)$.

DEFINITION 2.1.11 ([**Ih90**] and [**Ih92**]). The **stable derivation algebra** $\mathfrak{D}_{.}$ is the graded Lie subalgebra $\mathfrak{D}_{.}$ of $Der\mathbb{L}_{.}$ which has the following presentation: $\mathfrak{D}_{\bullet} = \underset{w\geqslant 1}{\oplus} \mathfrak{D}_w$, where

$$\mathfrak{D}_w = \{D_f \in Der\mathbb{L}_{.} |\ f \in \mathbb{L}_w \text{ satisfies (0)} \sim \text{(iii) below.}\}$$

$$\begin{cases} (0)\ \ f \in [\mathbb{L}_{\bullet}, \mathbb{L}_{\bullet}] \ \ (:= \overset{\infty}{\underset{a=2}{\oplus}} \mathbb{L}_a) \\ (\mathrm{i})\ \ f(A,B) + f(B,A) = 0 \\ (\mathrm{ii})\ \ f(A,B) + f(B,C) + f(C,A) = 0 \qquad \text{for } A+B+C=0 \\ (\mathrm{iii})\ \ \sum_{i\in\mathbf{Z}/5} f(X_{i,i+1}, X_{i+1,i+2}) = 0 \ \text{ in } \mathfrak{P}_{.}^{(5)} \ \text{(see } \textit{Note2.1.3}). \end{cases}$$

Here, for any Lie algebra H and $\alpha, \beta \in H$, $f(\alpha, \beta)$ denotes the image of $f \in \mathbb{L}_{\bullet}$ by the homomorphism $\mathbb{L}_{\bullet} \to H$ defined by $A \mapsto \alpha$ and $B \mapsto \beta$.

REMARK 2.1.12. (1) Each derivation $D \in \mathfrak{D}_{\bullet}$ determines a unique element $f \in [\mathbb{L}_{\bullet}, \mathbb{L}_{\bullet}]$ such that $D = D_f$.
(2) The relation (iii) implies (i).
(3) The completion by degree $\mathfrak{D}_{.}^{\wedge} = \underset{w\geqslant 1}{\widehat{\oplus}} \mathfrak{D}_w$ of Ihara's stable derivation algebra is equal to the pro-Lie algebra $\mathfrak{grt}_1(\mathbf{Q})$ [**Dr**] which is the Lie algebra *Lie* $\underline{GRT}_1$ of the graded Grothendieck-Teichmüller group.

On the structure of $\mathfrak{D}_{.}$, there is a standard conjecture in [**Ih02**] which is related to conjectures on the associated graded Lie algebra of the image of the Galois representation on $\pi_1^{(l)}(\mathbf{P}^1_{\overline{\mathbf{Q}}} - \{0,1,\infty\})$ by Ihara([**Ih90**]) and P. Deligne ([**De**]).

CONJECTURE 2.1.13 ([**Ih02**]). $\mathfrak{D}_{.}$ is a free Lie algebra generated by f_m, where f_m is a suitable element of $\mathfrak{D}_m$ $(m = 3, 5, 7, \cdots)$.

Furthermore, Ihara proposed the following problem:

PROBLEM 2.1.14 ([**Ih02**]). Construct f_m explicitly. Is there any canonical choice?

By his consideration of f_m $(m = 3, 5, 7, \cdots)$, each f_m must be of *depth* 1 (see [**Ih02**]Ch II).

REMARK 2.1.15. M. Matsumoto and H. Tsunogai ([**Tsu**]) calculated the dimensions of graded pieces of the stable derivation algebra on the lower weights. Especially Tsunogai verified *Conjecture* 2.1.13 up to weight 14.

Tsunogai posed the following problem (which arises from his computation table).

PROBLEM 2.1.16 (H. Tsunogai). On the defining relations (*Definition* 2.1.11) of the stable derivation algebra, does 5-cycle relation (iii) imply 3-cycle relation (ii)?

2.2. On $\underline{GT}$. §2.2.1 is devoted to reviewing the definition of the pro-algebraic group $\underline{GT}$ [**Dr**]§4. In §2.2.2, we will endow the regular function ring $\mathcal{O}(\underline{GT}_1)$ of its unipotent part $\underline{GT}_1$ with a weight filtration (§1.2) and show that it is equipped with a structure of filtered Hopf algebra.

2.2.1. *The Grothendieck-Teichmüller group.* Let F_2 be the free group of rank 2 generated by x and y and denote its Malcev completion (see, for example [**HM**] §A.1) by $\underline{F_2}$. Let k be any $\mathbf{Q}$-algebra.

DEFINITION 2.2.1 ([**Dr**]§4). The Grothendieck-Teichmüller (pro-algebraic) group is the pro-linear algebraic group $\underline{GT}$ defined over $\mathbf{Q}$ whose set of k-valued points is the subset of that of $\mathbb{G}_m \times \underline{F_2}$ defined as follows:

$$\underline{GT}(k) = \{(\lambda, f) \in k^\times \times \underline{F_2}(k) \ \big| \ (\lambda, f) \text{ satisfies (0)}\sim \text{(iii) below.}\}$$

$$\begin{cases} \text{(0)} \ \ f \in [\underline{F_2}, \underline{F_2}](k) \\ \text{(i)} \ \ f(x,y)f(y,x) = 1 \\ \text{(ii)} \ \ f(z,x)z^m f(y,z)y^m f(x,y)x^m = 1 \ \text{ for } \ xyz = 1, m = \frac{\lambda-1}{2} \\ \text{(iii)} \ \ f(x_{1,2}, x_{2,3})f(x_{3,4}, x_{4,5})f(x_{5,1}, x_{1,2})f(x_{2,3}, x_{3,4})f(x_{4,5}, x_{5,1}) = 1 \\ \qquad\qquad\qquad\qquad \text{in} \quad \underline{P_5}(k) \quad (\text{ see } Note 2.2.2.) \ . \end{cases}$$

The multiplication map [3] m of the Grothendieck-Teichmüller group is given as follows:

$$\begin{aligned} m : \underline{GT}(k) \times \underline{GT}(k) &\longrightarrow \underline{GT}(k) \\ (\lambda_2, f_2) \times (\lambda_1, f_1) &\longmapsto (\lambda_2, f_2) \circ (\lambda_1, f_1) \quad , \end{aligned}$$

where $(\lambda_2, f_2) \circ (\lambda_1, f_1) := \Big(\lambda_1\lambda_2 \ , f_2(x,y) f_1\big(x^{\lambda_2}, f_2(x,y)^{-1} y^{\lambda_2} f_2(x,y)\big)\Big)$.

NOTE 2.2.2. Here for any unipotent group scheme $\underline{H}$ over $\mathbf{Q}$ and morphism of group schemes $\Phi : \underline{F_2} \to \underline{H}$ with $\Phi(x) := \alpha$ and $\Phi(y) := \beta \in \underline{H}(\mathbf{Q})$, we denote the image of $f \in \underline{F_2}(\mathbf{Q})$ by $\Phi(f) \in \underline{H}(\mathbf{Q})$. We note that in (ii), x^m, y^m and z^m also make sense since $\underline{F_2}$ is unipotent. In the condition (iii), $\underline{P_5}$ means the Malcev completion of the pure sphere 5-braid group P_5 and $x_{i,j}$ $(1 \leqslant i, j \leqslant 5)$ stand for standard generators of P_5 (cf. [**Ih91**]). The relation (iii) implies (i).

V. G. Drinfel'd constructed $\underline{GT}$ and $\underline{GRT}$ as deformation groups of quasi-triangular quasi-Hopf quantized universal enveloping algebras. They act in a different way on the classifying space of these algebras. In [**Dr**], it is shown that their actions are commutative to each other.

DEFINITION 2.2.3 ([**Dr**]§5). The unipotent Grothendieck-Teichmüller (pro-algebraic) group $\underline{GT}_1$ is the unipotent sub-pro-algebraic group of $\underline{GT}$ whose set of k-valued points is

$$\underline{GT}_1(k) := \{f \in \underline{F_2}(k) \ \big| \ (1, f) \in \underline{GT}(k)\}.$$

Note that we get the following exact sequence of pro-algebraic group

$$\begin{aligned} 1 \longrightarrow \underline{GT}_1 \longrightarrow \underline{GT} &\longrightarrow \mathbb{G}_m \longrightarrow 1 \\ (\lambda, f) &\longmapsto \lambda \end{aligned} \qquad (2.2.1)$$

LEMMA 2.2.4. *The pro-algebraic group $\underline{GT}$ is a negatively weighted extension (*Definition *1.1.3) of $\mathbb{G}_m$ by the unipotent pro-algebraic group $\underline{GT}_1$ with respect to the central cocharacter* [4] $\frac{1}{\varpi} : \mathbb{G}_m \to \mathbb{G}_m$ *defined by* $x \mapsto x^{-1}$.

---

[3] For our convenience, we reverse the original definition of the multiplication of $\underline{GT}$ in [**Dr**]

[4] On the contrary, $\underline{GRT}$ was a negatively weighted extension with respect to ϖ.

PROOF . It follows from a calculation slightly more complicated than *Lemma* 2.1.6 (however, by combining *Lemma* 2.1.6 with *Proposition* 5.1.2, we can find another easier proof.). □

REMARK 2.2.5. By [**HM**]*Proposition A.8*, there exists the pro-linear algebraic group $\underline{AutF_2}$ defined over $\mathbf{Q}$ which represents the functor $K \mapsto \underline{AutF_2} \times K$ from field extensions over $\mathbf{Q}$ to groups (for the definition of $\underline{AutF_2} \times K$, see [**HM**]). By the correspondence $x \mapsto x^\lambda$ and $y \mapsto f^{-1}y^\lambda f$, where $(\lambda, f) \in \underline{GT}(K)$, $\underline{GT}$ can be regarded as a sub-algebraic group of $\underline{AutF_2}$.

2.2.2. *The filtered Hopf algebra* ($\mathcal{O}(\underline{GT}_1), W$). In this subsection, we will see that the regular function ring $\mathcal{O}(\underline{GT}_1)$. of $\underline{GT}_1$ is naturally equipped with a structure of filtered Hopf algebra.

As in the case of $\mathcal{O}(\underline{GT}_1)$ (§2.1.2) we can show that $\mathcal{O}(\underline{GT}_1)$ is equipped with a weight filtration $W = \{W_n\mathcal{O}(\underline{GT}_1)\}_{n\in\mathbf{Z}}$ satisfying properties (a)∼(c) (*Proposition* 1.3.2) with respect to $\frac{1}{\varpi}$ (*Lemma* 2.2.4). By *Proposition* 1.3.2,

PROPOSITION 2.2.6. (1) *The Hopf algebra* $\mathcal{O}(\underline{GT}_1)$ *is equipped with a weight filtration* W.
(2) *The pair* $(\mathcal{O}(\underline{GT}_1), W)$ *becomes a filtered Hopf algebra.*

Remark that we do not know any natural structure of graded Hopf algebra instead of that of filtered Hopf algebras on $\mathcal{O}(\underline{GT}_1)$, since we do not know any natural section of the exact sequence (2.2.1) like *Remark* 2.1.5. We remark that $\lambda \mapsto (\lambda, 1)$ is not a section because generally $z^m y^m x^m \neq 1$.

2.3. On $\underline{M}$ and φ_{KZ}. We shall review the definition of the pro-torsor $\underline{M}$ ([**Dr**]) in §2.3.1 and the original definition of the Drinfel'd associator φ_{KZ} ([**Dr**]) in §2.3.2 which plays a role to prove the main theorem in §3.

2.3.1. *The middle Grothendieck-Teichmüller torsor.* Let k be any $\mathbf{Q}$-algebra.

DEFINITION 2.3.1 ([**Dr**]§4). The middle Grothendieck-Teichmuller torsor is the pro-variety $\underline{M}$ defined over $\mathbf{Q}$ whose set of k-valued points is defined as follows:

$$\underline{M}(k) = \{(\mu, \varphi) \in k^\times \times k\langle\langle A, B\rangle\rangle \mid (\mu, \varphi) \text{ satisfies (0)}\sim \text{(iii) below.}\}$$

$$\begin{cases} \text{(0)} \ \ \varphi \in exp[\mathbb{L}_k^\wedge, \mathbb{L}_k^\wedge] \\ \text{(i)} \ \ \varphi(A,B)\varphi(B,A) = 1 \\ \text{(ii)} \ \ e^{\frac{\mu}{2}A}\varphi(C,A)e^{\frac{\mu}{2}C}\varphi(B,C)e^{\frac{\mu}{2}B}\varphi(A,B) = 1 \qquad \text{for } A+B+C=0 \\ \text{(iii)} \ \ \varphi(X_{1,2},X_{2,3})\varphi(X_{3,4},X_{4,5})\varphi(X_{5,1},X_{1,2})\varphi(X_{2,3},X_{3,4})\varphi(X_{4,5},X_{5,1}) = 1 \\ \qquad\qquad\qquad \text{in} \quad \underline{U\mathfrak{P}^{(5)}_{\cdot}}(k) \quad (\text{ see } Note 2.1.3). \end{cases}$$

The right $\underline{GT}$-action and the left $\underline{GRT}$-action on $\underline{M}$ are defined as follows [5]:

The right $\underline{GT}$-action:

$$\underline{M}(k) \times \underline{GT}(k) \longrightarrow \underline{M}(k)$$
$$(\mu, \varphi) \times (\lambda, f) \longmapsto (\mu, \varphi) \circ (\lambda, f) \ ,$$

$$\text{where } (\mu,\varphi)\circ(\lambda,f) := \Big(\mu\lambda \ , \varphi(A,B)\ f\big(e^A \ , \varphi(A,B)^{-1}e^B\varphi(A,B)\big)\Big).$$

[5]The right and left are opposite to the original ones in [**Dr**]. Because we turned over the direction of its multiplication of $\underline{GRT}(k)$ (resp. $\underline{GT}(k)$) in *Definition* 2.1.2 (resp. *Definition* 2.2.1).

The left $\underline{GRT}$-action:

$$\underline{GRT}(k) \times \underline{M}(k) \longrightarrow \underline{M}(k)$$
$$(c,g) \times (\mu,\varphi) \longmapsto (c,g)\circ(\mu,\varphi) \ ,$$

where $(c,g)\circ(\mu,\varphi) := \Big(\frac{\mu}{c}\ , g(A,B)\ \varphi(\frac{A}{c}\ , g(A,B)^{-1}\cdot\frac{B}{c}\cdot g(A,B))\Big)$.

REMARK 2.3.2. It can be checked easily that the relation (iii) implies (i).

Drinfel'd showed the followings:

PROPOSITION 2.3.3 ([**Dr**]*Proposition 5.1*). *The right action of* $\underline{GT}(k)$ *on* $\underline{M}(k)$ *is free and transitive.*

PROPOSITION 2.3.4 ([**Dr**]*Proposition 5.5*). *The left action of* $\underline{GRT}(k)$ *on* $\underline{M}(k)$ *is free and transitive.*

Therefore $\underline{M}$ has a structure of bi-torsor by the right $\underline{GT}$-action and the left $\underline{GRT}$-action. In [**Dr**]§4, Drinfel'd considered the following sub-bi-torsor.

DEFINITION 2.3.5 ([**Dr**]§4). The pro-variety $\underline{M}_1$ is the pro-subvariety of $\underline{M}$ defined over $\mathbf{Q}$ whose set of k-valued points is as follows:

$$\underline{M}_1(k) = \{\varphi \in k\langle\langle A,B\rangle\rangle \big|\ (1,\varphi)\in\underline{M}(k)\ \ \}$$

By restricting the right action of $\underline{GT}$ and the left action of $\underline{GRT}$ to their unipotent parts, $\underline{GT}_1$ and $\underline{GRT}_1$, respectively, we see that $\underline{M}_1$ has a structure of bi-torsor. The following was one of the main theorems in [**Dr**], which was proved in [**Dr**] §5.

PROPOSITION 2.3.6 ([**Dr**] Theorem A′). $\underline{M}_1(\mathbf{Q}) \neq \emptyset$.

REMARK 2.3.7. Z. Wojtkowiak constructed a pro-algebraic group G and a G-torsor TG in [**Wo**]*Appendix* §*A*. In fact, his G-torsor TG is isomorphic to Drinfel'd's $\underline{GRT}$-torsor $\underline{M}$.

2.3.2. *The Drinfel'd associator.* Consider the *Knizhnik-Zamolodchikov equation* (*KZ equation* for short)

$$\text{(KZ)}\qquad \frac{\partial g}{\partial u}(u) = \frac{1}{2\pi i}(\frac{A}{u}+\frac{B}{u-1})\cdot g(u) \qquad ,$$

where $g(u)$ is an analytic function in complex variable u with values in $\mathbf{C}\langle\langle A,B\rangle\rangle$, where 'analytic' means each coefficient is analytic. The equation (KZ) has singularities only at $0,1$ and ∞. Let $\mathbf{C}'$ be the complement of the union of the real half-lines $(-\infty,0]$ and $[1,+\infty)$ in the complex plane, which is a simply-connected domain. The equation (KZ) has a unique analytic solution on $\mathbf{C}'$ having a specified value at any given point on $\mathbf{C}'$. Moreover, at the singular points 0 and 1, there exist unique solutions $g_0(u)$ and $g_1(u)$ of (KZ) such that

$$g_0(u) \approx u^{\frac{A}{2\pi i}}\ (u\to 0), \qquad g_1(u)\approx (1-u)^{\frac{B}{2\pi i}}\ (u\to 1),$$

where $\approx$ means that $g_0(u)\cdot u^{-\frac{A}{2\pi i}}$ (resp. $g_1(u)\cdot(1-u)^{-\frac{B}{2\pi i}}$) has an analytic continuation in a neighborhood of 0 (resp. 1) in $\mathbf{C}$ with value 1 at 0 (resp. 1). Here, $u^a := exp(a\cdot log u) := 1+\frac{(a\cdot log u)}{1!}+\frac{(a\cdot log u)^2}{2!}+\frac{(a\cdot log u)^3}{3!}+\cdots$ and $log u := \int_1^u \frac{dt}{t}$ in $\mathbf{C}'$. In the same way, $(1-u)^b$ can be defined on $\mathbf{C}'$. Since $g_0(u)$ and $g_1(u)$ are both invertible unique solutions of (KZ) with the specified asymptotic behaviors, they must coincide with each other up to multiplication from the right by an invertible element of $\mathbf{C}\langle\langle A, B\rangle\rangle$.

DEFINITION 2.3.8. The **Drinfel'd associator** is the element $\varphi_{KZ}(A,B)$ of $\mathbf{C}\langle\langle A, B\rangle\rangle$ which is defined by

$$g_0(u) = g_1(u) \cdot \varphi_{KZ}(A,B) \quad .$$

In [**Dr**], the following is shown:

PROPOSITION 2.3.9 ([**Dr**]). *The pair* $(1, \varphi_{KZ})$ *satisfies (0)~ (iii) in Definition2.3.1, i.e.* $\varphi_{KZ} \in \underline{M}_1(\mathbf{C})$ *(for definition, see Definition 2.3.5).*

3. Hodge Side

We shall make a brief review on MZV's (multiple zeta values) in §3.1. We shall construct a canonical embedding from the spectrum of the $\mathbf{Q}$-algebra generated by all MZV's modulo the ideal generated by π^2 into the graded Grothendieck-Teichmüller group $\underline{GRT}_1$ in §3.2, which is one of our main results in this paper. In §3.3, we shall make another analogous embedding into the middle Grothendieck-Teichmüller torsor $\underline{M}_1$.

3.1. MZV's.

We make a short review on MZV's.

DEFINITION 3.1.1. For each (multi-)index $\mathbf{k} = (k_1, k_2, \ldots, k_m)$ of positive integers with $k_1, \cdots, k_{m-1} \geqslant 1$, $k_m > 1$, the corresponding **multiple zeta value** (MZV for short) $\zeta(\mathbf{k})$ is, by definition, the real number defined by the convergent series:

$$\zeta(\mathbf{k}) = \sum_{\substack{0<n_1<\cdots<n_m \\ n_i \in \mathbf{N}}} \frac{1}{n_1^{k_1} \cdots n_m^{k_m}} \quad .$$

The *weight* of $\mathbf{k}$: $wt(\mathbf{k})$ is defined as $wt(\mathbf{k}) = k_1 + \cdots + k_m$. For each natural number w, let Z_w be the $\mathbf{Q}$-vector subspace of $\mathbf{R}$ generated by all MZV's of indices with weight w : $Z_w = \langle \zeta(\mathbf{k}) | wt(\mathbf{k}) = w \rangle_{\mathbf{Q}} \subseteq \mathbf{R}$, and put $Z_0 = \mathbf{Q}$. Define $Z_\bullet$ as the *formal direct sum* of Z_w for all $w \geqslant 0$: $Z_\bullet = \bigoplus_{w \geqslant 0} Z_w$.

On the dimension of the $\mathbf{Q}$-vector space of MZV's at each weight, we have the following conjecture.

DIMENSION CONJECTURE . ([**Za**]) $dim_{\mathbf{Q}} Z_w$ is equal to d_w, which is given by the Fibonacci-like recurrence $d_w = d_{w-2} + d_{w-3}$, with initial values $d_0 = 1, d_1 = 0, d_2 = 1$.

More details on MZV's and the above conjecture were discussed in the author's previous article [**F**]. T. Terasoma [**Te**] and P. Deligne, A. Goncharov [**DG**] showed its upper-bound; $dim Z_w \leqslant d_w$ for all $w \geqslant 0$, by the theory of mixed Tate motives. On the contrary, to show its lower bound; $dim Z_w \geqslant d_w$ $(w \geqslant 1)$, seems to be quite difficult because we need to show their linear independency over $\mathbf{Q}$, which might be a difficult problem in transcendental number theory.

3.2. Main result. We shall construct a canonical embedding from $Spec\ Z_{\cdot}/(\pi^2)$ into $\underline{GRT}_1$.

PROPERTY 3.2.1. *The graded* $\mathbf{Q}$*-vector space* $Z_{\bullet}$ *has a structure of graded* $\mathbf{Q}$*-algebra, i.e.* $Z_a \cdot Z_b \subseteq Z_{a+b}$ *for* $a, b \geqslant 0$.

This follows from definitions of MZV's (for example, see [**F**]*Property 1.2.2.*). We call $Z_{\cdot}$ the MZV algebra.

NOTATION 3.2.2. Let $Z_{\cdot}/(\pi^2) = Z_{\cdot}/\pi^2 Z_{\cdot}$ be the MZV-algebra $Z_{\cdot}$ modulo the principal homogeneous ideal $(\pi^2) := \pi^2 Z_{\cdot}$ generated by $\pi^2 = 6\zeta(2) \in Z_2$. This is the graded $\mathbf{Q}$-algebra whose grading is given by $Z_{\cdot}/(\pi^2) = \underset{w\geqslant 0}{\oplus} \left(Z_{\cdot}/(\pi^2)\right)_w$, where

$$(3.2.1)\qquad \left(Z_{\cdot}/(\pi^2)\right)_w := \begin{cases} \mathbf{Q} & w = 0 \\ 0 & w = 1 \\ Z_w/(\pi^2 \cdot Z_{w-2}) & w \geqslant 2 \end{cases} .$$

REMARK 3.2.3. As far as the author knows, no point is known in $Spec\ Z_{\cdot}/(\pi^2)$ except one which is determined by the maximal ideal $Z_{>0}/\pi^2 Z_{\bullet}$, since it is related to problems in transcendental number theory.

NOTATION 3.2.4. Let $\mathbb{A}_{\bullet} = \underset{w\geqslant 0}{\oplus} \mathbb{A}_w = \mathbf{Q}\langle A, B\rangle$ be the non-commutative graded polynomial ring over $\mathbf{Q}$ with two variables A and B with $degA = degB = 1$, where $\mathbb{A}_w$ is the homogeneous degree w part of $\mathbb{A}_{\bullet}$.

THEOREM 3.2.5. *There is a surjection*

$$(3.2.2)\qquad \Phi^{\sharp}_{DR} : \mathcal{O}(\underline{GRT}_1)_{\bullet} \twoheadrightarrow Z_{\cdot}/(\pi^2)$$

of graded $\mathbf{Q}$*-algebras, which associates an embedding of schemes*

$$(3.2.3)\qquad \Phi_{DR} : Spec\ Z_{\cdot}/(\pi^2) \hookrightarrow \underline{GRT}_1 .$$

PROOF . Put

$$(3.2.4)\qquad \Phi_{KZ}(A,B) = 1 + \sum_{W:\text{words}} I(W)W := \varphi_{KZ}(2\pi i A, 2\pi i B) \in \mathbf{C}\langle\langle A, B\rangle\rangle,$$

where φ_{KZ} is the Drinfel'd associator (*Definition* 2.3.8). For each word W with $deg(W) = w$, $I(W)$ lies in Z_w (see [**F**]*Property* I §3.2.), i.e. $\Phi_{KZ} \in \underset{w\geqslant 0}{\widehat{\oplus}} (\mathbb{A}_w \otimes_{\mathbf{Q}} Z_w)$. By *Proposition* 2.3.9, it satisfies

$$\begin{cases} (0)\ log\,\Phi_{KZ}(A,B) := \sum_{n=1}^{\infty} \frac{(-1)^{n-1}}{n}\{\sum I(W)W\}^n \in [\mathbb{L}^{\wedge}_{\mathbf{C}}, \mathbb{L}^{\wedge}_{\mathbf{C}}] \left(= \underset{a\geqslant 2}{\widehat{\oplus}} (\mathbb{L}_a \otimes_{\mathbf{Q}} \mathbf{C})\right) \\ (\mathrm{I})\ \Phi_{KZ}(A,B)\Phi_{KZ}(B,A) = 1 \\ (\mathrm{II})\ e^{\pi i A}\Phi_{KZ}(C,A)e^{\pi i C}\Phi_{KZ}(B,C)e^{\pi i B}\Phi_{KZ}(A,B) = 1 \\ \qquad\qquad \text{for} \quad A + B + C = 0 \\ (\mathrm{III})\ \Phi_{KZ}(X_{1,2}, X_{2,3})\Phi_{KZ}(X_{3,4}, X_{4,5})\Phi_{KZ}(X_{5,1}, X_{1,2}) \\ \qquad \Phi_{KZ}(X_{2,3}, X_{3,4})\Phi_{KZ}(X_{4,5}, X_{5,1}) = 1 \quad \text{in} \quad \underline{U\mathfrak{P}^{(5)}_{\cdot}}(\mathbf{C}) . \end{cases}$$

For each word W with $deg(W) = w$, denote the quotient class of $I(W) \in Z_w$ in $\left(Z/(\pi^2)\right)_w$ by $\overline{I(W)}$ and put

$$\overline{\Phi_{KZ}}(A,B) := 1 + \sum_{W:\text{words}} \overline{I(W)}W \in \underset{w\geqslant 0}{\widehat{\oplus}} \left(\mathbb{A}_w \otimes_{\mathbf{Q}} \left(Z/(\pi^2)\right)_w\right).$$

Then the above four formulae imply

$$\begin{cases} (0)\ log\,\overline{\Phi_{KZ}(A,B)} \in \widehat{\bigoplus_{w\geqslant 2}}\left(\mathbb{L}_w \otimes_{\mathbf{Q}} \left(Z/(\pi^2)\right)_w\right) \\ (\mathrm{I})\ \overline{\Phi_{KZ}}(A,B)\overline{\Phi_{KZ}}(B,A) = 1 \qquad \text{in} \quad \widehat{\bigoplus_{w\geqslant 0}}\left(\mathbb{A}_w \otimes_{\mathbf{Q}} \left(Z/(\pi^2)\right)_w\right) \\ (\mathrm{II})\ \overline{\Phi_{KZ}}(C,A)\overline{\Phi_{KZ}}(B,C)\overline{\Phi_{KZ}}(A,B) = 1 \\ \qquad \text{for} \quad A+B+C=0 \quad \text{in} \quad \widehat{\bigoplus_{w\geqslant 0}}\left(\mathbb{A}_w \otimes_{\mathbf{Q}} \left(Z/(\pi^2)\right)_w\right) \\ (\mathrm{III})\ \overline{\Phi_{KZ}}(X_{1,2},X_{2,3})\overline{\Phi_{KZ}}(X_{3,4},X_{4,5})\overline{\Phi_{KZ}}(X_{5,1},X_{1,2}) \\ \qquad \overline{\Phi_{KZ}}(X_{2,3},X_{3,4})\overline{\Phi_{KZ}}(X_{4,5},X_{5,1}) = 1 \quad \text{in} \quad \widehat{\bigoplus_{w\geqslant 0}}\left(U\mathfrak{P}^{(5)}_w \otimes_{\mathbf{Q}} \left(Z/(\pi^2)\right)_w\right) \end{cases}.$$

So $\overline{\Phi_{KZ}}(A,B)$ determines a $Z_{\bullet}/(\pi^2)$-valued point of $\underline{GRT}_1$, i.e. $\overline{\Phi_{KZ}}(A,B) \in \underline{GRT}_1(Z_{\bullet}/(\pi^2))$. Thus we obtain the algebra homomorphism (3.2.2) by sending each x_W (*Notation* 2.1.7) to $\overline{I(W)}$. Since x_W's (resp. $\overline{I(W)}$'s) are algebraic generators of the graded algebra $\mathcal{O}(\underline{GRT}_1)$ (resp. $Z_{\bullet}/(\pi^2)$) whose degree is equal to $deg\ W$, $\Phi^{\sharp}_{DR}$ is a surjective algebra homomorphism preserving their degrees. From this surjective algebra homomorphism $\Phi^{\sharp}_{DR}$, we obtain the embedding (3.2.3) of schemes.

□

3.3. Related embedding into $\underline{M}_1$. In §3.2, we get an embedding from *Spec* $Z_{\bullet}/(\pi^2)$ into the graded Grothendieck-Teichmüller group $\underline{GRT}_1$. But on the other hand, we construct in this subsection a related embedding from the spectrum of a modified algebra of the MZV algebra into the middle Grothendieck-Teichmüller torsor $\underline{M}_1$.

DEFINITION 3.3.1. For each index $\mathbf{k} = (k_1, k_2, \ldots, k_m)$ of positive integers with $k_1, \cdots .k_{m-1} \geqslant 1$, $k_m > 1$, we define the corresponding **modified multiple zeta value** by

$$\widetilde{\zeta}(\mathbf{k}) := \frac{1}{(2\pi i)^{wt\ \mathbf{k}}}\zeta(\mathbf{k}).$$

For each natural number w, let $\mathfrak{Z}_{\leqslant w}$ be the $\mathbf{Q}$-vector subspace of $\mathbf{C}$ generated by all $\widetilde{\zeta}(\mathbf{k})$'s with $wt(\mathbf{k}) \leqslant w$: $\mathfrak{Z}_{\leqslant w} := \langle \widetilde{\zeta}(\mathbf{k}) | wt(\mathbf{k}) \leqslant w \rangle_{\mathbf{Q}} \subseteq \mathbf{C}$, and put $\mathfrak{Z}_{\leqslant 0} := \mathbf{Q}$. Define $\mathfrak{Z}$ to be the $\mathbf{Q}$-vector subspace of $\mathbf{C}$ generated by all $\widetilde{\zeta}(\mathbf{k})$'s.

NOTATION 3.3.2. By *Property* 3.2.1, $\mathfrak{Z}$ becomes a filtered $\mathbf{Q}$-algebra with ascending filtration $W = \{\mathfrak{Z}_{\leqslant a}\}_{a\geqslant 0}$, i.e. $\mathfrak{Z}_{\leqslant a} \cdot \mathfrak{Z}_{\leqslant b} \subseteq \mathfrak{Z}_{\leqslant a+b}$ $(a, b \geqslant 0)$. Let $Gr^W_{\bullet}\mathfrak{Z}$ denote the associated graded $\mathbf{Q}$-algebra of $\mathfrak{Z}$: $Gr^W_{\bullet}\mathfrak{Z} = \bigoplus_{a\geqslant 0} V_a$ where $V_a = \mathfrak{Z}_{\leqslant a}/\mathfrak{Z}_{\leqslant a-1}$ for $a \geqslant 1$ and $V_0 = \mathfrak{Z}_{\leqslant 0} = \mathbf{Q}$.

For each $a \geqslant 0$, let $f_a : \mathfrak{Z}_{\leqslant a} \to Z_a$ denote the $\mathbf{Q}$-linear map defined by sending each $\widetilde{\zeta}(\mathbf{k})$ with $wt(\mathbf{k}) \leqslant a$, to $Re\{(2\pi i)^a \widetilde{\zeta}(\mathbf{k})\} \in Z_a \subset \mathbf{R}$, where Re stands for the real parts. If $a \geqslant 2$ and $wt(\mathbf{k}) \leqslant a-1$, then $f_a(\widetilde{\zeta}(\mathbf{k})) \in \pi^2 Z_{a-2} \subset \mathbf{R}$. Thus f_a induces a $\mathbf{Q}$-linear map $g_a : Gr^W_a\mathfrak{Z} \to \left(Z_{\bullet}/(\pi^2)\right)_a$.

PROPOSITION 3.3.3. *The* $\mathbf{Q}$*-linear maps* $\{g_a\}_{a\geqslant 0}$ *induce the following canonical isomorphism of graded* $\mathbf{Q}$*-algebras:*

$$g := \bigoplus_{a\geqslant 0} g_a :\ Gr^W_{\bullet}\mathfrak{Z} \overset{\sim}{\to} Z_{\bullet}/(\pi^2) \quad .$$

PROOF . The surjectivity of g_a ($a \geqslant 0$) is trivial. The injectivity of g_a is trivial for $a = 0, 1$. Suppose that $a \geqslant 2$ and $g_a(\sum_{i=1}^{m} r_i\widetilde{\zeta}(\mathbf{k}_i)) \equiv 0$ for $m \in \mathbf{N}$, $r_i \in \mathbf{Q}$ and $wt(\mathbf{k}_i) = a$ $(1 \leqslant i \leqslant m)$. Then

$$f_a(\sum_{i=1}^{m} r_i\widetilde{\zeta}(\mathbf{k}_i)) \in \pi^2 Z_{a-2} \quad .$$

$$\therefore \quad \sum_{i=1}^{m} r_i\zeta(\mathbf{k}_i) \in \pi^2 Z_{a-2} \quad .$$

$$\therefore \quad \sum_{i=1}^{m} r_i\widetilde{\zeta}(\mathbf{k}_i) \in \mathfrak{Z}_{\leqslant a-2} \quad .$$

Therefore g_a is injective for $a \geqslant 2$. To check that the linear map g is a homomorphism of graded $\mathbf{Q}$-algebras is immediate. □

The following proposition is an analogue of *Theorem* 3.2.5.

PROPOSITION 3.3.4. *There is a surjection*

$$\Phi^{\sharp}_{Hod} : \mathcal{O}(\underline{M}_1) \twoheadrightarrow \mathfrak{Z} \tag{3.3.1}$$

of $\mathbf{Q}$*-algebras, which associates an embedding of schemes*

$$\Phi_{Hod} : Spec\ \mathfrak{Z} \hookrightarrow \underline{M}_1 \ . \tag{3.3.2}$$

PROOF . By imitating the proof of *Theorem* 3.2.5, we can construct the surjection $\Phi^{\sharp}_{Hod}$ thanks to *Proposition* 2.3.9. □

It can be verified directly that, in fact, the surjection $\Phi^{\sharp}_{Hod} : \mathcal{O}(\underline{M}_1) \twoheadrightarrow \mathfrak{Z}$ is strictly compatible with the weight filtration of $\mathcal{O}(\underline{M}_1)$ (§6.1) and that of $\mathfrak{Z}$ (*Notation* 3.3.2), i.e. $\Phi^{\sharp}_{Hod}(W_a\mathcal{O}(\underline{M}_1)) = \mathfrak{Z}_{\leqslant a}$ for $a \geqslant 0$.

Here *Hod* stands for Hodge. The relationship between *Theorem* 3.2.5 and *Proposition* 3.3.4 will be discussed in *Proposition* 6.2.4.

3.4. *Spec* $Z_{\cdot}/(\pi^2) = \underline{GRT}_1$?. We will discuss the conjecture that the embedding Φ_{DR} in *§3.2* might be an isomorphism.

NOTATION 3.4.1. For a graded vector space $V_{\cdot} = \bigoplus_{a\in\mathbf{Z}} V_a$, we denote its completion by degree by $V_{\cdot}^{\wedge} := \widehat{\bigoplus}_{a\in\mathbf{Z}} V_a$ and its graded dual vector space by $V_{\cdot}^{*} := \bigoplus_{a\in\mathbf{Z}} V_a^{*}$, where V_a^{*} is the dual vector space of V_a.

For any two formal power series $P(t)$, $Q(t)$ in $\mathbf{Q}[[t]]$, we express $P(t) \geqslant Q(t)$ when all coefficients of the formal power series $P(t) - Q(t)$ are all non-negative.

PROPOSITION 3.4.2. *Assume the generatedness part of Conjecture2.1.13 which is equivalent to saying that* $\mathfrak{D}$ *is a Lie algebra generated by one element in each degree* m *(*$m = 3, 5, 7, \cdots$*) and assume the lower bound part of Dimension conjecture (§3.1), which is equivalent to saying that* $dim Z_w \geqslant d_w$ *holds for all* $w \geqslant 0$*. Then the embedding* $\Phi_{DR} : Spec\ Z_{\cdot}/(\pi^2) \hookrightarrow \underline{GRT}_1$ *must be an isomorphism.*

PROOF . By taking the differential of Φ_{DR} at the point e which corresponds to the unit element of $\underline{GRT}_1$, we get an embedding

$$(d\Phi_{DR})_e : (NZ_{\cdot}^{*})^{\wedge} \hookrightarrow Lie\ \underline{GRT}_1 \quad .$$

Here $(NZ_{\bullet}^{*})^{\wedge}$ is the completion by degree of the dual vector space of the *new-zeta space* $NZ_{\bullet} = \underset{w\geqslant 2}{\oplus} NZ_w := \left(Z_{>2}/Z_{>0}^2\right)$ (see [**F**]§*1.3.*). Recall that $\mathfrak{D}_{\bullet}^{\wedge} = \underset{w\geqslant 1}{\widehat{\oplus}} \mathfrak{D}_w \simeq$ *Lie* $\underline{GRT}_1$ (see *Remark* 2.1.12). Thus we get an embedding of graded vector spaces

(3.4.1) $$(d\Phi_{DR})_e : (NZ_{\bullet}^{*})^{\wedge} = \underset{w\geqslant 1}{\widehat{\oplus}} NZ_w^{*} \hookrightarrow \mathfrak{D}_{\bullet}^{\wedge} = \underset{w\geqslant 1}{\widehat{\oplus}} \mathfrak{D}_w \quad .$$

Then we have

$$\sum_{w=0}^{\infty} d_w t^w \leqslant \sum_{w=0}^{\infty} dim_{\mathbf{Q}} Z_w \cdot t^w$$

by the above second assumption,

$$\leqslant \frac{1}{1-t^2} \prod_{w=1}^{\infty} \frac{1}{(1-t^w)^{dim_{\mathbf{Q}} NZ_w}}$$

by the definition of $NZ_{\bullet}$,

$$\leqslant \frac{1}{1-t^2} \prod_{w=1}^{\infty} \frac{1}{(1-t^w)^{dim_{\mathbf{Q}} \mathfrak{D}_w}}$$

by (3.4.1) and

$$\leqslant \sum_{w=0}^{\infty} d_w t^w$$

by [**F**]*Lemma4.3.6* combined with the first assumption.

Therefore all above inequalities must be equalities, which implies that $Z_{\bullet}$ is a polynomial algebra and $(d\Phi_{DR})_e : \left(Z_{>2}/Z_{>0}^2\right)^{*\wedge} \hookrightarrow \mathfrak{D}_{\bullet}^{\wedge}$ is an isomorphism. Thus

(3.4.2) $$\sum_{w=0}^{\infty} dim_{\mathbf{Q}} \big(Z/(\pi^2)\big)_w \cdot t^w = \prod_{w=1}^{\infty} \frac{1}{(1-t^w)^{dim_{\mathbf{Q}} \mathfrak{D}_w}} \quad .$$

On the other hand, from the surjectivity (*Theorem*3.2.5) of $\Phi_{DR}^{\sharp} : \mathcal{O}(\underline{GRT}_1)_{\bullet} \twoheadrightarrow Z_{\bullet}/(\pi^2)$, we get

(3.4.3) $$dim_{\mathbf{Q}} \big(Z/(\pi^2)\big)_w \leqslant dim_{\mathbf{Q}}\ \mathcal{O}(\underline{GRT}_1)_w \qquad \text{for all} \quad w.$$

Put $\mathfrak{m}_e = \underset{w>0}{\oplus} \mathcal{O}(\underline{GRT}_1)_w$. Then this graded vector space $\mathfrak{m}_e$ is the defining ideal of the unit element e in $\underline{GRT}_1$. Since $\mathfrak{D}_{\bullet}^{\wedge} \simeq$ *Lie* $\underline{GRT}_1$, $\mathfrak{D}_{\bullet}^{*} = \underset{w\geqslant 0}{\oplus} \mathfrak{D}_w^{*}$ is canonically isomorphic to the graded vector space $\mathfrak{m}_e / (\mathfrak{m}_e)^2$. Thus

(3.4.4) $$\sum_{w=0}^{\infty} dim_{\mathbf{Q}}\ \mathcal{O}(\underline{GRT}_1)_w \cdot t^w \leqslant \prod_{w=1}^{\infty} \frac{1}{(1-t^w)^{dim_{\mathbf{Q}} \mathfrak{D}_w}} \quad .$$

From (3.4.2) $\sim$ (3.4.4), it follows that

$$dim_{\mathbf{Q}} \big(Z/(\pi^2)\big)_w = dim_{\mathbf{Q}}\ \mathcal{O}(\underline{GRT}_1)_w \qquad \text{for all} \quad w.$$

Thus $\Phi_{DR}^{\sharp}$ must be an isomorphism, which means that the embedding Φ_{DR} (3.2.2) of the pro-algebraic group must be an isomorphism. □

Therefore it may be natural to pose

CONJECTURE A . The above embedding Φ_{DR} of the affine scheme is an isomorphism, i.e.

$$\Phi_{DR}: \ Spec\ Z_{\bullet}/(\pi^2) \cong \underline{GRT}_1$$

REMARK 3.4.3. (1) This conjecture claims that $Spec\ Z_{\bullet}/(\pi^2)$ is naturally equipped with a structure of non-commutative group scheme, which is equivalent to saying that the quotient algebra $Z_{\bullet}/(\pi^2)$ has a structure of non-co-commutative Hopf algebra.
(2) A unipotent algebraic group is isomorphic to its Lie algebra (as varieties), hence to an affine space. Together with this, the above conjecture would imply that $Z_{\bullet}/(\pi^2)$ must be a polynomial algebra. By taking *Remark* 3.2.3 into account, it looks difficult to show this last statement, due to problems in transcendental number theory.
(3) In §6.2, we shall see that *Conjecture* A is equivalent to saying that Φ_{Hod} is an isomorphism (*Conjecture* A′).

4. Galois Side

We will introduce pro-algebraic groups $\underline{Gal}^{(l)}_{\mathbf{Q}}$ and $\underline{Gal}^{(l)}_{\mathbf{Q}_l}$, and discuss their relationship to the Grothendieck-Teichmüller pro-algebraic group $\underline{GT}_1$.

4.1. The pro-l Galois representation. The *absolute Galois group* $Gal(\overline{\mathbf{Q}}/\mathbf{Q})$ acts on the algebraic fundamental group $\pi_1(\mathbf{P}^1_{\overline{\mathbf{Q}}} - \{0,1,\infty\}, \overrightarrow{01})$ of the projective line minus 3 points, where $\overrightarrow{01}$ means the tangential base point (see [**De**]§15). Let l be a prime. With this representation, we can associate the following continuous group homomorphism into the automorphism group of the free pro-l group $\widehat{F_2}^{(l)}$ of rank 2

$$p^{(l)}_1 : Gal\left(\overline{\mathbf{Q}}/\mathbf{Q}(\mu_{l^\infty})\right) \to Aut\ \widehat{F_2}^{(l)}$$

for each prime l, where μ_{l^∞} stands for the group of all l-powerth roots of unity. By [**HM**] *Corollary A.10*, there exists a natural topological group homomorphism into $\underline{AutF_2}(\mathbf{Q}_l)$ (cf. *Remark* 2.2.5)

$$p^{(l)}_2 : Aut\ \widehat{F_2}^{(l)} \to \underline{AutF_2}(\mathbf{Q}_l) \ .$$

By combining these two homomorphisms, we get the following Galois representation

$$\varphi_l := p^{(l)}_2 \circ p^{(l)}_1 : Gal\left(\overline{\mathbf{Q}}/\mathbf{Q}(\mu_{l^\infty})\right) \to \underline{AutF_2}(\mathbf{Q}_l) \ .$$

By imitating the construction of the embedding $Gal(\overline{\mathbf{Q}}/\mathbf{Q}) \hookrightarrow \widehat{GT}$ in [**Ih94**], we can show that its image is contained in the following pro-l group version of the Grothendieck-Teichmüller group $\widehat{GT}^{(l)}_1$.

LEMMA 4.1.1.

$$Im\ p^{(l)}_1 \subseteq \widehat{GT}^{(l)}_1 .$$

Here $\widehat{GT}_1^{(l)} := \left\{ \sigma \in Aut\ \widehat{F_2}^{(l)} \;\middle|\; \begin{array}{l} \sigma(x) = x,\ \sigma(y) = f^{-1}yf \ \text{ for } {}^{\exists} f \in \widehat{F_2}^{(l)} \\ \text{which satisfies (0)} \sim \text{(iii) below.} \end{array} \right\}.$

$$\begin{cases} (0)\ \ f \in [\widehat{F_2}^{(l)}, \widehat{F_2}^{(l)}] \\ (\mathrm{i})\ \ f(x,y)f(y,x) = 1 \\ (\mathrm{ii})\ \ f(z,x)f(y,z)f(x,y) = 1 \quad \text{for} \quad xyz = 1 \\ (\mathrm{iii})\ \ f(x_{1,2},x_{2,3})f(x_{3,4},x_{4,5})f(x_{5,1},x_{1,2})f(x_{2,3},x_{3,4})f(x_{4,5},x_{5,1}) = 1 \\ \qquad\qquad\qquad\qquad \text{in} \quad \widehat{P_5}^{(l)} \qquad \text{(see } \textit{Note } 4.1.2\text{)} \quad . \end{cases}$$

NOTE 4.1.2. Here $[\widehat{F_2}^{(l)}, \widehat{F_2}^{(l)}]$ means the topological commutator subgroup of $\widehat{F_2}^{(l)}$ and $\widehat{P_5}^{(l)}$ is the pro-l completion of the pure sphere braid group P_5 and $x_{i,j}$'s $(1 \leqslant i, j \leqslant 5)$ are its standard generators ([**Ih91**]). Note that $\sigma \in \widehat{GT}_1^{(l)}$ determines $f \in \widehat{F_2}^{(l)}$ uniquely because of the condition (0).

Since there exists a natural topological group homomorphism $\widehat{F_2}^{(l)} \to \underline{F_2}(\mathbf{Q}_l)$ with Zariski dense image, it follows from definitions

LEMMA 4.1.3.

$$p_2^{(l)}(\widehat{GT}_1^{(l)}) \subseteq \underline{GT}_1(\mathbf{Q}_l) \ \ (\subseteq \underline{AutF_2}(\mathbf{Q}_l)\ ) \quad .$$

But it seems not so clear that

PROBLEM 4.1.4. Is $p_2^{(l)}(\widehat{GT}_1^{(l)})$ Zariski dense in $\underline{GT}_1(\mathbf{Q}_l)$ or not?

By *Lemma 4.1.1* and *Lemma 4.1.3*, it follows

PROPOSITION 4.1.5.

$$Im\ \varphi_l \ \subseteq \underline{GT}_1(\mathbf{Q}_l)\ .$$

4.2. $\underline{Gal}_{\mathbf{Q}}^{(l)} = \underline{GT}_1$**?.** In this subsection, we will construct a pro-algebraic group version of the Galois image $Im\varphi_l$.

NOTATION 4.2.1. Let k be a field extension of $\mathbf{Q}$. For any pro-algebraic group $\mathcal{G} = \varprojlim_r G^{(r)}$ over $\mathbf{Q}$, we define the **k-structure** of $\mathcal{G}$ to be the pro-algebraic group $\mathcal{G} \times_{\mathbf{Q}} k := \varprojlim_r (G^{(r)} \times_{\mathbf{Q}} k)$ over k.

DEFINITION 4.2.2. The pro-algebraic group $\underline{Gal}_{\mathbf{Q}}^{(l)}$ over $\mathbf{Q}$ is the smallest pro-sub-variety of $\underline{GT}_1$ `defined over` $\mathbf{Q}$ whose set of $\mathbf{Q}_l$-rational points contains the image of $Im\ \varphi_l$.

Note that $\underline{Gal}_{\mathbf{Q}}^{(l)}$ has a structure of pro-linear algebraic group defined over $\mathbf{Q}$.

DEFINITION 4.2.3. The pro-algebraic group $\underline{Gal}_{\mathbf{Q}_l}^{(l)}$ over $\mathbf{Q}_l$ is the smallest pro-sub-variety of $\underline{GT}_1 \times_{\mathbf{Q}} \mathbf{Q}_l$ `defined over` $\mathbf{Q}_l$ whose set of $\mathbf{Q}_l$-rational points contains the image of $Im\ \varphi_l$.

Note that $\underline{Gal}_{\mathbf{Q}_l}^{(l)}$ has a structure of pro-unipotent algebraic group defined over $\mathbf{Q}_l$.

NOTATION 4.2.4. From these definitions, we get the following embeddings of pro-algebraic groups.

$$T_l : \underline{Gal}^{(l)}_{\mathbf{Q}_l} \hookrightarrow \underline{Gal}^{(l)}_{\mathbf{Q}} \times_{\mathbf{Q}} \mathbf{Q}_l \tag{4.2.1}$$

$$\Phi^{(l)}_{\mathbf{Q}} : \underline{Gal}^{(l)}_{\mathbf{Q}} \hookrightarrow \underline{GT}_1 \tag{4.2.2}$$

$$\varPhi^{(l)}_{\mathbf{Q}_l} : \underline{Gal}^{(l)}_{\mathbf{Q}_l} \hookrightarrow \underline{GT}_1 \times_{\mathbf{Q}} \mathbf{Q}_l \tag{4.2.3}$$

Note that $\varPhi^{(l)}_{\mathbf{Q}_l} = \left(\Phi^{(l)}_{\mathbf{Q}} \times id_{\mathbf{Q}_l}\right) \circ T_l$.

On these embeddings it may be natural to state the following two conjectures which are related to those of [**De**] and [**Ih02**].

CONJECTURE B . The embedding $\Phi^{(l)}_{\mathbf{Q}}$ (4.2.2) of the pro-algebraic group is an isomorphism, i.e.

$$\Phi^{(l)}_{\mathbf{Q}} : \underline{Gal}^{(l)}_{\mathbf{Q}} \cong \underline{GT}_1 \quad \text{for all prime } l$$

CONJECTURE C . The embedding T_l (4.2.1) of the pro-algebraic group is an isomorphism, i.e.

$$T_l : \underline{Gal}^{(l)}_{\mathbf{Q}_l} \cong \underline{Gal}^{(l)}_{\mathbf{Q}} \times_{\mathbf{Q}} \mathbf{Q}_l$$

It seems possible to deduce from the result in [**DG**] that there exists a sub-algebraic groups $\mathcal{M}$ of $\underline{AutF_2}$ defined over $\mathbf{Q}$ such that

$$\underline{Gal}^{(l)}_{\mathbf{Q}_l} \cong \mathcal{M} \times_{\mathbf{Q}} \mathbf{Q}_l$$

for all prime l. We remark that especially this implies the validity of *Conjecture C.*

4.3. The l-adic Galois image Lie algebra. In this subsection, we make a short review of the l-adic Galois image Lie algebra $\mathfrak{g}^{(l)}_{\bullet}$ [**Ih90**] and give a relationship among *Conjecture B*, *Conjecture C* and Ihara's conjecture (see *Conjecture* 4.3.2 below) [**Ih02**].

NOTATION 4.3.1. Let $\{\widehat{F_2}^{(l)}(m)\}_{m\in\mathbf{N}}$ be the lower central series of the free pro-l group $\widehat{F_2}^{(l)}$ of rank 2 which is defined inductively by $\widehat{F_2}^{(l)}(1) := \widehat{F_2}^{(l)}$, $\widehat{F_2}^{(l)}(m+1) := [\widehat{F_2}^{(l)}, \widehat{F_2}^{(l)}(m)]$ $(m \geqslant 1)$, where $[\ ,]$ means the topological commutator.

For each filtered vector space $\left(V, \{W_nV\}_{n\in\mathbf{Z}}\right)$ where $\{W_nV\}_{n\in\mathbf{Z}}$ is the ascending family of sub-vector spaces of V, we define the graded vector space $Gr^W_{\bullet}V := \underset{n\in\mathbf{Z}}{\oplus} Gr^W_nV$, where $Gr^W_nV := W_nV \,/\, W_{n-1}V$.

Let

$$p^{(l)}_1(m) : Gal\left(\overline{\mathbf{Q}}/\mathbf{Q}(\mu_{l^\infty})\right) \to Aut\left(\widehat{F_2}^{(l)} \Big/ \widehat{F_2}^{(l)}(m+1)\right)$$

be the induced homomorphisms from $p^{(l)}_1$ (§4.1) $(m \geqslant 1)$. We denote $\mathbf{Q}^{(l)}(m)$ by the field corresponding to $Ker\ p^{(l)}_1(m)$. The formal direct sum

$$\mathfrak{g}^{(l)}_{\bullet} := \underset{m\geqslant 1}{\oplus} \mathfrak{g}^{(l)}_m \quad \text{where} \quad \mathfrak{g}^{(l)}_m := Gal\left(\mathbf{Q}^{(l)}(m+1)/\mathbf{Q}^{(l)}(m)\right) \underset{\mathbf{Z}_l}{\otimes} \mathbf{Q}_l$$

has a structure of the graded Lie algebra over $\mathbf{Q}_l$ by taking the commutator as the bracket (see [**Ih90**]). This l-adic graded Lie algebra $\mathfrak{g}^{(l)}_{\bullet}$ is called the **l-adic Galois image Lie algebra**. In [**Ih90**], it was shown that there is a natural (graded) embedding from the l-adic Galois image Lie algebra $\mathfrak{g}_{\bullet}^{(l)}$ into the l-adic stable derivation (Lie-)algebra $\mathfrak{D}_{\bullet} \underset{\mathbf{Q}}{\otimes} \mathbf{Q}_l$ (*Definition* 2.1.11)

$$(4.3.1) \quad \Psi_l \left(= \underset{w \geqslant 2}{\oplus} \Psi_{(l),w} \right) : \mathfrak{g}_{\bullet}^{(l)} \left(= \underset{w \geqslant 2}{\oplus} \mathfrak{g}_w^{(l)} \right) \hookrightarrow \mathfrak{D}_{\bullet} \underset{\mathbf{Q}}{\otimes} \mathbf{Q}_l \left(= \underset{w \geqslant 2}{\oplus} \mathfrak{D}_w \underset{\mathbf{Q}}{\otimes} \mathbf{Q}_l \right) .$$

On this embedding of l-adic graded Lie algebras, we have the following

CONJECTURE 4.3.2 ([**Ih02**] *Conjecture 1*). The above embedding Ψ_l of l-adic graded Lie algebras is an isomorphism, i.e.

$$\Psi_l : \mathfrak{g}_{\bullet}^{(l)} \cong \mathfrak{D}_{\bullet} \underset{\mathbf{Q}}{\otimes} \mathbf{Q}_l \qquad \text{for all prime } l.$$

In fact, the validity of *Conjecture* 4.3.2 is equivalent to the validity of *Conjecture B* and *Conjecture C*:

PROPOSITION 4.3.3. *The embedding* $\Phi_{\mathbf{Q}_l}^{(l)} : \underline{Gal}_{\mathbf{Q}_l}^{(l)} \hookrightarrow \underline{GT}_1 \times_{\mathbf{Q}} \mathbf{Q}_l$ (4.2.3) *of pro-algebraic groups is an isomorphism if and only if the embedding* $\Psi_l : \mathfrak{g}_{\bullet}^{(l)} \hookrightarrow \mathfrak{D}_{\bullet} \underset{\mathbf{Q}}{\otimes} \mathbf{Q}_l$ (4.3.1) *of l-adic graded Lie algebras is an isomorphism.*

PROOF . Recall that $\underline{GT}_1$ is a unipotent pro-algebraic group. Therefore

(4.3.2) the embedding $\Phi_{\mathbf{Q}_l}^{(l)} : \underline{Gal}_{\mathbf{Q}_l}^{(l)} \hookrightarrow \underline{GT}_1 \times \mathbf{Q}_l$ is an isomorphism $\Longleftrightarrow$
the embedding of Lie algebras $d\Phi_{\mathbf{Q}_l}^{(l)} : Lie\ \underline{Gal}_{\mathbf{Q}_l}^{(l)} \hookrightarrow Lie\ (\underline{GT}_1 \times \mathbf{Q}_l)$
is an isomorphism.

We regard $Lie\ \underline{GT}_1$ as a $\underline{GT}$-module by the adjoint action. Since $\underline{GT}$ is a negatively weighted extension with respect to $\frac{1}{\varpi} : \mathbb{G}_m \to \mathbb{G}_m$ (*Lemma* 2.2.4), the pro Lie algebra $Lie\ \underline{GT}_1$ is naturally equipped with a weight filtration ([**HM**]*Proposition 4.5*). With this filtration, $Lie\ \underline{GT}_1$ becomes a filtered Lie algebra ([**HM**]*Proposition 3.4*). Since the weight grade functor $Gr_{\bullet}^W$ is exact ([**HM**] *Theorem 3.12*), it follows that

(4.3.3)
the embedding $d\Phi_{\mathbf{Q}_l}^{(l)} : Lie\ \underline{Gal}_{\mathbf{Q}_l}^{(l)} \hookrightarrow Lie\ (\underline{GT}_1 \times \mathbf{Q}_l)$ is an isomorphism $\Longleftrightarrow$
its associated embedding $Gr_{\bullet}^W d\Phi_{\mathbf{Q}_l}^{(l)} : Gr_{\bullet}^W Lie\ \underline{Gal}_{\mathbf{Q}_l}^{(l)} \hookrightarrow Gr_{\bullet}^W Lie\ (\underline{GT}_1 \times \mathbf{Q}_l)$
is an isomorphism

It was shown that $Gr_{\bullet}^W Lie\ \underline{Gal}_{\mathbf{Q}_l}^{(l)} \cong \mathfrak{g}_{\bullet}^{(l)}$ in [**HM**] *Theorem 8.4* and that $Gr_{\bullet}^W Lie\ \underline{GT}_1 \cong \mathfrak{D}_{\bullet}$ in [**Dr**] *Theorem 5.6*. Together with these two isomorphisms, the embedding $d\Phi_{\mathbf{Q}_l}^{(l)}$ yields Ψ_l. Namely

$$(4.3.4) \qquad Gr_{\bullet}^W d\Phi_{\mathbf{Q}_l}^{(l)} = \Psi_l .$$

From (4.3.2) $\sim$ (4.3.4), the statement of the proposition follows. □

5. $\underline{GT}_1 = \underline{GRT}_1$ Part I

In §5 and §6, we shall discuss a relationship between $\underline{GT}_1$ and $\underline{GRT}_1$ and then compare "Galois Side" (§4) and "Hodge Side" (§3) via these two Grothendieck-Teichmüller pro-algebraic groups.

In this section, we prove two kinds of isomorphism between $\underline{GT}$ and $\underline{GRT}$ in §5.1. In §5.3, we shall explain Figure 1 (§0).

5.1. Comparison between $\underline{GT}$ and $\underline{GRT}$. Here we shall prove a few propositions, which may be regarded as corollaries of Drinfel'd's results.

5.1.1. *Non-canonical isomorphism between their* $\mathbf{Q}$*-structures.* Let k be any field of characteristic 0.

PROPOSITION 5.1.1. *Between k-structures (*Notation *4.2.1) of $\underline{GT}$ and $\underline{GRT}$, there exists an isomorphism of group schemes*

$$S_\varphi : \underline{GT} \times k \overset{\sim}{\to} \underline{GRT} \times k \tag{5.1.1}$$

which arises from each point of φ of $\underline{M}(k)$.

PROOF . Take any point φ on $\underline{M}(k)$. Recall that its existence follows from *Proposition* 2.3.6. Assume that R is an arbitrary k-algebra. Then it is immediate that each φ determines an isomorphism $S_\varphi : \underline{GT}(R) \overset{\sim}{\to} \underline{GRT}(R)$ of groups such that $\varphi \circ f = S_\varphi(f) \circ \varphi$ for all $f \in \underline{GT}(R)$, by regarding $\varphi \in \underline{M}(R)$. □

Especially by taking a rational point φ of $\underline{M}$, we obtain an isomorphism between their $\mathbf{Q}$-structures. This isomorphism is non-canonical since it depends on the choice of φ.

5.1.2. *Standard isomorphism over* $\mathbf{C}$. Recall that φ_{KZ} is a standard point on $\underline{M}_1(\mathbf{C})$ (*Proposition* 2.3.9). Especially by taking the above (*Proposition* 5.1.1) point φ by $\varphi_{KZ} \in \underline{M}_1(\mathbf{C})$, we get

PROPOSITION 5.1.2. *Between* $\mathbf{C}$*-structures of $\underline{GT}$ and $\underline{GRT}$, there exists a standard isomorphism*

$$p = S_{\varphi_{KZ}} : \underline{GT} \times_{\mathbf{Q}} \mathbf{C} \overset{\sim}{\to} \underline{GRT} \times_{\mathbf{Q}} \mathbf{C}.$$

If we identify their groups of $\mathbf{C}$*-valued points by p, their subgroups of* $\mathbf{Q}$*-rational points are conjugate to each other in this common* $\mathbf{C}$*-structure. Namely, $p\,(\underline{GT}(\mathbf{Q})) = a \circ (\underline{GRT}(\mathbf{Q})) \circ a^{-1} \subseteq \underline{GRT}(\mathbf{C})$ for some $a \in \underline{GRT}(\mathbf{C})$. Moreover, we can take this a as an element of $\underline{GRT}_1(\mathbf{C})$.*

Thus by restricting p to unipotent parts, we get Figure 4.

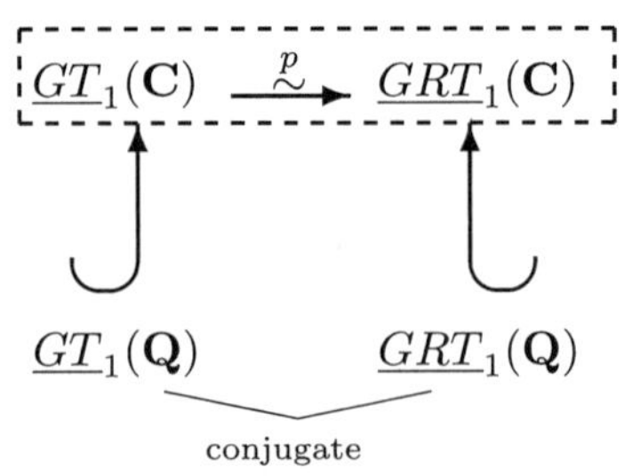

FIGURE 4

PROOF . Choose any rational point $\varphi \in \underline{M}_1(\mathbf{Q})$ (cf. *Proposition* 2.3.6). By *Proposition* 2.3.4, there should exist a unique $\mathbf{C}$-valued point $a \in \underline{GRT}_1(\mathbf{C})$ such that $\varphi_{KZ} = a \circ \varphi$. Thus

$$\varphi_{KZ} \circ f = p(f) \circ \varphi_{KZ} = (p(f) \circ a) \circ \varphi$$
$$\varphi_{KZ} \circ f = a \circ \varphi \circ f = (a \circ s_\varphi(f)) \circ \varphi.$$

for each $f \in \underline{GT}(\mathbf{Q})$. This implies that $p(f) = a \circ s_\varphi(f) \circ a^{-1}$ for all $f \in \underline{GT}(\mathbf{Q})$. Therefore $p(\underline{GT}(\mathbf{Q})) = a \circ s_\varphi(\underline{GT}(\mathbf{Q})) \circ a^{-1}$. By combining it with $s_\varphi(\underline{GT}(\mathbf{Q})) = \underline{GRT}(\mathbf{Q})$, we get $p(\underline{GT}(\mathbf{Q})) = a \circ \underline{GRT}(\mathbf{Q}) \circ a^{-1}$. □

We remark that this isomorphism p does not descend to that of $\mathbf{Q}$-structure, since there appear periods in each coefficient of $\varphi_{KZ}(A, B)$ such as $\frac{\zeta(3)}{(2\pi i)^3}$, $\frac{\zeta(2,3)}{(2\pi i)^5}$, ..., which do not belong to $\mathbf{Q}$.

5.2. Digression on a candidate of a canonical free basis of the stable derivation algebra. In this subsection, we make a few extra remarks on Ihara's Problem 2.1.14.

NOTATION 5.2.1. Denote the logarithmic isomorphism of $\underline{GRT}_1(\mathbf{C})$ by

$$Log : \underline{GRT}_1(\mathbf{C}) \stackrel{\sim}{\to} (\mathfrak{D}_\bullet \underset{\mathbf{Q}}{\otimes} \mathbf{C})^\wedge,$$

where $(\mathfrak{D}_\bullet \underset{\mathbf{Q}}{\otimes} \mathbf{C})^\wedge$ means the completion by degree of the stable derivation algebra $\mathfrak{D}_\bullet = \underset{w \geqslant 1}{\oplus} \mathfrak{D}_w$ (§2.1.3) tensored with $\mathbf{C}$.

Since $\underline{M}_1$ is defined over $\mathbf{Q}$, the complex conjugate $\overline{\varphi_{KZ}}$ of φ_{KZ} also lies in $\underline{M}_1(\mathbf{C})$. Thus by *Proposition* 2.3.4, there should exist a unique element $g \in \underline{GRT}_1(\mathbf{C})$ such that $g \circ \varphi_{KZ} = \overline{\varphi_{KZ}}$. In the proof of **[Dr]** *Proposition* 6.3, Drinfel'd get an element $\tilde{\psi} := Log\ g \in (\mathfrak{D}_\bullet \underset{\mathbf{Q}}{\otimes} \mathbf{C})^\wedge$, which has the following presentation:

$$\tilde{\psi} = \sum_{m \geqslant 3:\ \text{odd}} \psi_m \qquad \text{where} \qquad \psi_m = \frac{2\zeta(m)}{(2\pi i)^m}(adA)^{m-1}(B) + \cdots \in \mathfrak{D}_m \otimes \mathbf{C}.$$

The essentially same element was also obtained in **[Ra]**. It can be checked that ψ_m ($m \geqslant 3$: odd) lie on $\mathfrak{D}_m \underset{\mathbf{Q}}{\otimes} \frac{1}{(2\pi i)^m} Z_m$. Since each ψ_m is an element of $\mathfrak{D}_m \otimes \mathbf{C}$ with depth 1 in the sense of **[Ih02]** *Lecture II §2*, these ψ_m's might be a candidate of canonical free basis of $\mathfrak{D}_\bullet \otimes \mathbf{C}$ asked by Ihara in *Problem* 2.1.14.

PROBLEM 5.2.2. These ψ_m ($m \geqslant 3$: odd) generate a free Lie subalgebra of $\mathfrak{D}_\bullet \otimes \mathbf{C}$?

But Ihara asked for the free basis of the $\mathbf{Q}$-structure of the stable derivation algebra. On the above elements, we cannot expect that all $\psi'_m := \frac{(2\pi i)^m}{2\zeta(m)} \psi_m$ lies in $\mathfrak{D}_m$. Since, for example, to show that ψ'_{11} lies in $\mathfrak{D}_{11}$, we must show the *linear dependency* of 9 MZV's on the list of Z_{11} on **[F]** *Example* 4.3.4, which are predicted to be linearly independent by Dimension conjecture (§3.1).

5.3. Comparison between Galois Side and Hodge Side. By combining *Proposition* 5.1.2 with the embedding Φ_{DR} (3.2.3) and $\Phi^{(l)}_{\mathbf{Q}}$ (4.2.2), we get Figure 1 in §0.

Validities of *Conjecture A* (§3.4), *Conjecture B* (§4.2) and *Conjecture C* (§4.2) would imply the following relationship between Galois Side and Hodge Side.

> *" In Galois Side, there should exist a common $\mathbf{Q}$-structure of $\underline{Gal}^{(l)}_{\mathbf{Q}_l}$ for all prime l, i.e. $\underline{Gal}^{(l)}_{\mathbf{Q}_l} = \underline{GT}_1 \times_{\mathbf{Q}} \mathbf{Q}_l$. On the other hand, in Hodge Side, Spec $Z_\bullet/(\pi^2)$ should be equipped with a structure of pro-algebraic group and it should provide a different $\mathbf{Q}$-structure*

of the above common **Q***-structure. After tensored with* **C***, these two* **Q***-structures should be isomorphic. Moreover their two* **Q***-structures should be conjugate with each other in this common* **C***-structure."*

6. $\underline{GT}_1 = \underline{GRT}_1$ Part II

In §6.1, we shall introduce a weight filtration on the regular function ring $\mathcal{O}(\underline{M}_1)$. §6.2 will be devoted to a relationship between $\underline{M}_1$ and $\underline{GRT}_1$. In §6.3, we shall discuss a relationship between $\underline{GT}_1$ and $\underline{GRT}_1$ which is different from that of §5 and then compare "Galois Side" and "Hodge Side" via these two Grothendieck-Teichmüller pro-algebraic groups again in §6.5 and §6.6. We shall also explain Figure 2 (§0) in §6.5.

6.1. Weight filtration of $\mathcal{O}(\underline{M}_1)$. We introduce two kinds of weight filtration on the regular function ring $\mathcal{O}(\underline{M}_1)$ of $\underline{M}_1$ (*Definition* 2.3.5) W_{r_φ} and W_{l_φ} by the right $\underline{GT}_1$-action and the left $\underline{GRT}_1$-action respectively for each element $\varphi \in \underline{M}_1(\mathbf{Q})$. And then in *Proposition* 6.1.1 and *Proposition* 6.1.2, we shall show that their filtration are canonical (do not depend on the choice of $\varphi \in \underline{M}_1(\mathbf{Q})$) and coincide with each other.

Fix a point φ on $\underline{M}_1(\mathbf{Q})$. By *Proposition* 2.3.3 and *Proposition* 2.3.4, we get isomorphisms of schemes $r_\varphi : \underline{GT} \to \underline{M}$ and $l_\varphi : \underline{GRT} \to \underline{M}$ such that $r_\varphi(f) = \varphi \circ f$ for all $f \in \underline{GT}(A)$ and $l_\varphi(g) = g \circ \varphi$ for all $g \in \underline{GRT}(A)$, where A is an arbitrary $\mathbf{Q}$-algebra. Note that $S_\varphi = l_\varphi^{-1} \circ r_\varphi$ (5.1.1). Especially by restricting r_φ (resp. l_φ) to $\underline{GT}_1$ (resp. $\underline{GRT}_1$), we get an isomorphism of $\mathbf{Q}$-algebras $r_\varphi^\sharp : \mathcal{O}(\underline{M}_1) \to \mathcal{O}(\underline{GT}_1)$ (resp. $l_\varphi^\sharp : \mathcal{O}(\underline{M}_1) \to \mathcal{O}(\underline{GRT}_1)$). Denote $r_\varphi^\sharp \circ (l_\varphi^{-1})^\sharp : \mathcal{O}(\underline{GRT}_1) \to \mathcal{O}(\underline{GT}_1)$ by $S_\varphi^\sharp$. By transporting the weight filtration $W = \{W_n\mathcal{O}(\underline{GT}_1)\}_{n \in \mathbf{Z}}$ (§2.2.2) on $\mathcal{O}(\underline{GT}_1)$ via $r_\varphi^\sharp$, we get a weight filtration $W_{r_\varphi} = \{W_{n,r_\varphi}\mathcal{O}(\underline{M}_1)\}_{n \in \mathbf{Z}}$ on $\mathcal{O}(\underline{M}_1)$ such that $r_\varphi^\sharp(W_{n,r_\varphi}\mathcal{O}(\underline{M}_1)) = W_n\mathcal{O}(\underline{GT}_1)$ for $n \in \mathbf{Z}$.

Similarly we get another filtration $W_{l_\varphi} = \{W_{n,l_\varphi}\mathcal{O}(\underline{M}_1)\}_{n \in \mathbf{Z}}$ on $\mathcal{O}(\underline{M}_1)$ from $W = \{W_n\mathcal{O}(\underline{GRT}_1)\}_{n \in \mathbf{Z}}$ (§2.1.2) by $l_\varphi^\sharp$.

PROPOSITION 6.1.1. $W_{r_\varphi} = W_{l_\varphi}$.

PROOF . Recall that the weight filtration on $\mathcal{O}(\underline{GRT}_1)$ (resp. on $\mathcal{O}(\underline{GT}_1)$) was intrinsically defined by its adjoint action in §1.3. Since $\underline{GT}$ and $\underline{GRT}$ are isomorphic via S_φ (5.1.1) and the isomorphism S_φ is consistent with their central cocharacter (i.e. $S_\varphi^{-1} \circ s_0 \circ \varpi$ (*Remark* 2.1.5 and *Lemma* 2.1.6) becomes a lift of $\frac{1}{\varpi}$ (*Lemma* 2.2.4)), we get $r_\varphi^\sharp \circ (l_\varphi^{-1})^\sharp(W_n\mathcal{O}(\underline{GRT}_1)) = S_\varphi^\sharp(W_n\mathcal{O}(\underline{GRT}_1)) = W_n\mathcal{O}(\underline{GT}_1)$ for each $n \in \mathbf{Z}$. Therefore $W_{n,r_\varphi}\mathcal{O}(\underline{M}_1) = W_{n,l_\varphi}\mathcal{O}(\underline{M}_1)$ for $n \in \mathbf{Z}$. □

PROPOSITION 6.1.2. *The filtration W_{l_φ} on $\mathcal{O}(\underline{M}_1)$ does not depend on the choice of $\varphi \in \underline{M}_1(\mathbf{Q})$.*

PROOF . Take any two points φ, φ' of $\underline{M}_1(\mathbf{Q})$. By *Proposition* 2.3.4, there should exist a unique element $a \in \underline{GRT}_1(\mathbf{Q})$ such that $\varphi' = a \circ \varphi$. Denote $R_a : \underline{GRT}_1 \to \underline{GRT}_1$ to be the right action, deduced from $x \mapsto x \circ a$ for $x \in \underline{GRT}_1(A)$, where A is an arbitrary $\mathbf{Q}$-algebra. Then $l_{\varphi'} = l_\varphi \circ R_a$. Thus for each $n \in \mathbf{Z}$,

$$(l_{\varphi'}^{-1})^\sharp(W_n\mathcal{O}(\underline{GRT}_1)) = (l_\varphi^{-1})^\sharp(R_{a^{-1}})^\sharp(W_n\mathcal{O}(\underline{GRT}_1)). \tag{6.1.1}$$

Recall that $\mathcal{O}(\underline{GRT}_1)$ is a graded Hopf algebra whose grading is given by $deg\ x_W = deg\ W$ for each word W (*Note* 2.1.9), from which we can deduce that

$$(R_{a^{-1}})^\sharp(W_n\mathcal{O}(\underline{GRT}_1)) = W_n\mathcal{O}(\underline{GRT}_1) \tag{6.1.2}$$

for each $n \in \mathbf{Z}$. Thus by (6.1.1) and (6.1.2)

$$\begin{aligned} W_{n,l_{\varphi'}}\mathcal{O}(\underline{M}_1) &= (l_{\varphi'}^{-1})^\sharp(W_n\mathcal{O}(\underline{GRT}_1)) \\ &= (l_{\varphi}^{-1})^\sharp(W_n\mathcal{O}(\underline{GRT}_1)) = W_{n,l_\varphi}\mathcal{O}(\underline{M}_1) \end{aligned}$$

for $n \in \mathbf{Z}$. □

NOTATION 6.1.3. From now on, we denote briefly this filtration W_{l_φ} (or equivalently W_{r_φ}) by W.

From the algebra isomorphism $l_\varphi^\sharp$ and the construction of this filtration W, it immediately follows that W is compatible with the structure of $\mathbf{Q}$-algebra of $\mathcal{O}(\underline{M}_1)$. Namely $(\mathcal{O}(\underline{M}_1), W)$ becomes a filtered $\mathbf{Q}$-algebra.

6.2. Comparison between $\underline{M}_1$ and $\underline{GRT}_1$. In this subsection, we introduce a relationship among filtered algebras $\mathcal{O}(\underline{GRT}_1)$ and $\mathcal{O}(\underline{M}_1)$ in *Theorem* 6.2.2 and make a remark on a relationship between *Theorem* 3.2.5 and *Proposition* 3.3.4.

NOTATION 6.2.1. We denote by $Gr^W_\cdot\mathcal{O}(\underline{M}_1) := \bigoplus_{n\geqslant 0} Gr^W_n\mathcal{O}(\underline{M}_1)$ the graded algebra of the regular function ring $\mathcal{O}(\underline{M}_1)$ with respect to the filtration W (*Notation* 6.1.3).

Since $(\mathcal{O}(\underline{M}_1), W)$ is a filtered $\mathbf{Q}$-algebra, $Gr^W_\cdot\mathcal{O}(\underline{M}_1)$ is a graded $\mathbf{Q}$-algebra. Recall that $\mathcal{O}(\underline{GRT}_1)_\cdot$ is also a graded Hopf algebra over $\mathbf{Q}$ (§2.1.2).

THEOREM 6.2.2. *Between two $\mathbf{Q}$-structures of graded algebra $Gr^W_\cdot\mathcal{O}(\underline{M}_1)$ and $\mathcal{O}(\underline{GRT}_1)_\cdot$, there exists a canonical isomorphism*

$$r^\sharp : Gr^W_\cdot\mathcal{O}(\underline{M}_1) \xrightarrow{\sim} \mathcal{O}(\underline{GRT}_1)_\cdot . \tag{6.2.1}$$

PROOF . At first, fix any rational point $\varphi \in \underline{M}_1(\mathbf{Q})$. By the definition of weight filtration W on $\mathcal{O}(\underline{M}_1)$ and $l_\varphi^\sharp$ (§6.1), we obtain the isomorphism of graded $\mathbf{Q}$-algebras

$$Gr^W_\cdot l_\varphi^\sharp : Gr^W_\cdot\mathcal{O}(\underline{M}_1) \xrightarrow{\sim} Gr^W_\cdot\mathcal{O}(\underline{GRT}_1)_\cdot .$$

We put $r^\sharp = \tilde{s_0}\circ Gr^W_\cdot l_\varphi^\sharp$ (for $\tilde{s_0}$, see (2.1.2)). Then the theorem follows from the following lemma.

LEMMA 6.2.3. *The above isomorphism $Gr^W_\cdot l_\varphi^\sharp$ is canonical in the sense that it does not depend on the choice of rational points $\varphi \in \underline{M}_1(\mathbf{Q})$.*

PROOF . We proceed the way as in the proof on *Proposition* 6.1.2. Take another rational point $\varphi' \in \underline{M}_1(\mathbf{Q})$. Then there exists a unique element $a \in \underline{GRT}_1(\mathbf{Q})$ such that $l'_\varphi = l_\varphi \circ R_a$. Thus it is enough to show that the graded algebra homomorphism $Gr^W_\cdot R_a^\sharp : Gr^W_\cdot\mathcal{O}(\underline{GRT}_1) \to Gr^W_\cdot\mathcal{O}(\underline{GRT}_1)$ does not depend on the choice of $a \in \underline{GRT}_1(\mathbf{Q})$, where $Gr^W_\cdot R_a^\sharp$ means the associated graded quotient map of $R_a^\sharp : \mathcal{O}(\underline{GRT}_1) \to \mathcal{O}(\underline{GRT}_1)$ (cf. (6.1.2)). Recall that for all $a \in \underline{GRT}_1(\mathbf{Q})$, the first term $x_1(a)$ of its word expansion is always equal to 1 (*Notation* 2.1.7). By combining it with the fact that $\mathcal{O}(\underline{GRT}_1) = \mathbf{Q}[x_1, x_W]_{W:\text{words}}$ is equipped with a structure of graded Hopf algebra whose grading is given by $deg\ x_1 = 0$ and $deg\ x_w = deg\ W$ for each word W, we see that $Gr^W_\cdot R_a^\sharp$ induces the map on each graded component of $Gr^W_\cdot\mathcal{O}(\underline{GRT}_1)$, which is independent from $a \in \underline{GRT}_1(\mathbf{Q})$. □

Put $\underline{GrM}_1 := Spec\ Gr^W_{\bullet}\mathcal{O}(\underline{M}_1)$. Then the map (6.2.1) induces the following isomorphism of schemes over $\mathbf{Q}$.

$$r : \underline{GRT}_1 \overset{\sim}{\to} \underline{GrM}_1 \ . \tag{6.2.2}$$

We relate the two embeddings $\Phi_{DR} : Spec\ Z_{\bullet}/(\pi^2) \hookrightarrow \underline{GRT}_1$ (3.2.3) and $\Phi_{Hod} : Spec\ \mathfrak{Z} \hookrightarrow \underline{M}_1$ (3.3.2). It is easy to see that the surjective algebra homomorphism $\Phi^{\sharp}_{Hod} : \mathcal{O}(\underline{M}_1) \twoheadrightarrow \mathfrak{Z}$ (3.3.1) is strictly compatible with the filtrations of $\mathcal{O}(\underline{M}_1)$ (§6.1) and $\mathfrak{Z}$ (*Notation* 3.3.2). Therefore we obtain the surjective graded $\mathbf{Q}$-algebra homomorphism

$$Gr^W_{\bullet}\Phi^{\sharp}_{Hod} : Gr^W_{\bullet}\mathcal{O}(\underline{M}_1) \twoheadrightarrow Gr^W_{\bullet}\mathfrak{Z} \ . \tag{6.2.3}$$

Thus by *Proposition* 3.3.3, we get the associated embedding of schemes

$$Gr\Phi_{Hod} : Spec\ Z_{\bullet}/(\pi^2) \hookrightarrow \underline{GrM}_1 \ . \tag{6.2.4}$$

It is easy to see the following proposition.

PROPOSITION 6.2.4. $\Phi_{DR} = r^{-1} \circ Gr\Phi_{Hod}$.

Therefore *Conjecture A* (§3.4) is equivalent to the following:

CONJECTURE A$'$. The embedding Φ_{Hod} of affine schemes is an isomorphism, i.e.

$$\Phi_{Hod} : \ Spec\ \mathfrak{Z} \cong \underline{M}_1$$

REMARK 6.2.5. (1) This conjecture claims that all $\mathbf{Q}$-linear relations among modified MZV's (*Definition* 3.3.1) must be deduced from the defining equations (0)$\sim$(iii) of $\underline{M}_1$ (*Definition* 2.3.5) (Remark 2.3.1 and Tsunogai's problem 2.1.16 more strongly suggest that (0) and (iii) should be enough). In particular, it may be our interesting problem to try to check whether all various algebraic relations among (modified) MZV's, which was found by many mathematicians (for example, see [**HO**], [**IKZ**] and [**O**]) could be deduced from (0)$\sim$(iii) or not.

(2) In contrast, G. Racinet introduced a certain pro-algebraic group $\underline{DMR}_0$ over $\mathbf{Q}$ and a $\underline{DMR}_0$-torsor $\underline{DMR}_1$ in [**Ra**]. With his result, we also associate the embedding Φ_{Rac} : $Spec\ \mathfrak{Z} \hookrightarrow \underline{DMR}_1$. The conjecture ([**Ra**]) by D. Zagier and M. Kontsevich can be reformulated that Φ_{Rac} is an isomorphism. But it is not clear whether his $\underline{DMR}_1$ is isomorphic to $\underline{M}_1$ or not, which is left to our future research.

6.3. Comparison between $\underline{GT}_1$ and $\underline{GRT}_1$.

NOTATION 6.3.1. Denote by $Gr^W_{\bullet}\mathcal{O}(\underline{GT}_1) := \underset{n\geqslant 0}{\oplus} Gr^W_n\mathcal{O}(\underline{GT}_1)$ the graded algebra of the regular function ring $\mathcal{O}(\underline{GT}_1)$ by the filtration W (§2.2.2).

Since $(\mathcal{O}(\underline{GT}_1), W)$ has a structure of filtered Hopf algebra over $\mathbf{Q}$ (*Proposition* 2.2.6), $Gr^W_{\bullet}\mathcal{O}(\underline{GT}_1)$ becomes a graded Hopf algebra over $\mathbf{Q}$. Recall that $\mathcal{O}(\underline{GRT}_1)_{\bullet}$ is also a graded Hopf algebra over $\mathbf{Q}$ (*Proposition* 2.1.8).

THEOREM 6.3.2. *Between two* $\mathbf{Q}$*-structures of graded Hopf algebra* $Gr^W_{\bullet}\mathcal{O}(\underline{GT}_1)$ *and* $\mathcal{O}(\underline{GRT}_1)_{\bullet}$*, there exists a canonical isomorphism of graded Hopf algebras*

$$q^{\sharp} : Gr^W_{\bullet}\mathcal{O}(\underline{GT}_1) \overset{\sim}{\to} \mathcal{O}(\underline{GRT}_1)_{\bullet} \ . \tag{6.3.1}$$

PROOF . At first, pick up and fix any rational point $\varphi = \big(1, \varphi(A,B)\big) \in \underline{M}_1(\mathbf{Q}) \subseteq \underline{M}(\mathbf{Q})$. By restricting the map S_φ (5.1.1) to unipotent parts, we get the isomorphism of unipotent pro-algebraic groups $S_\varphi : \underline{GT}_1 \xrightarrow{\sim} \underline{GRT}_1$. Denote by $S_\varphi^\sharp : \mathcal{O}(\underline{GRT}_1) \to \mathcal{O}(\underline{GT}_1)$ the isomorphism of Hopf algebra (over $\mathbf{Q}$) induced by S_φ. The following two diagrams are commutative for each $g \in \underline{GT}(\mathbf{Q})$. (for τ, see *Notation* 1.3.1).

$$\begin{array}{ccccc} \underline{GT} & \longrightarrow & \mathbb{G}_m & \xleftarrow{\frac{1}{\varpi}} & \mathbb{G}_m \\ {\scriptstyle S_\varphi}\downarrow\simeq & & \frac{1}{\varpi}\downarrow & & \mathrm{id}\downarrow \\ \underline{GRT} & \longrightarrow & \mathbb{G}_m & \xleftarrow{\varpi} & \mathbb{G}_m \end{array} \qquad \begin{array}{ccc} \mathcal{O}(\underline{GT}_1) & \xrightarrow{\tau(g)=(\iota_g^{-1})^\sharp} & \mathcal{O}(\underline{GT}_1) \\ {\scriptstyle (S_\varphi^\sharp)^{-1}}\downarrow\simeq & & {\scriptstyle (S_\varphi^\sharp)^{-1}}\downarrow\simeq \\ \mathcal{O}(\underline{GRT}_1) & \xrightarrow{\tau(S_\varphi(g))=(\iota_{S_\varphi(g)}^{-1})^\sharp} & \mathcal{O}(\underline{GRT}_1) \end{array}$$

Namely we get a morphism of weighted modules ([**HM**]§*3.3.*)

$$(\mathcal{O}(\underline{GT}_1), \mathbb{G}_m, \frac{1}{\varpi}, \underline{GT}\ ) \longrightarrow (\mathcal{O}(\underline{GRT}_1), \mathbb{G}_m, \varpi, \underline{GRT}\ ) \quad ,$$

which means that $(S_\varphi^\sharp)^{-1} : \mathcal{O}(\underline{GT}_1) \to \mathcal{O}(\underline{GRT}_1)$ is an isomorphism of modules preserving weight filtrations of $\mathcal{O}(\underline{GT}_1)$ and $\mathcal{O}(\underline{GRT}_1)$. Since $S_\varphi^\sharp : \mathcal{O}(\underline{GRT}_1) \to \mathcal{O}(\underline{GT}_1)$ is a natural isomorphism of Hopf algebras, $(S_\varphi^\sharp)^{-1}$ is an isomorphism of filtered Hopf algebras. By taking the graded quotient of $(S_\varphi^\sharp)^{-1}$ by these weight filtrations, we get an isomorphism

$$Gr_\cdot^W (S_\varphi^\sharp)^{-1} : Gr_\cdot^W \mathcal{O}(\underline{GT}_1) \longrightarrow Gr_\cdot^W \mathcal{O}(\underline{GRT}_1)$$

of graded Hopf algebras.

LEMMA 6.3.3. *The above isomorphism $Gr_\cdot^W (S_\varphi^\sharp)^{-1}$ is canonical in the sense that it does not depend on the choice of rational points $\varphi \in \underline{M}_1(\mathbf{Q})$.*

PROOF . Take another rational point $\varphi' \in \underline{M}_1(\mathbf{Q})$. By *Proposition* 2.3.4, there should exist a unique element $a \in \underline{GRT}_1(\mathbf{Q})$ such that $\varphi' = a \circ \varphi$. It follows that $S_{\varphi'} = \tau_a \circ S_\varphi$, which implies that

$$(S_{\varphi'}^\sharp)^{-1} = (\tau_{a^{-1}})^\sharp \circ (S_\varphi^\sharp)^{-1} : \mathcal{O}(\underline{GT}_1) \overset{(S_\varphi^\sharp)^{-1}}{\longrightarrow} \mathcal{O}(\underline{GRT}_1) \overset{(\tau_{a^{-1}})^\sharp}{\longrightarrow} \mathcal{O}(\underline{GRT}_1)\ .$$

Since the $\underline{GRT}_1$-action on $Gr_\cdot^W \mathcal{O}(\underline{GRT}_1)$ by τ is trivial by *Proposition* 1.2.3.(b),

$$Gr_\cdot^W (S_\varphi^\sharp)^{-1} = Gr_\cdot^W (S_{\varphi'}^\sharp)^{-1} : Gr_\cdot^W \mathcal{O}(\underline{GT}_1) \xrightarrow{\sim} Gr_\cdot^W \mathcal{O}(\underline{GRT}_1). \ .$$

Thus we see that this isomorphism $Gr_\cdot^W (S_\varphi^\sharp)^{-1}$ does not depend on the choice of rational points of $\underline{M}_1(\mathbf{Q})$. □

The rest of the proof of theorem 6.3.2. : By composing the above isomorphism $Gr_\cdot^W (S_\varphi^{-1})^\sharp$ (*Lemma* 6.3.3) with the natural isomorphism $\tilde{s_0} : Gr_\cdot^W \mathcal{O}(\underline{GRT}_1) \to \mathcal{O}(\underline{GRT}_1)$ (2.1.2), we finally obtain a canonical isomorphism $q^\sharp$ of graded Hopf algebras over $\mathbf{Q}$

$$q^\sharp := \tilde{s_0} \circ Gr_\cdot^W (S_\varphi^{-1})^\sharp : Gr_\cdot^W \mathcal{O}(\underline{GT}_1) \xrightarrow{\sim} \mathcal{O}(\underline{GRT}_1). \ .$$

□

Put $\underline{GrGT}_1 := Spec\ Gr_\cdot^W \mathcal{O}(\underline{GT}_1)$. Then (6.3.1) induces the following isomorphism

(6.3.2) $$q : \underline{GRT}_1 \xrightarrow{\sim} \underline{GrGT}_1$$

of pro-algebraic groups over $\mathbf{Q}$.

REMARK 6.3.4. We can relate the above canonical isomorphism $q : \underline{GRT}_1 \xrightarrow{\sim} \underline{GrGT}_1$ (6.3.2) with the isomorphism $p : \underline{GT}_1 \times_{\mathbf{Q}} \mathbf{C} \to \underline{GRT}_1 \times_{\mathbf{Q}} \mathbf{C}$ (*Proposition* 5.1.2) of pro-algebraic groups, as follows. Recall that the map p is induced from $\varphi_{KZ}(A,B) \in \underline{M}_1(\mathbf{C})$, i.e. $p = S_{\varphi_{KZ}}$. As we have seen in the proof of *Theorem* 6.3.2, $Gr^W_{\cdot} S^\sharp_\varphi$ does not depend on the choice of points φ on $\underline{M}_1$. Therefore $\tilde{s_0} \circ Gr^W_{\cdot} p^\sharp = q^\sharp \underset{\mathbf{Q}}{\otimes} id_{\mathbf{C}}$.

6.4. Three embeddings on Galois Side. We describe here the prescription how to associate embeddings of pro-algebraic groups with T_l, $\Phi^{(l)}_{\mathbf{Q}}$ and $\Phi^{(l)}_{\mathbf{Q}_l}$ (§4.2).

(1) Suppose that $\mathcal{O}(\underline{Gal}^{(l)}_{\mathbf{Q}})$ (resp. $\mathcal{O}(\underline{Gal}^{(l)}_{\mathbf{Q}_l})$) is the regular function ring of the unipotent pro-algebraic group $\underline{Gal}^{(l)}_{\mathbf{Q}}$ (cf. *Definition* 4.2.2) over $\mathbf{Q}$ (resp. $\underline{Gal}^{(l)}_{\mathbf{Q}_l}$ (cf. *Definition* 4.2.3) over $\mathbf{Q}_l$), which is a Hopf algebra over $\mathbf{Q}$ (resp. over $\mathbf{Q}_l$). With the embeddings T_l, $\Phi^{(l)}_{\mathbf{Q}}$ and $\Phi^{(l)}_{\mathbf{Q}_l}$ (cf. (4.2.1)∼(4.2.3)) of unipotent pro-algebraic groups, we obtain the following surjective homomorphisms of Hopf algebras.

$$T_l^\sharp : \mathcal{O}(\underline{Gal}^{(l)}_{\mathbf{Q}}) \underset{\mathbf{Q}}{\otimes} \mathbf{Q}_l \twoheadrightarrow \mathcal{O}(\underline{Gal}^{(l)}_{\mathbf{Q}_l}) \tag{6.4.1}$$

$$\Phi^{(l)\sharp}_{\mathbf{Q}} : \mathcal{O}(\underline{GT}_1) \twoheadrightarrow \mathcal{O}(\underline{Gal}^{(l)}_{\mathbf{Q}}) \tag{6.4.2}$$

$$\Phi^{(l)\sharp}_{\mathbf{Q}_l} : \mathcal{O}(\underline{GT}_1) \underset{\mathbf{Q}}{\otimes} \mathbf{Q}_l \twoheadrightarrow \mathcal{O}(\underline{Gal}^{(l)}_{\mathbf{Q}_l}) \tag{6.4.3}$$

Notice that $\Phi^{(l)\sharp}_{\mathbf{Q}_l} = T_l^\sharp \circ \left(\Phi^{(l)\sharp}_{\mathbf{Q}} \underset{\mathbf{Q}}{\otimes} id_{\mathbf{Q}_l}\right)$.

(2) By the surjection $\Phi^{(l)\sharp}_{\mathbf{Q}}$ (resp. $\Phi^{(l)\sharp}_{\mathbf{Q}_l}$), the Hopf algebra $\mathcal{O}(\underline{Gal}^{(l)}_{\mathbf{Q}})$ over $\mathbf{Q}$ (resp. $\mathcal{O}(\underline{Gal}^{(l)}_{\mathbf{Q}_l})$ over $\mathbf{Q}_l$) is equipped with a structure of filtered Hopf algebra over $\mathbf{Q}$ (resp. over $\mathbf{Q}_l$) by the filtration $\left\{W_n\mathcal{O}(\underline{Gal}^{(l)}_{\mathbf{Q}})\right\}_{n\in\mathbf{Z}}$ (resp. $\left\{W_n\mathcal{O}(\underline{Gal}^{(l)}_{\mathbf{Q}_l})\right\}_{n\in\mathbf{Z}}$) induced from that of $\mathcal{O}(\underline{GT}_1)$(§2.2.2). Let $Gr^W_{\cdot}\mathcal{O}(\underline{Gal}^{(l)}_{\mathbf{Q}})$, $Gr^W_{\cdot}\mathcal{O}(\underline{Gal}^{(l)}_{\mathbf{Q}_l})$ and $Gr^W_{\cdot}\mathcal{O}(\underline{GT}_1)$ be the associated graded algebras of $\mathcal{O}(\underline{Gal}^{(l)}_{\mathbf{Q}})$, $\mathcal{O}(\underline{Gal}^{(l)}_{\mathbf{Q}_l})$ and $\mathcal{O}(\underline{GT}_1)$ by these filtrations respectively. Note that they are equipped with structures of graded Hopf algebra. Thus from (6.4.1)∼(6.4.3), we obtain surjective homomorphisms of graded Hopf algebras as follows:

$$Gr^W_{\cdot}T_l^\sharp : Gr^W_{\cdot}\mathcal{O}(\underline{Gal}^{(l)}_{\mathbf{Q}}) \underset{\mathbf{Q}}{\otimes} \mathbf{Q}_l \twoheadrightarrow Gr^W_{\cdot}\mathcal{O}(\underline{Gal}^{(l)}_{\mathbf{Q}_l}) \tag{6.4.4}$$

$$Gr^W_{\cdot}\Phi^{(l)\sharp}_{\mathbf{Q}} : Gr^W_{\cdot}\mathcal{O}(\underline{GT}_1) \twoheadrightarrow Gr^W_{\cdot}\mathcal{O}(\underline{Gal}^{(l)}_{\mathbf{Q}}) \tag{6.4.5}$$

$$Gr^W_{\cdot}\Phi^{(l)\sharp}_{\mathbf{Q}_l} : Gr^W_{\cdot}\mathcal{O}(\underline{GT}_1) \underset{\mathbf{Q}}{\otimes} \mathbf{Q}_l \twoheadrightarrow Gr^W_{\cdot}\mathcal{O}(\underline{Gal}^{(l)}_{\mathbf{Q}_l}) \tag{6.4.6}$$

Notice that $Gr^W_{\cdot}\Phi^{(l)\sharp}_{\mathbf{Q}_l} = Gr^W_{\cdot}T_l^\sharp \circ \left(Gr^W_{\cdot}\Phi^{(l)\sharp}_{\mathbf{Q}} \underset{\mathbf{Q}}{\otimes} id_{\mathbf{Q}_l}\right)$.

(3) Suppose that $\underline{GrGal}^{(l)}_{\mathbf{Q}}$ (resp. $\underline{GrGal}^{(l)}_{\mathbf{Q}_l}$) is the pro-algebraic group over $\mathbf{Q}$ (resp. over $\mathbf{Q}_l$) whose regular function ring is isomorphic to the graded Hopf algebra $Gr^W_{\cdot}\mathcal{O}(\underline{Gal}^{(l)}_{\mathbf{Q}})$ over $\mathbf{Q}$ (resp. $Gr^W_{\cdot}\mathcal{O}(\underline{Gal}^{(l)}_{\mathbf{Q}_l})$ over $\mathbf{Q}_l$). Recall that $\underline{GrGT}_1$ is isomorphic to $\underline{GRT}_1$ by q (6.3.2). Therefore, from

(6.4.4)$\sim$(6.4.6), we get the following embeddings of unipotent pro-algebraic groups:

(6.4.7) $$GrT_l : \underline{GrGal}^{(l)}_{\mathbf{Q}_l} \hookrightarrow \underline{GrGal}^{(l)}_{\mathbf{Q}} \times_{\mathbf{Q}} \mathbf{Q}_l$$

(6.4.8) $$Gr\Phi^{(l)}_{\mathbf{Q}} : \underline{GrGal}^{(l)}_{\mathbf{Q}} \hookrightarrow \underline{GRT}_1$$

(6.4.9) $$Gr\varPhi^{(l)}_{\mathbf{Q}_l} : \underline{GrGal}^{(l)}_{\mathbf{Q}_l} \hookrightarrow \underline{GRT}_1 \times_{\mathbf{Q}} \mathbf{Q}_l$$

Notice that $Gr\varPhi^{(l)}_{\mathbf{Q}_l} = \left(Gr\Phi^{(l)}_{\mathbf{Q}} \times id_{\mathbf{Q}_l}\right) \circ GrT_l$.

REMARK 6.4.1. Note that *Conjecture B* (§4.2) implies that $Gr\varPhi^{(l)}_{\mathbf{Q}}$ is an isomorphism and *Conjecture C* (§4.2) implies that GrT_l is an isomorphism.

6.5. Comparison between Galois Side and Hodge Side. By combining *Theorem* 6.3.2 with the embeddings $Gr\varPhi^{(l)}_{\mathbf{Q}}$ (6.4.8) and $\varPhi_{DR}$ (3.2.3), we get Figure 2 in §0. Note that we can compare two objects, $\underline{GrGal}^{(l)}_{\mathbf{Q}}$ in Galois Side and *Spec* $Z_{\bullet}/(\pi^2)$ in Hodge Side, in the same box $\underline{GRT}_1$. In contrast, we could not make a direct comparison between two objects in Galois Side and in Hodge Side inside a common box (over $\mathbf{Q}$) in §5.3.

Validities of *Conjecture A* (§3.4), *Conjecture B* (§4.2) and *Conjecture C* (§4.2) would imply the following relationship between Galois Side and Hodge Side.

> *" In Galois Side, $\underline{GRT}_1$ should be a common structure of $\underline{GrGal}^{(l)}_{\mathbf{Q}_l}$ for all prime l, i.e. $\underline{GrGal}^{(l)}_{\mathbf{Q}_l} \cong \underline{GRT}_1 \times_{\mathbf{Q}} \mathbf{Q}_l$. On the other hand, in Hodge Side, Spec $Z_{\bullet}/(\pi^2)$ should become a pro-algebraic group $\underline{GRT}_1$. Therefore Spec $Z_{\bullet}/(\pi^2)$ is the common $\mathbf{Q}$-structure of $\underline{GrGal}^{(l)}_{\mathbf{Q}_l}$, i.e. $\underline{GrGal}^{(l)}_{\mathbf{Q}_l} \cong$ Spec $Z_{\bullet}/(\pi^2) \times_{\mathbf{Q}} \mathbf{Q}_l$ for all prime l."*

Roughly speaking, MZV might be a regular function of the graded Galois image.

REMARK 6.5.1. The two surjections $\varPsi_{DR} : \mathfrak{D}^*_{\bullet} \twoheadrightarrow NZ_{\bullet}$ (for $NZ_{\bullet}$, see *Proposition* 3.4.2) and $\varPsi^*_l : \mathfrak{D}^*_{\bullet} \underset{\mathbf{Q}}{\otimes} \mathbf{Q}_l \twoheadrightarrow \mathfrak{g}^{(l)*}_{\bullet}$ of graded vector spaces are concerned in [**F**]§5 (see the picture below).

Galois Side		Hodge Side

$$\mathfrak{g}^{(l)*}_{\bullet} \overset{\varPsi^*_l}{\twoheadleftarrow} \mathfrak{D}^*_{\bullet} \underset{\mathbf{Q}}{\otimes} \mathbf{Q}_l$$

$$\mathfrak{D}^*_{\bullet} \overset{\varPsi_{DR}}{\twoheadrightarrow} NZ_{\bullet}$$

We can deduce these surjections from $\varPhi_{DR}$ (3.2.3) and $\varPhi^{(l)}_{\mathbf{Q}_l}$ (4.2.3) in the following way:

Take the dual of the differential (i.e. the cotangent space) of the embedding $\varPhi_{DR}$: *Spec* $Z_{\bullet}/(\pi^2) \hookrightarrow \underline{GRT}_1$ at the unit element e of $\underline{GRT}_1$. Then by (3.4.1) we get a surjection of graded vector spaces $(d\varPhi_{DR})^{\sharp}_e : \mathfrak{D}^*_{\bullet} \twoheadrightarrow NZ_{\bullet}$, which is equal to the above map $\varPsi_{DR}$. On the other hand, take the dual of the differential of the embedding $Gr\varPhi^{(l)}_{\mathbf{Q}_l} : \underline{GrGal}^{(l)}_{\mathbf{Q}_l} \hookrightarrow \underline{GRT}_1 \times_{\mathbf{Q}} \mathbf{Q}_l$ (6.4.9) at the unit element e of

$\underline{GRT}_1$. Then by (4.3.4) we get a surjection of graded vector spaces $(dGr\Phi^{(l)}_{\mathbf{Q}_l})^{\sharp}_e$: $\mathfrak{D}^*_{\bullet} \underset{\mathbf{Q}}{\otimes} \mathbf{Q}_l \twoheadrightarrow \mathfrak{g}^{(l)*}_{\bullet}$, which is equal to the above map Ψ^*_l.

6.6. Chase. Here we present a few relationships between Galois Side and Hodge Side deduced from *Figure* 2. §6.6.2 and §6.6.3 are just special cases of §6.6.1.

6.6.1. *Non-abelian case.*

$\underline{GRT}_1$: Recall that $\underline{GRT}_1(\mathbf{Q})$ is defined as a subset of the non-commutative formal power series ring $\mathbf{Q}\langle\langle A,B\rangle\rangle$ (*Definition* 2.1.4). Thus we obtain an embedding $g_1 : \underline{GRT}_1(\mathbf{Q}) \hookrightarrow \mathbf{Q}\langle\langle A,B\rangle\rangle$. This embedding determines an invertible element $\Phi_{GRT} = 1 + \sum_{W:\text{words}} d_W W \in \Big(\mathcal{O}(\underline{GRT}_1)_{\bullet}\langle\langle A,B\rangle\rangle\Big)^{\times}$ of the $\mathcal{O}(\underline{GRT}_1)_{\bullet}$-coefficient non-commutative formal power series ring with two variables.

$\underline{GT}_1$: Recall that $\underline{GT}_1(\mathbf{Q})$ is defined as a subset of the pro-algebraic group $\underline{F_2}(\mathbf{Q})$ (*Definition* 2.2.3). By sending free generators X and Y into e^A and e^B respectively, we get an embedding $g_2 : \underline{F_2}(\mathbf{Q}) \hookrightarrow \mathbf{Q}\langle\langle A,B\rangle\rangle$.

Galois Side: With the composition of embeddings $g_2 \circ \Phi^{(l)}_{\mathbf{Q}} : \underline{Gal}^{(l)}_{\mathbf{Q}}(\mathbf{Q}) \hookrightarrow \mathbf{Q}\langle\langle A,B\rangle\rangle$, we associate an invertible element $\Phi^{(l)}_{Ih} = 1 + \sum_{W:\text{words}} \kappa^{(l)}_W W \in \Big(\mathcal{O}(\underline{Gal}^{(l)}_{\mathbf{Q}})\langle\langle A,B\rangle\rangle\Big)^{\times}$. We call it as the **$l$-adic Ihara associator** [6]. Note that for each word W with $deg\ W = n$, $\kappa^{(l)}_W$ lies in $W_n\mathcal{O}(\underline{Gal}^{(l)}_{\mathbf{Q}})$. Denote by $Gr\ \kappa^{(l)}_W$ its quotient class in $W_n\mathcal{O}(\underline{Gal}^{(l)}_{\mathbf{Q}}) \big/ W_{n-1}\mathcal{O}(\underline{Gal}^{(l)}_{\mathbf{Q}})$. Put $Gr\Phi^{(l)}_{Ih} := 1 + \sum_{W:\text{words}} Gr\kappa^{(l)}_W\ W \in \Big(Gr^W_{\bullet}\mathcal{O}(\underline{Gal}^{(l)}_{\mathbf{Q}})\langle\langle A,B\rangle\rangle\Big)^{\times}$, which is also an invertible element associated with the composition $g_1 \circ Gr\Phi_l$.

Hodge Side: With the composition of embeddings $g_1 \circ \Phi_{DR}$, we obtain an invertible element of $\Big(Z_{\bullet}/(\pi^2)\Big)\langle\langle A,B\rangle\rangle$. From the definition of Φ_{DR} (see *Theorem* 3.2.5), It is equal to $\Phi_{KZ} \mod \pi^2$, which stands for the image of $\Phi_{KZ}(A,B) = \varphi_{KZ}(2\pi i A, 2\pi i B) \in Z_{\bullet}\langle\langle A,B\rangle\rangle$ (cf. *Definition* 2.3.8 and (3.2.4)) by the natural projection $Z_{\bullet}\langle\langle A,B\rangle\rangle \twoheadrightarrow \Big(Z_{\bullet}/(\pi^2)\Big)\langle\langle A,B\rangle\rangle$.

PROPOSITION 6.6.1. *Let* $Gr\Phi^{(l)\sharp}_{\mathbf{Q}\langle\langle A,B\rangle\rangle} : \mathcal{O}(\underline{GRT}_1)_{\bullet}\langle\langle A,B\rangle\rangle \twoheadrightarrow Gr^W_{\bullet}\mathcal{O}(\underline{Gal}^{(l)}_{\mathbf{Q}})\ \langle\langle A,B\rangle\rangle$ *be the surjection induced from* $Gr^W_{\bullet}\Phi^{(l)\sharp}_{\mathbf{Q}}$ (6.4.5) *and* $\Phi^{\sharp}_{DR\langle\langle A,B\rangle\rangle} : \mathcal{O}(\underline{GRT}_1)_{\bullet}\langle\langle A,B\rangle\rangle \twoheadrightarrow \Big(Z_{\bullet}/(\pi^2)\Big)\langle\langle A,B\rangle\rangle$ *be the surjection induced from* $\Phi^{\sharp}_{DR}$ (3.2.2). *Then*

$$Gr\Phi^{(l)\sharp}_{\mathbf{Q}\langle\langle A,B\rangle\rangle}(\Phi_{GRT}) = Gr\Phi^{(l)}_{Ih} \quad , \quad \Phi^{\sharp}_{DR\langle\langle A,B\rangle\rangle}(\Phi_{GRT}) = \Phi_{KZ} \mod \pi^2.$$

PROOF . This follows from definitions of $Gr\Phi^{(l)}_{Ih}$ and $\Phi_{KZ} \mod \pi^2$. □

Thus we get Figure 3(§0).

REMARK 6.6.2. Recall that *Conjecture B* and *Conjecture A* claim that both $Gr\Phi^{(l)\sharp}_{\mathbf{Q}\langle\langle A,B\rangle\rangle}$ and $\Phi^{\sharp}_{DR\langle\langle A,B\rangle\rangle}$ are isomorphisms. Therefore

---

[6]This should be an appropriate name since essentially the same series was studied in his series of papers [**AI88**], [**AI90**] and [**Ih86a**]~[**Ih00**].

" If Conjecture A and Conjecture B hold, then $Gr\Phi^{(l)}_{Ih}$ *must be identified with* $\Phi_{KZ} \mod \pi^2$ *via* $(\Phi^{\sharp}_{DR\langle\langle A,B\rangle\rangle}) \circ (Gr\Phi^{(l)\sharp}_{\mathbf{Q}\langle\langle A,B\rangle\rangle})^{-1}$ *for all prime* l*."*

6.6.2. *Meta-abelian case.*

DEFINITION 6.6.3. Suppose that R is a commutative ring. Let ψ be any invertible element of the non-commutative formal power series ring $R\langle\langle A, B\rangle\rangle$ with two variables. Decompose ψ^{-1} uniquely $\psi^{-1} = C_\psi + X_\psi A + Y_\psi B$ as in the way of [**Ih99**]. Here $C_\psi \in R$ and $X_\psi, Y_\psi \in R\langle\langle A, B\rangle\rangle$. The **meta-abelian quotient** of ψ is the image of $C_\psi + X_\psi A$ by the natural surjection $R\langle\langle A, B\rangle\rangle \twoheadrightarrow R[[A, B]]$ into the *commutative* formal power series ring.

$\underline{GRT}_1$: We denote by $B_{GRT} \in \mathcal{O}(\underline{GRT}_1).[[A, B]]$ the meta-abelian quotient of $\Phi_{GRT} \in \Big(\mathcal{O}(\underline{GRT}_1).\langle\langle A, B\rangle\rangle\Big)^\times$.

Galois Side: The meta-abelian quotient of the l-adic Ihara associator $\Phi^{(l)}_{Ih} \in \Big(\mathcal{O}(\underline{Gal}^{(l)}_{\mathbf{Q}})\langle\langle A, B\rangle\rangle\Big)^\times$ is equal to the **l-adic universal power series for Jacobi sums** $B^{(l)}(A, B) \in \mathcal{O}(\underline{Gal}^{(l)}_{\mathbf{Q}})[[A, B]]$(see [**Ih90**]§6.4.) [7], whose presentation is calculated in [**A89**], [**C**] and [**IKY**] as follows:

$$B^{(l)}(A, B) = exp\left[\sum_{m\geqslant 3:\mathrm{odd}} \frac{\kappa^{(l)*}_m}{m!}\{(A+B)^m - A^m - B^m\}\right] \in \mathcal{O}(\underline{Gal}^{(l)}_{\mathbf{Q}})[[A, B]].$$

Here $\kappa^{(l)*}_m \in \mathcal{O}(\underline{Gal}^{(l)}_{\mathbf{Q}})$ is the **l-adic m-th Soulé character** (see [**Sou**] and also [**F**]§5.) which is a kind of Euler system. We note that $\kappa^{(l)*}_m$ belongs to $W_m\mathcal{O}(\underline{Gal}^{(l)}_{\mathbf{Q}})$. We denote $Gr\kappa^{(l)*}_m$ by its quotient class in $W_m\mathcal{O}(\underline{Gal}^{(l)}_{\mathbf{Q}}) \Big/ W_{m-1}\mathcal{O}(\underline{Gal}^{(l)}_{\mathbf{Q}})$ and put

$$Gr^W_{\bullet} B^{(l)}(A, B) := exp\left[\sum_{m\geqslant 3:\mathrm{odd}} \frac{Gr\kappa^{(l)*}_m}{m!}\{(A+B)^m - A^m - B^m\}\right] \in Gr^W_{\bullet}\mathcal{O}(\underline{Gal}^{(l)}_{\mathbf{Q}})[[A, B]].$$

In fact, $Gr^W_{\bullet} B^{(l)}$ is the meta-abelian quotient of

$$Gr\Phi^{(l)}_{Ih} \in \Big(Gr^W_{\bullet}\mathcal{O}(\underline{Gal}^{(l)}_{\mathbf{Q}})\langle\langle A, B\rangle\rangle\Big)^\times.$$

Hodge Side: Recall that the *classical gamma function* $\Gamma(z)$ has a presentation $\Gamma(1-z) = exp\left\{\gamma z + \sum_{n=2}^{\infty} \frac{\zeta(n)}{n} z^n\right\}$ (γ: Euler constant), from which we can deduce the following formula of the **classical beta function** $B(x, y)$:

$$(1-x-y)\cdot B(1-x, 1-y) = exp\left[\sum_{m=2}^{\infty} \frac{\zeta(m)}{m}\{x^m + y^m - (x+y)^m\}\right].$$

By expanding this into **C**-coefficient commutative formal power series, we see that $(1-A-B)\cdot B(1-A, 1-B)$ naturally determines the element of

[7] According to the notation in [**Ih90**], $B^{(l)}(A, B) = \psi^{\mathrm{ab}}(e^A - 1, e^B - 1)$.

$Z_\bullet[[A,B]]$. We denote by $(1-A-B)\cdot B(1-A,1-B) \mod \pi^2$ its image by the natural projection $Z_\bullet[[A,B]] \twoheadrightarrow \Big(Z_\bullet/(\pi^2)\Big)[[A,B]]$, which is in fact equal to the meta-abelian quotient of $\Phi_{KZ} \mod \pi^2 \in \Big((Z_\bullet/(\pi^2))\langle\langle A,B\rangle\rangle\Big)^\times$ ([**Dr**]).

Therefore by taking the meta-abelian quotient part of *Proposition* 6.6.1, we get

COROLLARY 6.6.4. *Let* $Gr\Phi^{(l)\sharp}_{\mathbf{Q}[[A,B]]} : \mathcal{O}(\underline{GRT}_1)_\bullet[[A,B]] \twoheadrightarrow Gr^W_\bullet \mathcal{O}(\underline{Gal}^{(l)}_{\mathbf{Q}})\ [[A,B]]$ *be the surjection induced from* $Gr^W_\bullet\Phi^{(l)\sharp}_{\mathbf{Q}}$ (6.4.5) *and* $\Phi^\sharp_{DR[[A,B]]} : \mathcal{O}(\underline{GRT}_1)_\bullet[[A,B]] \twoheadrightarrow \Big(Z_\bullet/(\pi^2)\Big)[[A,B]]$ *be the surjection induced from* $\Phi^\sharp_{DR}$ (3.2.2). *Then*

$$Gr\Phi^{(l)\sharp}_{\mathbf{Q}[[A,B]]}(B_{GRT}) = Gr^W_\bullet B^{(l)}(A,B) \qquad ,$$
$$\Phi^\sharp_{DR[[A,B]]}(B_{GRT}) = (1-A-B)\cdot B(1-A,1-B) \mod \pi^2.$$

Galois Side				Hodge Side
$Gr^W_\bullet \mathcal{O}(\underline{Gal}^{(l)}_{\mathbf{Q}})_\bullet[[A,B]]$	$\overset{Gr\Phi^{(l)\sharp}_{\mathbf{Q}[[A,B]]}}{\twoheadleftarrow}$	$\mathcal{O}(\underline{GRT}_1)_\bullet[[A,B]]$	$\overset{\Phi^\sharp_{DR[[A,B]]}}{\twoheadrightarrow}$	$\Big(Z_\bullet/(\pi^2)\Big)[[A,B]]$
$Gr^W_\bullet B^{(l)}(A,B)$	$\longleftarrow$	$B_{GRT}(A,B)$	$\longmapsto$	$(1-A-B)\cdot B(1-A,1-B)$ mod π^2
<u>l-adic universal power series for Jacobi sums</u>				<u>classical beta function</u>

REMARK 6.6.5. (1) We recall that $\prod_{l:\text{prime}} B^{(l)}(A,B)$ was named the adelic beta function as an analogue of the classical beta function in [**A87**], and has been widely studied in [**A87**], [**A89**], [**C**], [**Ich**], [**Ih86a**]~[**Ih90**] and [**IKY**] (for a pro-finite group analogue, see also [**Ih99**] and [**Ih00**]).

(2) We recall that *Conjecture B* and *Conjecture A* claim that both $Gr\Phi^{(l)\sharp}_{\mathbf{Q}[[A,B]]}$ and $\Phi^\sharp_{DR[[A,B]]}$ are isomorphisms. Therefore

> *" If Conjecture A and Conjecture B hold, then* $Gr^W_\bullet B^{(l)}$ *must be identified with* $(1-A-B)\cdot B(1-A,1-B)$ mod π^2 *via* $(\Phi^\sharp_{DR[[A,B]]})\circ(Gr\Phi^{(l)\sharp}_{\mathbf{Q}[[A,B]]})^{-1}$ *for all prime* l."

We hope that this formulation is the desired direct correspondence.

6.6.3. *Special case.* Especially let $d_m := d_{A^{m-1}B} \in \mathcal{O}(\underline{GRT}_1)_\bullet$ be the coefficient of $W = A^{m-1}B$ of $\Phi_{GRT} \in \mathcal{O}(\underline{GRT}_1)_\bullet\langle\langle A,B\rangle\rangle$ (§6.6.1) for $m = 3,5,7,\cdots$. Then by restricting the correspondence in *Proposition* 6.6.1 to each coefficient of $A^{m-1}B$, we get

COROLLARY 6.6.6. *For* $m = 3,5,7,\cdots$,

$$Gr^W_\bullet\Phi^{(l)\sharp}_{\mathbf{Q}}(d_m) = \frac{1}{(m-1)!}Gr\ \kappa^{(l)*}_m \quad \textit{and} \quad \Phi^\sharp_{DR}(d_m) = -\zeta(m) \mod \pi^2 .$$

Galois Side | Hodge Side

$$
\begin{array}{ccccc}
Gr_{\cdot}^{W}\mathcal{O}(\underline{Gal_{\mathbf{Q}}^{(l)}}) & \overset{Gr_{\cdot}^{W}\Phi_{\mathbf{Q}}^{(l)\sharp}}{\twoheadleftarrow} & \mathcal{O}(\underline{GRT}_1). & \overset{\Phi_{DR}^{\sharp}}{\twoheadrightarrow} & Z./(\pi^2) \\
\frac{1}{(m-1)!}Gr\ \kappa_m^{(l)*} & \longleftarrow & d_m & \longmapsto & -\zeta(m) \mod \pi^2
\end{array}
$$

<u>l-adic Soulé character</u> | <u>Riemann zeta value</u>

Recall that *Conjecture B* and *Conjecture A* claim that both $Gr_{\cdot}^{W}\Phi_{\mathbf{Q}}^{(l)\sharp}$ and $\Phi_{DR}^{\sharp}$ are isomorphisms. Therefore

> *" If Conjecture A and Conjecture B hold, then Riemann zeta values are identified with Soulé characters via* $(Gr_{\cdot}^{W}\Phi^{(l)\sharp})\circ(\Phi_{DR}^{\sharp})^{-1}$
> ."

REMARK 6.6.7. We note that each coefficient of Φ_{KZ} is calculated in [**LM**] and [**F**] and remark that each coefficient of $\Phi_{Ih}^{(l)}$ can be also calculated in principle by Anderson-Ihara theory ([**AI90**]), from which we can get analogous correspondence to *Corollary* 6.6.6 associated with any given MZV's. Therefore the above correspondence is merely one of special corollaries of *Proposition* 6.6.1.

References

[A87] Anderson, G. W., The hyperadelic gamma function: a précis, Galois representations and arithmetic algebraic geometry (Kyoto, 1985/Tokyo, 1986), 1–19, Adv. Stud. Pure Math., 12, North-Holland, Amsterdam, 1987

[A89] ——— The hyperadelic gamma function, Invent. Math. 95 (1989), no. 1, 63–131.

[AI88] Anderson, G. W., Ihara, Y., Pro-l branched coverings of P^1 and higher circular l-units, Ann. of Math. (2) 128 (1988), no. 2, 271–293.

[AI90] ——— Pro-l branched coverings of P^1 and higher circular l-units. II, Internat. J. Math.1 (1990), no. 2, 119–148.

[C] Coleman, R. F., Anderson-Ihara theory: Gauss sums and circular units, Algebraic number theory, 55–72, Adv. Stud. Pure Math., 17, Academic Press, Boston, MA, 1989.

[De] Deligne, P., Le groupe fondamental de la droite projective moins trois points, Galois groups over Q (Berkeley, CA, 1987), 79–297, Math. Sci. Res. Inst. Publ., 16, Springer, New York-Berlin, 1989.

[DG] Deligne, P., Goncharov, A. B., Groupes fondamentaux motiviques de Tate mixte, Ann.Sci.Ecole Norm.Sup.(4)38 (2005), no.1, 1–56

[Dr] Drinfel'd, V. G., On quasitriangular quasi-Hopf algebras and a group closely connected with Gal($\overline{Q}/Q$), Leningrad Math. J. 2 (1991), no. 4, 829–860

[F] Furusho, H., The multiple zeta value algebra and the stable derivation algebra, Publ. Res. Inst. Math. Sci. Vol 39. no 4. (2003). 695-720.

[Gr] Grothendieck, A., Esquisse d'un programme, 1, 5–48, in [**SL**].

[HM] Hain, R., Matsumoto, M., Weighted Completion of Galois Groups and Some Conjectures of Deligne, Compositio Math. 139 (2003), no.2, 119–167.

[HM*] ——— The first version of [**HM**] available at `arXiv:math.AG/0006158v1`.

[HO] Hoffman, M. E., Ohno, Y., Relations of multiple zeta values and their algebraic expression, J. Algebra 262 (2003), 332-347.

[HS] Harbater, D., Schneps, L., Fundamental groups of moduli and the Grothendieck-Teichmüller group, Trans. Amer. Math. Soc. 352 (2000), no. 7, 3117–3148.

[Ich] Ichimura, H., On the coefficients of the universal power series for Jacobi sums, J. Fac. Sci. Univ. Tokyo Sect. IA Math. 36 (1989), no. 1, 1–7.

[Ih86a] Ihara, Y., Profinite braid groups, Galois representations and complex multiplications, Ann. of Math. (2) 123 (1986), no. 1, 43–106.

[Ih86b] ——— On Galois representations arising from towers of coverings of $P^1 - \{0, 1, \infty\}$, Invent. Math. 86 (1986), no. 3, 427–459.

[Ih90] ——— Braids, Galois groups, and some arithmetic functions, Proceedings of the International Congress of Mathematicians, Vol. I, II (Kyoto, 1990), 99–120, Math. Soc. Japan, Tokyo, 1991.

[Ih91] ——— Automorphisms of pure sphere braid groups and Galois representations, The Grothendieck Festschrift, Vol. II, 353–373, Progr. Math., 87, Birkhäuser Boston, Boston, MA, 1990.

[Ih92] ——— On the stable derivation algebra associated with some braid groups, Israel J. Math. 80 (1992), no. 1-2, 135–153.

[Ih94] ——— On the embedding of $\mathrm{Gal}(\overline{Q}/Q)$ into $\widehat{\mathrm{GT}}$, 289–321, in **[S]**.

[Ih99] ——— On beta and gamma functions associated with the Grothendieck-Teichmüller groups, Aspects of Galois theory, 144–179, London Math. Soc. Lecture Note Ser., 256, Cambridge Univ. Press, Cambridge, 1999.

[Ih00] ——— On beta and gamma functions associated with the Grothendieck-Teichmüller group II, J. Reine Angew. Math.527 (2000), 1–11.

[Ih02] ——— Some arithmetic aspects of Galois actions on the pro-p fundamental group of $\mathbb{P}^1 - \{0, 1, \infty\}$, Arithmetic fundamental groups and noncommutative algebra (Berkeley, CA, 1999), 247–273, Proc. Sympos. Pure Math., 70, Amer. Math. Soc., Providence, RI, 2002.

[IKY] Ihara, Y., Kaneko, M., Yukinari, A., On some properties of the universal power series for Jacobi sums, Galois representations and arithmetic algebraic geometry (Kyoto, 1985/Tokyo, 1986), 65–86, Adv. Stud. Pure Math., 12, North-Holland, Amsterdam, 1987.

[IKZ] Ihara, K., Kaneko, M. and Zagier, D.; Derivation and double shuffle relations for multiple zeta values, to appear in Compositio Math.

[LM] Le, T.T.Q., Murakami, J., Kontsevich's integral for the Kauffman polynomial, Nagoya Math. J. 142 (1996), 39–65.

[LNS] Lochak, P., Nakamura, H., Schneps, L., On a new version of the Grothendieck-Teichmüller group, C. R. Acad. Sci. Paris Sér. I Math. 325 (1997), no. 1, 11–16.

[Na] Nakamura, H., Galois rigidity of pure sphere braid groups and profinite calculus, J. Math. Sci.Univ. Tokyo 1 (1994), no. 1, 71–136.

[NS] Nakamura, H., Schneps, L., On a subgroup of the Grothendieck-Teichmüller group acting on the tower of profinite Teichmüller modular groups, Invent. Math. 141 (2000), no. 3, 503–560.

[O] Ohno, Y., A generalization of the duality and sum formulas on the multiple zeta values, J. Number Theory 74 (1999), no. 1, 39–43.

[Ra] Racinet, G., Séries génératrices non commutatives de polyzetas et associateurs de Drinfel'd, École Normale Supérieure Ph.D. thesis (2000).

[S] Schneps, L., The Grothendieck theory of dessins d'enfants, (Luminy, 1993), London Math. Soc. Lecture Note Ser., 200, Cambridge Univ. Press, Cambridge, 1994.

[SL] Schneps, L., Lochak, P., Geometric Galois actions. 1 and 2, London Mathematical Society Lecture Note Series, 242 and 243. Cambridge University Press, Cambridge, 1997.

[Sou] Soulé, C., On higher p-adic regulators, Algebraic K-theory, Evanston 1980 (Proc. Conf., Northwestern Univ., Evanston, Ill., 1980), pp. 372–401, Lecture Notes in Math., 854, Springer, Berlin-New York, 1981.

[Te] Terasoma, T., Mixed Tate motives and multiple zeta values, Invent. Math. 149 (2002), no. 2, 339–369.

[Tsu] Tsunogai, H., On ranks of the stable derivation algebra and Deligne's problem, Proc. Japan Acad. Ser. A Math. Sci. 73 (1997), no. 2, 29–31.

[Wo] Wojtkowiak, Z., Monodromy of iterated integrals and non-abelian unipotent periods, 219–289, in **[SL]** vol 2.

[Za] Zagier, D., Values of zeta functions and their applications, First European Congress of Mathematics, Vol. II (Paris, 1992), 497–512, Progr. Math., 120, Birkhäuser, Basel, 1994.

Graduate School of Mathematics, Nagoya University, Chikusa-ku, Furo-cho, Nagoya, 464-8602, Japan

E-mail address: `furusho@math.nagoya-u.ac.jp`

Contemporary Mathematics
Volume **416**, 2006

Asymptotics of q-difference equations

Stavros Garoufalidis and Jeffrey S. Geronimo

ABSTRACT. In this paper we develop an asymptotic analysis for formal and actual solutions of q-difference equations, under a regularity assumption, namely the non-collision and non-vanishing of the eigenvalues. In particular, evaluations of regular solutions of regular q-difference equations have an exponential growth rate which can be computed from the q-difference equation.

The motivation for the paper comes from a problem in Quantum Topology, the Hyperbolic Volume Conjecture, which states that a sequence on Laurent polynomials (the so-called colored Jones function of a knot), appropriately evaluated, becomes a sequence of complex numbers that grows exponentially. Moreover, the exponential growth rate is proportional to the volume of the knot complement.

The connection of the Hyperbolic Volume Conjecture with the paper comes from the fact that the colored Jones function of a knot is a solution of a q-difference equation, as was proven by TTQ. Le and the author.

CONTENTS

The authors were supported in part by the National Science Foundation.

1991 *Mathematics Classification.* Primary 57N10. Secondary 57M25.
Key words and phrases: knots, q-difference equations, asymptotics, colored Jones function, Hyperbolic Volume Conjecture, ϵ-difference equations, Birkhoff, Trjitzinsky, WKB, approximation schemes, recursion relations.

1. Introduction

1.1. The goal. The goal of the paper is to intiate an approach to the Hyperbolic Volume Conjecture, via asymptotics of solutions of difference equations with a small parameter. The Generalized Volume Conjecture links (conjecturally) the (colored) Jones polynomial of a knot to hyperbolic geometry of its complement.

Since the colored Jones polynomial is a specific solution to a linear q-difference equation, it follows that the generalized volume conjecture is the WKB limit of a specific solution of a linear difference equation with a small parameter.

Motivated by this, we study WKB asymptotics of formal and actual solutions of difference equations with a small parameter, under certain regularity asymptions.

1.2. The colored Jones function. A knot in 3-space is a smooth embedding of a circle, considered up to isotopy. Two of the simplest knots, the Trefoil (3_1) and the Figure Eight (4_1) are shown here:

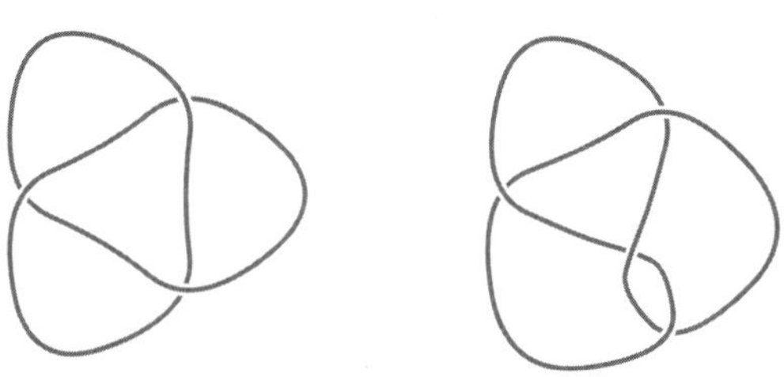

By the very definition, knots are flexible objects defined up to isotopy, which allows the embedding to move in a smooth and arbitrary way as long as it does not cross itself. In algebraic topology, a common way of studying knots (and more generally, spaces) is to associate computable numerical invariants (such as Euler characteristic, or Homology). Invariants are useful in deciding whether two knots are not the same. It is a much harder problem to construct computable invariants that separate knots.

The invariant that we will consider in this paper is the Jones polynomial of a knot; [**J**], which is a Laurent polynomial with integer coefficients, associated to each knot. The quantum nature of the Jones polynomial is apparent both in the original definition of Jones (using Temperley-Lieb algebras) and in the reformulation, due to Witten, in terms of the expectation value of a Quantum Field Theory; see [**J, Wt**].

The combinatorics associated to a planar projection of a knot show that the Jones polynomial is a computable invariant. However, it is hard to see from this point of view the relation between the Jones polynomial and Geometry. In Quantum Field Theory, one often reproduces Geometry by moving carefully chosen parameters of the theory to an appropriate limit.

In our case, we will introduce a new parameter, a natural number which roughly speaking corresponds to taking a connected n-parallel of a knot. The resulting invariant is no longer a Laurent polynomial, but rather a sequence of Laurent polynomials.

The *colored Jones function* of a knot K in 3-space is a sequence of Laurent polynomials

$$J_K : \mathbb{N} \longrightarrow \mathbb{Z}[q^{\pm}].$$

The first term in the above sequence, $J_K(1)$ is the Jones polynomial of K; see [**GL1**].

1.3. The Hyperbolic Volume Conjecture. Although knots are flexible objects, Thurston had the idea that their complements have a unique decomposition in pieces of unique "crystaline" shape. The shapes in question are the 8 different geometries in dimension 3, and the idea in question was termed the "Geometrization Conjecture". The most common of the 8 geometries is Hyperbolic Geometry, that is the existence of a complete, finite volume, constant curvature -1 Riemannian metric on knot complements. Thurston proved that unless the knot is torus or a satellite, then it carries a unique such metric; see [**Th**].

The *Hyperbolic Volume Conjecture* (HVC, in short) connects two very different views of knot: namely Quantum Field Theory and Riemannian Geometry. The HVC states for every hyperbolic knot K

$$\lim_{n \to \infty} \frac{\log |J_K(n)(e^{\frac{2\pi i}{n}})|}{n} = \frac{1}{2\pi} \operatorname{vol}(S^3 - K).$$

where $\operatorname{vol}(S^3 - K)$ is the *volume* of a complete hyperbolic metric in the knot complement $S^3 - K$. The conjecture was formulated in this form by Murakami-Murakami [**MM**] following an earlier version due to Kashaev, [**K**]. More generally, Gukov (see [**Gu**]) formulated a Generalized Hyperbolic Volume Conjecture that identifies the limit

$$\lim_{n \to \infty} \frac{\log |J_K(n)(e^{\frac{2\pi i \alpha}{n}})|}{n}$$

of a hyperbolic knot with known hyperbolic invariants (such as the volume of cone manifolds obtained by hyperbolic Dehn filling), for $\alpha \in (0,1] - \mathbb{Q}$ or $\alpha = 1$. Actually, the GHVC is stated for *complex* numbers α. For simplicity, we will study asympotics for real $\alpha \in [0,1]$.

At present, it is not known whether the limit in the HVC exists, let alone that it can be computed. Explicit finite multisum formulas for the colored Jones function of a knot exist; see for example [**GL1**]. From these formulas alone, it is difficult to study the above limit. In a sense, the question is to understand the sequence of Laurent polynomials that appears in the HVC. If the sequence is in some sense random, then it is hard to expect that the limit exists, or that it can be computed.

Since the first term of this sequence is the Jones polynomial, and since we know little about the possible values of the Jones polynomial, one would expect that there is even less to be said about the colored Jones function.

1.4. q-difference equations. Luckily, the colored Jones function behaves in a better way than its first term, namely the Jones polynomial. This can be quantified by recent work of TTQ Le and the first author, who proved that the colored Jones function of a knot satisfies a q-difference equation.

In other words, for every knot K there exist rational functions

$$b_1(u,v),\dots,b_d(u,v)\in\mathbb{Q}(u,v)$$

(which of course depend on K) such that for all $n\in\mathbb{N}$ we have:

$$\sum_{j=0}^{d} b_j(q^n,q)J_K(n+j)=0.$$

This opens the possibility of studying the q-difference equation rather than one of its solutions, namely the colored Jones function. Although the q-difference equation is not unique, it was shown by the first author in [**Ga1**] that one can choose a unique q-difference equation, which is a knot invariant. Moreover, it was conjectured in [**Ga1**] that the characteristic polynomial of this q-difference equation determines the characters of $\mathsf{SL}_2(\mathbb{C})$ representations of the knot complement, viewed from the boundary.

As was explained by the first author on several occasions, asymptotics of solutions of q-difference equations would have consequences on the HVC.

In this introductory article we review the history of asymptotics of solutions of q-difference equations.

1.5. Asymptotics of differential equations with a parameter. Excellent references for differential equations with a parameter are Olver's and Wasau's books; [**O**] and [**Wa**]. In 1837, Liouville and Green independently studied systematically existence of formal (i.e., perturbative) and actual solutions for second order differential equations with a parameter; see [**Gr, L**]. Second order equations are very important for classical and quantum physics.

In 1908 Birkhoff had the insight to introduce and study *arbitrary* order differential equation with a parameter (see [**B1**]):

$$y^{(n)}+\rho a_{n-1}(x,\rho)y^{(n-1)}+\cdots+\rho^n a_0(x,\rho)y=0 \tag{1}$$

where $y=y(x,\rho)$, $y^{(n)}$ means n-th derivative with respect to x (assumed to be restricted to a real interval), and ρ is a large complex parameter, and where the coefficient $a_j(x,\rho)$ are complex C^∞ functions with an expansion

$$a_j(x,\rho)=\sum_{s=0}^{\infty}a_{j,s}(x)\rho^{-s}$$

Birkhoff's working assumption was that the eigenvalues $\lambda_1(x), \dots, \lambda_n(x)$ of the characteristic equation

$$\lambda^n + a_{n-1,0}(x)\lambda^{n-1} + \cdots + a_{0,0}(x) = 0$$

were distinct but not necessarily nowhere vanishing.

In 1926, three theoretical physicists, Wentzel-Krammer-Brillouin studied the second order differential equation (1) under the assumption that its eigenvalues do not collide, and developed connection formulas linking solutions in the exponential region with those in the oscillatory region. Their method is often referred to as the WKB method.

1.6. Asymptotics of difference equations. As a motivation for our results, let us recall some fundamental results of Birkhoff and Trjitzinsky from 1930 on difference equations without a parameter; see [**B2**] and [**BT**].

A *difference equation* for a discrete function $f : \mathbb{N} \to \mathbb{C}$ has the form:

$$\sum_{j=0}^{d} a_j(n) f(n+j) = 0 \tag{2}$$

where $a_j : \mathbb{N} \longrightarrow \mathbb{C}$ are discrete functions so that $a_0(n)a_d(n) \neq 0$ for all $n \in \mathbb{N}$. We will assume the existence of asymptotic expansions of $a_j(n)$ around $n \to \infty$ for all $j = 1, \dots, d$:

$$a_j(n) \sim_{n\to\infty} n^{d_j/\omega}(a_{j,0} + a_{j,1}n^{-1/\omega} + a_{j,2}n^{-2/\omega} + \dots)$$

where $\omega \in \mathbb{N}$. This certainly holds for $\omega = 1$ if a_j are rational functions of n, as is often the case in combinatorial problems.

Due to the nowhere vanishing of $a_d \cdot a_0$, it follows that the set of solutions of (2) is a vector space of dimension d.

There are two main problems of difference equations:

- Existence of *formal series solutions* $\tilde{\psi}_1, \dots, \tilde{\psi}_d$ to (2).
- Existence of a basis $\{\psi_1, \dots, \psi_d\}$ of solutions so that $\psi_k(n)$ is asymptotic to $\tilde{\psi}_k(n)$ for large n.

In [**B2**], Birkhoff solved the existence of formal solutions in complete generality (that is, without any assumptions on the eigenvalues of the characteristic equation). In a sequel paper [**BT**], Birkhoff-Trjitzinsky solved the second problem in complete generality.

Among other things, the formal solutions of Birkhoff lead to the development of differential Galois theory, see [**vPS**].

Decades later, the results of Birkhoff and Trjitzinsky on difference equations have found applications to enumerative combinatorics and numerical analysis; see for example Wimp and Zeilberger in [**Wi, WZ**] and references therein. It is not surprising that difference equations are used in numerical analysis, since difference equations are numerical schemes of approximating differential equations. In enumerative combinatorics and complexity theory, difference equations appear in recursive computation. For example, the number $f(n)$ of involutions of $\{1, 2, \dots, n\}$ (that is, permutations which are a product of 1 and 2-cycles) is given by

$$f(n+2) = f(n+1) + (n+1)f(n)$$

with $f(1)=1$, $f(2)=2$. Using the results of Birkhoff-Trjitzinsky and the fact that $f(n)$ is monotone, it follows that

$$f(n) \sim_{n\to\infty} K n^{n/2} e^{-n/2+n^{1/2}} \left(1+\frac{c_1}{n^{1/2}}+\frac{c_2}{n}+\frac{c_3}{n^{3/2}}+\dots\right)$$

for nonzero constants c_i and some $K>0$. Actually, the c_k can be computed recursively from the difference equation; see [**WZ**, p.169].

1.7. Asymptotics of difference equations with a parameter. By some historical accident, asymptotics of solutions of difference equations with a parameter was not discussed a century ago. The first paper that discusses second order difference equation with a parameter appears to be due to Deift-McLaughlin (see [**DM**]) which was generalized by Costin-Costin to arbitrary order difference equations, [**CC**].

The purpose in this paper is to show that for regular q-difference equations, a regular solution has a well-defined and computable exponential growth rate in terms of a relative entropy of the characteristic polynomial of the q-difference equation; see Theorem 1 below.

This subject is classical and has been reinvented over the past hundred years by several groups, often unaware of each others results. In a sense, the problem of formal solutions of q-difference equations is a problem in differential Galois theory; [**vPS**], and a problem in numerical analysis; see for example [**CC**].

Our results are hardly new and are contained or can be obtained by minor modifications from results of Costin-Costin or from work of Birkhoff and collaborators, [**B1, BT, CC**].

Since the presentation in the above papers varies by time and taste, we have decided to give a self-contained account of the theory with complete proofs. Hopefully, this will benefit the researchers in Quantum Topology and in Analysis.

1.8. Statement of the results. In this paper, we will describe asymptotics of solutions of q-difference equations.

A *q-difference equation* for a sequence $(f(1), f(2), f(3), \dots)$ of smooth functions of q has the form:

$$\sum_{j=0}^{d} b_j(q^k, q) f(k+j, q) = 0 \tag{3}$$

where $b_j(v,u)$ are smooth functions and $f(k,q)=f(k)(q)$.

Before we proceed further, let us remark that q is a *variable* in (3) and not a complex number of absolute value less (or more) than 1. In the usual analytic theory of q-difference equations, q is a complex number inside or outside the unit disk.

Moreover, in the GHVC, we need to compute the nth term $f(n,q)$ in the above q-difference equation, and then evaluate it at $q_n = e^{2\pi i\alpha/n}$, for α fixed. In other words, in the GHVC, q_n is a complex number that varies with n in such a way that it stays in the unit circle and approaches 1 as $n\to\infty$.

With this in mind, ϵ-difference equations (defined below) are obtained from q-difference equations by the substitution $q=e^{2\pi i\epsilon}$ where ϵ is a small nonnegative real number, that plays the role of Planck's constant.

The *characteristic polynomial* of the q-difference equation (3) is

$$P(v,\lambda)=\sum_{j=0}^{d} b_j(v,1)\lambda^j$$

DEFINITION 1.1. We will say that (3) is *regular* if

$$\mathrm{Dsc}_\lambda P(v,\lambda)\cdot b_0(v,1)\cdot b_d(v,1)\neq 0$$

for all $v\in S^1$, where $\mathrm{Dsc}_\lambda P(v,\lambda)$ is the *discriminant* of $P(v,\lambda)$, which is a polynomial in the coefficients of $P(v,\lambda)$.

Let $\lambda_1(v),\dots,\lambda_d(v)$ denote the roots of the characteristic polynomial, which we call the *eigenvalues* of (3). It turns out that (3) is regular iff the eigenvalues $\lambda_1(v),\dots,\lambda_d(v)$ never collide and never vanish, for every $v\in S^1$. Moreover, it follows by the implicit function theorem that the roots are smooth functions of $v\in S^1$.

Since we are interested in asymptotics of solutions of q-difference equations which, as we shall see, are governed by the magnitude of the eigenvalues, we need to partition the circle according to the magnitudes of the eigenvalues.

Let $S^1=\cup_{p\in\mathcal{P}}I_p$ denote a partition of S^1 into a finite union of closed arcs (with nonoverlapping interiors), such that the magnitude of the eigenvalues does not change in each arc. In other words, for each $p\in\mathcal{P}$, there is a permutation σ_p of the set $\{1,\dots,d\}$ such that

$$|\lambda_{\sigma_p(1)}(v)|\geq|\lambda_{\sigma_p(2)}(v)|\geq\cdots\geq|\lambda_{\sigma_p(d)}(v)|\qquad\text{for all}\qquad v\in I_p.$$

The following definition introduces a locally fundamental set of solutions of q-difference equations.

DEFINITION 1.2. Fix a partition of I as above. A set $\{\psi_1,\dots,\psi_d\}$ is a *locally fundamental set of solutions* of (3) iff for every solution ψ for every $p\in\mathcal{P}$ and for every $m=1,\dots,d$ there exist smooth functions c^p_m such that

$$\psi(k,q)=c^p_1(q)\psi_{\sigma_p(1)}(k,q)+\cdots+c^p_d(q)\psi_{\sigma_p(d)}(k,q)\tag{4}$$

for all (k,q) such that $q^k\in I_p$.

THEOREM 1. *Assume that (3) is regular. Then, there exists a locally fundamental set of solutions $\{\psi_1,\dots,\psi_d\}$ such that*

- *For every $m=1,\dots,d$ and (k,q) such that $q^k\in I_p$ we have*

$$\psi_m\left(k,e^{\frac{2\pi i\alpha}{n}}\right)=\exp\left(\frac{n}{\alpha}\Phi_m\left(\frac{k\alpha}{n},\frac{\alpha}{n}\right)\right).$$

- *for some smooth functions Φ_m with uniform (with respect to $x\in I=[0,1]$)* asymptotic expansion

$$\Phi_m(x,\epsilon)\sim_{\epsilon\to 0}\sum_{s=0}^{\infty}\phi_{m,s}(x)\epsilon^s$$

where $\phi_{m,s}\in C^\infty(I)$ for all s

- *and leading term*

$$\phi_{m,0}(x)=\int_0^x\log(\lambda_m(e^{2\pi it}))dt\tag{5}$$

where we have chosen a branch for the logarithm of λ_m.

REMARK 1.3. For every $j = 1, \dots, d$ the smooth functions $\phi_{j,s}$ for positive s are uniquely determined from the coefficients $b_j(u, v)$ of (3) by a hierarchy of first-order differential equations along with specified initial conditions. On the other hand, the smooth functions Φ_m are not uniquely determined, since they are obtained by a smooth interpolation. Thus, the locally fundamental set of solutions is not uniquely determined from the q-difference equation, although its asymptotic behavior is.

It follows from Theorem 1 that each locally fundamental solution $\psi_m(n, q)$ of the q-difference equation (3) satisfies the GHVC in the sense that for every $\alpha \in [0, 1]$ we have:

$$\lim_{n\to\infty} \frac{\log|\psi_m(n, e^{\frac{2\pi i\alpha}{n}})|}{n} = \int_0^1 \log|\lambda_m(e^{2\pi i\alpha t})|dt$$

Fix a solution ψ of (3). Theorem 1 expresses ψ as a linear combination of ψ_m's in each arc I_p. For every $p \in P$, let

$$S_p = \{m \in \{1, \dots, d\} | \; c_m^p \neq 0\}. \tag{6}$$

Later (in Section 6.3) we will define the notion of a regular solution to a q-difference equation.

As a prototypical example, consider an q-difference equation that satisfies $|\lambda_1(v)| > |\lambda_j(v)|$ for all $j \neq 1$ and all $v \in S^1$. Then, any solution that satisfies $c_1(0) \neq 0$ (or more generally, c_1 has a nonvanishing derivative at 0) is regular.

REMARK 1.4. It is possible that $S_p \neq S_{p+1}$. In other words, the restriction of a fixed solution ψ to different intervals I_p may be a linear combination of different ψ_js. This is an important phenomenon, referred by the name of *Stokes phenomenon*; see [**Wa**].

Our next definition captures the growth rate of regular solutions to regular q-difference equations.

DEFINITION 1.5. Fix a collection $S = \{S_p | p \in \mathcal{P}\}$ of subsets of $\{1, \dots, d\}$. The *S-entropy*

$$\sigma_S : [0, 1] \to \mathbb{R}$$

of the q-difference equation (3) is defined by

$$\sigma_S(\alpha) = \int_0^1 \log \chi_S(e^{2\pi i\alpha t})dt.$$

where $\chi_S : [0, 1] \to \mathbb{R}$ is defined by

$$\chi_S(v) = \max_{j\in S_p} |\lambda_{\sigma_p(j)}(v)| \qquad \text{if} \qquad v \in I_p.$$

The *entropy* of (3) is the set of functions

$$\{\sigma_S : [0, 1] \to \mathbb{R} \quad | \quad S \subset \{1, \dots, d\}\}.$$

Notice that the entropy of a q-difference equation is not a real number, but rather a finite collection of functions.

The main result is the following

THEOREM 2. *If f is an S-regular solution of the regular q-difference equation* (3), *then for every $\alpha \in [0, 1]$ we have:*

$$\lim_{n\to\infty} \frac{\log|f(n)(e^{\frac{2\pi i\alpha}{n}})|}{n} = \sigma_S(\alpha)$$

Finally, let us define the J-entropy of a knot. In [**Ga1**] the first author showed that to every knot K one can associate a canonical q-difference equation of degree d, and a specific solution of it, namely the colored Jones function of K.

The q-difference equation itself is an invariant of a knot, which (by definition) is determined by the colored Jones function of the knot. Thus, any invariant of the q-difference equation is also an invariant of a knot, which is determined by the colored Jones function of the knot.

DEFINITION 1.6. The J-entropy of a knot is the entropy of its associated q-difference equation. We denote the J-entropy of a knot K by

$$\{\sigma^J_{S,K} : [0,1] \to \mathbb{R} \quad | \quad S \subset \{1, \dots, d\}\}.$$

1.9. What's next? The paper was written in the spring of 2004. Since then, a number of papers that discuss the asymptotics of the colored Jones function have appeared; see [**Ga2, Ga3, GL2, GL3**].

1.10. Acknowledgement. The results of this paper were announced in the JAMI 2003 meeting in Johns Hopkins. The first author wishes to thank J. Morava for the invitation, and P. Deligne who suggested the asymptotic behavior of solutions of q-difference equations. The first author wishes to thank D. Boyd for sharing and explaining his unpublished work and also A. Riese, T. Morley, and D. Zeilberger.

2. ϵ-difference equations

2.1. ϵ-difference equations. In this section, we will translate asymptotics of solutions of q-difference equation in terms of asymptotics of solutions of *ϵ-difference equations.* The latter are defined as follows.

Fix a positive number ϵ_0, a compact interval I of $\mathbb{R}$ and a natural number d. We will consider functions $\phi : \Delta_{\epsilon_0,I} \to \mathbb{C}$ with domain

$$\Delta_{\epsilon_0,I} := \{(k\epsilon, \epsilon) | \quad k \in \mathbb{N}, \epsilon \in (0, \epsilon_0], \quad k\epsilon \in I\}. \tag{7}$$

Consider the ϵ-difference equation for a function $\phi : \Delta_{\epsilon_0,I} \to \mathbb{C}$

$$\sum_{j=0}^{d} a_j(k\epsilon, \epsilon)\phi((k+j)\epsilon, \epsilon) = 0 \tag{8}$$

where $a_j \in C^\infty(I \times [0, \epsilon_0])$.

We will assume that for all $j = 0, \dots, d$, $a_j(x, \epsilon)$ has a uniformly (with respect to x) asymptotic expansion

$$a_j(x, \epsilon) \sim_{\epsilon \to 0} \sum_{s=0}^{\infty} a_{j,s}(x)\epsilon^s \tag{9}$$

where $a_{j,s} \in C^\infty(I)$.

As we mentioned before, ϵ-difference equations are obtained from q-difference equations by the substitution $q = e^{2\pi i\epsilon}$ where ϵ is a small nonnegative real number, that plays the role of Planck's constant.

The *characteristic polynomial* of (8) is

$$P(x, \lambda) = \sum_{j=0}^{d} a_j(x, 0)\lambda^j$$

DEFINITION 2.1. We will say that (8) *regular* if

$$\mathrm{Dsc}_\lambda P(x,\lambda)\cdot a_0(x,0)\cdot a_d(x,0)\neq 0$$

for all $x\in I$.

Let $\lambda_1(x),\dots,\lambda_d(x)$ denote the roots of the characteristic polynomial, which we call the *eigenvalues* of (8).

It turns out that (8) is regular iff the eigenvalues $\lambda_1(x),\dots,\lambda_d(x)$ never collide and never vanish, for every $x\in I$. Moreover, it follows by the implicit function theorem that the roots are smooth functions of $x\in I$.

Since we are interested in asymptotics of solutions of ϵ-difference equations which, as we shall see, are governed by the magnitude of the eigenvalues, we need to partition the interval I according to the magnitudes of the eigenvalues.

Let $I=\cup_{p\in\mathcal{P}}I_p$ denote a partition of I into a finite union of closed intervals (with nonoverlapping interiors), such that the magnitude of the eigenvalues does not change in each interval. In other words, for each $p\in\mathcal{P}$, there is a permutation σ_p of the set $\{1,\dots,d\}$ such that

$$|\lambda_{\sigma_p(1)}(x)|\geq|\lambda_{\sigma_p(2)}(x)|\geq\cdots\geq|\lambda_{\sigma_p(d)}(x)|\qquad\text{for all}\qquad x\in I_p.$$

The following definition introduces a locally fundamental set of solutions of ϵ-difference equations.

DEFINITION 2.2. Fix a partition of I as above. A set $\{\psi_1,\dots,\psi_d\}$ is a *locally fundamental set of solutions* of (8) iff for every solution $\psi:\Delta_{\epsilon,I}\to\mathbb{C}$, for every $p\in\mathcal{P}$ and for every $m=1,\dots,d$ there exist smooth functions $c^p_m\in C^\infty[0,\epsilon]$ such that

$$\psi(k\epsilon,\epsilon)=c^p_1(\epsilon)\psi_{\sigma_p(1)}(k\epsilon,\epsilon)+\cdots+c^p_d(\epsilon)\psi_{\sigma_p(d)}(k\epsilon,\epsilon)$$

for all $(k\epsilon,\epsilon)\in\Delta_{\epsilon,I}$. Here, the notation c^p_m does not indicate the pth power of c_m.

The next theorem summarizes the results of Costin-Costin.

THEOREM 3. *(*[**CC**]*) Assume that* (8) *is regular. Then, there exists a positive* $\epsilon'\leq\epsilon_0$ *and a locally fundamental set of solutions* $\{\psi_1,\dots,\psi_d\}$ *of* (8) *such that*

- *For every* $m=1,\dots,d$ *and* $(k\epsilon,\epsilon)\in\Delta_{e',I}$ *we have*

$$\psi_m(k\epsilon,\epsilon)=\exp\left(\epsilon^{-1}\Phi_m(k\epsilon,\epsilon)\right).$$

- *for some smooth functions* $\Phi_m\in C^\infty(I\times[0,\epsilon'])$ *with uniform (with respect to* $x\in I$*)* asymptotic expansion

$$\Phi_m(x,\epsilon)\sim_{\epsilon\to 0}\sum_{s=0}^\infty\phi_{m,s}(x)\epsilon^s \tag{10}$$

where $\phi_{m,s}\in C^\infty(I)$ *for all* s
- *and leading term*

$$\exp(\phi'_{m,0}(x))=\lambda_m(x). \tag{11}$$

Fix a solution ψ of (8). Theorem 3 expresses ψ as a linear combination of the ψ_j's in each interval I_p. For every $p\in P$, let

$$S_p=\{m\in\{1,\dots,d\}|\ c^p_m\neq 0\}. \tag{12}$$

Later (in Section 6.2) we will define the notion of a regular solution to an ϵ-difference equation.

As a prototypical example, consider an ϵ-difference equation that satisfies $|\lambda_1(x)| > |\lambda_j(x)|$ for all $j \neq 1$ and all $x \in I = [a, b]$. Then, any solution that satisfies $c_1(a) \neq 0$ (or more generally, c_1 has a nonvanishing derivative at a) is regular.

Our next definition captures the growth rate of regular solutions to regular ϵ-difference equations.

DEFINITION 2.3. Fix a collection $S = \{S_p | \, p \in \mathcal{P}\}$ of subsets of $\{1, \ldots, d\}$. The *S-entropy*

$$\sigma_S : I \to \mathbb{R}$$

of the ϵ-difference equation (8) is defined by

$$\sigma_S(x) = \int_0^x \log \chi_S(t) dt.$$

where $\chi_S : I \to \mathbb{R}$ is defined by

$$\chi_S(x) = \max_{j \in S_p} |\lambda_{\sigma_p(j)}(x)| \qquad \text{if} \qquad x \in I_p.$$

THEOREM 4. *If ψ is an S-regular solution to a regular ϵ-difference equation, and $x \in I$, we have:*

$$\lim_{\epsilon \to 0^+} \epsilon \log |\psi(x, \epsilon)| = \sigma_S(x)$$

The next remarks concern the uniqueness of a set of locally fundamental solutions to (8).

REMARK 2.4. For every $m = 1, \ldots, d$ the smooth functions $\phi_{m,s}$ for positive s are uniquely determined by (3) and the initial condition $\phi_{m,s}(0) = 0$. Indeed, applying Taylor series (with respect to ϵ) in (8) and collecting terms, we get for example:

$$\begin{aligned} \phi'_{m,1}(x) &= -\frac{1/2\phi''_{m,0}(x) \sum_{j=0}^d a_j(x,0) j^2 \lambda_m^j(x) + \sum_{j=0}^d \partial_\epsilon a_j(x,0) \lambda_m^j(x)}{\sum_{j=0}^d a_j(x,0) j \lambda_m^j(x)} \\ &= -\left(\frac{1}{2}\lambda' \frac{P_{\lambda\lambda}}{P_\lambda} + \frac{P_\epsilon}{P_\lambda \lambda}\right) |_{\lambda = \lambda_m(x)} \end{aligned}$$

where $f_x(x, \lambda) = \partial/\partial_x f(x, \lambda)$ and $f_\lambda(x, \lambda) = \partial/\partial_\lambda f(x, \lambda)$.

Similarly, for $s \geq 1$ we have:

$$\phi'_{m,s}(x) = -\frac{H_s(x)}{\sum_{j=0}^d a_j(x,0) j \lambda_m^j(x)}$$

where $H_s(x)$ is a function of derivatives of $a_j(x, 0)$ and $\phi_{m,t}$ for $t < s$. Notice that the denominator vanishes nowhere since the roots do not collide and do not vanish for every $x \in I$.

REMARK 2.5. If the coefficients $a_j(x, \epsilon)$ of the regular ϵ-difference equation (8) are analytic functions, then the functions $\phi_{m,s}$ of Theorem 3 are also analytic, for every m and s. This follows by induction from the differential hierarchy which these functions satisfy, and from the fact that the eigenvalues are analytic functions. Even though $\phi_{m,s}$ is analytic for every m and s, the series

$$\sum_{s=0}^{\infty} \phi_{m,s}(x) \epsilon^s$$

is in general divergent, and the functions $\Phi_{m,s}$ of Theorem 3 are not analytic.

REMARK 2.6. Even though the functions $\phi_{m,s}$ are uniquely determined by the ϵ-difference equation, the smooth functions F_m (and thus the locally fundamental set of solutions ψ_m) are not uniquely determined by the ϵ-difference equation. The problem is that smooth interpolation is not unique. Recently developed ideas of exponentially small corrections might construct a unique set of locally fundamental solutions when the coefficients of (8) are analytic functions. We will elaborate on this in a separate occasion.

2.2. Converting q-difference equations to ϵ-difference equations. The translation of q-difference equations to ϵ-difference equations is as follows. If f satisfies the q-difference equation

$$\sum_{j=0}^{d} b_j(q^k, q) f(k+j, q) = 0$$

then set

$$q = e^{2\pi i \epsilon}, \quad b_j(e^{2\pi i x}, e^{2\pi i \epsilon}) = a_j(x, \epsilon)$$

and consider the ϵ-difference equation for a function ϕ (with domain $\Delta_{\epsilon_0, I}$ for some $\epsilon_0 > 0$ and $I = [0, 2\pi]$):

$$\sum_{j=0}^{d} a_j(k\epsilon, \epsilon)\phi((k+j)\epsilon, \epsilon) = 0$$

The following lemma, although elementary, is the key to translating q-difference equations to ϵ-difference equations.

LEMMA 2.7. *For every $(k\epsilon, \epsilon) \in I \times [0, \epsilon_0]$ we have:*

$$\phi(k\epsilon, \epsilon) = f(k, e^{2\pi i \epsilon}). \tag{13}$$

Consequently, for every $\alpha \in (0, 1]$, we have:

$$\lim_{k \to \infty} \frac{1}{k} \log |f(k, e^{2\pi i \alpha / k})| = \alpha^{-1} \lim_{\epsilon \to 0} \epsilon \log |\phi(\alpha, \epsilon)|. \tag{14}$$

Thus, Theorem 3 implies Theorem 1.

PROOF. Observe that $a_j(k\epsilon, \epsilon) = b_j(e^{2\pi i k \epsilon}, e^{2\pi i \epsilon})$, thus $(k, \epsilon) \to \phi(k\epsilon, \epsilon)$ satisfies the equation

$$\sum_{j=0}^{d} b_j(e^{2\pi i k \epsilon}, e^{2\pi i \epsilon})\phi((k+j)\epsilon, \epsilon) = 0$$

and so does $(k, \epsilon) \to f(k, e^{2\pi i \epsilon})$. Since solutions with matching initial conditions are unique, (13) follows.

Equation (14) follows from equation (13) by the substitution $\epsilon = \alpha/k$:

$$\frac{1}{k} \log |f(k, e^{2\pi i \alpha / k})| = \frac{1}{k} \log |\phi(k\epsilon, \alpha/k)| = \alpha^{-1}\epsilon \log |\phi(\alpha, \epsilon)|.$$

□

3. Some linear algebra

In this section we will review some linear algebra. It is obvious that the complex roots of a monic polynomial uniquely determine it. It is also known [**GLR**] that the eigenvalues of a companion matrix uniquely determine it, in case they are distinct.

Consider a *companion* d by d matrix

$$A = \begin{pmatrix} 0 & 1 & 0 & \dots & 0 \\ 0 & 0 & 1 & \dots & 0 \\ 0 & 0 & 0 & \dots & 0 \\ 0 & 0 & 0 & \dots & \dots \\ \dots & \dots & \dots & \dots & 1 \\ -a_0 & -a_1 & -a_2 & \dots & -a_{d-1} \end{pmatrix}$$

The characteristic polynomial of A is

$$\lambda^d + \sum_{j=0}^{d-1} a_j \lambda^j$$

with roots $\lambda_1, \dots, \lambda_d$. Let $M = (\lambda_j^{i-1})_{i,j}$ be the Vandermonde matrix, and $D = \text{diag}(\lambda_1, \dots, \lambda_d)$ be the diagonal matrix with diagonal entries $\lambda_1, \dots, \lambda_d$.

LEMMA 3.1. *If a companion matrix has distinct eigenvalues, then with the above notation we have:*

$$A = MDM^{-1}$$

PROOF. Observe that $v_j = (1, \lambda_j, \dots, \lambda_j^{d-1})^T$ is an eigenvector of A with eigenvalue λ_j. Thus, $M = (v_1, \dots, v_d)$ and $AM = MD$. The result follows. □

Now, consider a companion matrix $A(u)$ whose entries in the bottom row are smooth functions in a variable u, with roots $\lambda_1(u), \dots, \lambda_d(u)$ which we assume are distinct for all u.

The next lemma is a key estimate for the norm of long products of slowly varying matrices. In the language of physics, $A(u)$ is the *transfer matrix* and $A(n) \dots A(2)A(1)$ is the *transition matrix*.

LEMMA 3.2. *Assume that the eigenvalues $\lambda_1(u), \dots, \lambda_d(u)$ of $A(u)$ are distinct for all u and*

$$\max_j \sup_u |\lambda_j(u)| \leq 1 + C\epsilon.$$

Then for $m \leq n$, $n\epsilon \in I$, we have

$$\|A(n)A(n-1)\dots A(m)\| \leq C'$$

PROOF. By Lemma 3.1, we have:

$$A(u) = M(u)D(u)M(u)^{-1}$$

If $m \leq n$, it follows that

$$\begin{aligned} A(n)A(n-1)\dots A(m) &= M(n)D(n)M(n)^{-1} \\ &\quad \cdot M(n-1)D(n-1)M(n-1)^{-1}\dots \\ &\quad \cdot M(m)D(m)M(m)^{-1}, \end{aligned}$$

which implies that

$$\begin{aligned}\|A(n)A(n-1)\dots A(m)\| &\leq \|M(n)\|M(m)^{-1}\| \cdot \|D(n)\|\dots\|D(m)\| \\ &\quad \|M(n)^{-1}M(n-1)\|\dots \\ &\quad \cdot\|M(m+1)^{-1}M(m)\|.\end{aligned}$$

Now, we have

$$\begin{aligned}\|D(k)\| &\leq 1+C\epsilon \qquad \text{for} \quad k=m,\dots,n \\ \|M(k)M(k-1)^{-1}\| &\leq 1+C'\epsilon \qquad \text{by Lemma 3.3.}\end{aligned}$$

If $I=[a,b]$, using the fact that $n\epsilon, m\epsilon \in I$, we obtain:

$$\begin{aligned}\|A(n)A(n-1)\dots A(m)\| &\leq (1+C\epsilon)^{n-m}(1+C'\epsilon)^{n-m} \\ &\leq (1+C''\epsilon)^{2(n-m)} \\ &\leq (1+C''\epsilon)^{2\frac{b-a}{\epsilon}} \\ &\leq e^{2C''(b-a)}.\end{aligned}$$

□

LEMMA 3.3. *If $M=(x_j^{i-1})_{i,j}$ and $N=(y_j^{i-1})_{i,j}$ are Vandermonde matrices, such that M is nonsingular, then*

$$(M^{-1}N)_{i,j} = \prod_{l\neq i} \frac{y_j - x_l}{x_i - x_l}.$$

4. Existence of formal solutions

In this section we will prove that (8) has a unique set of formal solutions. Let us define those first.

DEFINITION 4.1. A *formal series* $\tilde{\psi}(x,\epsilon)$ is one of the form

$$\tilde{\psi}(x,\epsilon) = \exp\left(\epsilon^{-1}\sum_{s=0}^{\infty}\phi_s(x)\epsilon^s\right) \tag{15}$$

where $\phi_s \in C^\infty(I)$ are smooth functions for all s.

Note that $\epsilon \log\tilde{\psi}(x,\epsilon)$ lies in the ring $C^\infty(I)[[\epsilon]]$ of formal power series with coefficients smooth functions on I.

Note further that if $\tilde{\psi}(x,\epsilon)$ is a formal series, so is $\tilde{\psi}(x+j\epsilon,\epsilon)$ for every $j\in\mathbb{Z}$, where the latter may be defined using the Taylor series of

$$\phi_s(x+j\epsilon) = \sum_{t=0}^{\infty}\frac{1}{t!}\phi_s^{(t)}(x)j^t\epsilon^t$$

as fommows:

$$\begin{aligned}\tilde{\psi}(x+j\epsilon,\epsilon) &= \exp\left(\epsilon^{-1}\sum_{s=0}^{\infty}\left(\sum_{t=0}^{s}\frac{1}{t!}\phi_{s-t}^{(t)}(x)\right)\epsilon^s\right) \\ &= \tilde{\psi}(x,\epsilon)\exp\left(\phi_0'(x)j + \left(\phi_1'(x)j+\frac{\phi_0''(x)}{2!}j^2\right)\epsilon\right. \\ &\quad \left.+\left(\phi_2'(x)j+\frac{\phi_1''(x)}{2!}j^2+\frac{\phi_0'''(x)}{3!}j^3\right)\epsilon^2+\dots\right)\end{aligned}$$

It follows that if $\tilde{\psi}(x,\epsilon)$ is a formal series, then the ratio $\tilde{\psi}(x+\epsilon,\epsilon)/\tilde{\psi}(x,\epsilon)$ lies in the ring $C^\infty(I)[[\epsilon]]$.

Using the language of *difference Galois theory* (see [**vPS**, p.4]) this implies that

LEMMA 4.2. *$C^\infty(I)[[\epsilon]]$ is a finite difference ring, under the map $x \to x+\epsilon$.*

DEFINITION 4.3. We say that a formal series $\tilde{\psi}$ of (15) is a *formal series solution* to (8) iff

$$\frac{1}{\tilde{\psi}(x,\epsilon)} \sum_{j=0}^{d} a_j(x,\epsilon)\tilde{\psi}(x+j\epsilon,\epsilon) = 0 \in C^\infty(I)[[\epsilon]]. \tag{16}$$

It is easy to see that if $\tilde{\psi}$ is a formal solution to (8), then the leading term ϕ_0 satisfies the equation

$$\exp(\phi_0'(x)) = \lambda(x) \tag{17}$$

where $\lambda(x)$ is an eigenvalue of (8).

PROPOSITION 4.4. *Assume that (8) is regular. Then, (8) has d unique formal series solutions $\tilde{\psi}_1,\dots,\tilde{\psi}_d$ with leading terms corresponding to the eigenvalues of (8).*

PROOF. First we need to show that (16) is indeed an equation in the power series ring $C^\infty(I)[[\epsilon]]$, i.e., that the terms involving negative powers of ϵ cancel.

Suppose that $\tilde{\psi}$ is given by (15). It follows from the calculation preceding Lemma 4.2 that for every $s \in \mathbb{N}$, we have:

$$\operatorname{coeff}\left(\epsilon^s, \frac{\tilde{\psi}(x+j\epsilon,\epsilon)}{\tilde{\psi}(x,\epsilon)}\right) = \begin{cases} \exp(j\phi_0'(x)) & \text{if } \; s=0 \\ j\exp(j\phi_0'(x))\phi_s'(x) + \text{terms}_s & \text{if } \; s>0 \end{cases} \tag{18}$$

where $\operatorname{coeff}(\epsilon^s, g(\epsilon))$ denotes the coefficient of ϵ^s in a power series $g(\epsilon)$, and where terms_s is a polynomial in the derivatives of ϕ_t for $t<s$.

Expand the terms of Equation (16) into power series in ϵ using the above equation and (9), and collect terms of powers of ϵ. It follows that (16) is equivalent to a hierarchy of first order differential equations:

$$\begin{aligned} \sum_{j=0}^{d} a_j(x,0)\exp(j\phi_0'(x)) &= 0 \\ \sum_{j=0}^{d} a_j(x,0) j \exp(j\phi_0'(x))\phi_s'(x) + \text{Terms}_s &= 0 \end{aligned}$$

where for positive s, Terms_s is a polynomial in the derivatives of ϕ_t and $a_j(x,0)$ for $t<s$.

Now fix an $m \in \{1,\dots,d\}$, and choose $\phi_{m,0}$ such that $\exp(\phi_{m,0}'(x)) = \lambda_m(x)$, where $\lambda_1(x),\dots,\lambda_d(x)$ are the eigenvalues of (8). Since (8) is regular, it follows that the roots $\lambda_1(x),\dots,\lambda_d(x)$ of the characteristic polynomial $P(x,\lambda)$ never collide, and never vanish for $x \in I$. Thus,

$$0 \neq \left(\lambda \frac{d}{d\lambda} P(x,\lambda)\right)_{\lambda=\lambda_m(x)} = \sum_{j=0}^{d} j a_j(x,0)\lambda_m^j(x)$$

for all $x \in I$. Thus, after we choose $\phi_{m,0}$, it follows that we can find functions $\phi_{m,s}$ for $s \geq 0$ that satisfy the above hierarchy. Moreover, for every m, the sequence of functions $\phi_{m,s}$ is uniquely determined by the above hierarchy and the initial conditions $\phi_{m,s}(0) = 0$. □

4.1. An alternative formal series. In this section we present an alternative, and slightly more general form, of formal series. In case of regular ϵ-difference equations this alternative form will not be needed. However, when eigenvalues collide or vanish, one must use this alternative form of formal series. Thus, in the present paper we will not use this alternative form of formal series, and the reader may skip this section.

DEFINITION 4.5. An *alternative formal series* $\tilde{\psi}(x,\epsilon)$ is one of the form

$$\tilde{\psi}(x,\epsilon) = \exp\left(\epsilon^{-1}\phi(x)\right)\sum_{s=0}^{\infty}\phi_s(x)\epsilon^s \tag{19}$$

where $\phi, \phi_s \in C^\infty(I)$ are smooth functions for all s, and $\phi_0(x) \neq 0$ for all $x \in I$.

In the remainder of this subsection, we will refer to alternative formal series simply by formal series.

Note that if $\tilde{\psi}(x,\epsilon)$ is a formal series, so is $\tilde{\psi}(x+j\epsilon,\epsilon)$ for any $j \in \mathbb{Z}$, where the latter may be defined using the Taylor series of $\phi_s(x+j\epsilon)$ and $\phi(x+j\epsilon)$ around x. It follows that

$$\begin{aligned}\tilde{\psi}(x+j\epsilon,\epsilon) &= \exp\left(\epsilon^{-1}\phi(x)\right)\\ &\cdot(\phi_0(x) + (\phi^{(1)}(x)\phi_0(x) + \phi_0^{(1)}(x) + \phi_1(x))\epsilon + \dots).\end{aligned}$$

Moreover, if $\tilde{\psi}(x,e)$ is a formal series, then the ratio $\tilde{\psi}(x+\epsilon,\epsilon)/\tilde{\psi}(x,\epsilon)$ lies in the ring $C^\infty(I)[[\epsilon]]$. This follows from (20) and the following computation, valid for every $j \in \mathbb{N}$:

$$\begin{aligned}\tilde{\psi}(x+j\epsilon,\epsilon) &= \exp(\epsilon^{-1}\phi(x+j\epsilon))(\phi_0(x+j\epsilon) + \phi_1(x+j\epsilon)\epsilon + O(\epsilon^2))\\ &= \exp\left(\epsilon^{-1}\phi(x) + \phi^{(1)}(x)j + \frac{\phi^{(2)}(x)}{2}j^2\epsilon + O(\epsilon^2)\right)\\ &\quad\cdot\left(\phi_0(x) + (\phi_0^{(1)}(x)j + \phi_1(x))\epsilon + O(\epsilon^2)\right)\\ &= \tilde{\psi}(x,\epsilon)\exp(\phi^{(1)}(x)j)\\ &\quad\cdot\left(1 + \left(\frac{\phi^{(2)}(x)}{2}j^2 + \frac{\phi_0^{(1)}(x)}{\phi_0(x)}\right)\epsilon + O(\epsilon^2)\right)\end{aligned}$$

In analogy with Lemma 4.2, this implies that

LEMMA 4.6. *$C^\infty(I)[[\epsilon]]$ is a finite difference ring, under the map $x \to x+\epsilon$.*

DEFINITION 4.7. We say that a formal series $\tilde{\psi}$ of (19) is a *formal series solution* to (8) iff

$$\frac{1}{\tilde{\psi}(x,\epsilon)}\sum_{j=0}^{d} a_j(x,\epsilon)\tilde{\psi}(x+j\epsilon,\epsilon) = 0 \in C^\infty(I)[[\epsilon]]. \tag{20}$$

It is easy to see that if $\tilde{\psi}$ is a formal solution to (8), then the leading term ϕ satisfies the equation

$$\exp(\phi'(x)) = \lambda(x) \tag{21}$$

where $\lambda(x)$ is an eigenvalue of (8).

In analogy with Proposition 4.4, we have the following:

PROPOSITION 4.8. *Assume that* (8) *is regular. Then,* (8) *has d unique formal series solutions $\tilde{\psi}_1, \dots, \tilde{\psi}_d$ with leading terms corresponding to the eigenvalues of* (8).

5. Proof of Theorem 3

In this section we prove Theorem 3. The strategy is to

(a) prove that there exists a solution ψ_1 with the stated properties where $\lambda_1(x)$ is an eigenvalue with maximum magnitude.
(b) use this solution ψ_1 to reduce Theorem 3 to the case of a ϵ-difference equation of degree one less than the original one.
(c) prove that the constructed set of solutions is a locally fundamental set.

Without loss of generality, we will assume that the eigenvalues of (8) satisfy the inequality:

$$|\lambda_1(x)| \geq |\lambda_2(x)| \geq \cdots \geq |\lambda_d(x)|$$

for all $x \in I$. Otherwise, we can partition I into subintervals where this is true.

5.1. Existence of a solution corresponding to the eigenvalue of maximum magnitude. Consider first a formal solution $\tilde{\psi}_1$ of (8) given in Proposition 4.4, which satisfies (15) and (17). Consider the smooth functions $\phi_{1,s} \in C^\infty(I)$ of (17).

The proof of the following lemma (due to Borel in case $\phi_{1,s}$ are constant functions, for all s) can be found in [**GG**, Lemma 2.5]:

LEMMA 5.1. *There exists a smooth function $\hat{\Phi}_1 \in C^\infty(I \times [0, \epsilon_0])$ such that we have (uniformly in $x \in I$):*

$$\hat{\Phi}_1(x, \epsilon) \sim_{\epsilon \to 0} \sum_{j=0}^{\infty} \phi_{1,s}(x)\epsilon^s.$$

Now, consider the unique solution ψ_1 of (8) with initial conditions

$$\psi_1(k\epsilon, \epsilon) = \exp(\epsilon^{-1}\hat{\Phi}_1(k\epsilon, \epsilon)) \quad \text{for} \quad k = 0, \dots, d-1$$

and for small enough ϵ, where without loss of generality, we assume that $I = [0, b]$.

Of course, for large k it may not be true that

$$\psi_1(k\epsilon, \epsilon) = \exp(\epsilon^{-1}\hat{\Phi}_1(k\epsilon, \epsilon)).$$

The next proposition estimates the error, uniformly with respect to k:

PROPOSITION 5.2. *There exists an $\epsilon' > 0$ and constants C_s such that for all $(k\epsilon, \epsilon) \in \Delta_{\epsilon', I}$ and all $s \in \mathbb{N}$, we have (uniformly in k):*

$$\left| \frac{\psi_1(k\epsilon, \epsilon)}{\exp(\epsilon^{-1}\hat{\Phi}_1(k\epsilon, \epsilon))} - 1 \right| < C_s \epsilon^s \tag{22}$$

PROOF. Let us make a change of variables:

$$\theta = \frac{\psi_1}{\hat{\psi}_1}, \tag{23}$$

where

$$\hat{\psi}_1(x,\epsilon) = \exp\left(\epsilon^{-1}\hat{\Phi}_1(x,\epsilon)\right).$$

We will show that for a fixed s_0, and for every s there exists a constant C_s such that for all $(k\epsilon, \epsilon) \in \Delta_{\epsilon_0, I}$ we have:

$$|\theta(k\epsilon,\epsilon) - 1| < C_s \epsilon^{s+1-s_0}, \tag{24}$$

Since ψ_1 satisfies (8), it follows that θ satisfies

$$\sum_{j=0}^{d} b_j(k\epsilon,\epsilon)\theta((k+j)\epsilon,\epsilon) = 0 \tag{25}$$

where

$$b_j(x,\epsilon) = a_j(x,\epsilon)\frac{\hat{\psi}_1(x+j\epsilon,\epsilon)}{\hat{\psi}_1(x,\epsilon)}.$$

It is easy to see that

- $b_j(x,\epsilon) \in C^\infty(I \times [0,\epsilon_0])$, has uniform (with respect to x) ϵ-asymptotic expansion as in (9),
- $b_{j,s}(x,0) = a_j(x,0)\lambda_1^j(x)$,
- and since $\tilde{\psi}$ is a formal series solution to (8) and $\hat{\Phi}_1$ is given by Lemma 5.1, if follows that for every s we have:

$$\sum_{j=0}^{d} b_j(x,\epsilon) = O(\epsilon^s) \tag{26}$$

The characteristic polynomial of (25) has roots $\mu_m(x) := \lambda_m(x)/\lambda_1(x)$ for $j = 1, \dots, d$. If (8) is regular, so is (25). Notice that $\lambda_1(x)$ vanishes nowhere since (25) is regular.

We now show (24). Let us write the difference equation (25) in matrix form

$$\Theta((k+1)\epsilon,\epsilon) = A(k\epsilon,\epsilon)\Theta(k\epsilon,\epsilon) \tag{27}$$

where

$$\Theta(x,\epsilon) = \begin{pmatrix} \theta(x,\epsilon) \\ \theta(x+\epsilon,\epsilon) \\ \theta(x+2\epsilon,\epsilon) \\ \dots \\ \dots \\ \theta(x+(d-1)\epsilon,\epsilon) \end{pmatrix}$$

and

$$A(x,\epsilon) = \begin{pmatrix} 0 & 1 & 0 & \dots & 0 \\ 0 & 0 & 1 & \dots & 0 \\ 0 & 0 & 0 & \dots & 0 \\ 0 & 0 & 0 & \dots & \dots \\ \dots & \dots & \dots & \dots & 1 \\ -c_0(x,\epsilon) & -c_1(x,\epsilon) & -c_2(x,\epsilon) & \dots & -c_{d-1}(x,\epsilon) \end{pmatrix}$$

and $c_j(x,\epsilon) = b_j(x,\epsilon)/b_d(x,\epsilon)$. Iterating, we obtain that

$$\Theta(k\epsilon,\epsilon) = A(k\epsilon,\epsilon)A((k-1)\epsilon,\epsilon)\dots A(\epsilon,\epsilon)\Theta(0,\epsilon)$$

for all $k \geq 1$, where $\Theta(0,\epsilon) = \mathbb{1}$, a column vector with all entries equal to 1.

Equation (26) gives:

$$A(k\epsilon,\epsilon)\mathbb{1} = \mathbb{1} + \epsilon^s \mathbb{E}_k(\epsilon) \tag{28}$$

where $\|\mathbb{E}_k(\epsilon)\| < C$ uniformly in k and ϵ. Feeding in the above equation, we obtain:

$$\Theta(k\epsilon,\epsilon) = \mathbb{1} + \epsilon^s\mathbb{E}_k(\epsilon) + \epsilon^s \sum_{j=1}^{k-1} A(k\epsilon,\epsilon)A((k-1)\epsilon,\epsilon)\dots A((j+1)\epsilon,\epsilon)\mathbb{E}_j(\epsilon). \tag{29}$$

Now, let us look at the roots $\mu_1(x,\epsilon),\dots,\mu_d(x,\epsilon)$ of

$$\sum_{j=0}^{d} b_j(x,\epsilon)\mu^j = 0.$$

Since $\max_j \sup_{x\in I} |\mu_j(x,0)| = 1$, it follows that

$$\max_j \sup_{x\in I} |\mu_j(x,\epsilon)| \leq 1 + C\epsilon.$$

Since $k\epsilon$ lies in I, a compact set, Lemma 3.2 and Equation (29) imply that

$$\|\Theta_k(\epsilon) - \mathbb{1}\| \leq kC'_s\epsilon^{s+1} \leq C_s\epsilon^{s-s_0+1} \tag{30}$$

for all $k\epsilon \in I$, where $s_0 = 1$. This completes the proof of (24).

Equations (24) and (22) differ in the presence of s_0. It is easy to see that if f is a function such that for a fixed s_0 and any $s \in \mathbb{N}$ we have:

$$|f(\epsilon) - \sum_{t=0}^{s+s_0} C_t\epsilon^t| < D_s\epsilon^{s+1},$$

then

$$|f(\epsilon) - \sum_{t=0}^{s} C_t\epsilon^t| < (D_s + |C_{s+1}| + \dots |C_{s+s_0}|)\epsilon^{s+1}.$$

This observation shows that (24) implies (22) and concludes the proof of the Proposition. □

PROPOSITION 5.3. (a) *There exists a smooth function* $\Phi_1 \in C^\infty(I\times[0,\epsilon'])$ *such that*

(a) *For all* $(k\epsilon,\epsilon) \in \Delta_{\epsilon',I}$, *we have:*

$$\psi_1(k\epsilon,\epsilon) = \exp\left(\epsilon^{-1}\Phi_1(k\epsilon,\epsilon)\right),$$

(b) Φ_1 *has an asymptotic expansion (uniform with respect to* x*):*

$$\Phi_1(x,\epsilon) \sim_{\epsilon\to 0} \exp\left(\epsilon^{-1}\sum_{s=0}^{\infty}\phi_{1,s}(x)\epsilon^s\right)$$

where $\phi_{1,s}$ *are as in Lemma 5.1. As a result, we have an asymptotic expansion (uniform with respect to* k*):*

$$\psi_1(k\epsilon,\epsilon) \sim_{\epsilon\to 0} \exp\left(\epsilon^{-1}\sum_{s=0}^{\infty}\phi_{1,s}(k\epsilon)\epsilon^s\right).$$

PROOF. Consider the change of variables θ as in (23).

Due to our choice of initial conditions it follows that for every fixed $k = 0, \dots, d-1$, the function $\epsilon \to \theta(k\epsilon, \epsilon)$ is smooth. Using this and the smoothness of the coefficients of (8), it follows that for every fixed k, the function $\epsilon \to \theta(k\epsilon, \epsilon)$ is smooth.

So far, the function θ is defined on $\Delta^{\text{seg}}_{\epsilon',I}$:

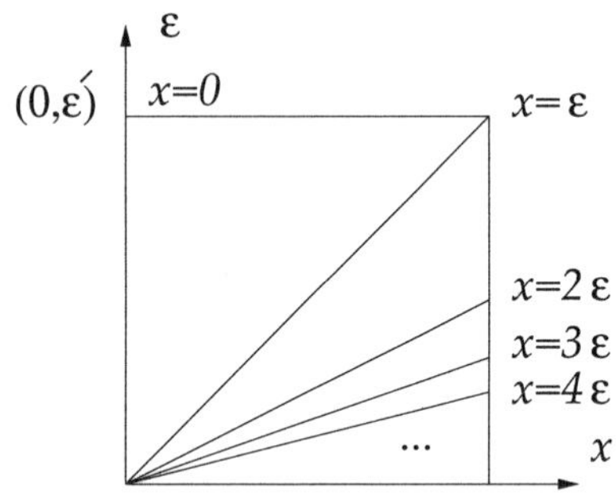

which is a union of line segments in a rectangle $I \times [0, \epsilon']$, and satisfies (22).

The complement of these line segments in $[0, \epsilon']$ consists of an infinite union of open triangles, together with the horizontal segment $I \times 0$. We can smoothly interpolate θ inside these open triangles so that it is defined on $I \times (0, \epsilon']$ and

$$|\theta(x, \epsilon) - 1| < C_s \epsilon^s \tag{31}$$

for all $(x, \epsilon) \in I \times (0, \epsilon']$ and all $s \in \mathbb{N}$.

Let us extend θ to $I \times [0, \epsilon']$ by defining $\theta(x, 0) = 1$ for all $x \in I$.

We claim that θ is smooth on $I \times [0, \epsilon']$. We need only to check this at the points $(x, 0)$ for $x \in I$. This follows easily from (31). For example, to check continuity at $(x, 0)$, consider a sequence (x_n, ϵ_n) such that $\lim_{n\to\infty}(x_n, \epsilon_n) = (x, 0)$. Then, (31) for $s = 1$ implies that $|\theta(x_n, \epsilon_n) - 1| < C_s \epsilon_n \sim_{n\to\infty} 0$, thus θ is continuous at $(x, 0)$. Using (31) for $s + 1$ it follows that $\partial^s / \partial \epsilon^s \theta(x, \epsilon)|_{\epsilon=0} = 0$ for all $s > 0$, and we find that θ has an ϵ-asymptotic expansion (uniform with respect to x):

$$\theta(x, \epsilon) \sim_{\epsilon\to 0} 1$$

Restricting further ϵ' if needed, we may assume that $|\theta(x, \epsilon)| > 0$ for all $(x, \epsilon) \in I \times [0, \epsilon']$; in other words $\log \theta(x, \epsilon)$ makes sense for all $(x, \epsilon) \in I \times [0, \epsilon']$.

Now, we can finish the proof of Proposition 5.3.

Let us define

$$\Phi_1(x, \epsilon) = \hat{\Phi}_1(x, \epsilon) + \epsilon \log \theta(x, \epsilon)$$

Then, (23) implies (a). Since $\theta(x, \epsilon)$ is asymptotic to 1 (uniformly on x), it follows that $\Phi_1(x, \epsilon)$ is asymptotic to $\hat{\Phi}_1(x, \epsilon)$. Using the asymptotic of $\hat{\Phi}_1$ given by Lemma 5.1, (b) follows. □

5.2. A reduction to an ϵ-difference equation of smaller degree. We will now prove Theorem 3 by induction on the degree d of the ϵ-difference equation. For $d = 1$, it follows from Proposition 5.3. The inductive step is the next Proposition.

PROPOSITION 5.4. *Assume that Theorem 3 holds for regular ϵ-difference equations of degree less than d. Then it holds for regular ϵ-difference equations of degree d.*

Proof. Consider a ϵ-difference equation (8) of degree d. We will use the solution ψ_1 of it constructed in Proposition 5.3 to reduce it to an equivalent equation of degree $d-1$, and an inhomogeneous ϵ-difference equation of degree 1.

Consider the dependent change of variables

$$\theta = \frac{\phi}{\psi_1} \tag{32}$$

This is well-defined since ψ_1 is nowhere zero. Then, ϕ satisfies (8) iff θ satisfies

$$\sum_{j=0}^{d} b_j(k\epsilon, \epsilon)\theta((k+j)\epsilon, \epsilon) = 0 \tag{33}$$

where

$$b_j(x,\epsilon) = a_j(x,\epsilon)\frac{\psi_1(x+j\epsilon,\epsilon)}{\psi_1(x,\epsilon)}.$$

The characteristic polynomials of (8) and (33) are related by

$$P_{(33)}(\lambda) = \lambda_1(x)^d P_{(8)}(\lambda/\lambda_1(x)).$$

As in the proof of Proposition 5.3, it is easy to see that (33) is a regular ϵ-difference equation. Moreover, it is easy to see that Theorem 3 holds for (8) iff it holds for (33). Indeed, check that the change of variables given by (32) preserves the asymptotics of the solutions of (8) and (33).

Thus, it suffices to work with (33). In that case, $\theta = 1$ is a solution of (33), since ψ is a solution of (8). It follows that

$$\sum_{j=0}^{d} b_j(x,\epsilon) = 0. \tag{34}$$

(Compare this with (26)). Let us define

$$\zeta(k\epsilon,\epsilon) = \theta((k+1)\epsilon,\epsilon) - \theta(k\epsilon,\epsilon). \tag{35}$$

Then, we get that ζ is a solution of the ϵ-difference equation

$$\sum_{j=0}^{d-1} c_j(k\epsilon,\epsilon)\zeta((k+j)\epsilon,\epsilon) = 0 \tag{36}$$

where

$$c_s(x,\epsilon) = \sum_{j=s+1}^{d} b_j(x,\epsilon).$$

The characteristic equations of (33) and (36) are related by

$$\sum_{j=0}^{d} b_j(x,0)\lambda^j = (\lambda-1)\sum_{j=0}^{d-1} c_j(x,0)\lambda^j$$

Since $c_0(x,\epsilon) = \sum_{j=1}^{d} b_j(x,\epsilon) = -b_0(x,\epsilon)$ (by (34)) and $c_d(x,\epsilon) = b_d(x,\epsilon)$, the same arguments of Proposition 5.3 imply that (36) is regular, assuming that (33) is regular.

By the induction hypothesis, it follows that (36) satisfies Theorem 3.

For the remainder of this section, fix a solution ζ of (36) which satisfies the properties of Theorem 3. In other words, ζ satisfies (33) and $\zeta(k\epsilon, \epsilon) = \exp(\epsilon^{-1}Z(k\epsilon, \epsilon))$ where Z is a smooth function with uniform (with respect to x) asymptotics:

$$Z(x, \epsilon) \sim_{\epsilon \to 0} \sum_{s=0}^{\infty} Z_s(x)\epsilon^s.$$

LEMMA 5.5. *There exists a formal solution*

$$\tilde{\theta}(x, \epsilon) = \exp(\epsilon^{-1}\Theta(x, \epsilon))$$

of (36) *such that*

$$\Theta(x, \epsilon) = \sum_{s=0}^{\infty} \Theta_s(x)\epsilon^s.$$

PROOF. We need to solve the formal power series Equation

$$\exp\left(\epsilon^{-1}\sum_{s=0}^{\infty}\Theta_s(x+\epsilon)\epsilon^s\right) - \exp\left(\epsilon^{-1}\sum_{s=0}^{\infty}\Theta_s(x)\epsilon^s\right) = \exp\left(\epsilon^{-1}\sum_{s=0}^{\infty}Z_s(x)\epsilon^s\right)$$

for Θ in terms of Z. Using the Taylor expansion $\Theta_0(x+\epsilon) = \Theta_0(x) + \Theta_0'(x)\epsilon + O(\epsilon^2)$ it is easy to see that the above equation equals to

$$\exp\left(\epsilon^{-1}\Theta_0(x) + O(1)\right) - \exp\left(\epsilon^{-1}\Theta_0(x) + O(1)\right) = \exp\left(\epsilon^{-1}Z_0(x) + O(1)\right)$$

from which follows that $\Theta_0 = Z_0$. Dividing the equation by $\exp(\epsilon^{-1}\Theta_0)$ we get an equation in formal power series with nonnegative powers of ϵ. Moreover, the coefficient of ϵ^s in that power series (for $s \geq 0$) equals to

$$(\exp(\Theta_0'(x)) - 1)\exp(\Theta_{s+1}(x))H_s(x)$$

where $H_s(x)$ is a function of Θ_j and Z_j for $j = 1, \dots, s$. Since $\exp(\Theta_0'(x)) = \exp(Z_0'(x))$ is an eigenvalue of (36), it is never equal to 1.

This and induction prove the lemma. □

LEMMA 5.6. (a) There exists a solution to the equation

$$\zeta(k\epsilon, \epsilon) = \theta((k+1)\epsilon, \epsilon) - \theta(k\epsilon, \epsilon) \tag{37}$$

for θ in terms of ζ with appropriate initial condition.
(b) For all $(k\epsilon, \epsilon) \in \Delta'_{\epsilon', I}$ we have:

$$\zeta(k\epsilon, \epsilon) \sim_{\epsilon \to 0} \exp(Z(k\epsilon, \epsilon))$$

where

$$\exp(Z(x, \epsilon)) = \exp(\Theta(x+\epsilon, \epsilon)) - \exp(\Theta(x, \epsilon))$$

PROOF. Fix $\epsilon > 0$ and let $I = [a, b]$. Consider a natural number k such that $k\epsilon \in I$ and $(k+1)\epsilon \in I$. This is equivalent to $k_1 \leq k \leq k_2$ where k_1 and k_2 are natural numbers that depend on ϵ and I, although we do not indicate this in our notation.

Then equation (37) implies that

$$\begin{aligned} \zeta(k_1\epsilon, \epsilon) &= \theta((k_1+1)\epsilon, \epsilon) - \theta(k_1\epsilon, \epsilon) \\ \zeta((k_1+1)\epsilon, \epsilon) &= \theta((k_1+2)\epsilon, \epsilon) - \theta((k_1+1)\epsilon, \epsilon) \\ \dots &= \dots \\ \zeta((k_2-1)\epsilon, \epsilon) &= \theta(k_2\epsilon, \epsilon) - \theta((k_2-1)\epsilon, \epsilon) \end{aligned}$$

Summing up, we obtain that

$$\theta(k_2\epsilon,\epsilon)=\theta(k_1\epsilon,\epsilon)+\sum_{j=k_1}^{k_2-1}\zeta(j\epsilon,\epsilon).$$

Choose initial conditions so that $\theta(k_1\epsilon,\epsilon)=\exp(Z(k_1\epsilon,\epsilon))$. This completes part (a).

Part (b) follows by a telescoping calculation. □

To finish the proof of Proposition 5.4, it suffices to show that the solution ζ of Lemma 5.6 is asymptotic to the formal solution $\tilde{\theta}$ of Lemma 5.5.

Since

$$\zeta(k\epsilon,\epsilon)\sim_{\epsilon\to 0}\exp(Z(k\epsilon,\epsilon))$$

and

$$\exp(Z(k\epsilon,\epsilon))=\exp(\Theta((k+1)\epsilon,\epsilon))-\exp(\Theta(k\epsilon,\epsilon))$$

it follows by the definition of θ given in Lemma 5.6 and by a telescopic sum, that:

$$\begin{aligned}
\theta(k\epsilon,\epsilon) &= \theta(k_1\epsilon,\epsilon)+\sum_{j=k_1}^{k-1}\zeta(j\epsilon,\epsilon)\\
&\sim_{\epsilon\to 0} \theta(k_1\epsilon,\epsilon)+\sum_{j=1}^{k-1}(\exp(\Theta((j+1)\epsilon,\epsilon))-\exp(\Theta(j\epsilon,\epsilon)))\\
&= \theta(k_1\epsilon,\epsilon)+\exp(\Theta(k\epsilon,\epsilon))-\exp(\Theta(k\epsilon,\epsilon))\\
&= \exp(\Theta(k\epsilon,\epsilon))
\end{aligned}$$

This concludes the proof of Proposition 5.4. □

5.3. The solutions form a locally fundamental set. Let us summarize what we have obtained so far.

Consider a partition $I=\cup_{p\in P}I_p$ of $I=[x_0,x_P]$ given by $I_p=[x_p,x_{p+1}]$ for $p=1,\dots,P-1$, and consider a permutation σ_p of $\{1,\dots,d\}$ such that

$$|\lambda_{\sigma_p(1)}(x)|\geq|\lambda_{\sigma_p(2)}(x)|\geq\cdots\geq|\lambda_{\sigma_p(d)}(x)| \qquad \text{for all} \qquad x\in I_p.$$

We have constructed solutions smooth functions Φ_m for $m=1,\dots,d$ with asymptotic expansion given by (10) and (11).

Let us define

$$\psi_m(x,\epsilon)=\exp\left(\epsilon^{-1}\Phi_m(x,\epsilon)\right), \tag{38}$$

where Φ_m are smooth functions with asymptotic expansions as in Equations (10) and (11).

Moreover, we have shown that for every interval I_p (as in the discussion prior to Theorem 3),

$$\{\psi_1(k\epsilon,\epsilon),\dots,\psi_d(k\epsilon,\epsilon)\}$$

is a set of solutions of (8) when $k\epsilon\in I_p$.

Fix a solution $\psi(k\epsilon,\epsilon)$ of (8) and an interval I_p. The following lemma certainly implies that $\{\psi_1,\dots,\psi_d\}$ is a locally fundamental set of solutions of (8). This concludes the proof of Theorem 3. □

In addition, the next lemma motivates the definition of a regular solution, given in the following section.

LEMMA 5.7. (a) Fix ψ and I_p as above. Then, there exist smooth functions c_m^p such that

$$\psi(k\epsilon, \epsilon) = c_1^p(\epsilon)\psi_{\sigma_p(1)}(k\epsilon, \epsilon) + \cdots + c_d^p(\epsilon)\psi_{\sigma_p(d)}(k\epsilon, \epsilon) \tag{39}$$

for all $k\epsilon \in I_p$.
(b) Moreover, for every p and $m = 1, \ldots, d$ we have

$$c_m^p(\epsilon) = \frac{\psi_{\sigma_{p-1}(m)}(x_p^\epsilon \epsilon - \epsilon, \epsilon)}{\psi_{\sigma_p(m)}(x_p^\epsilon \epsilon, \epsilon)} \gamma_m^p(\epsilon) \tag{40}$$

for some smooth functions γ_m^p, with the understanding that $\psi_{\sigma_{-1}(m)} = 1$. Here $[x]$ is the *largest integer smaller than* x, and

$$x_p^\epsilon = \left[\frac{x_p}{\epsilon}\right] + 1.$$

PROOF. Without loss of generality, let us assume that $\sigma_p(j) = j$ for $j = 1, \ldots, d$. Equation (39) is a linear equation in c_m^p, with solutions

$$c_m^p(\epsilon) = \frac{\det W_m(x_p^\epsilon, \epsilon)}{\det W(x_p^\epsilon, \epsilon)}$$

where $I_p = [x_p, x_{p+1}]$,

$$W(x,\epsilon) = \begin{pmatrix} \psi_1(x,\epsilon) & \ldots & \psi_m(x,\epsilon) & \ldots & \psi_d(x,\epsilon) \\ \psi_1(x+\epsilon,\epsilon) & \ldots & \psi_m(x+\epsilon,\epsilon) & \ldots & \psi_d(x+\epsilon,\epsilon) \\ \ldots & \ldots & \ldots & \ldots & \ldots \\ \psi_1(x+(d-1)\epsilon,\epsilon) & \ldots & \psi_m(x+(d-1)\epsilon,\epsilon) & \ldots & \psi_d(x+(d-1)\epsilon,\epsilon) \end{pmatrix}$$

and

$$W_m(x,\epsilon) = \begin{pmatrix} \psi_1(x,\epsilon) & \ldots & \psi(x,\epsilon) & \ldots & \psi_d(x,\epsilon) \\ \psi_1(x+\epsilon,\epsilon) & \ldots & \psi(x+\epsilon,\epsilon) & \ldots & \psi_d(x+\epsilon,\epsilon) \\ \ldots & \ldots & \ldots & \ldots & \ldots \\ \psi_1(x+(d-1)\epsilon,\epsilon) & \ldots & \psi(x+(d-1)\epsilon,\epsilon) & \ldots & \psi_d(x+(d-1)\epsilon,\epsilon) \end{pmatrix}$$

where the ψ's are in the mth column of W_m.

We will show that for small enough ϵ, $W(x, \epsilon)$ is nonsingular.

Using Equations (38), (10) and (11), it follows that

$$\begin{aligned} \frac{1}{\psi_1(x,\epsilon)\ldots\psi_d(x,\epsilon)} \det(W(x,\epsilon)) &= \det(\exp(j\phi_m'(x))) + O(\epsilon) \\ &= \det(\lambda_m(x)^j) + O(\epsilon) \\ &= \pm\prod_{i\neq j}(\lambda_i(x) - \lambda_j(x)) + O(\epsilon) \end{aligned}$$

uniformly in x, where λ_m are the eigenvalues of (8). Since (8) is regular, its eigenvalues never collide.

Similarly, using Equation (38), we have:

$$\frac{1}{\psi_1(x_p^\epsilon\epsilon,\epsilon)\ldots\widehat{\psi_m(x_p^\epsilon\epsilon,\epsilon)}\ldots\psi_d(x_p^\epsilon\epsilon,\epsilon)} \det(W_m(x_p^\epsilon\epsilon,\epsilon)) = \det B_{m,0}(x_p^\epsilon\epsilon,\epsilon) + O(\epsilon)$$

where

$$B_{m,j}(x,\epsilon) = \begin{pmatrix} \lambda_1(x)^{-j} & \ldots & \psi(x,\epsilon) & \ldots & \lambda_d(x)^{-j} \\ \lambda_1(x)^{1-j} & \ldots & \psi(x+\epsilon,\epsilon) & \ldots & \lambda_d(x)^{1-j} \\ \ldots & \ldots & \ldots & \ldots & \ldots \\ \lambda_1(x)^{d-1-j} & \ldots & \psi(x+(d-1)\epsilon,\epsilon) & \ldots & \lambda_d(x)^{d-1-j} \end{pmatrix}$$

where the ψ's are in the mth column of $B_{m,j}$. Thus,

$$c_m^p(\epsilon) = \frac{1}{\psi_m(x_p^\epsilon\epsilon,\epsilon)} \frac{\det(B_{m,0}(x_p^\epsilon\epsilon,\epsilon))}{\prod_{j\neq m}\lambda_j(x_p^\epsilon\epsilon)-\lambda_m(x_p^\epsilon\epsilon)} + O(\epsilon). \tag{41}$$

The idea now is to move the recursion relation backwards d times. Using the solution $\psi(k\epsilon,\epsilon)$ for $k\epsilon \in I_{p-1}$ will allow us to compute the smooth functions c_m^p.

In detail, consider the matrix $B_{m,0}(x_p^\epsilon\epsilon,\epsilon)$ and move the recursion relation backwards once. Using Equation (27) and the fact that the the jth column of $B_{m,j}(x_p^\epsilon,\epsilon)$ for $j\neq m$ is an eigenvector of $A(x_p^\epsilon\epsilon,\epsilon)$ (up to $O(\epsilon)$) with eigenvalue $\lambda_j(x_p^\epsilon\epsilon)$, it follows that

$$B_{m,0}(x_p^\epsilon\epsilon,\epsilon) = A(x_p^\epsilon\epsilon,\epsilon)B_{m,1}(x_p^\epsilon\epsilon-\epsilon,\epsilon)+O(\epsilon).$$

Iterating $d-1$ more times, it follows that

$$B_{m,0}(x_p^\epsilon\epsilon,\epsilon) = A(x_p^\epsilon\epsilon,\epsilon)\dots A(x_p^\epsilon\epsilon-(d-1)\epsilon,\epsilon)B_{m,d}(x_p^\epsilon\epsilon-d\epsilon,\epsilon)+O(\epsilon).$$

Since $x_p^\epsilon\epsilon - d\epsilon \in I_{p-1}$, it follows (as in the computation of $W_m(x,\epsilon)$ above) that:

$$\begin{aligned}
\det(B_m(x_p^\epsilon\epsilon-d\epsilon,\epsilon)) &= \psi_{\sigma_{p-1}(m)}(x_p^\epsilon\epsilon-\epsilon,\epsilon) \\
&\cdot\det\begin{pmatrix}
\lambda_1(x_p^\epsilon\epsilon-\epsilon)^{1-d} & \dots & \frac{\psi_{\sigma_{p-1}(m)}(x_p^\epsilon\epsilon-d\epsilon,\epsilon)}{\psi_{\sigma_{p-1}(m)}(x_p^\epsilon\epsilon-\epsilon,\epsilon)} & \dots & \lambda_d(x_p^\epsilon\epsilon-\epsilon)^{1-d} \\
\lambda_1(x_p^\epsilon\epsilon-\epsilon)^{2-d} & \dots & \frac{\psi_{\sigma_{p-1}(m)}(x_p^\epsilon\epsilon+(1-d)\epsilon,\epsilon)}{\psi_{\sigma_{p-1}(m)}(x_p^\epsilon\epsilon-\epsilon,\epsilon)} & \dots & \lambda_d(x_p^\epsilon\epsilon-\epsilon)^{2-d} \\
\dots & \dots & \dots & \dots & \dots \\
\lambda_1(x_p^\epsilon\epsilon-\epsilon)^{d-d} & \dots & \frac{\psi_{\sigma_{p-1}(m)}(x_p^\epsilon\epsilon-\epsilon,\epsilon)}{\psi_{\sigma_{p-1}(m)}(x_p^\epsilon\epsilon-\epsilon,\epsilon)} & \dots & \lambda_d(x_p^\epsilon\epsilon-\epsilon)^{d-d}
\end{pmatrix} \\
&= \psi_{\sigma_{p-1}(m)}(x_p^\epsilon\epsilon-\epsilon,\epsilon)\prod_{i\neq j, j\neq m}(\lambda_i^{-1}(x_p^\epsilon\epsilon-\epsilon)-\lambda_j^{-1}(x_p^\epsilon\epsilon-\epsilon)) \\
&\qquad\cdot\prod_{i\neq m}(\lambda_i^{-1}(x_p^\epsilon\epsilon-\epsilon)-\lambda_{\sigma_{p-1}(m)}^{-1}(x_p^\epsilon\epsilon-\epsilon))+O(\epsilon).
\end{aligned} \tag{42}$$

This, together with Equation (41) proves (40). □

6. Regular solutions and their asymptotics

In this section we discuss regular solutions of q and ϵ-difference equations and their asymptotics.

6.1. Regular solutions to ϵ-difference equations.

According to Lemma 5.7, a solution ψ to (8) determines a collection $S = \{S_p | p \in \mathcal{P}\}$ of subsets of $\{1,\dots,d\}$, where

$$S_p = \{m \in \{1,\dots,d\} | \gamma_m^p \neq 0\}.$$

DEFINITION 6.1. Fix a collection $S = \{S_p | p \in \mathcal{P}\}$ of subsets of $\{1,\dots,d\}$. We say that a solution ψ of a regular equation (8) is *S-regular* iff for every $p \in \mathcal{P}$ we have:

- $\gamma_m^p = 0$ if $m \notin S_p$.
- γ_m^p are *not flat* at 0 for all $m \in S_p$. That is, some derivative of γ_m^p at $\epsilon = 0$ does not vanish.

- For every $p \in \mathcal{P}$ there exists an element $\eta(p) \in S_p$ such that

  $|\lambda_{\eta(p)}(x)| > |\lambda_j(x)| \quad \text{for all} \quad j \in S_p - \{\eta(p)\}, \quad x \in \text{Interior}(I_p).$

  In other words, in the interior of the interval I_p, and among the eigenvalues $\lambda_j(x)$ for $j \in S_p$, there is a unique eigenvalue of strictly maximum magnitude.

We will say that a solution to (3) is *regular* if it is S-regular for some S.

6.2. Asymptotics of regular solutions of ϵ-difference equations.

PROOF. (of Theorem 4) Let ψ be an S-regular solution to (8). Let us assume that $S = \{1, \dots, d\}$, and $|\lambda_1(x)| > |\lambda_m(x)|$ for $m = 2, \dots, d$ and all x in the interior Interior(I) of the closed interval $I = [0, b]$. Fix an $x \in I$.

Then, we have:

$$\psi(x, \epsilon) = c_1(\epsilon)\psi_1(x, \epsilon) + \cdots + c_d(\epsilon)\psi_d(x, \epsilon).$$

for $x = k\epsilon$, where $c_1(\epsilon) = c_1\epsilon^{n_1} + O(\epsilon^{n_1+1})$, and $c_1 \neq 0$.

Then, we have:

$$\psi(x, \epsilon) = c_1(\epsilon)\psi_1(x, \epsilon)\left(1 + \sum_{m=2}^{d} \frac{c_j(\epsilon)}{c_1(\epsilon)} \frac{\psi_m(x, \epsilon)}{\psi_1(x, \epsilon)}\right). \tag{43}$$

Recall from Theorem 3 that

$$\begin{aligned} \psi_m(x, \epsilon) &= \exp(\epsilon^{-1}\Phi_m(x, \epsilon)) \\ \Phi_m(x, \epsilon) &= \Phi_{m,0}(x) + O(\epsilon) \\ \Phi'_{m,0}(x) &= \log \lambda_m(x). \end{aligned}$$

Thus,

$$\text{Re}(\Phi_{m,0})'(x) = \text{Re}(\log \lambda_m(x)) = \log |\lambda_m(x)|.$$

Combined with $|\lambda_m(x)| < |\lambda_1(x)|$ for $m \geq 2$, and $\Phi_{m,0}(0) = 0 = \Phi_{1,0}(0)$, it follows that

$$\text{Re}(\Phi_{m,0})(x) < \text{Re}(\Phi_{1,0})(x)$$

for all $x \in I$ and

$$\text{Re}(\Phi_{1,0})(x) = \int_0^x \log |\lambda_1(t)| dt.$$

Therefore,

$$\lim_{\epsilon \to 0^+} \left(1 + \sum_{m=2}^{d} \frac{c_j(\epsilon)}{c_1(\epsilon)} \frac{\psi_m(x, \epsilon)}{\psi_1(x, \epsilon)}\right) = 1.$$

and

$$\lim_{\epsilon \to 0^+} \epsilon \log |c_1(\epsilon)| = 0.$$

Thus, Equation (43) implies that

$$\begin{aligned} \lim_{\epsilon \to 0^+} \epsilon \log |\psi(x, \epsilon)| &= \lim_{\epsilon \to 0^+} \epsilon \log |\psi_1(x, \epsilon)| \\ &= \lim_{\epsilon \to 0^+} \epsilon \log |\exp(\epsilon^{-1}\Phi_m(x, \epsilon))| \\ &= \lim_{\epsilon \to 0^+} \epsilon \log |\exp(\epsilon^{-1}\Phi_m(x, 0) + O(1))| \\ &= \text{Re}(\Phi_{1,0}(x)) \\ &= \int_0^x \log |\lambda_1(t)| dt \end{aligned}$$

The result follows.

In the general case, we partition the interval I_p as in the discussion prior to Theorem 3 and repeat the above proof using Equation (44). The result follows. $\square$

6.3. Asymptotics of regular solutions of q-difference equations. First, we need to define what is a regular solution to a q-difference equation.

Consider a solution ψ of a q-difference equation and a partition of S^1 as in Section 1.8. Then, at each interval I_p, we can write the solution as a linear combination of fundamental solutions, as in Equation (4). Let S_p be the indexing set of the fundamental solutions that we use in each interval I_p; see (6).

Suppose that the partition of S^1 is given by $I_p = [e^{2\pi i x_p}, e^{2\pi i x_{p+1}}]$ for $p = 0, \dots, P-1$, with $q_0 = 1$, $q_P = e^{2\pi i}$.

Then, with $q = e^{2\pi i \epsilon}$ it turns out that for every p and $m = 1, \dots, d$ we have

$$c_m^p(q) = \frac{\psi_{\sigma_{p-1}(m)}(x_p^\epsilon \epsilon - \epsilon, \epsilon)}{\psi_{\sigma_p(m)}(x_p^\epsilon \epsilon, \epsilon)} \gamma_m^p(q) \tag{44}$$

for some smooth functions γ_m^p, with the understanding that $\psi_{\sigma_{-1}(m)} = 1$.

DEFINITION 6.2. Fix a collection $S = \{S_p | \, p \in \mathcal{P}\}$ of subsets of $\{1, \dots, d\}$. We say that a solution ψ of a regular equation (3) is *S-regular* iff for every $p \in \mathcal{P}$ we have:

- $\gamma_m^p = 0$ if $m \notin S_p$.
- γ_m^p are *not flat* at 0 for all $m \in S_p$. That is, some derivative of γ_m^p at $\epsilon = 0$ does not vanish.
- For every $p \in \mathcal{P}$ there exists an element $\eta(p) \in S_p$ such that

  $$|\lambda_{\gamma(p)}(v)| > |\lambda_j(v)| \quad \text{for all} \quad j \in S_p - \{\gamma(p)\}, \quad v \in \text{Interior}(I_p).$$

  In other words, in the interior of the interval I_p, and among the eigenvalues $\lambda_j(v)$ for $j \in S_p$, there is a unique eigenvalue of strictly maximum magnitude.

We will say that a solution to (3) is *regular* if it is S-regular for some S.

PROOF. (of Theorem 2) It follows from Equation (14) of Lemma 2.7 and Theorem 4. $\square$

7. Applications to Quantum Topology

7.1. The A-polynomial of a knot and its noncommutative version. In this section we discuss general features of q-difference equations for the colored Jones function of a knot.

The coefficients of the q-difference equations are rational functions of q and $q^n = Q$. In order to simplify the typesetting, we will give the q-difference equation

$$\sum_{j=0}^{d} b_j(q^n, q) f(n+j) = 0$$

in operator form

$$Pf = 0 \tag{45}$$

where

$$P = \sum_{j=0}^{d} b_j(Q, q) E^j$$

and where the operators E, Q and q, act on a discrete function $f : \mathbb{N} \to \mathbb{Z}[q^{\pm}]$ by

$$(qf)(n) = qf(n) \qquad (Qf)(n) = q^n f(n) \qquad (Ef)(n) = f(n+1).$$

Note that q commutes with Q and E, and that $EQ = qQE$.

It follows by definition that the characteristic polynomial $\mathrm{ch}P(v, \lambda)$ of (45) is obtained from P by setting $q = 1$, replacing (E, Q) by (λ, v). In other words, we have:

$$\mathrm{ch}P(v, \lambda) = \sum_{j=0}^{d} b_j(v, 1)\lambda^j.$$

In [**Ga1**], the first author showed that the colored Jones function J_K of a knot K satisfies an essentially unique smallest degree q-difference equation $P_K J_K = 0$ where the coefficients $a_j(u, v)$ of P_K are rational functions of u and v with rational coefficients.

In [**Ga1**], the operator P_K was called the *non-commutative A-polynomial* of K. In [**Ga1**], it was conjectured that:

CONJECTURE 1. *(AJ Conjecture) Up to a multiplication by a polynomial in v, we have*

$$\mathrm{ch}P_K(\lambda, v) = A_K(L, M)|_{(L,M^2)=(\lambda,v)}$$

where A_K is the A-polynomial of K, defined by [**CCGLS**].

The A-polynomial of K parametrizes the moduli space of characters of $\mathrm{SL}_2(\mathbb{C})$ representations of $\pi_1(S^3 - K)$, restricted to the boundary torus ∂M. The A-polynomial of a knot is an important ingredient to the Geometrization of the knot complement and its Dehn fillings.

The A-polynomial A_K of a knot K in S^3 satisfies symmetries, which we will list here, and refer to [**CCGLS**] and [**CL**] for proofs.

- **(S1)** It has integer coefficients and even powers of M, that is $A_K(L, M) \in \mathbb{Z}[L, M^2]$.
- **(S2)** It is reciprocal, that is, $A_K(L^{-1}, M^{-1}) = \pm L^k M^l A(L, M)$.
- **(S3)** It is tempered, that is the edge polynomials of its Newton polygon are cyclotomic.
- **(S4)** It specializes to
$$A_K(L, 1) = \pm(L-1)^{n_+}(L+1)^{n_-}$$
for some integers $n_{\pm}$.
- **(S5)** $L - 1$ is always a factor of A_K, that corresponds to $U(1)$ representations.

If the colored Jones function of a knot was an S-regular solution to a regular q-difference equation, and if the AJ Conjecture were true, then it follows that for every $\alpha \in [0, 1]$ we have:

$$\lim_{n\to\infty} \frac{\log |J_K(n)(e^{\frac{2\pi i \alpha}{n}})|}{n} = \sigma^J_{S,K}(\alpha) = \sigma^A_{S,K}(\alpha) \tag{46}$$

where $\sigma^A_{S,K}$ is the *A-entropy* of a knot, defined as follows.

DEFINITION 7.1. For a knot K in S^3, let $L_j(t)$ for $j = 1, \ldots, d$ denote the roots of the equation

$$A_K(L_j(t), e^{it/2}) = 0$$

for $t \in [0, 2\pi]$, where d is the L-degree of A_K. Fix a partition $\cup_{p \in \mathcal{P}} I_p$ of $[0, 2\pi]$ by closed intervals with nonoverlapping interiors and a permutation σ_p of the set $\{1, \dots, d\}$ such that

$$|L_{\sigma_p(1)}(t)| \geq |L_{\sigma_p(2)}(t)| \geq \cdots \geq |L_{\sigma_p(d)}(t)| \qquad \text{for all} \qquad t \in I_p.$$

For every collection $S = \{S_p | p \in \mathcal{P}\}$ of subsets of $\{1, \dots, d\}$, we can define the *S-entropy*

$$\sigma^A_{S,K} : [0,1] \to \mathbb{R}$$

by

$$\sigma^A_{S,K}(\alpha) = \frac{1}{2\pi} \int_0^{2\pi} \log \chi_S(\alpha t) dt,$$

where $\chi_S : [0,1] \to \mathbb{R}$ is defined by

$$\chi_S(t) = \max_{j \in S_p} |L_{\sigma_p(j)}(t)| \qquad \text{if} \qquad t \in I_p.$$

It is natural to ask how the A-entropy of a hyperbolic knot (evaluated at $\alpha = 1$) compares to the Hyperbolic Volume. The answer to this question is essentially contained in work of D. Boyd, [**Bo**], which we quote without proof here. We urge the reader to look in [**Bo**] for beautiful and suggestive calculations.

Boyd introduced and studied another invariant of a knot, the *Mahler measure*

$$m_K = \int_{S^1 \times S^1} \log |A_K(x,y)| dx dy$$

Using Jensen's formula, and the symmetry (S3), it follows that

$$2\pi m_K = \sum_{j=0}^{d} \int_0^{2\pi} \log^+ |L_j(t)| dt$$

where $\log^+(a) = \log \max\{1, a\}$.

Using (S2), it follows that $1/L_j(t)$ is an eigenvalue for every eigenvalue $L_j(t)$.

More generally, among the roots $L_j(t)$ there is a distinguished one, L_1, corresponding to the discrete faithful representation when $t = 2\pi$. Let $L_d(t) = 1$ denote the eigenvalue corresponding to the $U(1)$ representations. Boyd informs us that for 2-bridge knots K (in particular, for the 3_1 and 4_1 knots), it is true that

$$\sigma^A_{\{1,d\},K}(1) = \mathrm{vol}_K.$$

It follows by (S4) that the eigenvalues collide at $t = 0$. Moreover, Boyd informs us that for 2-bridge knots there exists a $t_0 \in (0, \pi)$ such that $|L_j(t)| = 1$ for all j and all $t : t_0 < t < \pi$.

Thus, if we want to apply Theorem 1 to the GHVC, we need to deal with irregular q-difference equations. We will discuss this topic in detail in a later publication.

Meanwhile, let us discuss some examples, taken from [**Ga1**].

7.2. Examples: The 3_1 and 4_1 knots. In this section we discuss in detail q-difference equation of the the colored Jones function of the two simplest knots, namely the trefoil 3_1 and the figure eight 4_1. The former is not hyperbolic, and the latter is.

In [**Ga1**], the first author computed that the colored Jones function J_{3_1} (resp. J_{4_1}) satisfies the second (resp. third) order q-difference equation

$$P_{3_1} J_{3_1} = 0 \qquad \text{resp.} \qquad P_{4_1} J_{4_1} = 0$$

where the noncommutative A-polynomials P_{3_1} and P_{4_1} are given by:

$$\begin{aligned} P_{3_1} &= \frac{q^3 Q^2 (q^2 - q^2 Q)}{q^3 - q^4 Q^2} \\ &+ \frac{(q - q^2 Q)(q + q^2 Q)(q^4 - q^5 Q + q^6 Q^2 - q^7 Q^2 - q^7 Q^3 + q^8 Q^4)}{q^2 Q (q - q^4 Q^2)(q^3 - q^4 Q^2)} E \\ &+ \frac{-1 + q^2 Q}{Q(q - q^4 Q^2)} E^2 \\ P_{4_1} &= \frac{q^5 Q(-q^3 + q^3 Q)}{(q^2 + q^3 Q)(-q^5 + q^6 Q^2)} \\ &-(q^2 - q^3 Q) \\ &\cdot \frac{(q^8 - 2q^9 Q + q^{10} Q - q^9 Q^2 + q^{10} Q^2 - q^{11} Q^2 + q^{10} Q^3 - 2q^{11} Q^3 + q^{12} Q^4)}{q^5 Q (q + q^3 Q)(q^5 - q^6 Q^2)} E \\ &+(-q + q^3 Q) \\ &\cdot \frac{(q^4 + q^5 Q - 2q^6 Q - q^7 Q^2 + q^8 Q^2 - q^9 Q^2 - 2q^{10} Q^3 + q^{11} Q^3 + q^{12} Q^4)}{q^4 Q (q^2 + q^3 Q)(-q + q^6 Q^2)} E^2 \\ &+ \frac{q^4 Q(-1 + q^3 Q)}{(q + q^3 Q)(q - q^6 Q^2)} E^3 \end{aligned}$$

If we wish, we may clear denominators in P_{3_1} and P_{4_1}. It follows that the characteristic polynomials are given by:

$$\begin{aligned} \mathrm{ch} P_{3_1}(L, M) &= -\frac{(L-1)(L+M^3)}{M(1+M)} \\ \mathrm{ch} P_{4_1}(L, M) &= \frac{(L-1)(L - LM - M^2 - 2LM^2 - L^2 M^2 - LM^3 + LM^4)}{M(1+M)^2} \end{aligned}$$

Inspection shows that P_{3_1} and P_{4_1} are not regular. Nevertheless, let us try to compute the S-entropy.

For the case of 3_1, we have $|L_j(t)| = 1$ for $j = 1, 2, 3$ and in this case

$$\sigma^A_{S, 3_1}(1) = \mathrm{vol}_{3_1} = 0$$

for all S.

For 4_1 knot, we have 3 eigenvalues $L_1(t)$, $L_2(t) = 1/L_1(t)$ and $L_3(t) = 1$. Assuming appropriate choices for the branches of the eigenvalues, the plot of $\log |L_1(t)|$ and $\log |L_2(t)| = -\log |L_1(t)|$ for $t \in [0, 2\pi]$ is given by:

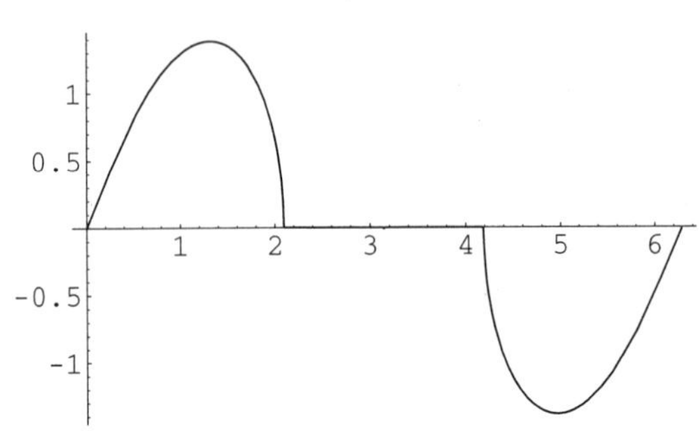

and

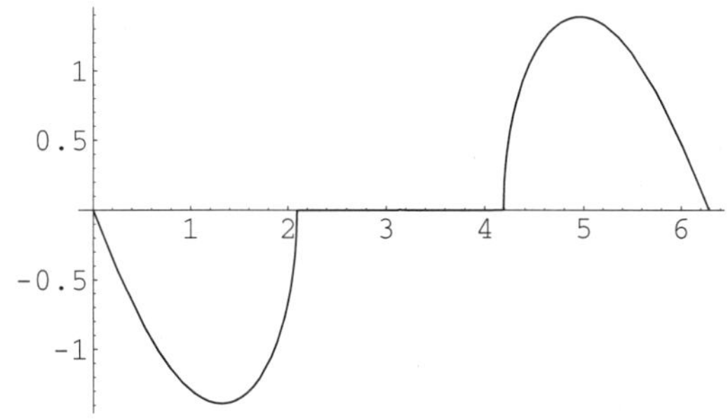

It follows that

$$\sigma^A_{\{1,3\},4_1}(1) = \mathrm{vol}(4_1) = 2.029883 \qquad \sigma^A_{\{1,2,3\},4_1}(1) = 2\mathrm{vol}_{4_1} = 4.05977.$$

Since the HVC is true for the 4_1 knot, it suggests that the colored Jones function lies in a strictly smaller subspace of the vector space of solutions to the q-difference equation $P_K J_K = 0$. Using work of Murakami [**Mu**], one can figure out exactly the selection principle; that is which locally fundamental solutions contribute to the colored Jones function.

Note that the associated q-difference equation of the 4_1 knot has the following features: of *collision*, *resonance* and *vanishing*:

- The eigenvalues collide at $t = 0$ (since $L_1(0) = L_1(0) = -1$), at $t = \pi/3$ (since $L_1(\pi/3) = L_1(\pi/3) = -1$) and by symmetry at $t = 2\pi/3$ and $t = 2\pi$.
- There is resonance on the interval $[\pi/3, 2\pi/3]$ where all three eigenvalues have equal magnitude.
- There is vanishing of the coefficients at $t = \pi/2$ (since the denominator $M + 1$ of the coefficients is singular at $t = \pi/2$).

Moreover, there is an additional difficult problem of *selection principle*.

We plan to study these problems in later publications.

References

[B1] G. Birkhoff, *On the asymptotic character of the solutions of certain linear differential equations containing a parameter*, Trans. Amer. Math. Soc. **9** (1908) 219–231.

[B2] ———, *Formal theory of irregular difference equations*, Acta Math. **54** (1930) 205–246.

[BT] ——— and W. Trjitzinsky, *Analytic theory of singular difference equations*, Acta Math. **60** (1932) 1–89.

[Bo] W. Boyd, *Mahler's measure and invariants of hyperbolic manifolds*, Number theory for the millennium, I (Urbana, IL, 2000), 127–143, A K Peters, Natick, 2002.

[CC] O. Costin, and R. Costin, *Rigorous WKB for finite-order linear recurrence relations with smooth coefficients*, SIAM J. Math. Anal. 27 (1996), no. 1, 110–134.

[CCGLS] D. Cooper, D, M. Culler, H. Gillet, D. Long and P. Shalen, *Plane curves associated to character varieties of* 3-*manifolds*, Invent. Math. **118** (1994) 47–84.

[CL] ——— and D. Long, *Remarks on the A-polynomial of a knot*, J. Knot Theory Ramifications **5** (1996) 609–628.

[DM] P. Deift and K. McLaughlin, *A continuum limit of the Toda lattice*, Mem. Amer. Math. Soc. **131** (1998), no. 624.

[GL1] S. Garoufalidis and T.T.Q. Le, *The colored Jones function is q-holonomic* Geom. and Topology **9** (2005) 1253–1293.

[GL2] ——— and ———, *Asymptotics of the colored Jones function*, preprint 2005 `math.GT/0508100`.

[GL3] ——— and ———, *Gevrey asymptotics in quantum topology*, preprint 2005.

[Ga1] ———, *On the characteristic and deformation varieties of a knot*, Proceedings of the CassonFest, Geometry and Topology Monographs **7** (2004) 291–309.

[Ga2] ———, *Difference and differential equations for the colored Jones function*, preprint 2003 `math.GT/0306229`.

[Ga3] ———, *Gevrey asymptotics of difference equations with a parameter*, preprint 2005.

[GG] M. Golubitsky and V. Guillemin, *Stable mappings and their singularities*, Graduate Text in Mathematics **14** Springer-Verlag, 1973.

[GLR] I. Gohberg, P. Lancaster, and L. Rodman, *Matrix Polynomials*, Academic Press, NY, 1982.

[Gr] G. Green, *On the motion of waves in a variable canal of small depth and width*, Trans. Cambridge Philos. Soc. **6** (1837) 457–462.

[Gu] S. Gukov, *Three-Dimensional Quantum Gravity, Chern-Simons Theory, and the A-Polynomial*, Comm. Math. Phys. **255** (2005) 577–627.

[K] R. Kashaev, *The hyperbolic volume of knots from the quantum dilogarithm*, Modern Phys. Lett. A **39** (1997) 269–275.

[L] J. Liouville, *Sur le développement des fonctions ou parties de fonctions en séries*, J. Math. Pures Appl. **2** (1837) 16–35.

[J] V. Jones, *Hecke algebra representation of braid groups and link polynomials*, Annals Math. **126** (1987) p. 335–388.

[MM] H. Murakami and J. Murakami, *The colored Jones polynomials and the simplicial volume of a knot*, Acta Math. **186** (2001) 85–104.

[Mu] ———, *Some limits of the colored Jones polynomials of the figure-eight knot*, Kyungpook Math. J. **44** (2004) 369–383.

[O] F. Olver, *Asymptotics and special functions*, Reprint. AKP Classics. A K Peters, Ltd., Wellesley, MA, 1997.

[vPS] M. van der Put and M. Singer, *Galois theory of difference equations*, Lecture Notes in Mathematics, **1666** Springer-Verlag 1997.

[Th] W. Thurston, *The geometry and topology of 3-manifolds*, 1979 notes, available from MSRI.

[Wa] W. Wasow, *Asymptotic expansions for ordinary differential equations*, Reprint of the 1976 edition. Dover Publications, New York, 1987.

[WZ] J. Wimp, and D. Zeilberger, *Resurrecting the asymptotics of linear recurrences*, J. Math. Anal. Appl. **111** (1985) 162–176.

[Wi] ———, *Computation with recurrence relations*, Applicable Mathematics Series, Pitman, Boston 1984.

[Wt] E.Witten, *Quantum field theory and the Jones polynomial*, Commun. Math. Physics. **121** (1989) 360–376.

[Z] D. Zeilberger, *A holonomic systems approach to special functions identities*, J. Comput. Appl. Math. **32** (1990) 321–368.

School of Mathematics, Georgia Institute of Technology, Atlanta, GA 30332-0160, USA

E-mail address: `stavros@math.gatech.edu`, URL: `http://www.math.gatech.edu/~stavros`

School of Mathematics, Georgia Institute of Technology, Atlanta, GA 30332-0160, USA

E-mail address: `geronimo@math.gatech.edu`

Contemporary Mathematics
Volume **416**, 2006

The mapping class group acts reducibly on $\mathsf{SU}(n)$-character varieties

William M. Goldman

ABSTRACT. When G is a connected compact Lie group, and π is a closed surface group, then $\mathsf{Hom}(\pi, G)/G$ contains an open dense $\mathsf{Out}(\pi)$-invariant subset which is a smooth symplectic manifold. This symplectic structure is $\mathsf{Out}(\pi)$-invariant and therefore defines an invariant measure μ, which has finite volume. The corresponding unitary representation of $\mathsf{Out}(\pi)$ on $L^2(\mathsf{Hom}(\pi, G)/G, \mu)$ contains no finite-dimensional subrepresentations besides the constants. This paper proves the remark that the representation $L^2(\mathsf{Hom}(\pi, G)/G, \mu)$ contains other invariant subspaces, for $G = \mathsf{SU}(n)$.

Let $G = \mathsf{SU}(n)$ and π be the fundamental group of a closed oriented surface Σ. Let $\mathsf{Hom}(\pi, G)$ be the space of representations $\pi \longrightarrow G$. The group $\mathsf{Aut}(\pi) \times \mathsf{Aut}(G)$ acts on $\mathsf{Hom}(\pi, G)$. Let $\mathsf{Hom}(\pi, G)/G$ be the quotient of $\mathsf{Hom}(\pi, G)$ by $\{1\} \times \mathsf{Inn}(G)$. Then $\mathsf{Out}(\pi) := \mathsf{Aut}(\pi)/\mathsf{Inn}(\pi)$ acts on $\mathsf{Hom}(\pi, G)/G$. The $\mathsf{Out}(\pi)$-action preserves a symplectic structure on $\mathsf{Hom}(\pi, G)/G$ which determines a finite invariant smooth measure μ (on an invariant dense open subset which is a smooth manifold). When G is compact, the total measure is finite (Jeffrey-Weitsman [**6, 7**], Huebschmann [**5**]). There results a unitary representation of $\mathsf{Out}(\pi)$ on the Hilbert space

$$\mathcal{H} := L^2(\mathsf{Hom}(\pi, G)/G, \mu).$$

Let $C \subset \mathcal{H}$ denote the subspace corresponding to the constant functions.

The following theorem is proved in Goldman [**3**] for $n = 2$ and Pickrell-Xia [**8, 9**] in general:

THEOREM. *The only finite dimensional* $\mathsf{Out}(\pi)$*-invariant subspace of* $\mathcal{H}$ *is* C.

According to [**3**], the only finite-dimensional invariant subspace of $\mathcal{H}$ consists of constants. Let $\mathcal{H}_0$ denote the orthocomplement of the constants in $\mathcal{H}$, that is, the set of $f \in \mathcal{H}$ such that $\int f d\mu = 0$.

The goal of this note is an elementary proof of the following:

THEOREM. *The representation of* $\mathsf{Out}(\pi)$ *on* $\mathcal{H}_0$ *is reducible.*

In general for a compact connected Lie group G, the components of $\mathsf{Hom}(\pi, G)$ and $\mathsf{Hom}(\pi, G)/G$ are indexed by $\pi_1(G')$ where G' is the commutator subgroup. In that case the ergodicity/mixing results of [**3, 8, 9**] imply that the only invariant

---

Partially supported by NSF grants DMS0405605 and DMS-0103889.

finite dimensional subspaces are subspaces of the space of locally constant functions, a vector space of dimension $|vert\pi_1(G')|$. The method of proving reducibility requires a nontrivial element of the center of G. For simplicity, we only discuss the case $G = \mathsf{SU}(n)$ in this paper.

I am grateful to Steve Zelditch, David Fisher, Charlie Frohman, and Jeff Hakim for helpful conversations.

1. The center of $\mathsf{SU}(n)$

Let $\mathfrak{Z} \cong \mathbb{Z}/n$ denote the center of G, the group consisting of all scalar matrices $\zeta\mathbb{I}$ where $\zeta^n = 1$. Then $\mathsf{Hom}(\pi, \mathfrak{Z}) \cong H^1(M; \mathbb{Z}/n)$ acts on $\mathsf{Hom}(\pi, G)/G$ by pointwise multiplication: If $\rho \in \mathsf{Hom}(\pi, G)$ and $u \in \mathsf{Hom}(\pi, \mathfrak{Z})$, then define the action $u \cdot \rho$ of u on ρ by:

$$\gamma \overset{u\cdot\rho}{\longmapsto} \rho(\gamma)u(\gamma). \tag{1}$$

Recall the definition [**2**] of the symplectic structure on $\mathsf{Hom}(\pi, G)/G$. Suppose that $\rho \in \mathsf{Hom}(\pi, G)$ is an irreducible representation. By Weil [**11**] (compare also Raghunathan [**10**]), the Zariski tangent space to $\mathsf{Hom}(\pi, G)$ at ρ identifies with the space $Z^1(\pi, \mathfrak{g}_{\mathsf{Ad}\rho})$ of cocycles where $\mathfrak{g}_{\mathsf{Ad}\rho}$ is the π-module defined by the composition

$$\pi \xrightarrow{\rho} G \xrightarrow{\mathsf{Ad}} \mathsf{Aut}(\mathfrak{g})$$

and $\mathfrak{g}$ is the Lie algebra of G. Since ρ is irreducible, G acts locally freely and $\mathsf{Hom}(\pi, G)/G$ has a smooth structure in a neighborhood of $[\rho]$. The tangent space to the orbit $G \cdot \rho$ equals the space of coboundaries $B^1(\pi, \mathfrak{g}_{\mathsf{Ad}\rho})$. The tangent space to $\mathsf{Hom}(\pi, G)/G$ at $[\rho]$ identifies with the cohomology group $H^1(\pi, \mathfrak{g}_{\mathsf{Ad}\rho})$.

Let $u \in \mathsf{Hom}(\pi, \mathfrak{Z})$. Since $\mathsf{Ad}(\mathfrak{Z})$ is trivial, the action of u induces an identification of π-modules $\mathfrak{g}_{\mathsf{Ad}\rho} \longrightarrow \mathfrak{g}_{\mathsf{Ad}(u\cdot\rho)}$ and hence of tangent spaces

$$\begin{aligned} T_{[\rho]}\mathsf{Hom}(\pi, G)/G &= H^1(\pi, \mathfrak{g}_{\mathsf{Ad}\rho}) \\ &\longrightarrow \quad T_{[u\cdot\rho]}\mathsf{Hom}(\pi, G)/G = H^1(\pi, \mathfrak{g}_{\mathsf{Ad}(u\cdot\rho)}). \end{aligned}$$

The symplectic form ω_ρ at $[\rho]$ is defined by the cup product

$$H^1(\pi, \mathfrak{g}_{\mathsf{Ad}\rho}) \times H^1(\pi, \mathfrak{g}_{\mathsf{Ad}\rho}) \longrightarrow H^2(\pi; \mathbb{R})$$

using the pairing of π-modules induced by an Ad-invariant nondegenerate symmetric bilinear form on $\mathfrak{g}$ as a coeefficient pairing. Evidently the symplectic form ω, and therefore the corresponding measure μ, are $\mathsf{Hom}(\pi, \mathfrak{Z})$-invariant.

2. Trace functions

Suppose that $\gamma \in \pi$. The function

$$\begin{aligned} \mathsf{Hom}(\pi, G) &\longrightarrow \mathbb{C} \\ \rho &\longmapsto \mathsf{tr}(\rho(\gamma)) \end{aligned}$$

is $\mathsf{Inn}(G)$-invariant and defines a function

$$\mathsf{Hom}(\pi, G)/G \xrightarrow{t_\gamma} \mathbb{C}.$$

We extend this definition to products of more than one element of π: Let $\gamma = (\gamma_1, \dots, \gamma_s) \in \pi^s$. Define the *trace function* of γ as the product:

$$t_\gamma([\rho]) := t_{\gamma_1}([\rho]) \dots t_{\gamma_s}([\rho])$$

Using the definition (1) of the action of $\mathsf{Hom}(\pi, \mathfrak{Z})$ on $\mathsf{Hom}(\pi, G)/G$, the action of $u \in \mathsf{Hom}(\pi, \mathfrak{Z})$ on the trace function t_γ is given by $u \cdot t_\gamma$, defined by:

$$\rho \overset{u \cdot t_\gamma}{\longmapsto} u(\gamma)^{-1} t_\gamma(\rho) \tag{2}$$

since

$$\begin{aligned}(u \cdot t_\gamma)(\rho) &:= t_\gamma(u^{-1} \cdot \rho) \\ &= \mathsf{tr}\big((u^{-1} \cdot \rho)(\gamma)\big) \\ &= \mathsf{tr}\big(u(\gamma)^{-1} \rho(\gamma)\big) \\ &= u(\gamma)^{-1} \mathsf{tr}(\rho(\gamma)) \\ &= u(\gamma)^{-1} t_\gamma(\rho)\end{aligned}$$

Since $\mathfrak{Z} \cong \mathbb{Z}/n$, the evaluation $u(\gamma)$ depends only on the *total homology class*

$$[\gamma] \;:=\; [\gamma_1] + \cdots + [\gamma_s] \;\in\; H_1(\Sigma; \mathbb{Z}/n),$$

defined as the sum of the $\mathbb{Z}/n$-homology classes of the γ_i. The evaluation of $u \in \mathsf{Hom}(\pi, \mathfrak{Z})$ on γ is just the natural pairing

$$\begin{aligned}\mathsf{Hom}(\pi, \mathfrak{Z}) \times H_1(\Sigma; \mathbb{Z}/n) &\longrightarrow \mathbb{Z}/n \cong Z \\ (u, [\gamma]) &\longmapsto u \cdot [\gamma]\end{aligned}$$

where $[\gamma]$ is the total homology class of γ.

LEMMA 1. *If* $[\gamma] \neq 0 \in H_1(\Sigma; \mathbb{Z}/n)$, *then*

$$\int t_\gamma d\mu = 0.$$

PROOF. Since the homology class of γ in $H_1(\Sigma, \mathbb{Z}/n)$ is nonzero, $u(\gamma) \neq 1$ for some $u \in \mathsf{Hom}(\pi, \mathfrak{Z})$. Since $(u^{-1})_* \mu = \mu$,

$$\int t_\gamma \, d\mu \;=\; \int (t_\gamma \circ u^{-1}) \, d\Big((u^{-1})_* \mu \Big) \;=\; u(\gamma)^{-1} \int t_\gamma d\mu$$

by (2). Since $u(\gamma)^{-1} \neq 1$, the integral is zero as claimed. □

An element $\gamma \in \pi$ is *nonseparating simple* if it contains a nonseparating simple loop which is homotopically nontrivial and not homotopic to $\partial\Sigma$. Since such an element has nontrivial homology class $[\gamma] \in H_1(\Sigma, \mathbb{Z}/n)$, Lemma 1 implies:

COROLLARY 2. *Suppose* γ *is nonseparating simple. Then* $t_\gamma \in \mathcal{H}_0$.

For $n = 2$, Frohman and Kania-Bartoszynska [**1**] have calculated $\int t_\gamma d\mu$ for separating simple loops γ in terms of Bernoulli numbers and $6j$-symbols.

3. Invariant subspaces

Let $\mathcal{W} \subset \mathcal{H}_0$ denote the closure of the span of all t_α, where α is nonseparating simple loops. (Alternatively, let α run over all elements of π with nonzero homology class modulo n.) Clearly $\mathcal{W}$ is $\mathsf{Out}(\pi)$-invariant.

PROPOSITION 3. *$\mathcal{W}$ is a proper subspace of $\mathcal{H}_0$.*

PROOF. Choose $\gamma \in \pi^s$ such that:

- t_γ is not constant;
- γ has trivial homology class in $H_1(\Sigma, \mathbb{Z}/n)$.

Such elements are easy to find: for example if γ consists of one nontrivial separating simple loop, or if γ consists of one n-th power, or if $\gamma = (\gamma_1, \dots, \gamma_1)$ where s is a positive multiple of n and γ_1 is nontrivial.

Normalize t_γ so that it lies in $\mathcal{H}_0$:

$$\hat{t}_\gamma := t_\gamma - \int t_\gamma d\mu.$$

We claim that $\hat{t}_\gamma$ is a nonzero vector in $\mathcal{H}_0$ orthogonal to $\mathcal{W}$. It is nontrivial since t_γ is nonconstant. Furthermore if $\alpha \in \pi$ has nonzero homology class $[\alpha] \in H_1(\Sigma, \mathbb{Z}/n)$ (for example if it nonseparating simple), then t_α is orthogonal to $\hat{t}_\gamma$.

Since $\rho(\alpha)$ is unitary, $\overline{t_\alpha} = t_{\alpha^{-1}}$. Thus

$$\begin{aligned} \langle \hat{t}_\gamma, t_\alpha \rangle &= \int \hat{t}_\gamma \overline{t_\alpha} d\mu = \int \hat{t}_\gamma t_{\alpha^{-1}} d\mu \\ &= \int t_\gamma t_{\alpha^{-1}} d\mu \text{ (by Corollary 2)} \\ &= \int t_\eta d\mu \end{aligned}$$

where $\eta = (\gamma_1, \dots, \gamma_s, \alpha^{-1})$. Since

$$[\eta] = [\gamma] - [\alpha] \neq 0,$$

Lemma 1 implies that $\langle \hat{t}_\gamma, t_\alpha \rangle = 0$, as desired. □

When π is a free group and $n = 2$, then ergodicity and weak-mixing of the action of $\mathsf{Out}(\pi)$ on $\mathsf{Hom}(\pi, G)/G$ is proved in Goldman [**4**]. Exactly the same argument as above (with $\mathcal{W}$ replaced by the closure of the span of t_γ where γ is an element of a free generating set of π) shows that the unitary representation of $\mathsf{Out}(\pi)$ on $L^2(\mathsf{Hom}(\pi, G)/G, \mu)$ contains invariant (necessarily infinite-dimensional) subspaces other than the constants.

References

1. Frohman, C. and Kania-Bartoszynska, J., *Shadow world evaluation of the Yang-Mills measure,* Algebraic and Geometric Topology, **4** (2001), 311–332.
2. Goldman, W., *The symplectic nature of fundamental groups of surfaces,* Adv. Math. **54** (1984), 200–225.
3. ______, *Ergodic theory on moduli spaces,* Ann. Math. **146** (1997), 475–507.
4. ______, *An ergodic action of the outer automorphism group of a free group,* `math.DG/0506401` Geometric and Functional Analysis (to appear)
5. Huebschmann, J., *Symplectic and Poisson structures of certain moduli spaces,* Duke Math J. **80** (1995) 737-756.

6. Jeffrey,L. and Weitsman, J., *Bohr-Sommerfeld orbits and the Verlinde dimension formula,* Commun. Math. Phys. 150 (1992) 593-630
7. ———, *Toric structures on the moduli space of flat connections on a Riemann surface: Volumes and the moment map,* Adv. Math. **109,** 151-168 (1994).
8. Pickrell, D. and Xia, E., *Ergodicity of Mapping Class Group Actions on Representation Varieties, I. Closed Surfaces,* Comment. Math. Helv. **77** (2001), 339–362.
9. ———, *Ergodicity of Mapping Class Group Actions on Representation Varieties, II. Surfaces with Boundary,* Transformation Groups **8** (2003), no. 4, 397–402.
10. Raghunathan, M., "Discrete Subgroups of Lie Groups," Ergebniße der Math. **58**, Springer-Verlag Berlin-Heidelberg-New York (1972).
11. Weil, A., *Remarks on the cohomology of groups,* Ann. Math. **80** (1964), 149–157.

MATHEMATICS DEPARTMENT, UNIVERSITY OF MARYLAND, COLLEGE PARK, MD 20742 USA
E-mail address: wmg@math.umd.edu

Contemporary Mathematics
Volume **416**, 2006

Pro-p link groups and p-homology groups

Jonathan Hillman, Daniel Matei and Masanori Morishita

Contents

Introduction

Let L be a tame link in the 3-sphere S^3 consisting of n knots $K_1, \cdots, K_n$ and let G_L be the link group $\pi_1(X_L)$, $X_L = S^3 \setminus L$. For a prime number p, let $\widehat{G_L}$ denote the pro-p completion of the group G_L, $\widehat{G_L} = \varprojlim G_L/N$ where N runs over normal subgroups of G_L having p-power indices. By a theorem of J. Milnor [Mi], it is shown that $\widehat{G_L}$ has the following simple presentation as a pro-p group

$$\widehat{G_L} = \langle x_1, \cdots, x_n \mid [x_1, y_1] = \cdots = [x_n, y_n] = 1 \rangle$$

where x_i and y_i represent the meridian and longitude around K_i respectively (Theorem 1.1.4). The purpose of this paper is to use the pro-p link group $\widehat{G_L}$ and the associated group-theoretic invariants for the study of the p-homology groups of p^m-fold cyclic branched covers of S^3 along L, following the analogies between link theory and number theory [Mo1$\sim$4],[Rez1,2]. The invariants we derive from $\widehat{G_L}$ are the p-adic Milnor invariants and the completed Alexander module over the formal power series ring $\widehat{\Lambda_n} = \mathbf{Z}_p[[X_1, \cdots, X_n]]$ with coefficients in the ring $\mathbf{Z}_p$ of p-adic integers. The tool involved here is the Fox differential calculus on a free pro-p group [Ih]. Although these invariants are simply p-adic analogues of the usual Milnor invariants and Alexander modules, it is natural to work over $\widehat{\Lambda_n}$ since the completed Alexander module can be presented over $\widehat{\Lambda_n}$ by a sort of *universal p-adic higher linking matrix* $\widehat{T_L}$, called the *p-adic Traldi matrix*. This is defined in terms of the p-adic Milnor numbers and we can derive from $\widehat{T_L}$ systematically the "p-primary" information on the homology of p^m-fold branched covers of L. This is an idea analogous to Iwasawa theory [Iw] which may also be regared as a p-adic strengthening of the method employed by W. Massey [Mas] and L. Traldi [T]. We note that the method using the truncated Traldi matrices was considered in [Mat] to study the homology of unbranched covers.

The homology of cyclic branched covers of a link L is one of the basic invariants of L and has been extensively investigated by many authors. The Betti number and the order have been determined in terms of the Alexander (Hosokawa) polynomial ([HK],[MM],[S1] etc) and further the (Galois) module structure has been studied ([Da],[HS],[S2] etc), however most results are concerned mainly with the part which is prime to the covering degree. In [Rez1,2], A. Reznikov studied the p-homology of p-fold branched covers after the model of the classical problem on p-ideal class groups in number theory (see also [Mo1]). In this paper, we push this line of study in *arithmetic topology* further and determine the Galois module structure of the p-homology of a p-fold branched cover along a link completely in terms of the p-adic higher linking matrices. To be precise, let M be the p-fold cyclic branched cover of S^3 along L obtained from the completion of the p-fold total linking cover of X_L and let σ denote a generator of the Galois group of M over S^3. The homology group $H_1(M, \mathbf{Z}_p) = H_1(M, \mathbf{Z}) \otimes_{\mathbf{Z}} \mathbf{Z}_p$ is then a module over the complete discrete valuation ring $\widehat{\mathcal{O}} := \mathbf{Z}_p[\langle\sigma\rangle]/(\sigma^{p-1} + \cdots + \sigma + 1) = \mathbf{Z}_p[\zeta]$, $\zeta := \sigma \bmod (\sigma^{p-1} + \cdots + \sigma + 1)$. Assume that $H_1(M, \mathbf{Z})$ is finite. Then the p-primary part $H_1(M, \mathbf{Z}_p)$ has the $\mathfrak{p}$-rank $n-1$ ([Mo1],[Rez2]) so that it has form

$$H_1(M, \mathbf{Z}_p) = \bigoplus_{i=1}^{n-1} \widehat{\mathcal{O}}/\mathfrak{p}^{a_i} \quad (a_i \geq 1)$$

as $\widehat{\mathcal{O}}$-module where $\mathfrak{p} := (\zeta - 1)$ is the maximal ideal of $\widehat{\mathcal{O}}$. Hence the determination of the Galois module structure of $H_1(N, \mathbf{Z}_p)$ is equivalent to that of the $\mathfrak{p}^k$-rank

$$e_k := \#\{i \mid a_i \geq k\} \quad (k \geq 1).$$

Our main result is to give formulas for e_k's in terms of the higher linking matrices obtained by specializing the truncated p-adic Traldi matrices at $X_1 = \cdots = T_n = \zeta - 1$ (Theorem 4.1.3). For the simplest case of $k = 2$, our formula reads

$$e_2 = n - 1 - \operatorname{rank}_{\mathbf{F}_p}(C \bmod p)$$

where $C = (C_{ij})$ is the linking matrix defined by $C_{ij} = \operatorname{lk}(K_i, K_j)$ for $i \neq j$ and $C_{ii} = -\sum_{j \neq i} \operatorname{lk}(K_i, K_j)$. In view of the analogy between the linking number and the power residue symbols [Mo2,3], this is seen as a link-theoretic analog of L. Rédei's formula for the 4-rank of the class group of a quadratic field ([Réd1]), and our general result was partly suggested by the relation between Rédei's triple symbol and the 8-rank of a class group [Réd2]. In fact, the whole argument here can be translated into arithmetic [Mo5]. In the last section, we study the asymptotic behavior of the order $|H_1(M_m, \mathbf{Z}_p)|$ for the p^m-fold cyclic branched cover M_m as $m \to \infty$, following Iwasawa theory on $\mathbf{Z}_p$-extensions [Iw]. Though our results obtained in this paper are rather elementary, they seem to indicate further possibilities of our arithmetic approach to link theory.

Acknowledgement. We would like to thank K. Murasugi for helpful communication.

Notation. Throughout this paper, we fix a prime number p. We denote by $\mathbf{F}_p$ the field with p elements and by $\mathbf{Z}_p$ the ring of p-adic integers. Let ord_p denote the additive p-adic valuation extended on the algebraic closure $\overline{\mathbf{Q}_p}$ of the p-adic field

$\mathbf{Q}_p$ with $\mathrm{ord}_p(p) = 1$ and set $|x|_p = p^{-\mathrm{ord}_p(x)}$, $x \in \overline{\mathbf{Q}_p}$. We use the letter q to denote p or 0. For a topological (possibly discrete) group G, we denote by $G^{(k,q)}$ the k-th term of lower central q-series defined by $G^{(1,q)} = G, G^{(k+1,q)} = (G^{(k,q)})^q[G^{(k,q)}, G]$ where for closed subgroups A, B of G, $[A, B]$ stands for the closed subgroup of G generated by $[a, b] = aba^{-1}b^{-1}, a \in A, b \in B$. We simply write $G^{(k)}$ for $G^{(k,0)}$, the k-th term of the lower central series of G. For a pro-finite group G and a complete local ring R, we denote by $R[[G]]$ the completed group ring of G over R [Ko,§7].

1. Pro-p completion of a link group

1.1. *The pro-p completion of a link group.* Let L be a tame link in the 3-sphere S^3 consisting of n component knots $K_1, \cdots, K_n$ and let G_L be the link group $\pi_1(X_L)$, $X_L = S^3 \setminus L$. After the work of K.T. Chen, J. Milnor [Mi] derived the following information about the presentation of the nilpotent quotient $G_L/G_L^{(k,q)}$. Let F be the free group on the n words $x_1, \cdots, x_n$ where x_i represents the meridian m_i around K_i and let $\pi : F \to G_L$ be the meridianal homomorphism defined by $\pi(x_i) = m_i$ $(1 \le i \le n)$.

Theorem 1.1.1 ([Mi]). *For each $k \ge 1$ and i $(1 \le i \le n)$, there is a word $y_i^{(k)}$ in $x_1, \cdots, x_n$ representing the image of the i-th longitude in the quotient $G_L/G_L^{(k,q)}$ such that*

$$y_i^{(k)} \equiv y_i^{(k+1)} \ \ mod \ \ F^{(k,q)} \tag{1.1.2}$$

and such that $\pi : F \to G_L$ induces the isomorphism

$$F/N_k F^{(k,q)} \ \simeq \ G_L/G_L^{(k,q)} \tag{1.1.3}$$

where N_k is the subgroup of F generated normally by $[x_i, y_i^{(k)}]$ $(1 \le i \le n)$.

Let $\widehat{G_L}$ be the pro-p completion of G_L, namely the inverse limit $\varprojlim G_L/N$ of the tower of quotients G_L/N which are finite p-groups. Since the quotients by the lower central p-series of G_L are cofinal in this tower, we have

$$\widehat{G_L} \ = \ \varprojlim_k G_L/G_L^{(k,p)}.$$

Since $\{y_i^{(k)} F^{(k,p)}\}_{k \ge 1}$ forms an inverse system in $\{F/F^{(k,p)}\}_{k \ge 1}$ by (1.1.2), we define the pro-p word y_i to be $(y_i^{(k)} F^{(k,p)})$ in the free pro-p group $\widehat{F} := \varprojlim F/F^{(k,p)}$ which represents the i-th "longitude" in $\widehat{G_L}$ under the map $\widehat{\pi} : \widehat{F} \to \widehat{G_L}$ induced by π. By taking the inverse limit with respect to k in the isomorphism (1.1.3) of finite p-groups for $q = p$, we have the following

Theorem 1.1.4. *The map $\widehat{\pi}$ induces the isomorphism of pro-p groups*

$$\widehat{F}/\widehat{N} \ \simeq \ \widehat{G_L}$$

where $\widehat{N}$ is the closed subgroup of $\widehat{F}$ generated normally by $[x_i, y_i]$ $(1 \le i \le n)$. In particular, we have $\widehat{G_K} \ \simeq \ \mathbf{Z}_p$ for a knot K.

Remark 1.1.5. (1) By the construction above, we note $y_i \equiv y_i^{(k)}$ mod $\widehat{F}^{(k,q)}$ (F is embedded in $\widehat{F}$).

(2) In view of the analogy between knots and primes, the pro-p link group $\widehat{G_L}$ is regarded as an analog of the maximal pro-p Galois group over the rational number field $\mathbf{Q}$ unramified outside prime numbers $p_1, \cdots, p_n$, $p_i \equiv 1 \bmod p$ [Mo2].

Theorem 1.1.4 tells us that from the group-theoretic point of view, any link of n components looks like a pure braid link with n strings after the pro-p completion. In particular, by applying the method of D. Anick [A] to determine the graded quotients of the lower central series of a pure braid link group to our pro-p link group $\widehat{G_L}$, we see that the pro-p analog of Murasugi's conjecture holds (cf. [L]). We define the *mod p linking diagram* of L to be the graph with vertices the components of L and an edge joining K_i and K_j if and only if the linking number $\mathrm{lk}(K_i, K_j) \not\equiv 0$ mod p.

Theorem 1.1.6. *If the mod p linking diagram of L is connected, we have the isomorphisms*

$$\widehat{G_L}^{(q)}/\widehat{G_L}^{(q+1)} \simeq \widehat{F}_1^{(q)}/\widehat{F}_1^{(q+1)} \times \widehat{F}_{n-1}^{(q)}/\widehat{F}_{n-1}^{(q+1)} \quad for \ q \geq 1,$$

where $\widehat{F}_r$ denotes the free pro-p group of rank r.

1.2. *The p-goodness of a link group.* Let G be a group and $\widehat{G}$ be the pro-p completion of G. We then call G *p-good* if the natural map $G \to \widehat{G}$ induces the isomorphisms on cohomology $H^q(\widehat{G}, M) \ \overset{\sim}{\to} \ H^q(G, M)$ for all $q \geq 0$ for any finite p-primary $\widehat{G}$-modules M (cf. [Se]).

Theorem 1.2.1. *A link group G_L is p-good.*

Proof. We shall say that a subgroup G of finite index in G_L is *open* if $[G_L : G]$ is a power of p. Let M be a finite p-primary $\widehat{G_L}$-module. We shall show by induction on the length of M that if G is an open subgroup of G_L then there is a smaller open subgroup G_1 such that restriction from $H^2(G, M)$ to $H^2(G_1, M)$ is trivial.

Suppose first that $M = \mathbf{F}_p$ with trivial G_L-action and let $H^*(G)$ denote $H^*(G, \mathbf{F}_p)$ for ease of reading. Since $[G_L : G]$ is finite and $G_L/G_L^{(2)} \cong \mathbf{Z}^n$ there is an epimorphism $\tau : G \to C = \mathbf{Z}/p\mathbf{Z}$. Then $K = \mathrm{Ker}(\tau)$ is another open subgroup of G_L. The Hochschild-Serre spectral sequence for G as an extension of C by K has E_2 term $E_2^{p,q} = H^p(C, H^q(K))$, r^{th} differential d_r of bidegree $(r, 1-r)$ and converges to $H^*(G)$. Since $H^p(K) = 0$ for $p > 2$ there are only three nonzero rows, and since $H^*(G) = 0$ for $* > 2$ we see that $d_3^{p,2}$ is an isomorphism for all $p \geq 1$. The spectral sequence is an algebra over the ring $H^*(C) = E_2^{*,0}$. Since C has cohomological period 2, the cup product with a generator of $H^2(C) \cong \mathbf{F}_p$ induces isomorphisms $\gamma_2^{p,q} : E_2^{p,q} \cong E_2^{p+2,q}$ such that $d_2^{p+2,q}\gamma_2^{p,q} = \gamma_2^{p+2,q-1}d_2^{p,q}$ for all $p, q \geq 0$. Therefore we have the isomorphisms $\mathrm{Ker}(d_2^{p,q}) \simeq \mathrm{Ker}(d_2^{p+2,q})$, $\mathrm{Im}(d_2^{p,q}) \simeq \mathrm{Im}(d_2^{p+2,q})$ for any $p, q \geq 0$. In particular we have the isomorphisms $\gamma_3^{0,2} : E_3^{0,2} = \mathrm{Ker}(d_2^{0,2}) \cong E_3^{2,2} = \mathrm{Ker}(d_2^{2,2})$ and $\gamma_3^{3,0} : E_3^{3,0} = E_2^{3,0}/\mathrm{Im}(d_2^{1,1}) \cong E_3^{5,0} = E_2^{5,0}/\mathrm{Im}(d_2^{3,1})$ with $d_3^{2,2} \circ \gamma_3^{0,2} = \gamma_3^{2,1} \circ d_3^{0,2}$. It follows that $d_3^{0,2}$ is also an isomorphism, and so $E_\infty^{0,2} = 0$. But the edge homomorphism from $H^2(G)$ to $H^2(K)$ factors through $E_\infty^{0,2} \leq E_2^{0,2} = H^2(K)^C$, and so is 0.

In general, M has a finite composition series whose factors are copies of the simple module $\mathbf{F}_p$. Suppose that M_1 is a maximal proper submodule of M, with quotient $M/M_1 \cong \mathbf{F}_p$. Restriction from G to K induces a homomorphism from the exact sequences of cohomology corresponding to the coefficient sequence $0 \to M_1 \to M \to \mathbf{F}_p \to 0$. The result for $\mathbf{F}_p$ implies that the image of $H^2(G;M)$ lies in the image of $H^2(K, M_1)$. By the hypothesis of induction we may assume the result is true for M_1, and so there is an open subgroup $K_1 < K$ such that restriction from $H^2(K, M_1)$ to $H^2(K_1, M_1)$ is trivial. Hence restriction from $H^2(G, M)$ to $H^2(K_1, M)$ is also trivial. This establishes the inductive step.

In particular, restriction from $H^2(G_L, M)$ to $H^2(J, M)$ is trivial, for some open subgroup J, and so the result follows, as in Exercise 1 of Chapter I.§2.6 of [Se]. (This exercise is stated in terms of profinite completions, but extends easily to the pro-p case). □

Since the cohomological dimension $cd(G_L) \leq 2$, with equality if and only if L is nontrivial, Theorem 1.2.1 gives the corresponding bound for the pro-p completion $\widehat{G_L}$.

Corollary 1.2.2. *The cohomological p-dimension* $cd_p(\widehat{G_L}) \leq 2$.

If the Milnor invariants of L are all 0 mod p (cf. Section 2), then $\widehat{G_L}$ is a free pro-p group and so $cd_p(\widehat{G_L})$ may be strictly less than $cd(G_L)$. In particular, this is so if L is a nontrivial knot.

2. p-adic Milnor invariants

2.1. *The pro-p Fox differential calculus.* Let $\widehat{F}$ be the pro-p completion of the free group F on n generators $x_1, \cdots, x_n$. Y. Ihara [Ih] extended the Fox differential calculus on the abstract free group F ([F]) to that on $\widehat{F}$. The basic result is stated as the following

Theorem 2.1.1 ([Ih]). *There is a unique continuous* $\mathbf{Z}_p$*-homomorphism*

$$\partial_i = \frac{\partial}{\partial x_i} \; : \; \mathbf{Z}_p[[\widehat{F}]] \longrightarrow \mathbf{Z}_p[[\widehat{F}]]$$

for each i $(1 \leq i \leq n)$ *such that any element* $\alpha \in \mathbf{Z}_p[[\widehat{F}]]$ *is expressed uniquely in the form*

$$\alpha = \epsilon(\alpha)1 + \sum_{i=1}^{n} \partial_i(\alpha)(x_i - 1)$$

where ϵ *is the augmentation map* $\mathbf{Z}_p[[\widehat{F}]] \to \mathbf{Z}_p$.

The higher order derivatives are defined inductively by

$$\partial_{i_1} \cdots \partial_{i_r}(\alpha) = \partial_{i_1}(\partial_{i_2} \cdots \partial_{i_r}(\alpha)).$$

Here are some basic rules (cf. [Ih,2]).

(2.1.2) 1. If one restricts ∂_i to $\mathbf{Z}[F]$ under the natural embedding $\mathbf{Z}[F] \to \mathbf{Z}_p[[\widehat{F}]]$, we get the usual Fox derivative on $\mathbf{Z}[F]$ ([F]).
2. $\partial_i(x_j) = \delta_{ij}$ (Kronecker delta).
3. $\partial_i(\alpha\beta) = \partial_i(\alpha)\epsilon(\beta) + \alpha\partial_i(\beta)$ $(\alpha, \beta \in \mathbf{Z}_p[[\widehat{F}]])$.
4. $\partial_i(f^{-1}) = -f^{-1}\partial_i(f)$ $(f \in \widehat{F})$.
5. For $f \in \widehat{F}$ and $a \in \mathbf{Z}_p$, $\partial_i(f^a) = b\partial_i(f)$, where b is any element of $\mathbf{Z}_p[[\widehat{F}]]$ such that $b(f-1) = f^a - 1$.

6. Let $\widehat{F'}$ be another free pro-p group on $x'_1, \cdots, x'_m$ and let $\varphi : \widehat{F} \to \widehat{F'}$ be a continuous surjective homomorphism. Then one has $\partial'_i(\varphi(\alpha)) = \sum_{j=1}^{n} \varphi(\partial_j(\alpha))\partial'_i(\varphi(x_j))$, where $\partial'_i = \frac{\partial}{\partial x'_i}$, $\alpha \in \mathbf{Z}_p[[\widehat{F}]]$.
7. For $f \in \widehat{F}$, $\epsilon(\partial_i^r(f)) = \binom{a}{r}$ where $a = \epsilon(\partial_i(f))$ and $\binom{a}{r} = \frac{a(a-1)\cdots(a-r+1)}{r!} \in \mathbf{Z}_p$.

Let $\mathbf{Z}_p\langle\langle X_1, \cdots, X_n\rangle\rangle$ be the formal power series ring over $\mathbf{Z}_p$ in non-commuting variables $X_1, \cdots, X_n$ which is compact in the topology taking the ideals $I(r)$ of power series with homogeneous components of degree $\geq r$ as the system of neighborhood of 0. The pro-p Magnus embedding M of $\widehat{F}$ into $\mathbf{Z}_p\langle\langle X_1, \cdots, X_n\rangle\rangle^\times$ is defined by

$$M(x_i) = 1 + X_i,\ M(x_i^{-1}) = 1 - X_i + X_i^2 + \cdots$$

and it is extended to the isomorphism $\mathbf{Z}_p[[\widehat{F}]] \simeq \mathbf{Z}_p\langle\langle X_1, \cdots, X_n\rangle\rangle$ of compact $\mathbf{Z}_p$-algebras. The resulting expansion of $\alpha \in \mathbf{Z}_p[[\widehat{F}]]$ is given by the Fox derivatives:

$$\text{(2.1.3)} \qquad M(\alpha) = \epsilon(\alpha) + \sum_{I=(i_1\cdots i_r)} \epsilon_I(\alpha) X_{i_1} \cdots X_{i_r},$$
$$\epsilon_I(\alpha) = \epsilon(\partial_{i_1} \cdots \partial_{i_r}(\alpha)),\quad I = (i_1 \cdots i_r).$$

Finally, we recall the following fact ([Ko],7.14): Let $\widehat{J}_q$ be the two-sided ideal of $\mathbf{Z}_p[[\widehat{F}]]$ generated by q and the augmentation ideal $I_{\mathbf{Z}_p[[\widehat{F}]]}$. Then for $f \in \widehat{F}$ and $k \geq 1$, we have

(2.1.4)
$$f \in \widehat{F}^{(k,q)} \Longleftrightarrow f - 1 \in \widehat{J}_q$$
$$\Longleftrightarrow M(f) = 1 + (\text{term of degree} \geq k) \ \text{(for case } q = 0).$$

2.2. *p-adic Milnor invariants.* We keep the same notation as in Section 1. Let $\widehat{F}$ be the free pro-p group on $x_1, \cdots, x_n$ where each x_i represents the i-th meridian. For a multi-index $I = (i_1 \cdots i_r), r \geq 1$, we define the *$p$-adic Milnor number* $\hat{\mu}(I)$ by

$$\text{(2.2.1)} \qquad \hat{\mu}(I) := \epsilon_{I'}(y_{i_r}) \quad I' = (i_1 \cdots i_{r-1})$$
$$= \epsilon(\partial_{i_1} \cdots \partial_{i_{r-1}}(y_{i_r}))$$

where $y_j \in \widehat{F}$ represents the j-th "longitude" in $\widehat{G_L}$ (cf. Section 1.1). By convention, we set $\hat{\mu}(I) = 0$ for $|I| = r = 1$. We let $\hat{\Delta}(I)$ denote the ideal of $\mathbf{Z}_p$ generated by $\hat{\mu}(J)$ where J runs over all cyclic permutations of proper subsequences of I. We then define the *p-adic Milnor invariant* $\overline{\hat{\mu}}(I)$ by

$$\text{(2.2.2)} \qquad \overline{\hat{\mu}}(I) := \hat{\mu}(I) \bmod \hat{\Delta}(I).$$

Since the usual Milnor number $\mu(I), I = (i_1 \cdots i_r)$ is defined by $\epsilon(\partial_{i_1} \cdots \partial_{i_{r-1}}(y_{i_r}^{(r)}))$ by (2.1.2),1, Remark 1.1.5, (1) and (2.2.1) yield

$$\text{(2.2.3)} \qquad \hat{\mu}(I) = \mu(I) \ \text{ and } \ \hat{\Delta}(I) = \Delta(I)$$

as elements in $\mathbf{Z}_p$. Hence, we have

(2.2.4) $\overline{\hat{\mu}}(I)$'s are isotopy invariants of L and satisfy the same properties such as the cycle symmetry and Shuffle relations $\overline{\mu}(I)$'s enjoy ([Mi]).

Theorem 1.1.4 and (2.1.4) implies the following:

(2.2.5) All $\overline{\hat{\mu}}(I) = 0$ for $|I| < r$ if and only if $\hat{\pi} : \hat{F} \to \widehat{G_L}$ induces an isomorphism $\hat{F}/\hat{F}^{(r,q)} \simeq \widehat{G_L}/\widehat{G_L}^{(r,q)}$. In particular, if all p-adic Milnor invariants of L are zero, one has $\widehat{G_L} \simeq \hat{F}$. This is the case for boundary links.

Finally, we remark that (2.1.2), 7 implies

$$\hat{\mu}(\underbrace{i \cdots i}_{r} j) = \binom{\mathrm{lk}(K_i, K_j)}{r} \quad \text{for } i \neq j. \tag{2.2.6}$$

The Milnor invariant is also given by the Massey products in the cohomology of $\widehat{G_L}$. For the normalized Massey system in profinite group cohomology and sign convention, we refer to [Mo3]. Let $\xi_1, \cdots, \xi_n$ be the $\mathbf{Z}_p$-basis of $H^1(\widehat{G_L}, \mathbf{Z}_p)$ dual to the meridians m_i's, and let $\eta_j \in H_2(\widehat{G_L}, \mathbf{Z}_p)$ be the image of $[x_j, y_j]$ under the transgression $H_1(\hat{N}, \mathbf{Z}_p)_{\widehat{G_L}} \to H_2(\widehat{G_L}, \mathbf{Z}_p)$. Then for $I = (i_1 \cdots i_r), r \geq 2$, there is a normalized Massey system M for the product $\langle \xi_{i_1}, \cdots, \xi_{i_r} \rangle \in H_2(\widehat{G_L}, \mathbf{Z}_p/\hat{\Delta}(I))$ so that

$$\langle \xi_{i_1}, \cdots, \xi_{i_r} \rangle(\eta_j) = \begin{cases} (-1)^r \overline{\hat{\mu}}(I) & j = i_r \neq i_1, \\ (-1)^{r+1} \overline{\hat{\mu}}(I) & j = i_1 \neq i_r, \\ 0 & \text{otherwise.} \end{cases} \tag{2.2.7}$$

3. Completed Alexander modules

3.1. *The Alexander module of* $\widehat{G_L}$. The Alexander module of a finitely presented pro-p group was introduced in several modes in [Mo2]. As a particular case, the Alexander module of the pro-p link group $\widehat{G_L}$ is defined using the pro-p Fox free differential calculus as follows. We keep the same notation as in Section 1 and 2. Let $\hat{\psi}$ be the abelianization map $\widehat{G_L} \to \hat{H} := \widehat{G_L}/\widehat{G_L}^{(2)} = \mathbf{Z}_p^n$ and denote by the same $\hat{\psi}$ the continuous $\mathbf{Z}_p$-homomorphism $\mathbf{Z}_p[[\widehat{G_L}]] \to \mathbf{Z}_p[[\hat{H}]]$ on the completed group rings. We identify $\mathbf{Z}_p[[\hat{H}]]$ with the commutative formal power series ring $\widehat{\Lambda_n} := \mathbf{Z}_p[[X_1, \cdots, X_n]]$ over $\mathbf{Z}_p$ by setting $t_i := \hat{\psi} \circ \hat{\pi}(x_i) = 1 + X_i$. By Theorem 1.1.4, we call the matrix over $\widehat{\Lambda_n}$

$$\widehat{P_L} = \left(\hat{\psi} \circ \hat{\pi}(\partial_j([x_i, y_i]))\right) \tag{3.1.1}$$

the *Alexander matrix* of $\widehat{G_L}$. We then define the *Alexander module* $\widehat{A_L}$ of $\widehat{G_L}$ by the compact $\widehat{\Lambda_n}$-module presented by $\widehat{P_L}$:

$$\widehat{A_L} := \mathrm{Coker}(\widehat{\Lambda_n}^n \xrightarrow{\widehat{P_L}} \widehat{\Lambda_n}^n) \tag{3.1.2}$$

and call $\widehat{A_L}$ the *completed Alexander module* of L over $\widehat{\Lambda_n}$. Let $\hat{\Lambda} = \mathbf{Z}_p[[X]]$, the *Iwasawa algebra* and $\tau : \widehat{\Lambda_n} \to \hat{\Lambda}$ the reducing homomorphism defined by $\tau(X_i) = X$ $(1 \leq i \leq n)$. Then the *reduced completed Alexander module* $\widehat{A_L}^{red}$ is defined by the compact $\hat{\Lambda}$-module

$$\widehat{A_L}^{red} := \tau(\widehat{A_L}) = \widehat{A_L} \otimes_{\widehat{\Lambda_n}} \hat{\Lambda} \tag{3.1.3}$$

which is presented by $\tau(\widehat{P_L}) = \widehat{P_L}(X, \cdots, X)$. For a knot K, $\widehat{P_L} = O_n$ (zero matrix) and so we have

$$(3.1.4) \qquad \widehat{A_L} = \widehat{A_L}^{red} = \widehat{\Lambda}.$$

We also define the *i-th completed Alexander ideal* $\widehat{E_i}(L)$ of L by the i-th elementary ideal $E_i(\widehat{A_L})$ of $\widehat{A_L}$ and *i-th p-adic Alexander series* $\widehat{\Delta_i}(L)$ of L by the greatest common divisor $\Delta_i(\widehat{E_i}(L))$ of generators of the ideal $\widehat{E_i}(L)$:

$$(3.1.5) \qquad \widehat{E_i}(L) := E_i(\widehat{A_L}), \ \ \widehat{\Delta_i}(L) := \Delta_i(\widehat{E_i}(L))$$

The relation with the usual Alexander module is given as follows. Let $\psi : G_L \to H = H_1(X_L, \mathbf{Z})$ be the abelianization map which induces the ring homomorphism $\psi : \mathbf{Z}[G_L] \to \mathbf{Z}[H]$ on the group rings where $\mathbf{Z}[H]$ is identified with the Laurent polynomial ring $\Lambda_n = \mathbf{Z}[t_1^{\pm 1}, \cdots, t_n^{\pm 1}]$, $t_i = \psi \circ \pi(x_i)$. Given a presentation $G_L = \langle x_1, \cdots, x_m \mid r_1 = \cdots r_l = 1 \rangle$ $(m \geq n)$, the Alexander module of L is given as the Λ_n-module presented by the Alexander matrix $(\psi \circ \pi(\partial_j(r_i)))$. By Theorem 1.1.1, we can take $m = n$ and the relators to be $[x_i, y_i^{(k+1)}]$ $(1 \leq i \leq n)$ and some finite number of generators $f_i^{(k+1)}$ of $F^{(k+1,p)}$ $(k \geq 1)$. We let $J_p = \widehat{J_p} \cap \mathbf{Z}[F]$ and $J_{\Lambda_n,p} = \psi \circ \pi(J_p)$. Since $\partial_j(f_i^{(k+1)}) \in {J_p}^k$ by (2.1.4), passing to the quotients modulo $(J_{\Lambda_n,p})^k$, $A_L/(J_{\Lambda_n,p})^k A_L = A_L \otimes_{\Lambda_n} \Lambda_n/(J_{\Lambda_n,p})^k$ is presented by the matrix $\left(\psi \circ \pi(\partial_j([x_i, y_i^{(k+1)}])) \bmod (J_{\Lambda_n,p})^k\right)$. Here we see by Remark 1.1.5, (1) that the elements $\{\psi \circ \pi(\partial_j([x_i, y_i^{(k+1)}])) \bmod (J_{\Lambda_n,p})^k\}$ form an inverse system with respect to k and its limit is given as $\widehat{\psi} \circ \widehat{\pi}(\partial_j([x_i, y_i]))$ under the identification $\varprojlim_k \Lambda_n/(J_{\Lambda_n,p})^k = \widehat{\Lambda_n}$ defined by $t_i = 1 + X_i$. Hence by (3.1.2) we have

$$(3.1.6) \qquad \widehat{A_L} = \varprojlim_k A_L \otimes_{\Lambda_n} (\Lambda_n/(J_{\Lambda_n,p})^k) = A_L \otimes_{\Lambda_n} \widehat{\Lambda_n}.$$

Similarly, $\widehat{A_L}^{red}$ is related with the usual reduced Alexander module $A_L^{red} = \tau(A_L)$ by

$$(3.1.7) \qquad \widehat{A_L}^{red} = A_L \otimes_\Lambda \widehat{\Lambda}$$

where $\Lambda = \mathbf{Z}[t^{\pm 1}]$ is embedded into $\widehat{\Lambda}$ by $t = 1 + X$.

3.2. *The p-adic Traldi matrix.* The Alexander matrix $\widehat{P_L}$ of $\widehat{G_L}$ (3.1.1) is computed explicitly as a *universal p-adic higher linking matrix* in terms of p-adic Milnor numbers. This is regarded as a p-adic strengthening of the work by L. Traldi [Tr].

Definition 3.2.1. The *p-adic Traldi matrix* $\widehat{T_L} = (\widehat{T_L}(i,j))$ of L over $\widehat{\Lambda_n}$ is defined by

$$\widehat{T_L}(i,j) = \begin{cases} -\displaystyle\sum_{r \geq 1} \sum_{\substack{1 \leq i_1, \cdots, i_r \leq n \\ i_r \neq i}} \hat{\mu}(i_1 \cdots i_r i) X_{i_1} \cdots X_{i_r} & i = j \\ \hat{\mu}(ji) X_i + \displaystyle\sum_{r \geq 1} \sum_{1 \leq i_1 \cdots i_r \leq n} \hat{\mu}(i_1 \cdots i_r j i) X_i X_{i_1} \cdots X_{i_r} & i \neq j. \end{cases}$$

and we also define the *reduced p-adic Traldi matrix* $\widehat{T_L}^{red}$ of L over $\widehat{\Lambda}$ by

$$\widehat{T_L}^{red} := \tau(\widehat{T_L}) = \widehat{T_L}(X, \cdots, X).$$

Our theorem is then stated as

Theorem 3.2.2. *The p-adic Traldi matrix $\widehat{T_L}$ gives a presentation matrix for the completed Alexander module $\widehat{A_L}$ over $\widehat{\Lambda_n}$, and the reduced p-adic Traldi matrix $\widehat{T_L}^{red}$ gives a presentation matrix for the reduced completed Alexander module $\widehat{A_L}^{red}$ over $\widehat{\Lambda}$.*

Proof. By (3.1.1),(3.1.2) and (3.1.3), it suffices to show $\widehat{\psi}\circ\widehat{\pi}(\partial_j([x_i, y_i])) = \widehat{T_L}(i,j)$. By the rules $2 \sim 4$ of (2.1.2), we have

$$\partial_j([x_i, y_i]) = (1 - x_i y_i x_i^{-1})\delta_{ij} + x_i(1 - y_i x_i^{-1} y_i^{-1})\partial_j(y_i).$$

Here $y_i = 1 + \sum_{r\geq 1} \sum_{1\leq i_1,\cdots,i_r\leq n} \hat{\mu}(i_1 \cdots i_r i)(x_{i_1} - 1)\cdots(x_{i_r} - 1)$ by (2.1.3) and (2.2.1).
Hence we get

(3)

$$\begin{aligned}\widehat{\psi}\circ\widehat{\pi}\,(\partial_j([x_i, y_i])) = \delta_{i,j} &- \sum_{r\geq 1} \sum_{1\leq i_1,\cdots,i_r\leq n} \hat{\mu}(i_1 \cdots i_r i) X_{i_1} \cdots X_{i_r} \\ &+ \hat{\mu}(ji)X_i + \sum_{r\geq 1} \sum_{1\leq i_1,\cdots,i_r\leq n} \hat{\mu}(i_1 \cdots i_r j i) X_i X_{i_1} \cdots X_{i_r}.\end{aligned}$$

which yields the assertion. □

The following is an extension of (3.1.4).

Corollary 3.2.3. *For a link whose p-adic Milnor invariants are all zero, we have*

$$\widehat{A_L} \simeq \widehat{\Lambda_n}^n, \quad \widehat{A_L}^{red} \simeq \widehat{\Lambda}^n.$$

This is the case for boundary links.

Finally, we introduce the *truncated p-adic Traldi matrices.*

Definition 3.2.4. For $k \geq 2$, the *k-th truncated p-adic Traldi matrix* $\widehat{T_L}^{(k)} = (\widehat{T_L}^{(k)}(i,j))$ is defined by

$$\widehat{T_L}^{(k)}(i,j) = \begin{cases} -\displaystyle\sum_{r=1}^{k-1} \sum_{\substack{1\leq i_1,\cdots,i_r\leq n \\ i_r\neq i}} \hat{\mu}(i_1 \cdots i_r i) X_{i_1} \cdots X_{i_r} & i = j \\ \hat{\mu}(ji)X_i + \displaystyle\sum_{r=1}^{k-2} \sum_{1\leq i_1,\cdots,i_r\leq n} \hat{\mu}(i_1 \cdots i_r j i) X_i X_{i_1} \cdots X_{i_r} & i \neq j \end{cases}$$

and we also define the *k-th truncated reduced p-adic Traldi matrix* $\widehat{T_L}^{red,(k)}$ by

$$\widehat{T_L}^{red,(k)} := \tau(\widehat{T_L}^{(k)}) = \widehat{T_L}^{(k)}(X, \cdots, X).$$

We note that $\widehat{T_L}^{red,(2)}$ is the *linking matrix* multiplied by X, where the linking matrix $C = (C_{ij})$ is defined by $C_{ii} = -\sum_{j\neq i} \mathrm{lk}(K_i, K_j)$ and $C_{ij} = \mathrm{lk}(K_i, K_j)$ for

$i \neq j$. Thus the p-adic Traldi matrix $\widehat{T_L}$ is regarded as a *universal higher linking matrix* over $\widehat{\Lambda_n}$ which contains all information on the completed Alexander module. In the following section, we derive from $\widehat{T_L}$ the information on the p-homology groups of p^m-fold cyclic branched covers along L.

4. Galois module structure for the p-homology group of a p-fold cyclic branched cover

4.1. *Galois module structure of the p-homology of a p-fold cover.* Let X_∞ be the infinite cyclic cover of X_L associated to the kernel of the homomorphism $G_L \to \langle t \rangle$ sending each meridian to t. Let M be the the completion of the p-fold subcover of X_∞ over X_L so that M is a p-fold cyclic cover of S^3 branched along L. We set $\nu_d(t) = t^{d-1} + \cdots + t + 1$ for $d \geq 1$. Let $\phi : M \to S^3$ be the covering map and σ a generator of its Galois group. Since $\nu_p(\sigma) = tr \circ \phi_* = 0$ where $tr : H_1(S^3, \mathbf{Z}) \to H_1(M, \mathbf{Z})$ is the transfer, we can regard $H_1(M, \mathbf{Z})$ as a module over the Dedekind ring $\mathcal{O} = \mathbf{Z}[\langle\sigma\rangle]/(\nu_p(\sigma)) = \mathbf{Z}[\zeta]$, $\zeta = \sigma \bmod (\nu_p(\sigma))$ is a primitive p-th root of 1. Hence $H_1(M, \mathbf{Z}_p) = H_1(M, \mathbf{Z}) \otimes_{\mathbf{Z}} \mathbf{Z}_p$ is regarded as a module over the complete discrete valuation ring $\widehat{\mathcal{O}} = \mathcal{O} \otimes_{\mathbf{Z}} \mathbf{Z}_p = \mathbf{Z}_p[\zeta]$. Note that $\widehat{\mathcal{O}}$ is the completion of $\mathcal{O}$ with respect to the maximal ideal $\mathfrak{p}$ generated by the prime element $\pi := \zeta - 1$ and the residue field $\widehat{\mathcal{O}}/\mathfrak{p}$ is $\mathbf{F}_p$. By Theorem 2.2.1, we can derive the following information on a presentation matrix for the $\widehat{\mathcal{O}}$-module $H_1(M, \mathbf{Z}_p)$. Note that the evaluation of a power series $F(X) \in \mathbf{Z}_p[[X]]$ at $s = \pi$ makes sense in the $\mathfrak{p} = (\pi)$-adically complete ring $\widehat{\mathcal{O}}$.

Theorem 4.1.1. *A presentation matrix for $H_1(M, \mathbf{Z}_p) \oplus \widehat{\mathcal{O}}$ over $\widehat{\mathcal{O}}$ is given by $\widehat{T_L}^{red}(\pi)$. Further, for any integer $k \geq 2$, a presentation matrix for $(H_1(M, \mathbf{Z}_p) \otimes_{\widehat{\mathcal{O}}} \widehat{\mathcal{O}}/\mathfrak{p}^k) \oplus \widehat{\mathcal{O}}/\mathfrak{p}^k$ over $\widehat{\mathcal{O}}/\mathfrak{p}^k$ is given by $\widehat{T_L}^{red,(k)}(\pi)$. Here $\widehat{T_L}^{red}$ (resp. $\widehat{T_L}^{red,(k)}$) is the reduced (resp. reduced truncated) Traldi matrix defined in Section 3.2.*

Proof. Note that the well-known relation ([S1,Theorem 6])

$$H_1(M, \mathbf{Z}) \simeq H_1(X_\infty, \mathbf{Z})/\nu_p(t) H_1(X_\infty, \mathbf{Z})$$

is an $\widehat{\mathcal{O}}$-isomorphism since σ acts on the r.h.s by t. Hence we have the following isomorphisms over $\widehat{\mathcal{O}}$

$$\begin{aligned} H_1(M, \mathbf{Z}_p) &\simeq (H_1(X_\infty, \mathbf{Z})/\nu_p(t) H_1(X_\infty, \mathbf{Z})) \otimes_{\mathbf{Z}} \mathbf{Z}_p \\ &\simeq H_1(X_\infty, \mathbf{Z}) \otimes_\Lambda (\Lambda \otimes_{\mathbf{Z}} \mathbf{Z}_p)/(\nu_p(t))) \\ &\simeq H_1(X_\infty, \mathbf{Z}) \otimes_\Lambda (\widehat{\Lambda}/(\nu_p(1+X))) \\ &\simeq H_1(X_\infty, \mathbf{Z}) \otimes_\Lambda \widehat{\mathcal{O}}. \end{aligned}$$

Since $A_L^{red} \simeq H_1(X_\infty, \mathbf{Z}) \oplus \Lambda$ as Λ-module ([H, 5.4]), tensoring with $\widehat{\mathcal{O}}$ over Λ, we have an isomorphism of $\widehat{\mathcal{O}}$-modules

$$A_L^{red} \otimes_\Lambda \widehat{\mathcal{O}} \simeq H_1(M, \mathbf{Z}_p) \oplus \widehat{\mathcal{O}}.$$

Since the l.h.s is same as $\widehat{A_L}^{red} \otimes_{\widehat{\Lambda}} \widehat{\mathcal{O}}$, the first assertion follows from Theorem 3.2.4. The second assertion is obtained from the first one by taking modulo $\mathfrak{p}^k$. □

Now, we asuume that $H_1(M, \mathbf{Z})$ is finite so that $H_1(M, \mathbf{Z}_p)$ is the p-primary part of $H_1(M, \mathbf{Z})$. Using Theorem 3.1.1, we will see the $\widehat{\mathcal{O}}$-module structure of $H_1(M, \mathbf{Z}_p)$ more precisely. First, we recall the following result on the $\mathfrak{p}$-rank of $H_1(M, \mathbf{Z}_p)$ (cf. [Mo1],[Rez]).

Lemma 4.1.2. *$H_1(M, \mathbf{Z}_p) \otimes_{\widehat{\mathcal{O}}} \mathbf{F}_p$ has dimension $n-1$ over $\mathbf{F}_p$.*

Proof. By [Mo1], the map $\Phi : H_1(M, \mathbf{Z}) \to \mathbf{F}_p^n$ defined by

$$\Phi(c) := (\mathrm{lk}(\phi_*(c), K_1) \bmod p)$$

induces an isomorphism

$$H_1(M, \mathbf{Z})/(\sigma - 1)H_1(M, \mathbf{Z}) \simeq \{(\xi_i) \in \mathbf{F}_p^n \mid \sum_{i=1}^{n} \xi_i = 0\} \simeq \mathbf{F}_p^{n-1}$$

where the l.h.s is $H_1(M, \mathbf{Z}) \otimes_{\mathcal{O}} \mathcal{O}/\mathfrak{p} = H_1(M, \mathbf{Z}_p) \otimes_{\widehat{\mathcal{O}}} \mathbf{F}_p$, and hence we are done. □

By Lemma 4.1.2, $H_1(M, \mathbf{Z}_p)$ has the form

$$H_1(M, \mathbf{Z}_p) = \bigoplus_{i=1}^{n-1} \widehat{\mathcal{O}}/\mathfrak{p}^{a_i} \qquad (a_i \geq 1)$$

as $\hat{\mathcal{O}}$-module. Hence, the determination of $\widehat{\mathcal{O}}$-module structure of $H_1(M, \mathbf{Z}_p)$ is equivalent to the determination of the $\mathfrak{p}^k$-rank

$$e_k := \#\{i \mid a_i \geq k\} \ \ (k \geq 1).$$

By Theorem 4.1.1, we can describe the $\mathfrak{p}^k$-rank e_k in terms of $\widehat{T_L}^{red,(k)}(\pi)$.

Theorem 4.1.3. *For $k \geq 2$, let $\varepsilon_1^{(k)}, \cdots, \varepsilon_{n-1}^{(k)}, \varepsilon_n^{(k)} = 0$ be the elementary divisors of $\widehat{T_L}^{red,(k)}(\pi)$, where $\varepsilon_i^{(k)} | \varepsilon_{i+1}^{(k)}$ $(1 \leq i \leq n-1)$. Then we have*

$$e_k = \#\{i \,|\, \varepsilon_i^{(k)} \equiv 0 \bmod \mathfrak{p}^k\} - 1.$$

Here we may call $\widehat{T_L}^{red,(k)}(\pi)$ the *k-th higher linking matrix* in view of the following

Corollary 4.1.4. *For $k = 2$, we have*

$$e_2 = n - 1 - \ \mathrm{rank}_{\mathbf{F}_p}(C \bmod p)$$

where $C = (C_{ij})$ is the linking matrix of L defined by $C_{ii} = -\sum_{j \neq i} \mathrm{lk}(K_i, K_j)$ and $C_{ij} = \mathrm{lk}(K_i, K_j)$ for $i \neq j$.

Proof. In fact, we have, by definition,

$$\widehat{T_L}^{red,(2)}(\pi) = \pi C$$

and hence $e_2 = n - 1 - \mathrm{rank}_{\mathbf{F}_p}(C \bmod p)$ by Theorem 4.1.3. □

4.2. *2-component case.* We suppose $n = 2$ and keep to assume $H_1(M, \mathbf{Z})$ is finite. By Lemma 4.1.2, $H_1(M, \mathbf{Z}_p)$ has the $\mathfrak{p}$-rank 1 so that we have

$$H_1(M, \mathbf{Z}_p) = \widehat{\mathcal{O}}/\mathfrak{p}^a, \ \ a \geq 1.$$

Hence $e_k = 0$ or 1 for $k \geq 2$, and by Theorem 4.1.3 we have

$$e_k = 1 \Longleftrightarrow \widehat{T_L}^{red,(k)}(\pi) \equiv O_2 \bmod \pi^k.$$

As $\widehat{T_L}^{red,(k)}(1,2)(\pi) = -\widehat{T_L}^{red,(k)}(1,1)(\pi)$, $\widehat{T_L}^{red,(k)}(2,2)(\pi) = -\widehat{T_L}^{red,(k)}(2,1)(\pi)$, we have the following

Theorem 4.2.1. *Suppose $n = 2$. For each $k \geq 1$, assuming $e_k = 1$, we have*

$$e_{k+1} = 1 \Longleftrightarrow \begin{cases} \displaystyle\sum_{r=1}^{k} \sum_{i_1,\cdots,i_{r-1}=1,2} \hat{\mu}(i_1 \cdots i_{r-1}21)\pi^r \equiv 0 \bmod \pi^{k+1}, \\ \displaystyle\sum_{r=1}^{k} \sum_{i_1,\cdots,i_{r-1}=1,2} \hat{\mu}(i_1 \cdots i_{r-1}12)\pi^r \equiv 0 \bmod \pi^{k+1}. \end{cases}$$

We give the condition in Theorem 4.2.1 in more concise forms for lower k. In the following computation, we use simply the usual Milnor number $\mu(I)$ instead of $\hat{\mu}(I)$ by (2.2.3).

Example 4.2.2. e_2: Since $e_1 = 1$, we have by Theorem 4.2.1

(4.2.2.1) $$e_2 = 1 \Longleftrightarrow \mu(12)\pi \equiv 0 \bmod \pi^2 \Longleftrightarrow \mathrm{lk}(K_1, K_2) \equiv 0 \bmod p.$$

e_3: Assume $\mathrm{lk}(K_1, K_2) \equiv 0 \bmod p$. By Theorem 4.2.1, we have

$$e_3 = 1 \Longleftrightarrow \begin{cases} \mu(21)\pi + (\mu(121) + \mu(221))\pi^2 \equiv 0 \bmod \pi^3, \\ \mu(12)\pi + (\mu(112) + \mu(212))\pi^2 \equiv 0 \bmod \pi^3. \end{cases}$$

By cycle symmetry, $\mu(121) \equiv \mu(112)$, $\mu(221) \equiv \mu(212) \bmod \mu(12)$. Here $\mu(112) \equiv \mu(221) \equiv \binom{\mu(12)}{2} \bmod \mu(12)$ by (2.2.6). Thus we have $\mu(121) + \mu(221) \equiv \mu(112) + \mu(212) \equiv \mu(12) \equiv 0 \bmod p$. Hence, we have

(4.2.2.2) $$e_3 = 1 \Longleftrightarrow \mathrm{lk}(K_1, K_2) \equiv 0 \bmod p^2.$$

As one easily sees, this condition is also equivalent to

(4.2.2.3) $$\mu(112) \equiv \mu(221) \equiv 0 \bmod p.$$

e_4: Assume $\mathrm{lk}(K_1, K_2) \equiv 0 \bmod p^2$. By Theorem 4.2.1, we have

$$e_4 = 1 \Longleftrightarrow \begin{cases} \mu(21)\pi + (\mu(121) + \mu(221))\pi^2 \\ \quad +(\mu(1121) + \mu(1221) + \mu(2121) + \mu(2221))\pi^3 \equiv 0 \bmod \pi^4, \\ \mu(12)\pi + (\mu(112) + \mu(212))\pi^2 \\ \quad +(\mu(1112) + \mu(1212) + \mu(2112) + \mu(2212))\pi^3 \equiv 0 \bmod \pi^4. \end{cases}$$

As in case of e_3, we have $\mu(121)+\mu(221) \equiv \mu(12) \equiv 0 \bmod p^2$. Similarly, $\mu(1121) \equiv \mu(1112) \equiv \binom{\mu(12)}{3} \bmod \Delta(1121)$ and $\mu(2221) \equiv \binom{\mu(12)}{3} \bmod \Delta(2221)$ by (2.2.6). Since $\Delta(1121) \equiv \Delta(2221) \equiv 0 \bmod p$, $\mu(1121) + \mu(2221) \equiv \mu(12)/3 \equiv 0 \bmod p$. Finally, by shuffle relation, $\mu(1221) + \mu(2121) + \mu(2211) \equiv \mu(2121) + 2\mu(2211) \equiv 0 \bmod p$. Thus the first condition is equivalent to $\mu(21)\pi - \mu(2211)\pi^3 \equiv 0 \bmod \pi^4$. Similarly, we see that the second condition is equivalent to $\mu(12)\pi - \mu(1122)\pi^3 \equiv 0 \bmod \pi^4$ which is same as the first one. Hence, we obtain

$$e_4 = 1 \Longleftrightarrow \mathrm{lk}(K_1, K_2) - \mu(1122)\pi^2 \equiv 0 \bmod \pi^3.$$

For case $p = 2$, this is equivalent to the following condition*

$$(4.2.2.4) \quad \begin{cases} \mathrm{lk}(K_1, K_2) \equiv 0 \bmod 8, \ \mu(1122) \equiv 0 \bmod 2 \\ \text{or} \\ \mathrm{lk}(K_1, K_2) \equiv 0 \bmod 4, \mathrm{lk}(K_1, K_2) \not\equiv 0 \bmod 8, \mu(1122) \equiv 1 \bmod 2. \end{cases}$$

For example, the Whitehead link $L = K_1 \cup K_2$ satisfies $\mathrm{lk}(K_1, K_2) = 0, \mu(1122) = 1$ and so $e_3 = 1, e_4 = 0$, hence $H_1(M, \mathbf{Z}_p) = \widehat{\mathcal{O}}/\mathfrak{p}^3$. For the 2-bridge link of type (48,37), the latter condition of (4.2.2.4) is satisfied and so $H_1(M, \mathbf{Z}_2) = \mathbf{Z}/2^k\mathbf{Z}$, $k \geq 4$.

5. Iwasawa type formulas for the p-homology groups of p^m-fold cyclic branched covers

5.1. *Asymptotic formula for the p-homology of p^m-fold covers.* For $m \geq 1$, let M_m be the completion of the p^m-fold subcover of X_∞ over X_L so that M_m is a p^m-fold cyclic cover of S^3 branched along L. In this last Section, we are concerned with the asymptotic behavior of the order of $H_1(M_m, \mathbf{Z}_p)$ as $m \to \infty$ using the standard argument in Iwasawa theory. As in Section 4, we start again with the following isomorphisms

$$(5.1.1) \qquad \widehat{A_L}^{red} \simeq (H_1(X_\infty, \mathbf{Z}) \otimes_\Lambda \widehat{\Lambda}) \oplus \widehat{\Lambda},$$

$$(5.1.2) \qquad H_1(M_m, \mathbf{Z}_p) \simeq (H_1(X_\infty, \mathbf{Z}) \otimes_\Lambda \widehat{\Lambda})/(\nu_{p^m}(1 + X)).$$

From these, we get immediately an extension of a theorem of M. Dellomo [D] for a knot.

Proposition 5.1.3 *For a link L whose p-adic Milnor invariants are all zero, for example a boundary link, we have $H_1(M_m, \mathbf{Z}_p) = \mathbf{Z}_p^{(p^m-1)(n-1)}$ for $m \geq 1$. In particular, $H_1(M_m, \mathbf{Z}_p) = 0$ for $m \geq 1$ if L is a knot.*

Proof. In fact, $\widehat{A_L}^{red} = \widehat{\Lambda}^n$ for such a link L by Corollary 3.2.5. Hence $H_1(X_\infty, \mathbf{Z}) \otimes_\Lambda \widehat{\Lambda} = \widehat{\Lambda}^{n-1}$ by (5.1.1) and so $H_1(M_m, \mathbf{Z}_p) = \mathbf{Z}_p^{(p^m-1)(n-1)}$ by (5.1.2). □

In the following, we assume that $n \geq 2$. By (5.1.1), the 0-th elementary ideal $E_0(H_1(X_\infty, \mathbf{Z}) \otimes_\Lambda \widehat{\Lambda})$ over $\widehat{\Lambda}$ is same as the 1st completed Alexander ideal $\widehat{E_1}(L)$ (3.1.5). Note that the 1st p-adic Alexander series $\widehat{\Delta_1}(L)$ is given as the greatest common divisors of all $n-1$ minors of the reduced p-adic Traldi matrix and so it is written by the form

$$\widehat{\Delta_1}(L) = X^{n-1} \cdot \widehat{\nabla_L}$$

where we call $\widehat{\nabla_L}$ the *p-adic Hosokawa series* of L. Then by (5.1.2), we have the following formula on the order $|H_1(M_m, \mathbf{Z}_p)|$ which is seen as the p-primary part of the well known formula by S. Kinoshita and F. Hosokawa [KT] (See also [MM]). Here we interpret $|H_1(M_m, \mathbf{Z}_p)| = 0$ to mean $H_1(M_m, \mathbf{Z}_p)$ is infinite.

Proposition 5.1.4. $|H_1(M_m, \mathbf{Z}_p)| = p^{m(n-1)} \prod_{\substack{\zeta^{p^m}=1 \\ \zeta \neq 1}} |\widehat{\nabla_L}(\zeta - 1)|_p^{-1}.$

*K. Murasugi informed us of this condition and examples which are obtained by the relation between the Alexander polynomial and Milnor invariants [Mu].

Now, we assume $H_1(M_m, \mathbf{Z}_p)$ is finite for any m and see the asymptotic behaviour of the order $|H_1(M_m, \mathbf{Z}_p)|$ as $m \to \infty$. For this, we recall the following standard facts from Iwasawa theory. We call a polynomial $g(X) \in \mathbf{Z}_p[X]$ *distinguished* if $g(X) = X^d + a_{d-1}X^{d-1} + \cdots + a_0$, $a_i \equiv 0 \bmod p$ for $0 \leq i \leq d-1$.

Lemma 5.1.5 (p-adic Weierstrass preparation theorem [W, Theorem 7.3]). *A non-zero element $f(X) \in \widehat{\Lambda}$ is written uniquely as*

$$f(X) = p^{\mu} g(X) u(X)$$

where μ is a non-negative integer, $g(X)$ is a distinguished polynomial and $u(X) \in \widehat{\Lambda}^{\times}$.

Lemma 5.1.6 ([W,Theorem 7.14]). *Let $f(X) \in \widehat{\Lambda}$ and assume $f(\zeta - 1) \neq 0$ for any primitive p^m-th root ζ of 1 for $k \geq 1$. Write $f(X) = p^{\mu} g(X) u(X)$ according to Lemma 5.5 and define λ by the degree of $g(X)$. Then there is an integer ν independent of k such that we have the equality*

$$\mathrm{ord}_p\Big(\prod_{\substack{\zeta^{p^m}=1 \\ \zeta \neq 1}} f(\zeta - 1)\Big) = \lambda m + \mu p^m + \nu$$

for sufficiently large m.

For the convenience of the reader, we include herewith a proof of Lemma 5.1.6.

Proof of Lemma 5.1.6. Since $u(x) \in \widehat{\Lambda}^{\times}$, we have

$$\mathrm{ord}_p\Big(\prod_{\substack{\zeta^{p^m}=1 \\ \zeta \neq 1}} f(\zeta - 1)\Big) = (p^m - 1)\mu + \mathrm{ord}_p\Big(\prod_{\substack{\zeta^{p^m}=1 \\ \zeta \neq 1}} g(\zeta - 1)\Big).$$

Write $g(X) = X^{\lambda} + a_{\lambda-1}X^{\lambda-1} + \cdots + a_0$, $a_i \equiv 0 \bmod p$ $(0 \leq i \leq \lambda - 1)$. For a primitive p^l-th root ζ of 1 $(1 \leq l \leq k)$, one has the equality $(p) = (\zeta - 1)^{\phi(p^l)}$ of ideals of $\mathbf{Z}[\zeta]$ and so $\mathrm{ord}_p((\zeta - 1)^{\lambda}) = \frac{\lambda}{\phi(p^l)}$ where $\phi(x)$ is the Euler function. Therefore, if l is large enough, $\mathrm{ord}_p((\zeta - 1)^{\lambda}) < \mathrm{ord}_p(a_i(\zeta - 1)^i)$ for $0 \leq i \leq \lambda - 1$ and so $\mathrm{ord}_p(g(\zeta - 1)) = \mathrm{ord}_p((\zeta - 1)^{\lambda})$. Hence, there is a constant C independent of k such that for sufficiently large m, we have

$$\mathrm{ord}_p\Big(\prod_{\substack{\zeta^{p^m}=1 \\ \zeta \neq 1}} g(\zeta - 1)\Big) = \mathrm{ord}_p\Big(\prod_{\substack{\zeta^{p^m}=1 \\ \zeta \neq 1}} (\zeta - 1)^{\lambda}\Big) + C = \mathrm{ord}_p(p^{m\lambda}) + C = \lambda m + C. \square$$

To apply Lemma 5.1.6 to $\widehat{\nabla_L}$, write

$$\widehat{\nabla_L} = p^{\mu(L;p)} g(L;p) u(L;p)$$

where $\mu(L;p)$ is a nonnegative integer, $g(L;p)$ is a distinguished polynomial and $u(L,p) \in \widehat{\Lambda}^{\times}$ according to Lemma 5.1.5 and set $\lambda(L;p) = \deg(g(L;p))$. Then Proposition 5.1.4 and Lemma 5.1.6 yield the following

Theorem 5.1.7. *Notation and assumption being as above, there is a constant $\nu(L;p)$ depending only on L and p such that we have*

$$\mathrm{ord}_p(|H_1(M_m, \mathbf{Z}_p)|) = (n - 1 + \lambda(L;p))m + \mu(L;p)p^m + \nu(L;p)$$

for sufficiently large m.

We call the invariants $\lambda(L;p), \mu(L;p)$ the *Iwasawa* λ, μ-*invariants* of L with respect to p respectively after the model of the Iwasawa invariants in the theory of $\mathbf{Z}_p$-extensions [Iw].

5.2. *Examples.*

1. Let L be the Whitehead link. We then have

$$\widehat{T_L} = \begin{pmatrix} X_1X_2^2 & -X_1^2X_2 \\ X_1^2X_2 & -X_1X_2^2 \end{pmatrix}$$

and $\widehat{\nabla_L} = X^2$. Hence, we have

$$\lambda(L;p) = 2, \mu(L;p) = 0 \text{ and } \mathrm{ord}_p(H_1(M_m, \mathbf{Z}_p)) = 3m \text{ for } m \geq 1.$$

2. Let $L = K_1 \cup K_2 \cup K_3$ be the Borromean rings so that we can take $y_1 = [x_3, x_2], y_2 = [x_3, x_1], y_3 = [x_1, x_2]$. Then we can compute all Milnor number needed to get the reduced Traldi matrix

$$\widehat{T_L}^{red} = \begin{pmatrix} X + X^2 & -X^2 & -X \\ -X & X + X^2 & -X^2 \\ -X^2 & -X & X + X^2 \end{pmatrix}$$

and so $\widehat{\nabla_L} = 1 + X + X^2$. Hence, we have

$$\lambda(L;p) = \mu(L;p) = 0 \text{ and } \mathrm{ord}_p(H_1(M_m, \mathbf{Z}_p)) = 2m \text{ for } m \geq 1.$$

References

[A] D. Anick, *Inert sets and the Lie algebra associated to a group*, J. Algebra **111** (1987), 154–165.

[Da] J. Davis, *The homology of cyclic branched covers of S^3*, Math. Ann. **301** (1995), 507–518.

[De] M. Dellomo, *On the inverse limit of the branched cyclic covers associated with a knot*, J. Pure Appl. Algebra **40** (1986), 15–26.

[F] R.H. Fox, *Free differential calculus. I: Derivation in the free group ring*, Ann. of Math **57** (1953), 547-560.

[H] J.A. Hillman, *Algebraic invariants of links*, Series on Knots and Everything, **32**, World Scientific Publishing Co, 2002.

[HS] J. Hillman, M. Sakuma, *On the homology of finite abelian coverings of links*, Canad. Math. Bull. **40** (1997), 309-315.

[HK] F. Hosokawa, S. Kinoshita, *On the homology group of branched cyclic covering spaces of links*, Osaka Math. J. **12** (1960), 331–355.

[Ih] Y. Ihara, *On Galois representations arising from towers of coverings of* $\mathbf{P}^1 \setminus \{0, 1, \infty\}$, Invent. Math. **86** (1986), 427-459.

[Iw] K. Iwasawa, *On* Γ*-extensions of algebraic number fields*, Bull. Amer. Math. Soc. **65** (1959), 183–226.

[Ko] H. Koch, *Galoissche Theorie der p-Erweiterungen*, Springer-Verlag, Berlin-New York; VEB Deutscher Verlag der Wissenschaften, Berlin, 1970.

[L] J.P Labute, *The Lie algebra associated to the lower central series of a link group and Murasugi's conjecture*, Proc. Amer. Math. Soc. **109** (1990), 951–956.

[Mas] W. Massey, *Completion of link modules*, Duke Math. J. **47** (1980), 399-420.

[Mat] D. Matei, *Fundamental groups of links and arrangements: Characteristic varieties, resonance varieties and finite index subgroups*, Ph.D Thesis, Northeastern University (1999).

[MM] J.P. Mayberry, K. Murasugi, *Torsion-groups of abelian coverings of links*, Trans. Amer. Math. Soc. **271,** (1982), 143-173.

[Mi] J. Milnor, *Isotopy of links*, in Algebraic Geometry and Topology, A symposium in honour of S. Lefschetz (edited by R.H. Fox, D.S. Spencer and W. Tucker), Princeton Univ. Press, Princeton (1957), 280-306.

[Mo1] M. Morishita, *A theory of genera for cyclic coverings of links*, Proc. Japan Academy **77** (2001), 115-118.

[Mo2] ———, *On certain analogies between knots and primes*, J. Reine Angew. Math. **550** (2002), 141-167.

[Mo3] ———, *Milnor invariants and Massey products for prime numbers*, Compositio Math. **140** (2004), 69-83.

[Mo4] ———, *Analogies between knots and primes, 3-manifolds and number rings*, to appear in Sugaku Expositions.

[Mo5] ———, *Milnor invariants and l-class groups*, submitted (2004).

[Mu] K. Murasugi, *On Milnor's invariant for links*, Trans. Amer. Math. Soc. **124** (1966), 94-110.

[Réd1] L. Rédei, *Arithmetischer Beweis des Satzes über die Anzahl der durch vier teilbaren Invarianten der absoluten Klassengruppe im quadratischen Zahlkörper*, J. Reine Angew. Math. **171** (1934), 55-60.

[Réd2] ———, *Ein neues zahlentheoretisches Symbol mit Anwendungen auf die Theorie der quadratischen Zahlkörper, I*, J. Reine Angew. Math. **180** (1938), 1-43.

[Rez1] A. Reznikov, *Three-manifolds class field theory (homology of coverings for a nonvirtually b_1-positive manifold)*, Selecta Math. New ser. **3** (1997), 361–399.

[Rez2] ———, *Embedded incompressible surfaces and homology of ramified coverings of three-manifolds*, Selecta Math. New ser. **6** (2000), 1-39.

[S1] M. Sakuma, *Homology groups of abelian coverings of links*, Math. Seminar Note, Kobe Univ. **7** (1979), 515-530.

[S2] ———, *Homology of abelian coverings of links and spatial graphs*, Canad. J. Math. **47** (1995), 201-224.

[Se] J.-P. Serre, *Cohomologie Galoisienne*, Lecture Notes in Mathematics **5**, Springer-Verlag, Berlin - Heidelberg - New York, 1965.

[Tr] L. Traldi, *Milnor's invariants and the completions of link modules*, Trans. Amer. Math. Soc. **284** (1984), 401-424.

[Tu] V.G. Turaev, *Milnor's invariants and Massey products*, English transl. J. Soviet Math. **12** (1979), 128-137.

[W] L. Washington, *Introduction to cyclotomic fields*, GTM, **83**, Springer-Verlag, New York, 1982.

J.H: School of Mathematics and Statistics, The University of Sydney, Sydney, NSW 2006, Australia.

jonh@maths.usyd.edu.au

D.M: Institute of Mathematics of the Romanian Academy, P.O. BOX 1-764, RO-014700, Bucharest, Romania, and Graduate School of Mathematical Sciences, University of Tokyo, 3-8-1 Komaba, Meguro, Tokyo, 153-8914, Japan.

daniel.matei@imar.ro

M.M: Graduate School of Mathematics, Kyushu University, Fukuoka, 812-8581, Japan.

morisita@math.kyushu-u.ac.jp

Contemporary Mathematics
Volume **216**, 2006

A Quantum Introduction to Knot Theory

Hitoshi Murakami

ABSTRACT. This is an introduction to knot theory for non-specialists, focused on quantum invariants. I describe the homologies of covering spaces of a knot, the Alexander polynomial, the HOMFLY polynomial, the colored Jones polynomial, and the volume conjecture. There are essentially no new results in this article.

In this article I would like to introduce knot theory to non-specialists.

After V. Jones' discovery of his celebrated knot invariant [**20**], much attention has been paid to 'quantum' invariants corresponding to representations of quantum groups. But 'classical' knot invariants, essentially defined by using the fundamental group of the knot complement, are still quite interesting objects.

I start with calculation of the first homology group of the n-fold branched covering space of the three-sphere branched along a knot. It was shown by D. Silver and S. Williams that the growth rate of the order of the homology group with respect to n is given by the Mahler measure of the Alexander polynomial [**51**], the order of the first homology group of the infinite cyclic cover.

Next I will review a formal, recursive calculation of the Alexander polynomial due to J. Conway [**6**] and generalize it to introduce the HOMFLY polynomial, the quantum invariant corresponding to the fundamental representation of $\mathfrak{sl}_n$. I use a graphical approach to define it, and it turns out that it is a generalization of L. Kauffman's definition of the Jones polynomial [**26**]. In fact the Jones polynomial is the quantum invariant corresponding to the fundamental (2-dimensional) representation of $\mathfrak{sl}_2$.

There is another generalization of the Jones polynomial. For the N-dimensional irreducible representation of $\mathfrak{sl}_2$ one can define a knot invariant, which is called the N-dimensional colored Jones polynomial. We use a graphical way again to define it.

So far we know very little about geometric meanings of the quantum invariants compared with the Alexander polynomial. The N-dimensional colored Jones polynomial evaluated at the N-th root of unity would have an important geometric

---

2000 *Mathematics Subject Classification.* Primary 57M25, 57M27.

Key words and phrases. knot, cyclic cover, Alexander polynomial, HOMFLY polynomial, colored Jones polynomial, volume conjecture.

This research is partially supported by Grant-in-Aid for Scientific Research (B).

meaning. I will describe the Volume Conjecture by R. Kashaev [**24**], and J. Murakami and I [**36**] that would relate the asymptotic behavior of the colored Jones polynomial and the volume of the knot complement.

Here are the contents of this article. In Section 1 I describe how to construct cyclic covering spaces of a knot and show how to calculate their homologies. In Section 2 I define the Alexander polynomial and its relation to the homologies of cyclic coverings. I introduce the HOMFLY polynomial following [**38**] in Section 3. By using the Jones–Wenzl idempotent, I define the colored Jones polynomial in Section 4. Section 5 is devoted to the volume conjecture.

ACKNOWLEDGMENTS. I would like to thank the organizers of the JAMI 2002-2003 program 'Primes and Knots'; especially to Jack Morava. I also thank the Johns Hopkins University for its hospitality.

Part of this article was also reported in the talks at Université de Genève and Seoul National University. Thanks are also due to Rinat Kashaev, Claude Weber, Hyuk Kim and Andrey Vesnin, who invited me to their workshop/seminar.

1. Branched covering space of S^3 along a knot

For more detail about this and the next sections see [**4, 29, 45**].

Let K be a knot in the three-sphere S^3, that is, K is an embedded circle in S^3. Since the first homology group $H_1(S^3 \setminus K; \mathbb{Z})$ of the knot complement $S^3 \setminus K$ is $\mathbb{Z}$, there is the n-fold covering space $X_n(K)$ corresponding to the kernel of the following homomorphism:

$$\pi_1(S^3 \setminus K) \xrightarrow{\alpha} H_1(S^3 \setminus K)(\cong \mathbb{Z}) \to \mathbb{Z}/n\mathbb{Z},$$

where α is the abelianization map, and the second map is the reduction modulo n. We also allow the case where $n = \infty$ (in this case $\mathbb{Z}/n\mathbb{Z} := \mathbb{Z}$) and $X_\infty(K)$ is called the universal abelian covering space of K.

We can visualize $X_n(K)$ by using a surface bounding K. A compact, connected oriented surface F with $\partial F = K$ is called a *Seifert surface* of K, where ∂F is the boundary of F. See Figures 1 and 2 for the figure-eight knot and one of its Seifert surfaces. One can always construct a Seifert surface for any knot [**10, 48**]. For a proof see [**45**] for example. Let Y be the three-manifold obtained from S^3 by

FIGURE 1. The figure-eight knot and its Seifert surface.

cutting along F, whose boundary is $F_+ \cup F_-$, where $F_\pm$ is a copy of F. We make n copies of Y and denote them by Y_i with boundary $F_{i,+} \cup F_{i,-}$ $(i = 1, 2, \dots, n)$.

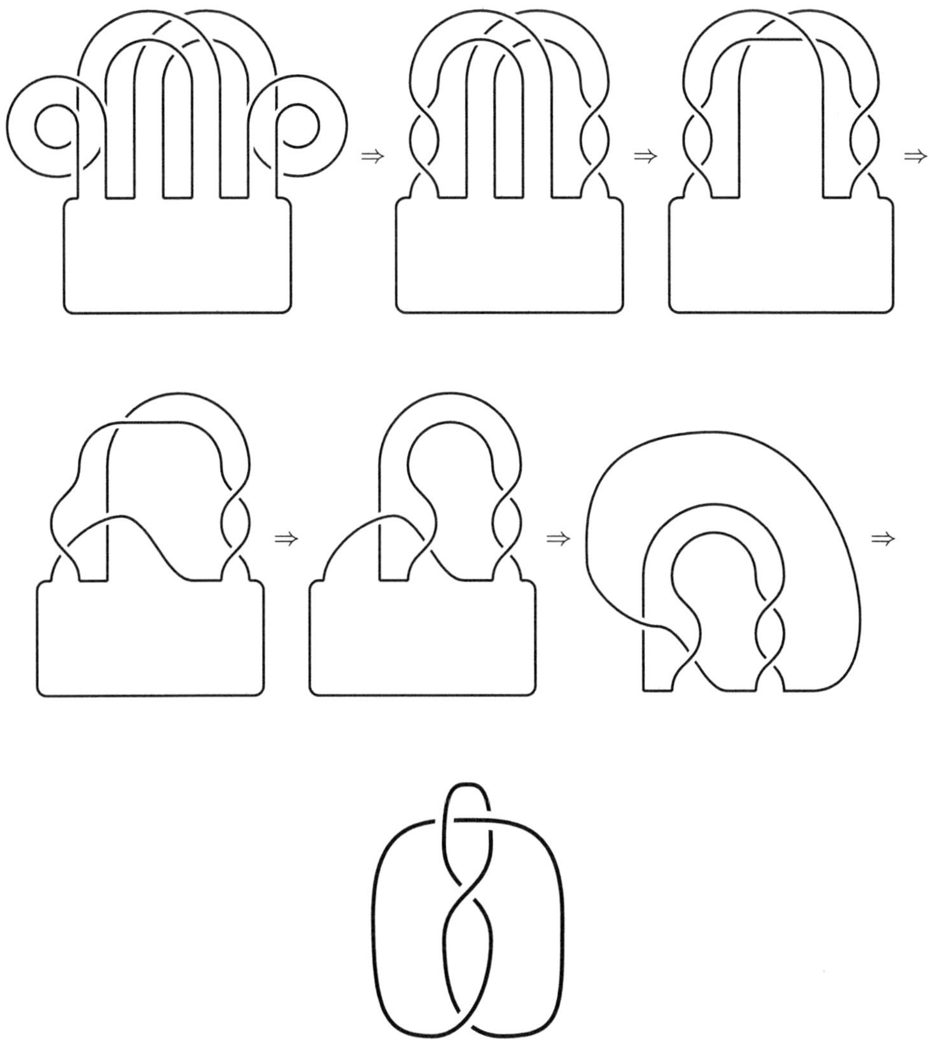

FIGURE 2. All of these represent the figure-eight knot.

Then X_n is obtained from $\bigcup_{i=1}^{n} Y_i$ by identifying $F_{1,+}$ and $F_{2,-}$, $F_{2,+}$ and $F_{3,-}$, $\cdots$, $F_{n-1,+}$ and $F_{n,-}$, and $F_{n,+}$ and $F_{1,-}$. See Figures 3–4.

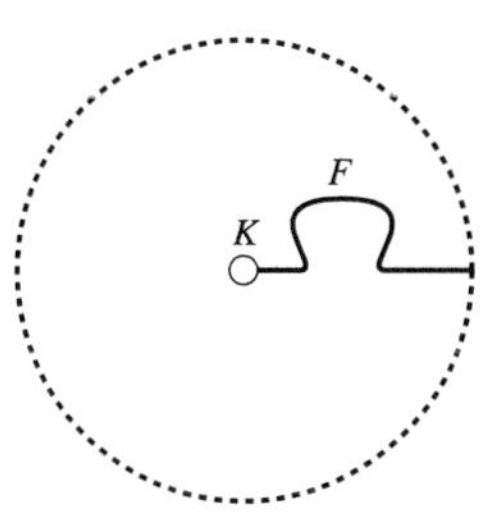

FIGURE 3. A knot K and its Seifert surface F.

Now suppose for simplicity that K is a fibered knot, that is, $S^3 \setminus K$ is a fiber bundle with the interior of an orientable surface $\operatorname{Int} F$ as the fiber and and with S^1 as the base space. Then Y is homeomorphic to the product $\operatorname{Int} F \times I$, where I is the unit interval $[0,1]$, and $S^3 \setminus K$ can be reconstructed as $\operatorname{Int} F \times I/(x,0) \sim (f(x),1)$, the space obtained from $\operatorname{Int} F \times I$ by identifying $F \times 0$ with $F \times 1$ such that $(x,0)$ is identified with $(f(x),1)$, for a homeomorphism $f\colon F \to F$. The induced homomorphism $f_*\colon H_1(F;\mathbb{Z}) \to H_1(F;\mathbb{Z})$ is called the monodromy map. So $X_n(K)$ is homeomorphic to $\operatorname{Int} F \times I/(x,0) \sim (f^n(x),1)$ and $H_1(X_n;\mathbb{Z}) = \operatorname{Coker}(A^n - E) \oplus \mathbb{Z}$, where A is a matrix presenting f and E is the identity matrix with the same size of A (Figure 4). (The summand $\mathbb{Z}$ comes from a cycle that goes around $X_n(K)$.) If we fill in $X_n(K)$ with a solid torus $D^2 \times S^1$, we have a closed three-manifold $\hat{X}_n(K)$ and $H_1(\hat{X}_n(K);\mathbb{Z}) = \operatorname{Coker}(A^n - E)$ which is called the n-fold cyclic branched covering space of S^3 branched along a knot (Figure 5).

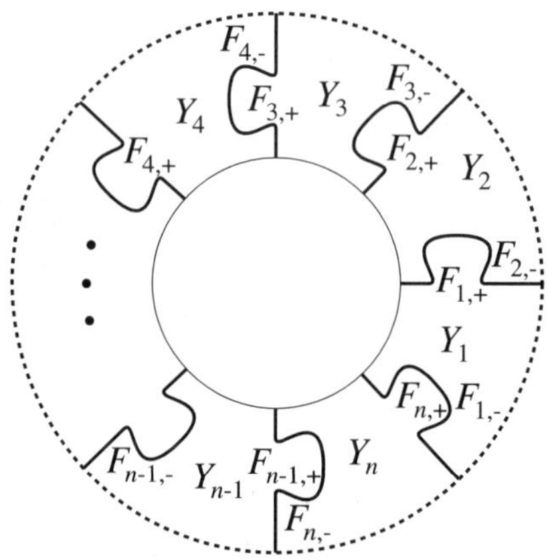

FIGURE 4. The n-fold cyclic (unbranched) covering space of a knot.

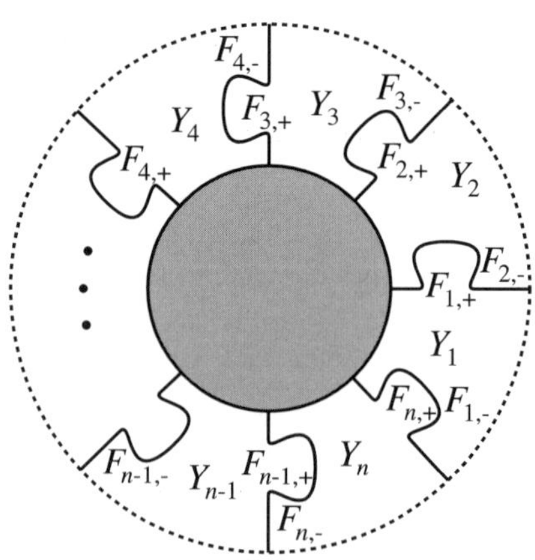

FIGURE 5. The n-fold cyclic branched covering space of a knot. (The shaded disk indicates $D^2 \times S^1$.)

2. Alexander polynomial

Let $\Delta(K;t)$ be the characteristic polynomial of the monodromy matrix A, that is, $\Delta(K;t) := \det(A - t\,I)$. The polynomial $\Delta(K;t)$ is called the Alexander polynomial of K and is an invariant of knots up to $\pm t^k$ ($k \in \mathbb{Z}$) [**2**]. The order $|H_1(\hat{X}_n;\mathbb{Z})|$ can be determined by the Alexander polynomial.

THEOREM 2.1 ([**9**]).

$$|H_1(\hat{X}_n(K);\mathbb{Z})| = \left|\prod_{k=1}^{n-1} \Delta\left(K; \exp\left(\frac{2\pi k\sqrt{-1}}{n}\right)\right)\right|,$$

where $|H_1(\hat{X}_n(K);\mathbb{Z})|$ is the number of elements in $H_1(\hat{X}_n(K);\mathbb{Z})$ if it is a finite group and zero otherwise. Note that the right hand side is equal to the resultant $|R(\Delta(K;t), t^n - 1)|$ and can be calculated as $\prod_{j=1}^{2g}({\xi_j}^n - 1)$ with ξ_j zeroes of $\Delta(K;t)$, since $\Delta(K;1) = 1$ and $\Delta(K;t)$ is monic.

REMARK 2.2. For any knot K, the Alexander polynomial $\Delta(K;t)$ is also defined as the 'order' of the first homology group $H_1(\hat{X}_\infty(K);\mathbb{Z})$ as a $\mathbb{Z}[t,t^{-1}]$-module, where t acts as the covering transformation. It is also true that $\Delta(K;1) = 1$ for any knot, but in general $\Delta(K;t)$ may not be monic. See Remark 3.6

EXAMPLE 2.3. Let 3_1 be the trefoil knot. (Knot theorists often use the symbol n_j that means the j-th knot with n minimal crossing in the knot table. For example the figure-eight knot is the unique knot with minimal crossing four, and so it is written as 4_1.) See Figure 6 for diagrams of 3_1 and its Seifert surface. It is well known that 3_1 is a fibered knot with the surface in the picture as a fiber. (See for example [**45**].) Let $\{a,b\}$ be a generator system of $H_1(F;\mathbb{Z})$ indicated in Figure 7. Let also $\{\alpha,\beta\}$ be its dual generator system of $H_1(S^3 \setminus F;\mathbb{Z})$, that is, it is chosen so that $\mathrm{lk}(a,\alpha) = 1$, $\mathrm{lk}(a,\beta) = 0$, $\mathrm{lk}(b,\alpha) = 0$, and $\mathrm{lk}(b,\beta) = 1$, where $\mathrm{lk}(x,y)$ denotes the linking number in S^3. If we push off a in the positive direction (to this side of the paper), then we get a 1-cycle $a^+ \in H_1(S^3 \setminus F;\mathbb{Z})$. We also get $a^- \in H_1(S^3 \setminus F;\mathbb{Z})$ pushing off a in the negative side. Similarly we get $b^\pm$. It is not hard to see

$$\begin{cases} a^+ &= -\alpha \\ a^- &= -\alpha + \beta \\ b^+ &= \alpha - \beta \\ b^- &= -\beta. \end{cases}$$

So the monodromy map sends $a \mapsto a^+ = -\alpha = a^- + b^- \mapsto a + b$ and $b \mapsto b^+ = \alpha - \beta = -a^- \mapsto -a$ it can be presented by the matrix $A = \begin{pmatrix} 1 & -1 \\ 1 & 0 \end{pmatrix}$. Then its Alexander polynomial is

$$\Delta(3_1;t) = \begin{vmatrix} 1-t & -1 \\ 1 & -t \end{vmatrix} = t^2 - t + 1.$$

Therefore the order of the first homology of the branched covering space of 3_1 is

$$|H_1(\hat{X}_n(3_1);\mathbb{Z})| = \left|\left(\exp(\pi\sqrt{-1}/3)^n - 1\right)\left(\exp(-\pi\sqrt{-1}/3)^n - 1\right)\right|$$

FIGURE 6. Diagrams of the right-hand trefoil (below) and its Seifert surface (above).

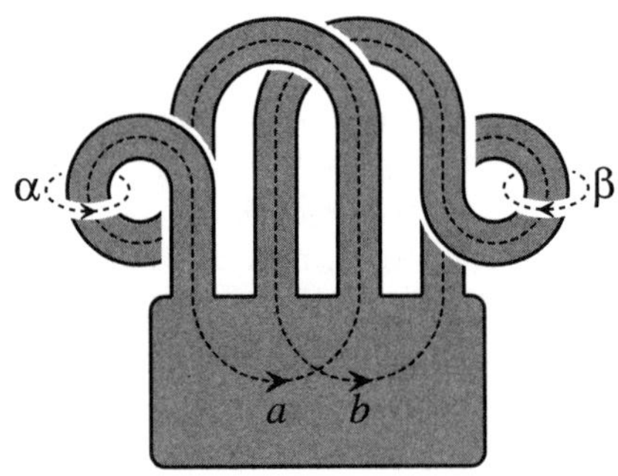

FIGURE 7. $\{a, b\}$ and $\{\alpha, \beta\}$ are generator systems of $H_1(F; \mathbb{Z})$ and $H_1(S^3 \setminus F, \mathbb{Z})$ respectively.

$$= \begin{cases} 0 & \text{if } n \equiv 0 \pmod 6 \\ 1 & \text{if } n \equiv 1 \pmod 6 \\ 3 & \text{if } n \equiv 2 \pmod 6 \\ 4 & \text{if } n \equiv 3 \pmod 6 \\ 3 & \text{if } n \equiv 4 \pmod 6 \\ 1 & \text{if } n \equiv 5 \pmod 6 \end{cases}.$$

In general, $H_1(\hat{X}_n(K);\mathbb{Z})$ is periodic with period k if and only if $\Delta(K;t)$ divides $t^k - 1$ [**14**].

EXAMPLE 2.4. Next we consider the figure-eight knot 4_1. Denote by $\{a, b\}$ a

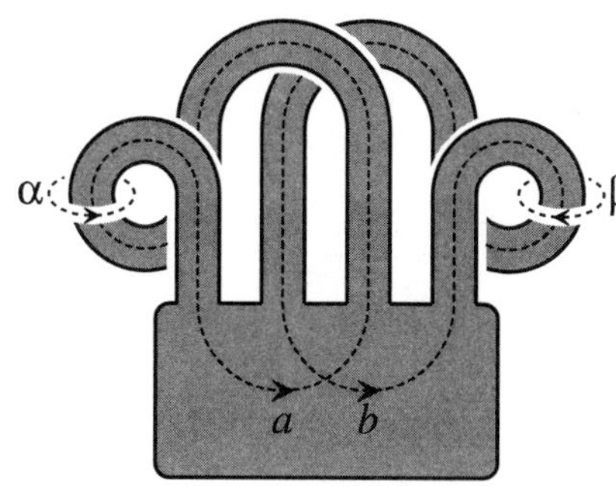

FIGURE 8. $\{a,b\}$ and $\{\alpha,\beta\}$ are generator systems of $H_1(F;\mathbb{Z})$ and $H_1(S^3 \setminus F, \mathbb{Z})$ respectively.

generator system of $H_1(F;\mathbb{Z})$ and by $\{\alpha,\beta\}$ a generator system of $H_1(S^3 \setminus F;\mathbb{Z})$. In this case since

$$\begin{cases} a^+ &= -\alpha \\ a^- &= -\alpha + \beta \\ b^+ &= \alpha + \beta \\ b^- &= \beta, \end{cases}$$

we have $A = \begin{pmatrix} 1 & -1 \\ -1 & 2 \end{pmatrix}$ and $\Delta(4_1;t) = t^2 - 3t + 1$. Since $\xi_1 := \frac{3+\sqrt{5}}{2}$ and $\xi_2 := \frac{3-\sqrt{5}}{2}$ are the roots of $t^2 - 3t + 1 = 0$, we have

$$|H_1(\hat{X}_n(4_1);\mathbb{Z})| = |(\xi_1^n - 1)(\xi_2^n - 1)| = |2 - \xi_1^n - \xi_2^n|.$$

Observe that

$$\lim_{n\to\infty} \frac{\log|H_1(\hat{X}_n;\mathbb{Z})|}{n} = \log \xi_1$$

since

$$\xi_1^n < |2 - \xi_1^n - \xi_2^n| < 4\xi_1^n.$$

Similarly let $\xi_1, \xi_2, \dots, \xi_l, \xi_{l+1}, \dots, \xi_{2g}$ be the eigenvalues of a monodromy matrix A of a knot K with $|\xi_i| > 1$ for $i = 1, 2, \dots, l$. Assume that no roots of $\Delta(K;t) = 0$ are on the unit circle in the complex plane. Then

$$|H_1(\hat{X}_n K;\mathbb{Z})| = \left|(\xi_1^n - 1)(\xi_2^n - 1)\cdots(\xi_l^n - 1)(\xi_{l+1}^n - 1)\cdots(\xi_{2g}^n - 1)\right|$$

and so

$$\lim_{n\to\infty} \frac{\log|H_1(X_n(K);\mathbb{Z})|}{n} = \sum_{i=1}^{l} \log|\xi_i|$$

since $\prod_{i=1}^{l}|\xi_i|$ is the (unique) biggest term in the expansion of $\left|(\xi_1^n - 1)\cdots(\xi_{2g}^n - 1)\right|$. See [**14**] for the case where $\Delta(K;t)$ has zeroes on the unit circle and some of them are not roots of unity.

In general D. Silver and S. Williams proved

THEOREM 2.5 ([**51**]). *Let $\hat{X}_n(K)$ be the n-fold branched cyclic cover of a knot K. Then*

$$\lim_{n\to\infty}\frac{\log|\operatorname{Tor}H_1(\hat{X}_n;\mathbb{Z})|}{n}=\log\mathbf{M}(\Delta(K;t)),$$

where $\operatorname{Tor}G$ *is the torsion part of a group* G, *and* $\log\mathbf{M}(f(t))$ *is the logarithmic Mahler measure of a polynomial* $f(t)$:

$$\log\mathbf{M}(f(t)):=\int_{S^1}\log|f(t)|\,dt=\int_0^1\log|f(\exp(2\pi\sqrt{-1}x))|\,dx.$$

REMARK 2.6. See also [**44, 13, 52**].

REMARK 2.7. It is well known that

$$\mathbf{M}(f(t))=|\text{the leading coefficient of } f| \times\prod|\text{zeroes of } f(t) \text{ whose absolute value is bigger than one}|$$

3. HOMFLY polynomial

In this section we follow [**38**] to define the HOMFLY polynomial for links.

It is well known that the Alexander polynomial for links satisfies the following recursive formula:

$$\Delta(L_+;t)-\Delta(L_-;t)=(t^{1/2}-t^{-1/2})\Delta(L_0;t) \tag{A1}$$

with the initial condition

$$\Delta(O;t)=1, \tag{A0}$$

where L_+, L_-, L_0 are three links (several circles in $\mathbb{R}^3$) that differ only in a small ball where they are as in Figure 9, and O is the unknot, that is, the knot bounding a disk. The relation (A1) is often called the skein relation and the triple (L_+, L_-, L_0)

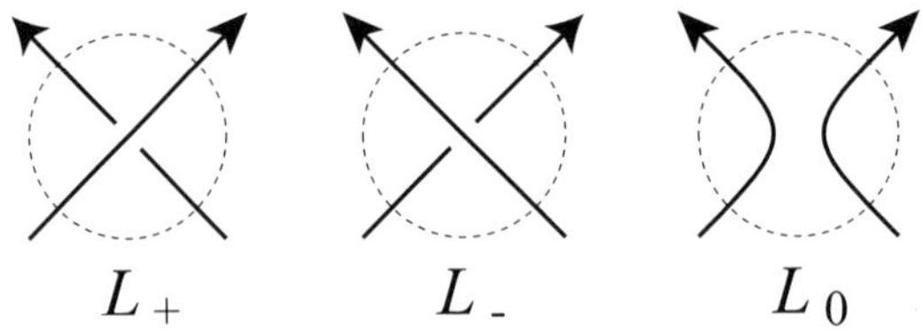

FIGURE 9. A skein triple L_+, L_-, and L_0.

is called a skein triple. This recursive formula of the Alexander polynomial was introduced by J. H. Conway [**6**]. The proof of the well-definedness was given by L. H. Kauffman [**26**]. See also [**12, 17**]. Note that here we normalize $\Delta(L;t)$ so that $\Delta(L;t^{-1})=\Delta(L;t)$ for any link L and that it is defined without ambiguity (not up to $\pm t^k$).

REMARK 3.1. Here is a subtle remark. Our skein relation (A1) *differs* from the Alexander polynomial defined by using Seifert surfaces by $(-1)^{\sharp(L)-1}$, where $\sharp(L)$ is the number of components in L.

We can use skein relations repeatedly to calculate $\Delta(L;t)$ for any oriented link L as follows.

EXAMPLE 3.2. First let us consider the following skein relation.

$\Delta($ 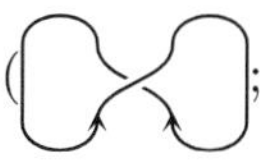 $;t) - \Delta($ 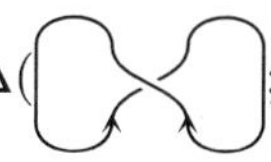 $;t) = (t^{1/2} - t^{-1/2})\Delta($ 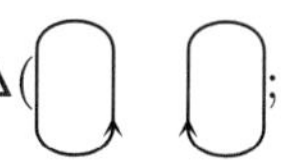 $;t).$

Since both of the knots in the left hand side are the unknot, from (A0) the Alexander polynomial of the trivial two-component link, which is depicted in the right hand side, is 0.

EXAMPLE 3.3. Next consider the following skein relation.

$\Delta($ 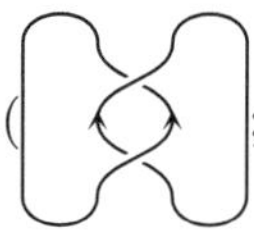 $;t) - \Delta($ 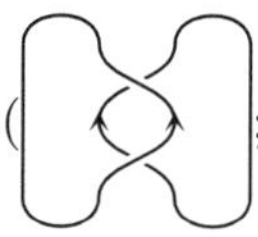 $;t) = (t^{1/2} - t^{-1/2})\Delta($ 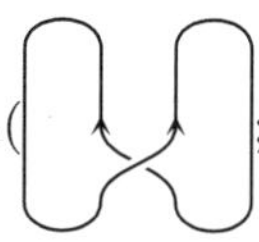 $;t).$

The second knot in the left hand side presents the two-component trivial link and its Alexander polynomial vanishes. Therefore the Alexander polynomial of the positive Hopf link, the first link in the left hand side, is $t^{1/2} - t^{-/2}$ since the knot in the right hand side is the unknot whose Alexander polynomial is 1 from (A0).

EXAMPLE 3.4. Now we calculate the Alexander polynomial of the trefoil again. In the skein relation

$\Delta($ 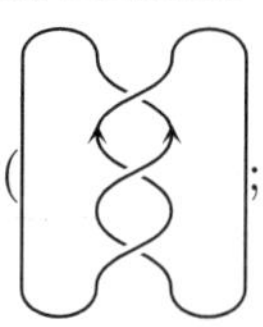 $;t) - \Delta($ 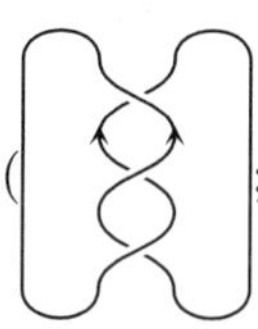 $;t) = (t^{1/2} - t^{-1/2})\Delta($ 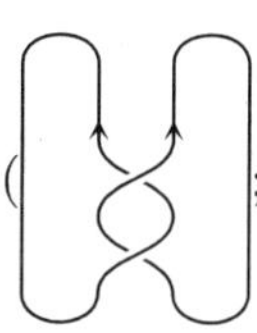 $;t),$

the second knot in the left hand side is the unknot and the figure in the right hand side presents the positive Hopf link. Therefore from Examples 3.2 and 3.3, we have

$\Delta($ 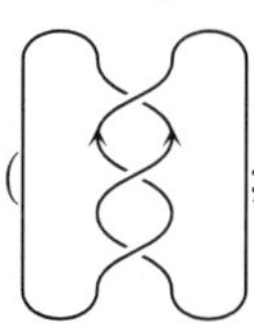 $;t) = \Delta($ 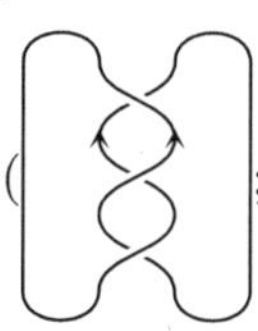 $;t) + (t^{1/2} - t^{-1/2})\Delta($ 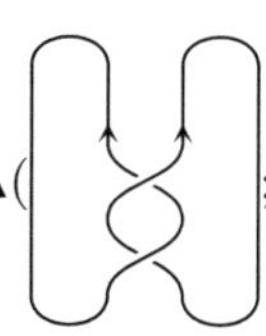 $;t)$

$$= 1 + (t^{1/2} - t^{-1/2})^2 = t - 1 + t^{-1},$$

which coincides with the Alexander polynomial calculated in the previous section up to a unit in $\mathbb{Z}[t, t^{-1}]$.

REMARK 3.5. One can prove combinatorially that the Alexander polynomial is an invariant for links by only using the recursive relations (A1) and (A0). If one does it, one also obtains a generalization of $\Delta(L;t)$, which will be described below.

REMARK 3.6. It is an easy exercise to calculate that the Alexander polynomial of the knot 5_2 shown in Figure 10 is $2t - 3 + 2t^{-1}$. Therefore 5_2 is not a fibered knot since the leading coefficient of the Alexander polynomial of any fibered knot, which equals the determinant of the monodromy matrix, is ± 1. See Remark 2.2.

We can generalize the skein relation (A1) to define the HOMFLY (or HOMFLYpt) polynomial $P_n(K;t) \in \mathbb{Z}[t^{1/2}, t^{-1/2}]$. The HOMFLY polynomial is defined by using the following two axioms.

$$\text{(P1)} \qquad t^{n/2} P_n(L_+;t) - t^{-n/2} P_n(L_-;t) = (t^{1/2} - t^{-1/2}) P_n(L_0;t)$$

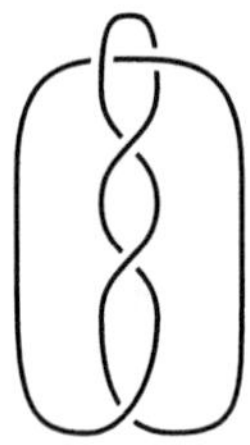

FIGURE 10. The knot 5_2, the second knot with five crossing.

with the initial condition

(P0) $$P_n(O;t) = 1,$$

where L_+, L_-, and L_0 are the skein triple.

The HOMFLY polynomial was defined in [**11, 40**], generalizing both the Alexander polynomial and the Jones polynomial [**20, 21**]. Note that $n = 0$ and $n = 2$ give the Alexander polynomial and the Jones polynomial respectively.

Observe that when $n = 1$, $P_n(L) = 1$ for any link L.

As noted in Remark 3.5, one can prove the well-definedness of $P_n(L;t)$ combinatorially by using the skein relation. But I will prove it by defining P_n directly and constructively.

To do that I will first introduce local moves on link diagrams [**41, 42**]. The Reidemeister move I removes/creates a small kink as described in Figure 11. The

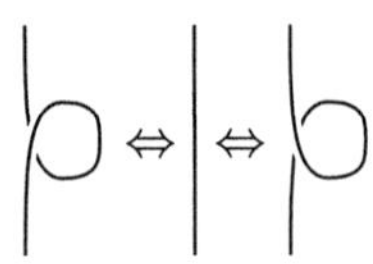

FIGURE 11. Reidemeister move I.

Reidemeister move II changes a pair parallel strands into two strands one of which crosses over the other and vice versa (Figure 12). Figure 13 shows the Reidemeister

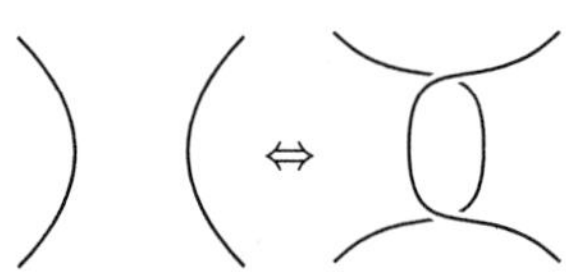

FIGURE 12. Reidemeister move II.

move III, where a strand crosses over a crossing.

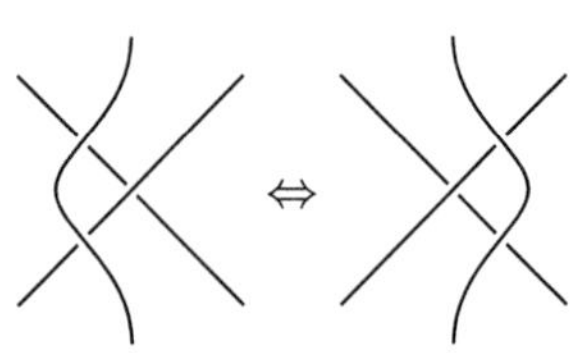

FIGURE 13. Reidemeister move III.

THEOREM 3.7 (Reidemeister's Theorem). *Two link diagrams define the same link if and only if one of them can be transformed into the other by a finite sequence of Reidemeister moves I, II, III.*

PROOF. For a proof see [**4**]. (One may need a piecewise-linear topology textbook, for example [**46**], to understand fully.) Here is a rough sketch.

First we replace links with polygons. Then equivalent polygonal links can be transformed to each other by a composition of *triangle moves*, that is, one replaces a segment of a polygonal link with two sides of a triangle, or vice versa, where the triangle does not intersect other part of the polygonal link. Now observe that the projection of such a triangle can be divided into small triangles so that in each triangle there is either, an arc with both ends on edges of the triangle (Reidemeister move II; see Figure 15), a crossing (Reidemeister move III; see Figure 16), or an arc with an end on a vertex of the triangle (Reidemeister move I; see Figure 14).

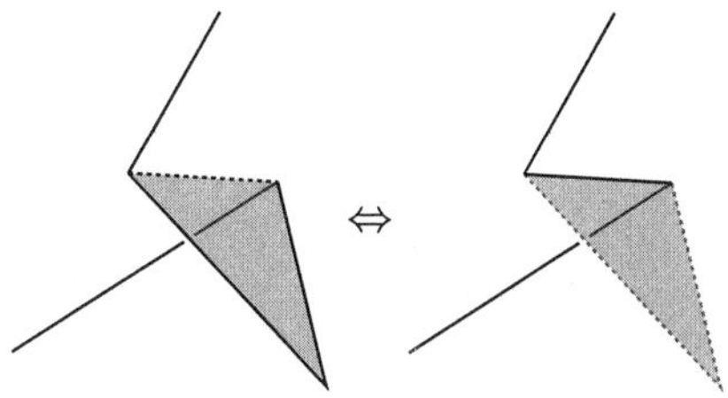

FIGURE 14. Triangle move corresponding to Reidemeister move I.

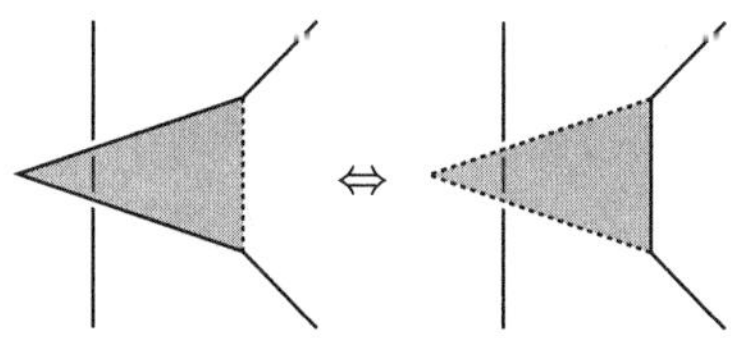

FIGURE 15. Triangle move corresponding to Reidemeister move II.

□

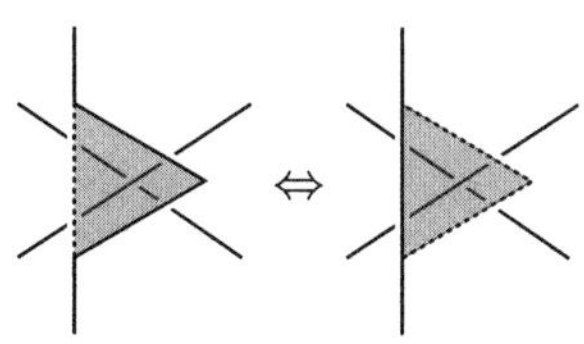

FIGURE 16. Triangle move corresponding to Reidemeister move III.

Now I will explain how to define P_n.

Let n be a positive integer greater than one. Put $\mathcal{N} := \{-(n-1)/2, -(n-3)/2, \dots, (n-3)/2, (n-1)/2\}$. (Note that $\mathcal{N}$ consists of integers if n is odd and half-integers if n is even.) Let G be an oriented trivalent plane graph such that each vertex looks like Figure 17. A *flow* on such a graph is an assignment of a positive integer less than or equal to n to each oriented edge such that at each vertex the sum

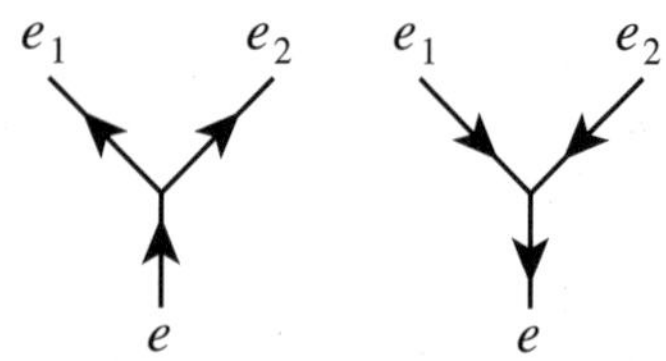

FIGURE 17. Two types of trivalent vertices.

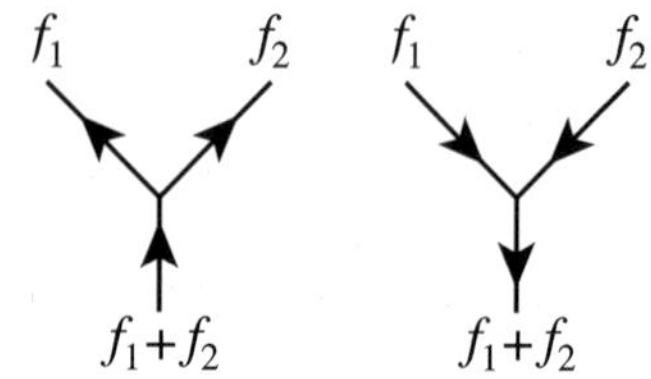

FIGURE 18. Graphs with flows.

of the integers coming in (going out, respectively) is equal to the integer going out (coming in, respectively) as in Figure 18 with $n \geq 5$. See Figure 19 for an example of a graph with flow. For a plane graph with flow f, a *state* σ is an assignment

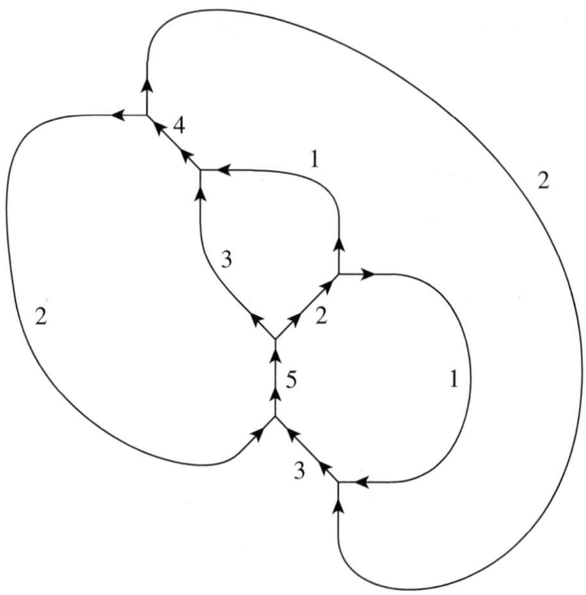

FIGURE 19. A trivalent graph with flow.

of a subset $\sigma(e) \subset \mathcal{N}$ to each edge e with flow $f(e)$ such that $\sharp\sigma(e) = f(e)$, where $\sharp\sigma(e)$ is the cardinality of $\sigma(e)$. We also require that at each vertex with edges e_1, e_2, and e, $\sigma(e_1) \cap \sigma(e_2) = \emptyset$ and that $\sigma(e_1) \cup \sigma(e_2) = \sigma(e)$, where e_1, e_2 and e are edges in the left, in the right and in the bottom respectively (see Figure 17). See Figure 20 for an example of a state on the graph in Figure 19.

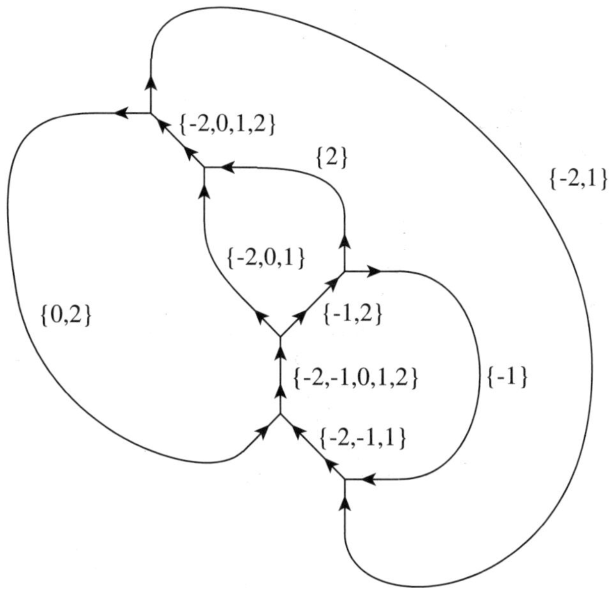

FIGURE 20. A state on the graph in Figure 19

Given a flow f and a state σ on a trivalent plane graph, we define the *weight* $\mathrm{wt}(v;\sigma)$ at a vertex v to be

$$\mathrm{wt}(v;\sigma) := t^{f(e_1)f(e_2)/4-\pi(\sigma(e_1),\sigma(e_2))/2},$$

where t is an indeterminate, e_1 (e_2, respectively) is the edge in the left (right, respectively) in the Figure 17, and

$$\pi(A_1, A_2) := \sharp\{(a_1, a_2) \in A_1 \times A_2 \mid a_1 > a_2\}.$$

For example for the upper left vertex of Figure 20, the weight is

$$t^{2\times 2/4-\pi(\{0,2\},\{-2,1\})/2} = t^{1-3/2} = t^{-1/2}.$$

Given a flow f and a state σ, we can decompose each edge e into $f(e)$ parallel segments and connect them to make circles so that each circle carries the same element in $\mathcal{N}$. We define the *rotation number*

$$\mathrm{rot}(\sigma) := \sum_C \sigma(C)\,\mathrm{rot}(C),$$

where C runs over all the circles defined above, and $\mathrm{rot}(C)$ for an oriented circle C is 1 if it is oriented counterclockwise and -1 otherwise. See Figure 21 how to calculate $\mathrm{rot}(\sigma)$ for the graph with state in Figure 20.

For an oriented, trivalent, plane graph G with flow, we define

$$\langle G\rangle_n := \sum_\sigma \left\{\prod_v \mathrm{wt}(v;\sigma)\right\} t^{\mathrm{rot}(\sigma)}.$$

We have the following local relations for $\langle G\rangle_n$.

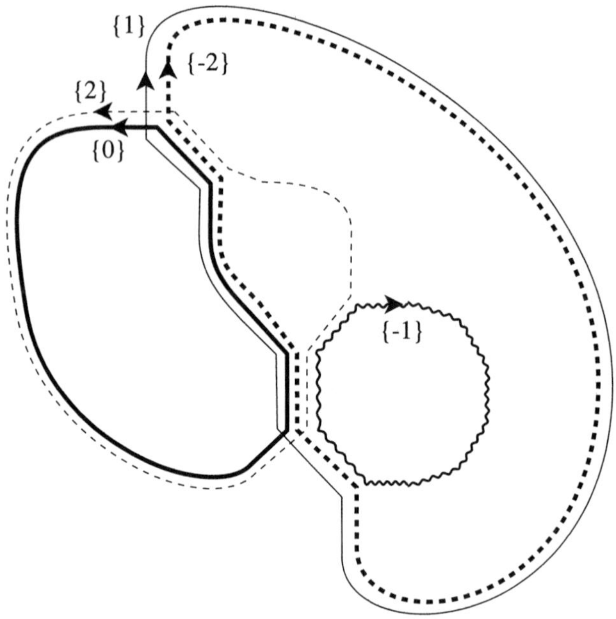

FIGURE 21. The thick circle carries 0, the thin circle carries 1, the thick broken circle carries -2, the thin broken circle carries 2, and the wavy circle carries -1. Therefore the rotation number is $0\times 1+1\times(-1)+(-2)\times(-2)+2\times 1+(-1)\times(-1)=6$.

PROPOSITION 3.8 ([**38**]). *Put* $[k] := (t^{k/2}-t^{-k/2})/(t^{1/2}-t^{-1/2})$. *Then the following formulas hold, where in each formula the diagrams outside the angle brackets are the same.*

(3.1)

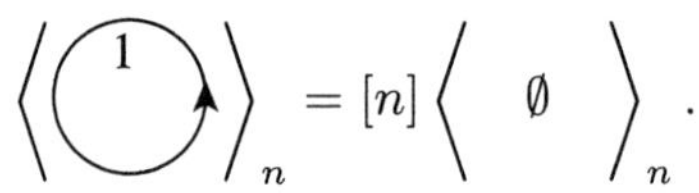

(3.2)

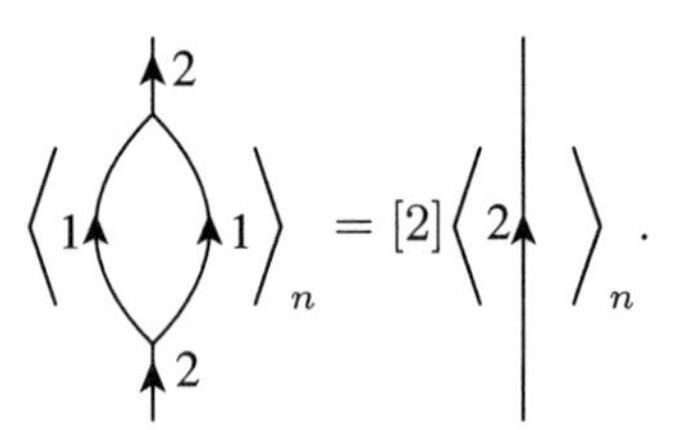

(3.3)

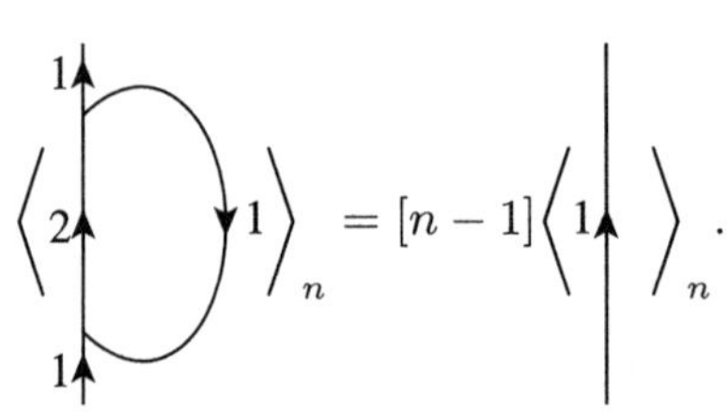

(3.4) 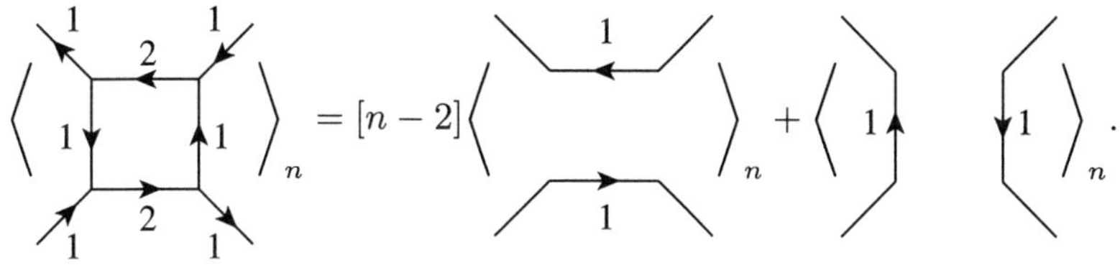

(3.5) 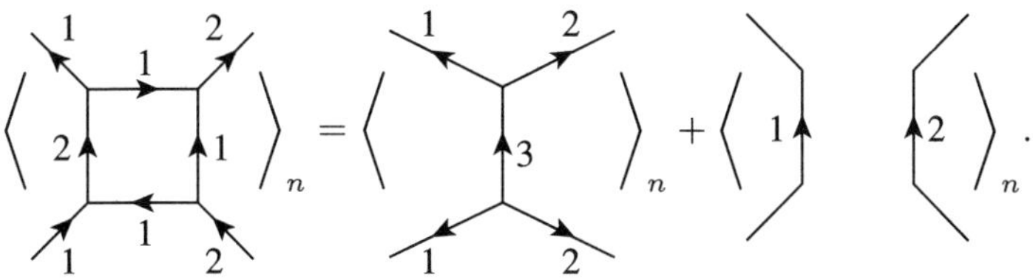

(3.6) 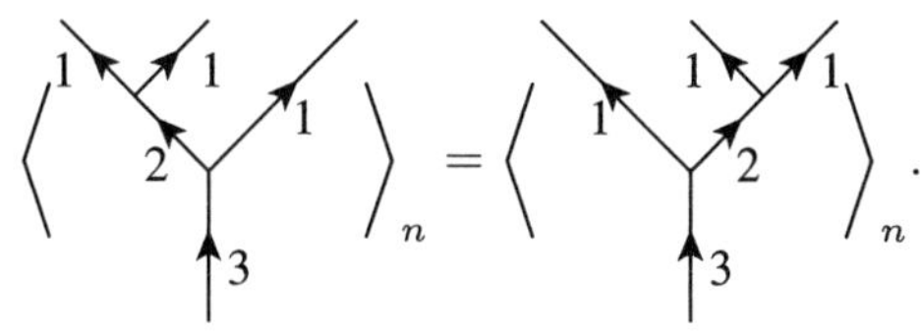

REMARK 3.9. If we put $t = 1$, then $\langle G\rangle_n$ is equal to the number of colorings of edges with n colors so that each edge e carries $f(e)$ different colors. In this case it is easy to prove the theorem since $[k] = k$ when $t = 1$.

REMARK 3.10. In a graph with flow, an edge with flow 1 corresponds to the fundamental (n-dimensional) representation V of the Lie algebra $\mathfrak{sl}_n$. An edge with flow k corresponds to its k-fold exterior power (antisymmetric power) $\wedge^k V$. A vertex in the left hand side of Figure 18 corresponds to the inclusion $\wedge^{f_1+f_2}V \to (\wedge^{f_1}V) \wedge (\wedge^{f_2}V)$ and that in the right hand side corresponds to the projection $(\wedge^{f_1}V) \wedge (\wedge^{f_2}V) \to \wedge^{f_1+f_2}V$.

Now we define $\langle D\rangle_n$ for an oriented link diagram D by

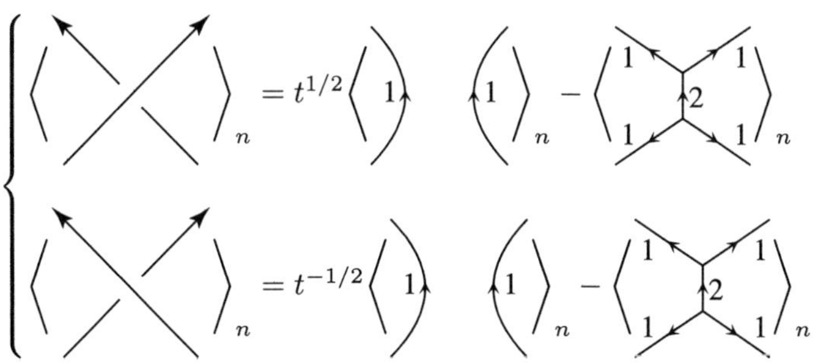

Then we can rather easily prove the following formulas.

PROPOSITION 3.11 ([**38**]). *The following formulas hold for any orientations of strings.*

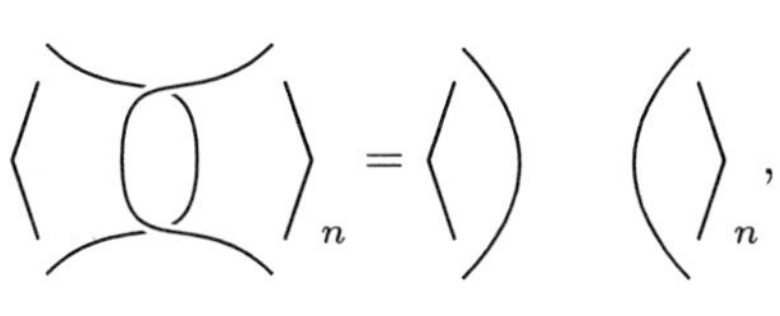

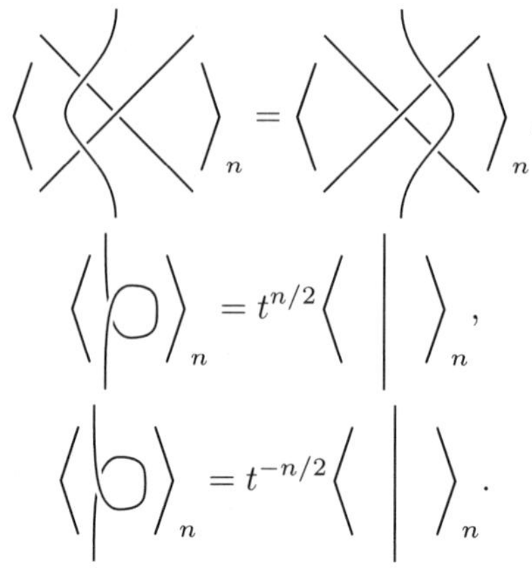

As a corollary if we define

$$\tilde{P}_n(D) := t^{-nw(D)/2}\langle D\rangle_n$$

for an oriented link diagram D, where $w(D)$ the *writhe* of D, that is, the sum of the signs of the crossings in D with

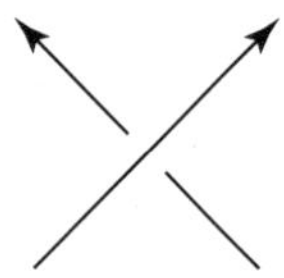

: +1,

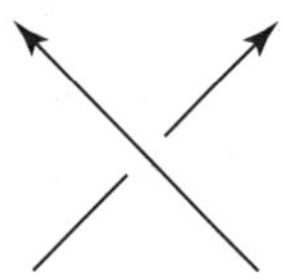

: −1,

then $\tilde{P}_n$ is a link invariant. So we can denote $\tilde{P}_n(D)$ by $\tilde{P}_n(L;t)$ if D represents a link L. Note that $\tilde{P}_n(O;t) = [n]$ and so if we normalize $\tilde{P}_n$ to define $P_n(L;t) := \tilde{P}_n(L;t)/[n]$, then $P_n(L;t)$ satisfies the axioms of the HOMFLY polynomial.

It can be easily proved that P_n is a Laurent polynomial in t for knots (and for links with odd number of components), and Laurent polynomial in $t^{1/2}$ for links with even number of components. Note that it is not clear from the graphical definition of P_n that it is a Laurent polynomial in t.

REMARK 3.12. One can calculate $\langle G\rangle_n$ for any plane graph G by only using Equations (3.1)–(3.6). But I do not know whether one can prove the well-definedness by only using these axioms.

REMARK 3.13. For $n = 2$, we get a version of the Kauffman bracket [**27**].

Since there is a unique state for an edge with flow 2, we can ignore edges with flow 2. Then a graph is decomposed into circles consisting of segments with arrows. Taking into account the weights, this gives the correspondence between our invariant for $n = 2$ and the Kauffman bracket.

REMARK 3.14. For $n = 3$, G. Kuperberg gave a definition via Equations (3.1)–(3.6) [**28**]. Note that in this case there is a duality changing an edge with flow 2 and an edge with flow 1 with opposite edge. Therefore Equations (3.2) and (3.3) become the same, and so do Equations (3.4) and (3.5). Equation (3.6) becomes trivial since one can ignore an edge with flow 3.

4. Colored Jones polynomial

In this section I will describe another generalization of the Jones polynomial, $P_2(L;t)$ in our notation.

Since $P_2(L;t)$ corresponds to the two-dimensional standard representation V of $\mathfrak{sl}_2$, one can expect a link invariant corresponding to the d-dimensional representation, which can be defined as the $(d-1)$-fold symmetric tensor. It is in fact called the colored Jones polynomial and one can define it graphically as follows.

First let us consider the following graphical object defined by

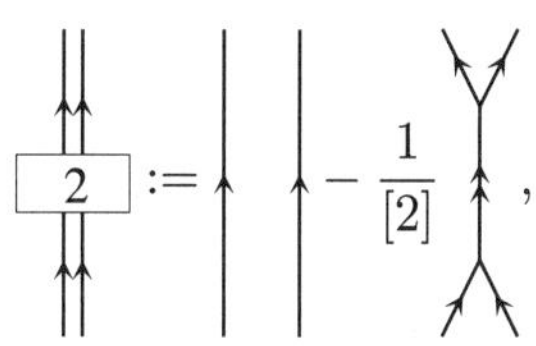

where in the right hand side a segment with single arrow means an edge of flow 1 corresponding to V and a segment with double arrow means an edge of flow 2 corresponding to the antisymmetric tensor product $V \wedge V$ as described in the previous section.

If $t = 1$ then $[2] = 2$ and the meaning of the equation is as follows. The second figure in the right hand side means the composition of the homomorphisms $V \otimes V \to V \wedge V \to V \otimes V$ defined by $a \otimes b \mapsto a \wedge b \mapsto a \otimes b - b \otimes a$. Therefore if we go up from the bottom to the top in the left hand side, we have a homomorphism $V \otimes V \to V \otimes V$ defined by the right hand side, that is:

$$(a,b) \mapsto a \otimes b - \frac{1}{2}(a \otimes b - b \otimes a) = \frac{1}{2}(a \otimes b + b \otimes a),$$

which is nothing but the definition of the symmetric tensor product.

In general we define 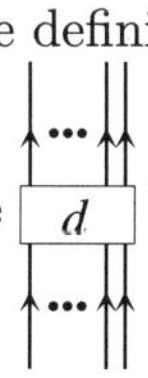 by

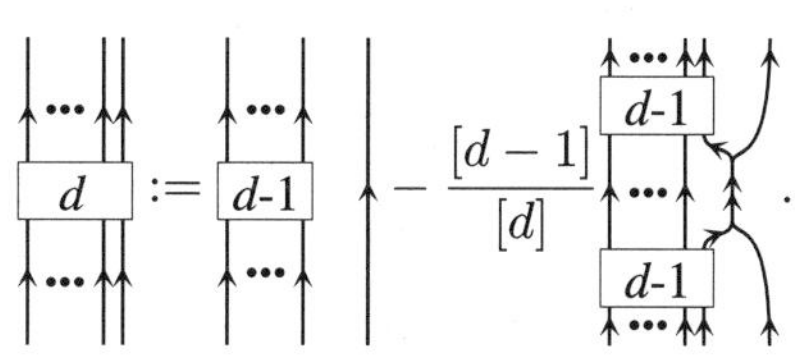

It can be also proved that 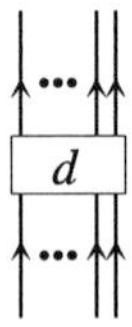 corresponds to the d-fold symmetric tensor of V if we put $t = 1$. Note that its dimension is $d - 1$. Therefore 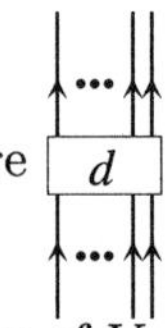 can be regarded as a deformation or quantization of d-fold symmetric tensor of V. It is not hard to prove that

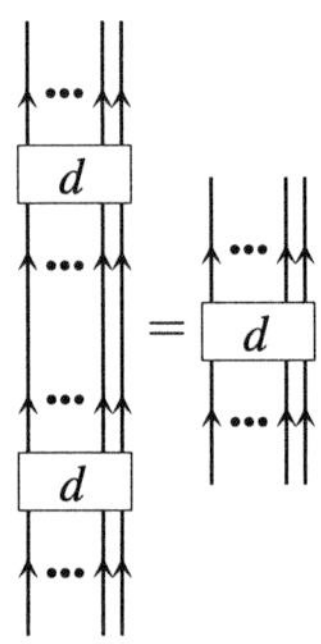

and 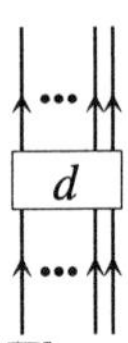 is often called the Jones–Wenzl idempotent [**56**].

The colored Jones polynomial $J_n(K;t)$ of dimension $d+1$ is defined by

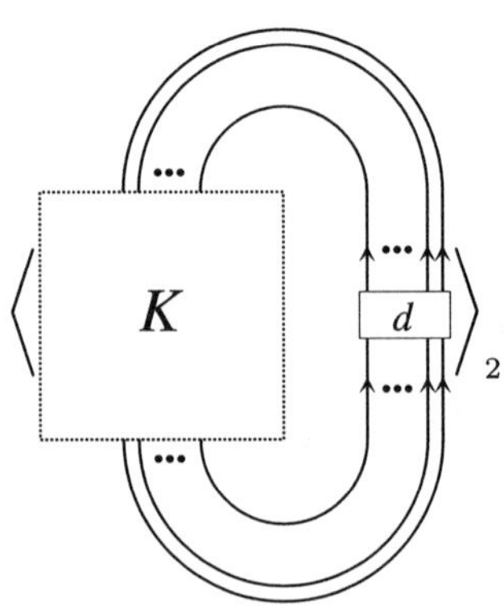
$/[d+1]$,

where 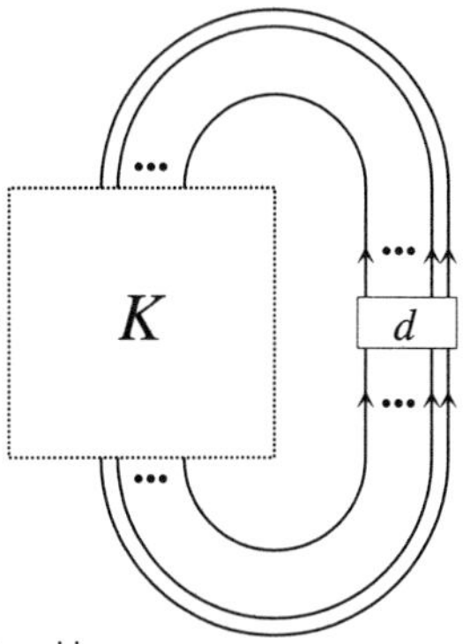

is obtained by replacing the string in 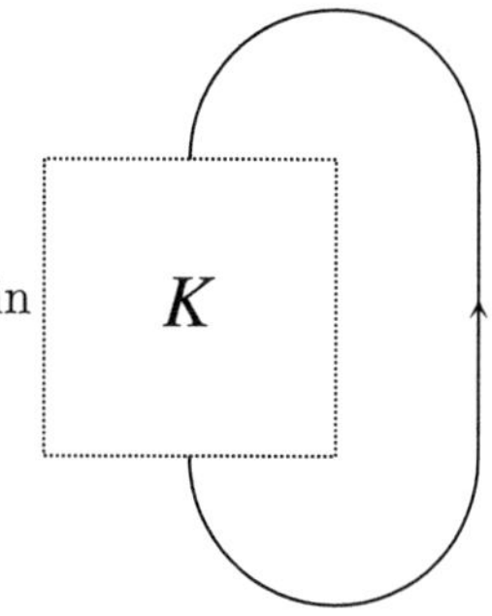

with 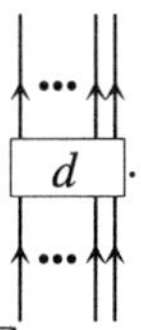

.

For example the $d+1$-dimensional colored Jones polynomial of the unknot $J_{d+1}(O;t)$ can be calculated as follows: First of all

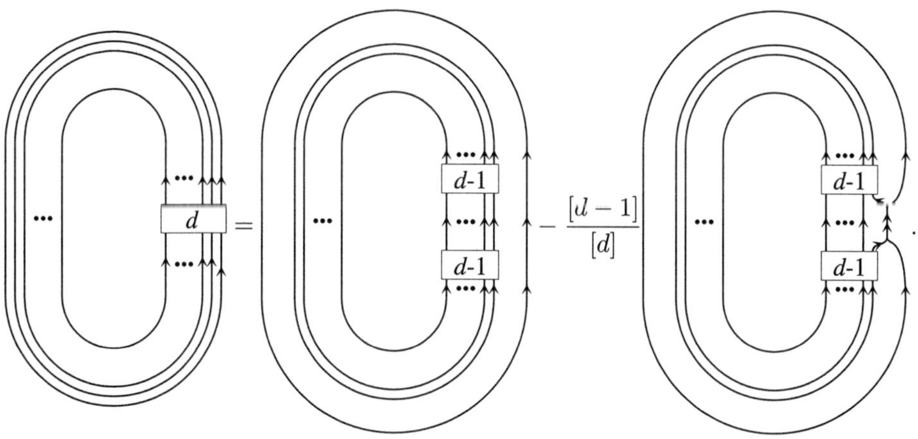

Since we can erase the edges with flow 2, and forget the orientations of other edges, we have

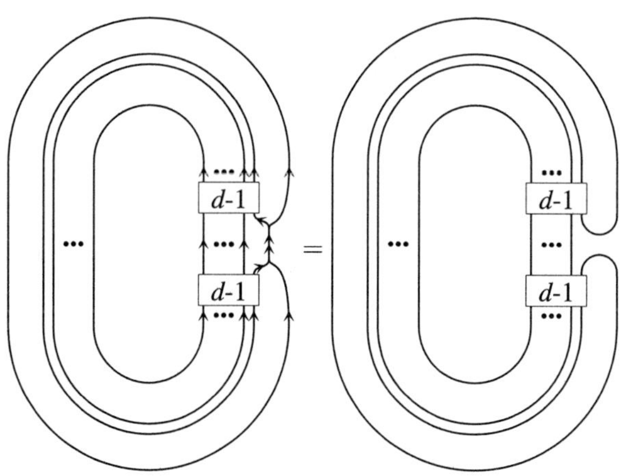

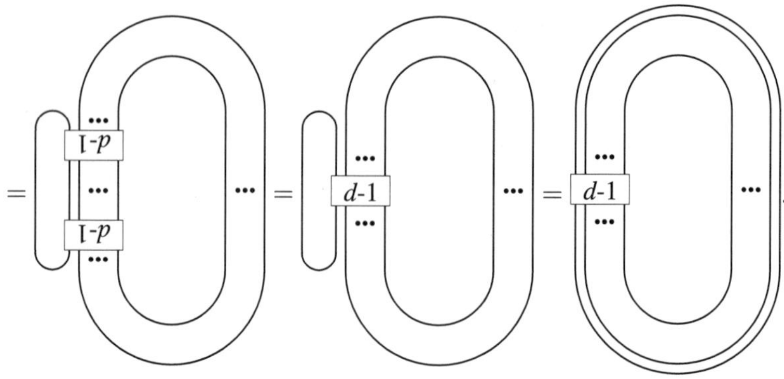

Therefore

$$
\begin{aligned}
J_{d+1}(O;t)[d+1] &= [2]J_d(O;t)[d] - \frac{[d-1]}{[d]} J_d(O;t)[d] \\
&= \frac{(t^{1/2}+t^{-1/2})(t^{d/2}-t^{-d/2}) - (t^{(d-1)/2}-t^{-(d-1)/2})}{t^{1/2}-t^{-1/2}} J_d(O;t) \\
&= [d+1]J_d(O;t),
\end{aligned}
$$

and by induction we have $J_{d+1}(O;t) = 1$. In general it can be proved that $J_d(L;t)$ is a Laurent polynomial in $t^{\pm 1/2}$ for any link L.

For example for the figure-eight knot 4_1, K. Habiro [**16**] and T. Le proved that

$$
J_d(4_1;t) = \sum_{k=0}^{d-1} \prod_{l=1}^{k} \left(t^{(d+l)/2} - t^{-(d+l)/2} \right) \left(t^{(d-l)/2} - t^{-(d-l)/2} \right). \tag{4.1}
$$

See also [**30**] for other examples.

5. Volume Conjecture

We want to investigate the asymptotic behavior of the following quantity:

$$
\frac{\log J_N\left(K; \exp(2\pi\sqrt{-1}/N)\right)}{N}. \tag{5.1}
$$

Let us calculate the $N \to \infty$ limit of Equation (5.1) for the figure-eight knot 4_1 following T. Ekholm. From Equation (4.1) we have

$$
J_N\left(4_1; \exp(2\pi\sqrt{-1}/N)\right) = \sum_{k=0}^{N} \prod_{l=1}^{k} \left(2\sin(l\pi/N)\right)^2.
$$

Putting $f(k) := \prod_{l=1}^{k} \left(2\sin(l\pi/N)\right)^2$, we have $J_N\left(4_1; \exp(2\pi\sqrt{-1}/N)\right) = \sum_{k=0}^{N} f(k)$. Now $f(k)$ takes its maximum when $l\pi/N = 5\pi/6$, roughly speaking. Therefore we have

$$
f(5N/6) \le \sum_{k=0}^{N-1} f(k) \le N f(5N/6).
$$

Taking logarithm, we have

$$
\frac{\log f(5N/6)}{N} \le \frac{\log J_N\left(4_1; \exp(2\pi\sqrt{-1}/N)\right)}{N} \le \frac{\log N f(5N/6)}{N} = \frac{\log N}{N} + \frac{\log f(5N/6)}{N}.
$$

Now since $\frac{\log N}{N} \to 0$ when $N \to \infty$, we have

$$\begin{aligned}\lim_{N\to\infty} \frac{\log J_N\left(4_1; \exp(2\pi\sqrt{-1}/N)\right)}{N} &= \lim_{N\to\infty} \frac{\log f(5N/6)}{N} \\ &= 2 \lim_{N\to\infty} \sum_{l=1}^{5N/6} \frac{\log\left(2\sin(l\pi/N)\right)}{N} \\ &= \frac{2}{\pi}\int_0^{5\pi/6} \log\left(2\sin(l\pi/N)\right)\,dx.\end{aligned}$$

Here we define the Lobachevsky function $Л(\theta)$ as follows.

$$Л(\theta) := -\int_0^{\theta} \log 2\sin x\,dx.$$

Then we have

$$\lim_{N\to\infty} \frac{\log J_N\left(4_1; \exp(2\pi\sqrt{-1}/N)\right)}{N} = -\frac{2}{\pi}\,Л(5\pi/6).$$

The Lobachevsky function satisfies the following properties (see for example [**32**] and [**54**, Chapter 7].)

(1) $Л(\theta)$ is an odd function.
(2) $Л(\theta)$ has period π.
(3) $Л(n\theta) = n\sum_{k \pmod{n}} Л(\theta + k\pi/n)$ for any integer n.

Therefore

$$Л(5\pi/6) = -\,Л(\pi/6)$$

and

$$\begin{aligned}Л(\pi/3) &= Л(2\cdot\pi/6) = 2\,Л(\pi/6) + 2\,Л(\pi/6+\pi/2) = 2\,Л(\pi/6) + 2\,Л(2\pi/3) \\ &= 2\,Л(\pi/6) - 2\,Л(\pi/3)\end{aligned}$$

and so we have

$$Л(5\pi/6) = -3\,Л(\pi/3)/2.$$

Now we finally have

$$\lim_{N\to\infty} \frac{\log J_N\left(4_1; \exp(2\pi\sqrt{-1}/N)\right)}{N} = -\frac{2}{\pi}\,Л(5\pi/6) = \frac{6\,Л(\pi/3)}{2\pi}.$$

What is this?

Here I will explain a little about hyperbolic geometry. See for example [**54, 55, 3**]. for more details.

Let us consider the upper half space model $\mathbf{H}^3$, which is topologically same as the upper half Euclidean xyz-space. A geodesic plane in $\mathbf{H}^3$ is either a (Euclidean) upper half plane perpendicular to the xy-plane or an upper hemisphere with center on the xy-plane. A geodesic line is the intersection of two geodesics planes; either upper half line perpendicular to the xy-plane or an semicircle perpendicular to the xy-plane. A hyperbolic tetrahedron has geodesic faces and an ideal hyperbolic tetrahedron means a tetrahedron with all the vertices on the 'infinite sphere', that is, xy-plane $\cup\,\infty$. By isometry of $\mathbf{H}^3$ one can assume that a vertex of any ideal hyperbolic tetrahedron is ∞. Then it looks like Figure 22. If we see from the above, it becomes a triangle (Figure 23). Two ideal hyperbolic tetrahedra are isometric if and only if their associated triangles are similar. Therefore ideal hyperbolic

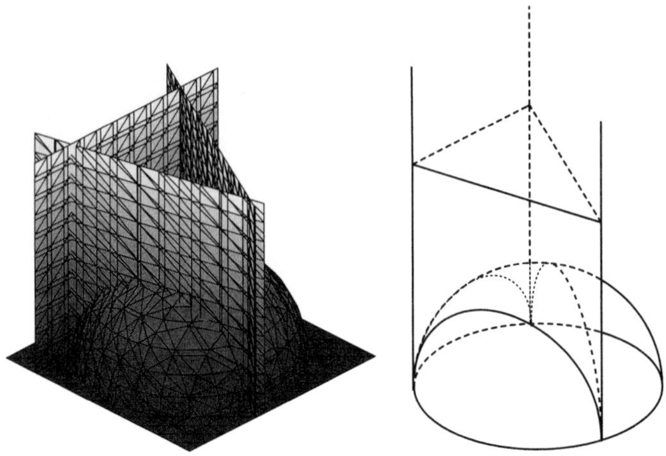

FIGURE 22. Ideal hyperbolic tetrahedron

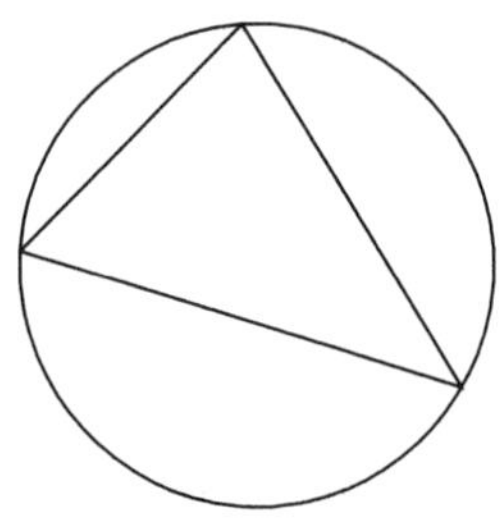

FIGURE 23. Top view of the ideal hyperbolic tetrahedron

tetrahedra are parameterized by three interior angles α, β and γ with $\alpha+\beta+\gamma=2\pi$, which are the dihedral angles of tetrahedra. (There are six dihedral angles but the opposite angles are the same.) It can be proved that the volume of the ideal hyperbolic tetrahedron with dihedral angles α, β, γ is given by $\text{Л}(\alpha)+\text{Л}(\beta)+\text{Л}(\gamma)$.

Roughly speaking a hyperbolic three-manifold can be obtained by pasting hyperbolic tetrahedra. In particular, it is well known that the complement of the figure-eight knot can be obtained from two regular ideal hyperbolic tetrahedra [**43**].

REMARK 5.1. Note that since an ideal tetrahedron is a tetrahedron with vertices removed, if we paste ideal tetrahedra we have a non-compact three-manifold without boundary. If we cut out small tetrahedra near the vertices of the ideal tetrahedron, we have a truncated tetrahedron. If we paste these truncated tetrahedra together in the same way as the ideal tetrahedra, we get a compact three-manifold with boundaries, each of which is homeomorphic to a torus.

Thus the volume of the complement of the figure-eight $\mathrm{Vol}(S^3\setminus 4_1)$ is equal to $6\,\text{Л}(\pi/3)$ and so we conclude that

$$2\pi\lim_{N\to\infty}\frac{\log J_N\left(4_1;\exp(2\pi\sqrt{-1}/N)\right)}{N}=\mathrm{Vol}(S^3\setminus 4_1). \tag{5.2}$$

REMARK 5.2. R. Kashaev observed by numerical calculations that for the knots 4_1, 5_2, and 6_1 (5.1) is equal to the volume of the complement divided by 2π. He also conjectured that this also holds for any hyperbolic knot (a knot whose complement has complete hyperbolic structure) [**24**].

He calculated the limits of these invariants by first approximating the summations by meromorphic functions and then applying the saddle point method. He consulted the table of [**1**] and found out that these limits coincide with the volumes. Unfortunately his computation was not rigorous but efficient enough to be applied to other knots and links [**37**].

REMARK 5.3. To be more precise, Kashaev conjectured that the limit of his link invariant defined in [**22, 23**] determines the hyperbolic volume. In [**36**], J. Murakami and I proved that Kashaev's invariant is the absolute value of the colored Jones polynomial evaluated at a root of unity.

In [**36**] J. Murakami and I conjectured that (5.2) holds for any knot if we replace the hyperbolic volume with a version of simplicial volume (or Gromov norm) [**15**]. Here the simplicial volume can be calculated as follows.

It is known [**18, 19**] that any link complement space can be uniquely decomposed by tori into hyperbolic spaces and Seifert fibered spaces (JSJ decomposition):

$$S^3 \setminus K = (H_1 \cup_{\text{torus}} H_2 \cup_{\text{torus}} \cdots \cup_{\text{torus}} H_k) \bigcup_{\text{torus}} (F_1 \cup_{\text{torus}} F_2 \cup_{\text{torus}} \cdots \cup_{\text{torus}} F_l),$$

where H_i is hyperbolic and F_j is Seifert fibered.

A Seifert fibered space is a generalization of a circle bundle over a surface with (possibly) singularities [**47**] (English translation can be found in [**50**]). Here a neighborhood of a 'singular fiber' is obtained from a cylinder $D^2 \times [0,1] \subset \mathbb{C} \times [0,1]$ ($D^2 := \{z \in \mathbb{C} \mid |z| \leq 1\}$) by identifying $D^2 \times \{0\}$ and $D^2 \times \{1\}$ by the map $(x, 0) \mapsto (x \exp(2p\pi\sqrt{-1}/q), 1)$ for coprime integers p and q. Note that this is homeomorphic to a solid torus $D^2 \times S^1$ ($S^1 := \{z \in \mathbb{C} \mid |z| = 1\}$).

EXAMPLE 5.4. The complement of the trefoil 3_1 is Seifert fibered. To see this I first note that 3_1 can be put on a standard torus (Figure 24). If we remove the

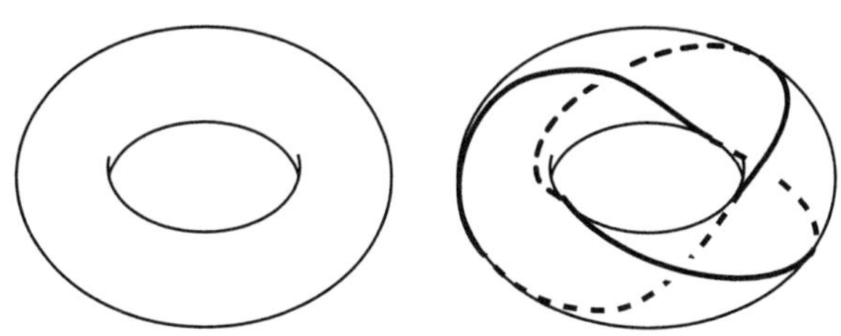

FIGURE 24. Standard torus and the trefoil on it.

standard torus from S^3 we have another standard torus. This is because we first decompose S^3 into two three-balls, dig a hole through a three-ball (remove a solid cylinder) to make a solid torus, and attach the cylinder to the other three-ball to make another solid torus. (See Figure 25.)

Then our trefoil is on the common boundary of the two solid tori. See Figure 26. Let us regard the solid torus on the left as $D^2 \times S^1 := D^2 \times [0,1]/(z,0) \sim (z,1)$ (the quotient space obtained from $D^2 \times [0,1]$ by identifying $(z,0)$ and $(z,1)$.) and that on the right as $S^1 \times D^2 := [0,1] \times D^2/(0,w) \sim (1,w)$. Then the trefoil in the

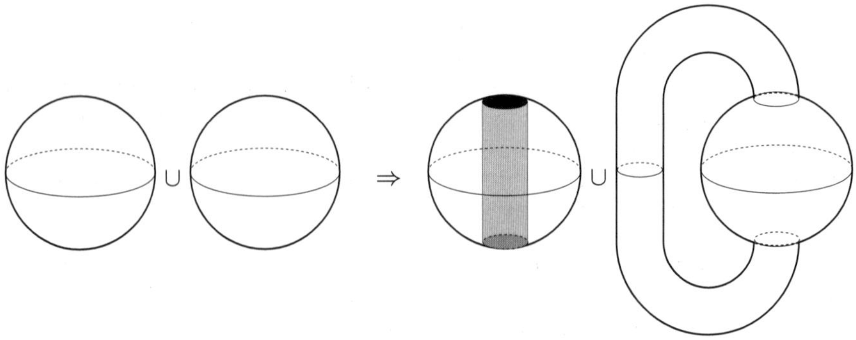

FIGURE 25. $S^3 = \text{ball} \cup \text{ball} = \text{ball without cylinder} \cup \text{ball with cylinder}$.

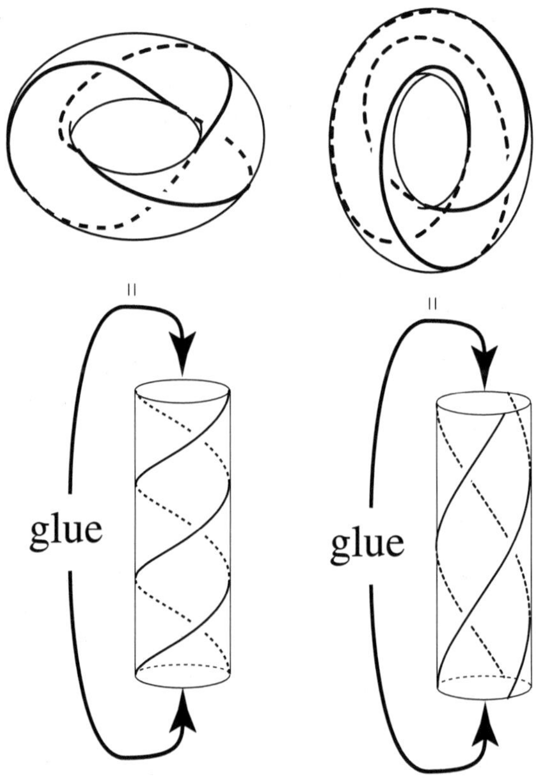

FIGURE 26. S^3 is decomposed into two solid tori and the trefoil is on the common boundary.

left of Figure 26 can be described as

$$\left(\exp\left(\frac{3}{2}\times 2\pi s\sqrt{-1}\right), s\right) \cup \left(-\exp\left(\frac{3}{2}\times 2\pi s\sqrt{-1}\right), s\right)$$

and that on the right as

$$\left(u, \exp\left(\frac{2}{3}\times 2\pi u\sqrt{-1}\right)\right)$$

$$\cup\left(u, \exp\left(\frac{1}{3}\times 2\pi\sqrt{-1}\right)\exp\left(\frac{2}{3}\times 2\pi u\sqrt{-1}\right)\right)$$
$$\cup\left(u, \exp\left(\frac{2}{3}\times 2\pi\sqrt{-1}\right)\exp\left(\frac{2}{3}\times 2\pi u\sqrt{-1}\right)\right).$$

Now we define maps $p_1 \colon D^2 \times S^1 \to D^2$ and $p_2 \colon S^1 \times D^2$ by $p_1\big((z,s)\big) \mapsto |z|\exp\big(2(\arg z - \frac{3}{2}\times 2\pi s)\sqrt{-1}\big)$ and $p_2\big((u,w)\big) \mapsto |w|\exp\big(3(\arg w - \frac{2}{3}\times 2\pi u)\sqrt{-1}\big)$. These maps are well defined as maps from the solid tori and also agree on their boundaries. It is easily seen that both $p_1\big|_{D^2\setminus\{0\}\times S^1}$ and $p_2\big|_{S^1\times D^2\setminus\{0\}}$ is a circle bundle. Therefore p_1 and p_2 define Seifert fibered space structure on $D^2 \times S^1$ and $S^2 \times D^2$ with singular fibers $\{0\} \times S^2$ and $S^2 \times \{0\}$ respectively. Note that each trefoil can be regarded as a regular (non-sigular) fiber.

Since p_1 and p_2 agree on the boundaries, they define a map from $p \colon S^3 = D^2 \times S^1 \cup S^1 \times D^2 \to D^2 \cup D^2 = S^2$. This also defines a Seifert fibered space structure on S^3 with two singular fibers. Since the trefoil in each solid torus is a regular fiber, we finally see that $S^3 \setminus$ trefoil is a Seifert fibered space with two singular fibers.

EXAMPLE 5.5. Let us consider the knot depicted in Figure 27, which is called the $(2,1)$-cable of the figure-eight knot since the knot goes along the figure-eight knot twice and around it once. The complement of this knot can be decomposed

FIGURE 27. $(2,1)$-cable of the figure-eight knot.

as follows: The thin line in the left hand side of Figure 28 shows the decomposing

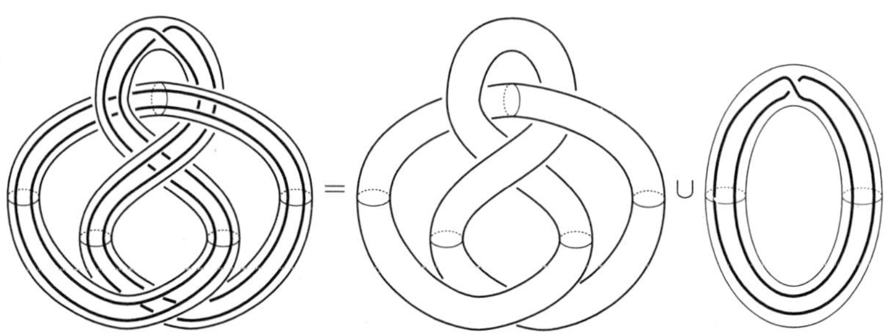

FIGURE 28. The JSJ decomposition of the $(2,1)$-cable of the figure-eight knot.

torus. The first figure of the right hand side shows the complement of the figure-eight knot and the second shows the solid torus minus $(2,1)$-cable, which runs twice along S^1 and once around D^2.

As noted in the above, the figure-eight knot complement is hyperbolic. The second one $D^2 \times S^1 \setminus (2,1)$-cable is Seifert fibered, which can be proved as before.

Now the simplicial volume of a knot complement is the sum of the hyperbolic volumes of the hyperbolic pieces:

$$\|S^3 \setminus K\| = \sum_{i=1}^{k} \operatorname{Vol}(H_i).$$

REMARK 5.6. The original definition of the simplicial volume $|M|$ for a closed three-manifold is as follows. Let $[M]$ be the fundamental class of the third homology $H_3(M;\mathbb{R})$. (So $[M]$ generates $H_3(M;\mathbb{R}) \cong \mathbb{R}$.) Then the Gromov norm is defined to be the infimum of the norm of three-chains that present $[M]$, where the norm of a chain $\sum_{i=1}^{m} r_i c_i$ is $\sum_{i=1}^{m} |r_i|$.

It is known that the Gromov norm is proportional to the hyperbolic volume; more precisely $\operatorname{Vol}(M) = v_3|M|$ with v_3 is the hyperbolic volume of the regular ideal hyperbolic tetrahedra (see for example [**54**, Chapter 6]).

In this paper I abuse the notation and call $\|M\| := v_3|M|$ the simplicial volume.

EXAMPLE 5.7. Since the complement of the trefoil is Seifert fibered, its simplicial volume is zero.

EXAMPLE 5.8. From Example 5.5, the $(2,1)$-cable of the figure-eight knot can be decomposed into two pieces, one hyperbolic and one Seifert fibered. Therefore its simplicial volume is the same as the figure-eight knot's.

Now the Volume Conjecture can be stated as follows.

CONJECTURE 5.9 (Volume Conjecture [**36**]). *For any knot K, the asymptotic behavior of $J_N(K;\exp(2\pi\sqrt{-1}))$ determines the simplicial volume of the complement. More precisely the following equality holds.*

$$2\pi \lim_{N\to\infty} \frac{\log\left|J_N\left(K;\exp(2\pi\sqrt{-1}/N)\right)\right|}{N} = \|S^3 \setminus K\|. \tag{5.3}$$

REMARK 5.10. If we drop the absolute value sign from the left hand side in Equation (5.3), we have a complex number whose real part is the volume. In [**37**] we also conjecture that the imaginary part gives the so called Chern–Simons invariant [**5, 31**]. It is known that (Volume$+\sqrt{-1}$Chern–Simons) gives in a sense an analytic function [**39, 62**].

REMARK 5.11. The Volume Conjecture was proved so far for

- The figure-eight knot.
- Torus knots (knots on the standardly embedded torus) by Kashaev and O. Tirkkonen [**25**]. Note that the complement of a torus knot is Seifert fibered and so its simplicial volume is zero.
- The Borromean rings by T. Le. (Interesting topics about the Borromean rings can be found in [**7**]. Visit also the website: http://www.liv.ac.uk/ spmr02/rings/.) The complement of the Borromean rings is hyperbolic. You can also enjoy a video: [**8**].

REMARK 5.12. Y. Yokota showed (inspired by D. Thurston's observation [**53**]) that one can obtain a decomposition by ideal tetrahedra corresponding to the limit of Kashaev's invariant for any hyperbolic knot complement whose parameters naturally defines a potential function that gives both volume and Chern–Simons invariant ([**59, 57, 58, 60, 61**]. See also [**34**].)

REMARK 5.13. For further development of the Volume Conjecture see [**33, 35**].

References

[1] C. C. Adams, *The knot book*, W. H. Freeman and Company, New York, 1994.

[2] J. W. Alexander, *Topological invariants of knots and links*, Trans. Amer. Math. Soc. **30** (1928), no. 2, 275–306.

[3] R. Benedetti and C. Petronio, *Lectures on hyperbolic geometry*, Universitext, Springer-Verlag, Berlin, 1992.

[4] G. Burde and H. Zieschang, *Knots*, de Gruyter Studies in Mathematics, vol. 5, Walter de Gruyter & Co., Berlin, 2003.

[5] S.-S. Chern and J. Simons, *Characteristic forms and geometric invariants*, Ann. of Math. (2) **99** (1974), 48–69.

[6] J. H. Conway, *An enumeration of knots and links, and some of their algebraic properties*, Computational Problems in Abstract Algebra (Proc. Conf., Oxford, 1967), Pergamon, Oxford, 1970, pp. 329–358.

[7] P. Cromwell, E. Beltrami, and M. Rampichini, *The Borromean rings*, Math. Intelligencer **20** (1998), no. 1, 53–62.

[8] D. Epstein and C. Gunn, *Not knot. VHS video (20 min.) with paperback supplement.*, Wellesley, MA: A. K. Peters. 48 p. $ 39.95 , 1991 (English).

[9] R. H. Fox, *Free differential calculus. III*, Ann. of Math. (2) **64** (1956), 407–419.

[10] F Frankl and L Pontrjagin, *Ein Knotensatz mit Anwendung auf die Dimensionstheorie*, Math. Ann. **102** (1930), 785–789.

[11] P. Freyd, D. Yetter, J. Hoste, W. B. R. Lickorish, K. Millett, and A. Ocneanu, *A new polynomial invariant of knots and links*, Bull. Amer. Math. Soc. (N.S.) **12** (1985), no. 2, 239–246.

[12] C. A. Giller, *A family of links and the Conway calculus*, Trans. Amer. Math. Soc. **270** (1982), no. 1, 75–109.

[13] F. González-Acuña and H. Short, *Cyclic branched coverings of knots and homology spheres*, Rev. Mat. Univ. Complut. Madrid **4** (1991), no. 1, 97–120.

[14] C. McA. Gordon, *Knots whose branched cyclic coverings have periodic homology*, Trans. Amer. Math. Soc. **168** (1972), 357–370.

[15] M. Gromov, *Volume and bounded cohomology*, Inst. Hautes Études Sci. Publ. Math. (1982), no. 56, 5–99 (1983).

[16] K. Habiro, *On the colored Jones polynomials of some simple links*, Sūrikaisekikenkyūsho Kōkyūroku (2000), no. 1172, 34–43.

[17] R. Hartley, *The Conway potential function for links*, Comment. Math. Helv. **58** (1983), no. 3, 365–378.

[18] W. H. Jaco and P. B. Shalen, *Seifert fibered spaces in 3-manifolds*, Mem. Amer. Math. Soc. **21** (1979), no. 220, viii+192.

[19] K. Johannson, *Homotopy equivalences of 3-manifolds with boundaries*, Lecture Notes in Mathematics, vol. 761, Springer, Berlin, 1979.

[20] V. F. R. Jones, *A polynomial invariant for knots via von Neumann algebras*, Bull. Amer. Math. Soc. (N.S.) **12** (1985), no. 1, 103–111.

[21] ———, *Hecke algebra representations of braid groups and link polynomials*, Ann. of Math. (2) **126** (1987), no. 2, 335–388.

[22] R. M. Kashaev, *Quantum dilogarithm as a 6j-symbol*, Modern Phys. Lett. A **9** (1994), no. 40, 3757–3768.

[23] ———, *A link invariant from quantum dilogarithm*, Modern Phys. Lett. A **10** (1995), no. 19, 1409–1418.

[24] ———, *The hyperbolic volume of knots from the quantum dilogarithm*, Lett. Math. Phys. **39** (1997), no. 3, 269–275.

[25] R. M. Kashaev and O. Tirkkonen, *A proof of the volume conjecture on torus knots*, Zap. Nauchn. Sem. S.-Peterburg. Otdel. Mat. Inst. Steklov. (POMI) **269** (2000), no. Vopr. Kvant. Teor. Polya i Stat. Fiz. 16, 262–268, 370.
[26] L. H. Kauffman, *The Conway polynomial*, Topology **20** (1981), no. 1, 101–108.
[27] ———, *State models and the Jones polynomial*, Topology **26** (1987), no. 3, 395–407.
[28] G. Kuperberg, *The quantum G_2 link invariant*, Internat. J. Math. **5** (1994), no. 1, 61–85.
[29] W. B. R. Lickorish, *An introduction to knot theory*, Graduate Texts in Mathematics, vol. 175, Springer-Verlag, New York, 1997.
[30] G. Masbaum, *Skein-theoretical derivation of some formulas of Habiro*, Algebr. Geom. Topol. **3** (2003), 537–556, arXiv:math.GT/0306345.
[31] R. Meyerhoff, *Density of the Chern–Simons invariant for hyperbolic 3-manifolds*, Low-dimensional topology and Kleinian groups (Coventry/Durham, 1984), Cambridge Univ. Press, Cambridge, 1986, pp. 217–239.
[32] J. Milnor, *Hyperbolic geometry: the first 150 years*, Bull. Amer. Math. Soc. (N.S.) **6** (1982), no. 1, 9–24.
[33] H. Murakami, *Optimistic calculations about the Witten-Reshetikhin-Turaev invariants of closed three-manifolds obtained from the figure-eight knot by integral Dehn surgeries*, Sūrikaisekikenkyūsho Kōkyūroku (2000), no. 1172, 70–79.
[34] ———, *Kashaev's invariant and the volume of a hyperbolic knot after Y. Yokota*, Physics and combinatorics 1999 (Nagoya), World Sci. Publishing, River Edge, NJ, 2001, pp. 244–272.
[35] ———, *Some limits of the colored Jones polynomials of the figure-eight knot*, Kyungpook Math. J. **44** (2004), no. 3, 369–383.
[36] H. Murakami and J. Murakami, *The colored Jones polynomials and the simplicial volume of a knot*, Acta Math. **186** (2001), no. 1, 85–104.
[37] H. Murakami, J. Murakami, M. Okamoto, T. Takata, and Y. Yokota, *Kashaev's conjecture and the Chern-Simons invariants of knots and links*, Experiment. Math. **11** (2002), no. 3, 427–435.
[38] H. Murakami, T. Ohtsuki, and S. Yamada, *Homfly polynomial via an invariant of colored plane graphs*, Enseign. Math. (2) **44** (1998), no. 3-4, 325–360.
[39] W. D. Neumann and D. Zagier, *Volumes of hyperbolic three-manifolds*, Topology **24** (1985), no. 3, 307–332.
[40] J. H. Przytycki and P. Traczyk, *Invariants of links of Conway type*, Kobe J. Math. **4** (1988), no. 2, 115–139.
[41] K. Reidemeister, *Knotentheorie*, Springer-Verlag, Berlin, 1974 (German), reprint.
[42] ———, *Knot theory*, BCS Associates, Moscow, Idaho, USA, 1983, Transl. from [**41**] and ed. by Leo F. Boron, Charles O. Christenson, and Bryan A. Smith.
[43] R. Riley, *A quadratic parabolic group*, Math. Proc. Cambridge Philos. Soc. **77** (1975), 281–288.
[44] ———, *Growth of order of homology of cyclic branched covers of knots*, Bull. London Math. Soc. **22** (1990), no. 3, 287–297.
[45] D. Rolfsen, *Knots and links*, Mathematics Lecture Series, vol. 7, Publish or Perish Inc., Houston, TX, 1990.
[46] C. P. Rourke and B. J. Sanderson, *Introduction to piecewise-linear topology*, Springer Study Edition, Springer-Verlag, Berlin, 1982.
[47] H. Seifert, *Topologie dreidimensionaler gefaserter Räume.*, Acta Math. **60** (1933), 147–238.
[48] ———, *Über das Geschlecht von Knoten.*, Math. Ann. **110** (1934), 571–592 (German).
[49] H. Seifert and W. Threlfall, *Lehrbuch der Topologie.*, Leipzig und Berlin: B. G. Teubner. VII, 353 S., 132 Fig. , 1934 (German).
[50] ———, *A textbook of topology*, Pure and Applied Mathematics, vol. 89, Academic Press Inc. [Harcourt Brace Jovanovich Publishers], New York, 1980, Transl. from [**49**] by Michael A. Goldman. Seifert, H.: Topology of 3-dimensional fibered spaces. Transl. by Wolfgang Heil. Ed. by Joan S. Birman and Julian Eisner.
[51] D. S. Silver and S. G. Williams, *Mahler measure, links and homology growth*, Topology **41** (2002), no. 5, 979–991.
[52] ———, *Mahler measure of Alexander polynomials*, J. London Math. Soc. (2) **69** (2004), no. 3, 767–782.

[53] D. Thurston, *Hyperbolic volume and the Jones polynomial*, Lecture notes, École d'été de Mathématiques 'Invariants de nœuds et de variétés de dimension 3', Institut Fourier - UMR 5582 du CNRS et de l'UJF Grenoble (France) du 21 juin au 9 juillet 1999.

[54] W. P. Thurston, *The Geometry and Topology of Three-Manifolds*, Electronic version 1.1 - March 2002, http://www.msri.org/publications/books/gt3m/.

[55] ———, *Three-dimensional geometry and topology. Vol. 1*, Princeton Mathematical Series, vol. 35, Princeton University Press, Princeton, NJ, 1997, Edited by Silvio Levy.

[56] H. Wenzl, *On sequences of projections*, C. R. Math. Rep. Acad. Sci. Canada **9** (1987), no. 1, 5–9.

[57] Y. Yokota, *On the volume conjecture for hyperbolic knots*, arXiv:math.QA/0009165.

[58] ———, *On the volume conjecture for hyperbolic knots*, Proceedings of the 47th Topology Symposium (Inamori Hall, Kagoshima University), July 2000, pp. 38–44.

[59] ———, *On the volume conjecture of hyperbolic knots*, Knot Theory – dedicated to Professor Kunio Murasugi for his 70th birthday (M. Sakuma, ed.), March 2000, pp. 362–367.

[60] ———, *On the potential functions for the hyperbolic structures of a knot complement*, Geom. Topol. Monogr. **4** (2002), 303–311.

[61] ———, *From the Jones polynomial to the A-polynomial of hyperbolic knots*, Interdiscip. Inform. Sci. **9** (2003), 11–21.

[62] T. Yoshida, *The η-invariant of hyperbolic 3-manifolds*, Invent. Math. **81** (1985), no. 3, 473–514.

DEPARTMENT OF MATHEMATICS, TOKYO INSTITUTE OF TECHNOLOGY, OH-OKAYAMA, MEGURO, TOKYO 152-8551, JAPAN

E-mail address: `starshea@tky3.3web.ne.jp`

Contemporary Mathematics
Volume **416**, 2006

Classical Knot Invariants and Elementary Number Theory

Kunio Murasugi

1. Introduction and Preliminaries

(1) **Knots, equivalent knots and knot types**

A knot K is a simple closed curve in a 3-space $\mathbb{R}^3$ or a 3-sphere S^3. (Since S^3 is obtained from $\mathbb{R}^3$ by adding one point ∞, there are no essential differences between knots in $\mathbb{R}^3$ and knots in S^3.) Further, in order to avoid unnecessary complications, we may assume that K is a polygon (with many sides). See Fig. 1.1. Both $\mathbb{R}^3$ and S^3 are oriented. A 3-space $\mathbb{R}^3$ (and S^3) is given a right handed orientation. See Fig. 1.2.

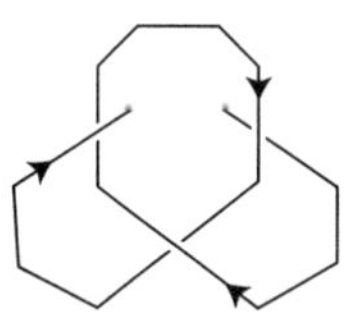

Figure 1.1

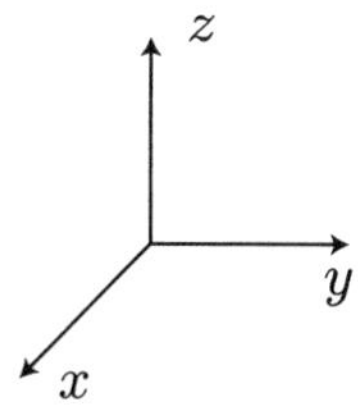

Figure 1.2

The orientation of a knot is usually denoted by an arrow. See Fig. 1.1. Such a knot is called an *oriented* knot. Two knots K_1 and K_2 (in $\mathbb{R}^3$ or S^3) are said to be *equivalent* (or *ambient isotopic*) if there is a homeomorphism f of $\mathbb{R}^3$ (or S^3) onto itself that maps K_1 to K_2, where f preserves orientations of $\mathbb{R}^3$ (or S^3). If K_1 and K_2 are oriented, then f sends K_1 to K_2 including orientations. Intuitively speaking, two knots are *equivalent* if we can move one knot to another knot in $\mathbb{R}^3$ (or S^3) without allowing any self-intersections. See Fig. 1.3. The simplest knot is a knot equivalent to a circle on the plane in $\mathbb{R}^3$. It is called a *trivial knot* or a *unknotted knot.* Fig. 1.4.

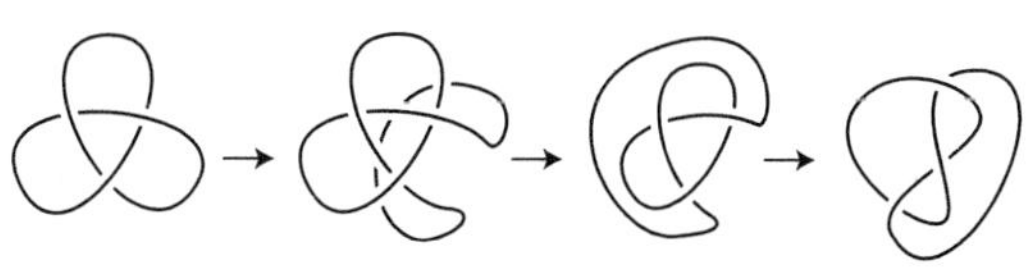

Figure 1.3

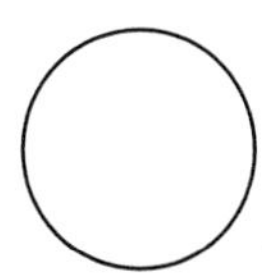

Figure 1.4

Two equivalent knots are said to be of the *same type.*

2000 *Mathematics Subject Classification.* Primary 57M25, Secondary 57M05.

Partially supported by NSERC Grant A 4034.

(2) **Invertibility**

Each knot can have two different orientations, one is opposite to another. See Fig. 1.5.

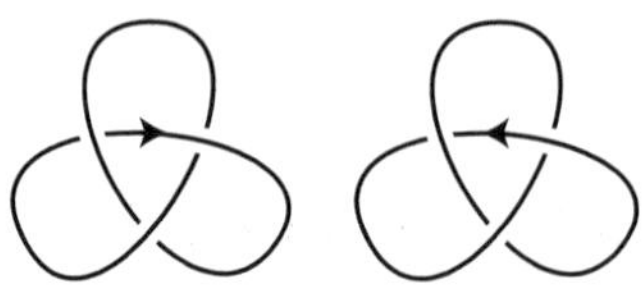

(a) (b)

Figure 1.5

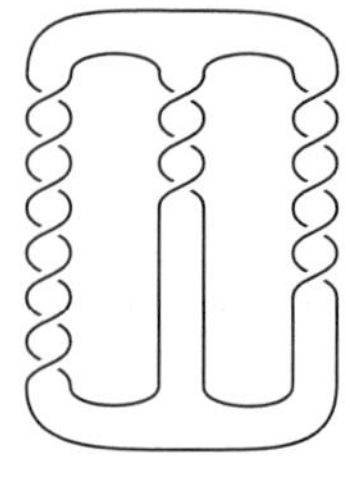

Figure 1.6

It is conceivable that two oriented knots with opposite orientations given to the same knot may not be equivalent. But it was 1964 when H.Trotter first proved that the knot in Fig. 1.6 is one of such knots [T3]. A knot is called *invertible* if two oriented knots with opposite orientations obtained from the same knot are equivalent. Many knots are invertible, but non-invertible knots are by no means rare.

(3) **Amphicheirality**

Given a knot K in $\mathbb{R}^3$, we can construct a mirror image of K. More precisely, let f be a reflection of $\mathbb{R}^3$, i.e. $f(x,y,z) = (x,y,-z)$. The image of K under f is called the *mirror* image of K. If K is equivalent to its mirror image K^*, then K is said to be *amphicheiral.* If an oriented knot K and K^* with the orientation induced from that of K are equivalent, we say an (oriented) knot K is $+$amphicheiral. However, if K and K^* with opposite orientation are equivalent, K is called $-$amphicheiral. For example, the knot in Fig. 1.1 is not amphicheiral, but the proof is rather complicated. The simplest non-trivial amphicheiral knot is shown in Fig. 1.7.

Figure 1.7

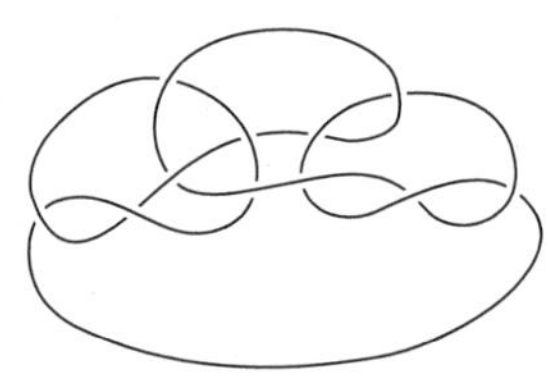

Figure 1.8

(4) **Knot invariant**

One of the fundamental problems in knot theory is a so-called the *classification problem.* The problem is to construct a list $\mathcal{L}$ of knots for which (1) any knot is equivalent to a knot in $\mathcal{L}$ and (2) two knots in $\mathcal{L}$ are not equivalent. The classification problem has been solved for certain families of knots which we discuss later. However, so far, there are no algorithms by which we can determine whether or not given two knots are equivalent. One of the effective methods to distinguish two knots is to compare their knot invariants. A quantity $\rho(K)$ assigned to a knot K is called a *knot invariant* if $\rho(K_1) = \rho(K_2)$ for equivalent knots K_1 and K_2. One of the earliest knot invariants is probably the Minkowski unit $C_p(K)$ for a knot K and an odd prime p. (See Section 6.) Later, Alexander [A2] introduced the most important knot invariant in knot theory called the *Alexander polynomial.*

The Alexander polynomial is an integer polynomial for any knot, and in particular, for a trivial knot, its value is 1. The Alexander polynomial is one of the major tools to distinguish two knots, however it is not a *complete invariant*. In fact, Kinoshita-Terasaka knot (or KT knot) depicted in Fig. 1.8 is not a trivial knot, but the Alexander polynomial is 1.

(5) **Links**

So far, we have only looked at knots, but many concepts can be extended to a link, a collection of knots. A *link* is a finite ordered collection of mutually disjoint knots $L = K_1 \cup K_2 \cup \cdots \cup K_n$, in $\mathbb{R}^3$ or S^3. Each knot K_i is called a *component* of L. If L consists of n knots, L is called an *n-component link*. If each component is oriented, L is called an oriented link. Two links $L = \{K_1, K_2, \dots, K_n\}$ and $L' = \{K_1', K_2', \dots, K_m'\}$ are *equivalent* if (1) $n = m$, and (2) there is a homeomorphism f that preserves the orientation of $\mathbb{R}^3$ (or S^3) and f maps K_1 to K_1', K_2 to $K_2', \dots, K_n$ to K_n' (since $m = n$). Strictly speaking, the equivalence of links should also be related to how we order the components. However, such a stringent condition is not necessary, since we may suitably reorder the components. Therefore (2) is usually replaced by the following (2A): (2A) There is a homeomorphism f that preserves the orientation of $\mathbb{R}^3$ (or S^3) and maps $K_1 \cup K_2 \cup \cdots \cup K_n$ to $K_1' \cup K_2' \cup \cdots \cup K_m'$. For a link, invertibility or amphicheirality is not a major problem. In fact, the Alexander polynomial is defined only for an oriented link, and if one component K_i reverses its orientation, the Alexander polynomial changes entirely. For a knot, however, the Alexander polynomial does not depend on its orientation. Therefore, the non-invertibility of a knot cannot be detected by the Alexander polynomial. Although these notable differences exist between knots and links, many properties of knots are extended to those of links in a straightforward manner. However, in this article, knots and links are carefully distinguished. Therefore, a link always means a link with more than one component.

2. Knot (or link) Group

(1) **Diagram**

We usually represent a knot K by a graph D with "broken edges" on a plane. See Fig. 1.1. This is called a *regular diagram* of K. A regular diagram gives us a sense of how the knot may in fact lie in 3-dimensions, i.e. it allows us to depict the knot as a spatial diagram on the plane. Since we consider exclusively regular diagrams, we call them simply *diagrams* of knots.

(2) **Knot group**

For a knot K, $\pi_1(\mathbb{R}^3 - K, *)(\simeq \pi_1(S^3 - K, *))$, the fundamental group of the complement of K in $\mathbb{R}^3$ (or S^3), is called the *group of knot* or *knot group* of K, where $*$ denotes a base point, but since two groups with different base points are conjugate, we usually omit $*$. We denote the knot group of K by $G(K)$. $G(K)$ is finitely presented, i.e. $G(K)$ has a presentation in which both the number of generators and relations are finite.

(3) **Wirtinger presentation**

Using a diagram, we can obtain one of finite presentations, called a *Wirtinger presentation*, of $G(K)$. In this presentation, each generator corresponds to an *over-passing* arc, defined below, and each relation corresponds to a crossing point. Let K be an oriented knot and D a diagram. Select one under-crossing point, P_1 say, and trace the arc (according to its orientation) until the next under-crossing point P_2.

See Fig. 2.1. The arc connecting P_1 and P_2 is denoted by s_1, called an *over-passing arc.* The next arc s_2 is the arc connecting P_2 and P_3, and so on. Note that s_2 is a long arc, but s_5 and s_6 are short arcs in Fig. 2.1. A short arc does not pass over any arc. Each arc s_i corresponds to a generator x_i of $G(K)$. See Fig. 2.2.

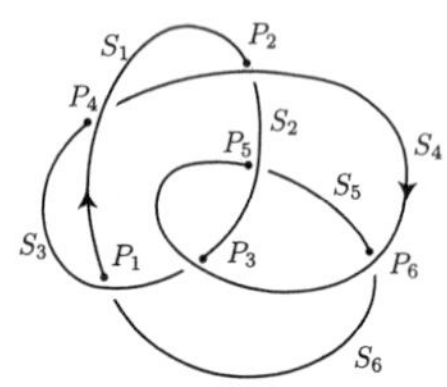

Figure 2.1

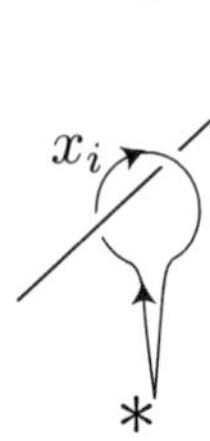

Figure 2.2

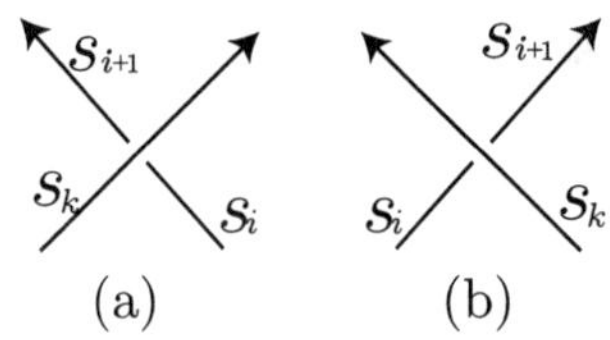

Figure 2.3

x_i is an oriented loop in $\mathbb{R}^3$ (or S^3) that circles once around s_i as is shown in Fig. 2.2. The set of these elements $\{x_1, x_2, \ldots, x_n\}$ is a set of generators of $G(K)$, where n is the number of crossing points in D. A relation of $G(K)$ is obtained as follows: To each crossing point, 3 arcs meet as in Fig. 2.3 (a) or (b). For (a), the corresponding relation is $R_i = x_i x_k x_{i+1}^{-1} x_k^{-1}$. For (b), it is $R_i = x_i x_k^{-1} x_{i+1}^{-1} x_k$.

Therefore, $G(K)$ has a presentation $G(K) = \langle x_1, x_2, \ldots, x_n | R_1 = 1, R_2 = 1, \ldots, R_n = 1\rangle$. It is known that any one of R_i is unnecessary, so that $G(K)$ is presented by n generators and $n-1$ relations, where n is the number of crossing points of D. This presentation of $G(K)$ is called a *Wirtinger presentation* of $G(K)$.

(4) **Examples**

Example 2.1 For the knot with the diagram D in Fig. 2.1,
$G(K) = \langle x_1, x_2, \ldots, x_6 | R_1 = 1, R_2 = 1, \ldots, R_6 = 1\rangle$, where $R_1 = x_1 x_4^{-1} x_2^{-1} x_4$, $R_2 = x_2 x_4 x_3^{-1} x_4^{-1}$, $R_3 = x_3 x_1^{-1} x_4^{-1} x_1$, $R_4 = x_4 x_2 x_5^{-1} x_2^{-1}$, $R_5 = x_5 x_4 x_6^{-1} x_4^{-1}$, $R_6 = x_6 x_3^{-1} x_1^{-1} x_3$.

Example 2.2 A Wirtinger presentation of K in Fig. 2.4 is
$G(K) = \langle x_1, x_2, x_3 | R_1 = 1, R_2 = 1\rangle$, where $R_1 = x_1 x_3 x_2^{-1} x_3^{-1}$, $R_2 = x_2 x_1 x_3^{-1} x_1^{-1}$. We can simplify this presentation as follows. First, $R_2 = 1$ yields $x_1^{-1} x_2 x_1 = x_3$, and hence x_3 can be eliminated. A new presentation has 2 generators x_1 and x_2, and one relation R_2' that comes from R_1: $R_1' = x_2 x_1 x_2^{-1} x_1^{-1} x_2^{-1} x_1$. If we set $a = x_2 x_1$ and $b = x_2 x_1 x_2$, then a and b generate $G(K)$ and one relation R_1' becomes $ab^{-2}a^2 = 1$, and thus we have a simple presentation $G(K) = \langle a, b | a^3 = b^2\rangle$.

(5) **Link group**

For an oriented link $L = K_1 \cup K_2 \cup \cdots \cup K_r$, the link group $G(L) = \pi_1(\mathbb{R}^3 - L)$ $(\simeq \pi_1(S^3 - L))$ has a Wirtinger presentation, where the set of generators is $\{x_{i,j} | i = 1, 2, \ldots, r; j = 1, 2, \ldots, m_i\}$, and the set of relations is $\{R_{i,j} = 1 | i = 1, 2, \ldots, r; j = 1, 2, \ldots, m_i\}$, and $R_{i,j}$ is of the form : $x_{i,j} x_{k,\ell}^{e_{i,j}} x_{i,j}^{-1} x_{k,\ell}^{-e_{i,j}}$, where $e_{i,j} = +1$ or-1. $x_{i,j}$ corresponds to an over-passing arc in the ith component K_i of L. As in the case of the knot group, at least one of the relations can be eliminated.

Example 2.3 Let L be a link consisting of n *completely separated* trivial knots. More precisely, L has a diagram consisting of n trivial knots that are separated by $n-1$ parallel lines. See Fig. 2.5 for $n = 3$. Then L is called a *trivial n-component link.*

The group of L is a free group of rank n freely generated by $x_{1,1}, x_{2,1}, \ldots, x_{n,1}$.

Example 2.4 Let L be a 2-component link in Fig. 2.6.
Then $G(L) = \langle x_{1,1} x_{2,1} | R_{1,1} = 1, R_{2,1} = 1\rangle$ where $R_{1,1} = x_{1,1} x_{2,1} x_{1,1}^{-1} x_{2,1}^{-1}$, and

$R_{2,1} = R_{1,1}^{-1}$. Therefore, $G(L)$ is a free abelian group of rank 2. The link L is called a *Hopf link*.

Remark 2.1 The knot (or link) group is an invariant of an *unoriented* knot (or link). Furthermore, the group of a knot (or link) K is isomorphic to that of the mirror image of K. Therefore, we cannot detect the invertibility and amphicheirality of K by its group alone.

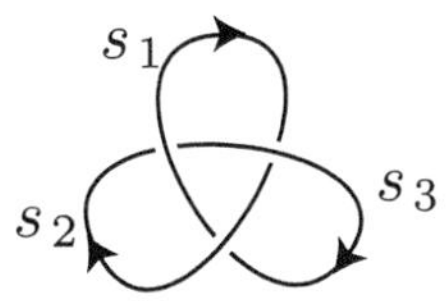

Figure 2.4

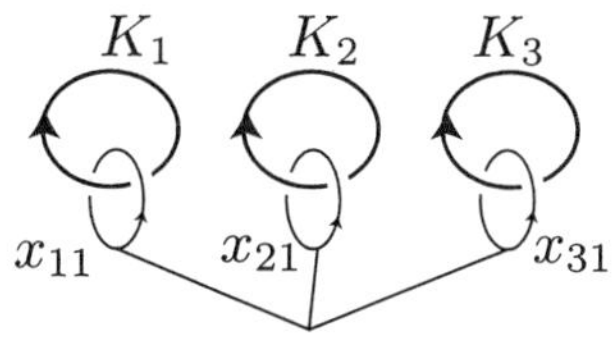

Figure 2.5

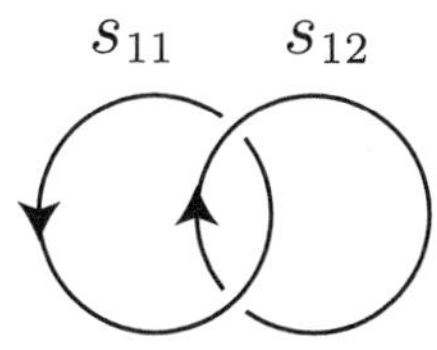

Figure 2.6

3. Braids

(1) **n-braids**

Take a cube B in $\mathbb{R}^3$, and plot n points $A_1, A_2, \ldots, A_n$ on the top of B and other n points $A'_1, A'_2, \ldots, A'_n$ on the base of B. For the sake of neatness, we specify the coordinates of these points. First, describe B as the set $\{(x,y,z)|0 \leq x,y,z \leq 1\}$, and define the coordinates of A_i as $(\frac{1}{2}, \frac{i}{n+1}, 1)$ and that of A'_i as $(\frac{1}{2}, \frac{i}{n+1}, 0)$. Now join these $A_1, \ldots, A_n$ to $A'_1, \ldots, A'_n$ by means of n (polygonal) arcs $u_1, u_2, \ldots, u_n$ inside B (except their end points) as follows: (1) u_i and u_j do not intersect if $i \neq j$. (2) u_i joins A_i to A'_k for some k. These arcs u_i are called *strings*. If any plane E parallel to the base of B either intersects each string at one and only one point, or it does not intersect at all, then the set of these n strings in B is called an *n-braid*. Given an n-braid, by projecting the braid onto the yz-plane, we obtain a (regular) diagram of a braid (as in the case of knots) . See Fig. 3.1 (a) or (b).

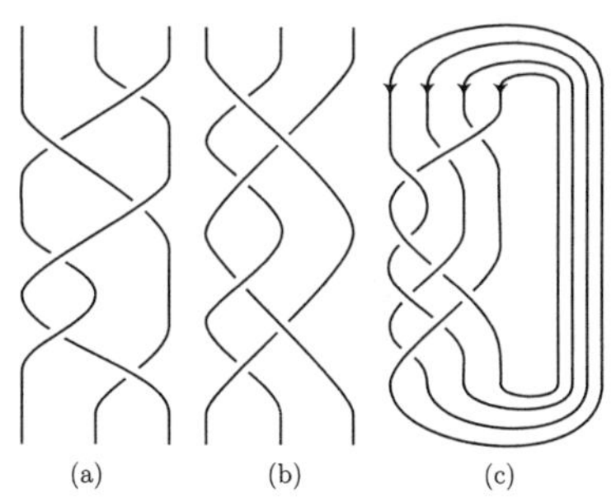

Figure 3.1

As is shown in Fig. 3.1 (b), if each A_i $(i = 1, 2, \ldots, n)$ is connected to A'_i by the ith string u_i, then an n-braid is called a *pure* n-braid. Intuitively, two braids (in a cube) whose end-points we keep fixed, is said to be *equivalent*, if we can continuously deform one to the other without causing any of the strings to intersect each other.

(2) **The n-braid group B_n**

Given two n-braids α and β, we can define the product $\alpha\beta$. First glue the base of the cube that contains α to the top face of the cube that contains β. The gluing together of two cubes produces a rectangular solid in which there exists an n-braid that has been created from α and β. This braid is the *product* $\alpha\beta$. It is well known that under this product, the set of all (equivalent classes of) n-braids forms a group, called the *n-braid group*, denoted by B_n. The group B_n is generated by $n-1$ n-braids, $\sigma_1, \sigma_2, \ldots, \sigma_{n-1}$. See Fig. 3.2.

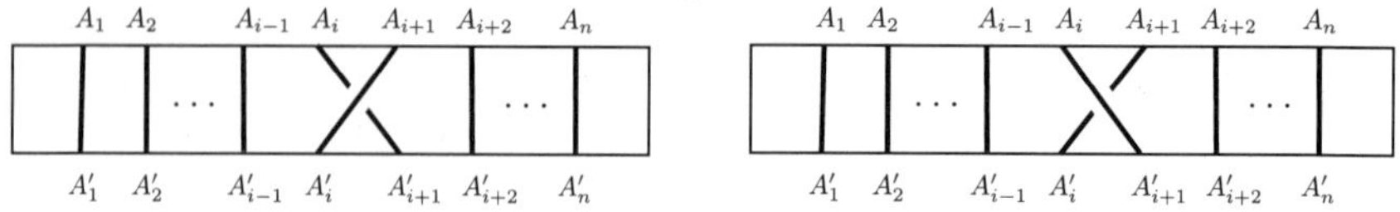

(a) σ_i (b) σ_i^{-1}

Figure 3.2

There are two types of relations for B_n. One is of the form (I) $\sigma_i\sigma_j = \sigma_j\sigma_i$ for $|i-j| > 1$, and another is of the form (II) $\sigma_i\sigma_{i+1}\sigma_i = \sigma_{i+1}\sigma_i\sigma_{i+1}$. These relations (I) and (II) form a complete set of relations for B_n.

(3) **Examples**

Example 3.1 The 3-braids shown in Fig. 3.1 (a) and (b) are written, respectively, as $\sigma_2\sigma_1^{-1}\sigma_2\sigma_1^2\sigma_2^{-1}$ and $\sigma_1^{-1}\sigma_2^{-1}\sigma_1^2\sigma_2^{-1}\sigma_1$.

Example 3.2 (1) $B_1 = \langle 1 \rangle$ is a trivial group. (2) $B_2 = \langle \sigma_1 | - \rangle$ is an infinite cyclic group.

(3) $B_3 = \langle \sigma_1, \sigma_2 | \sigma_1\sigma_2\sigma_1 = \sigma_2\sigma_1\sigma_2 \rangle$

(4) $B_4 = \langle \sigma_1, \sigma_2, \sigma_3 | \sigma_1\sigma_3 = \sigma_3\sigma_1, \sigma_1\sigma_2\sigma_1 = \sigma_2\sigma_1\sigma_2, \sigma_2\sigma_3\sigma_2 = \sigma_3\sigma_2\sigma_3 \rangle$

(4) **closed n-braid (or the closure of a braid)**

As is shown in Fig. 3.1 (c), we connect, by a set of parallel arcs that lie outside the square, the points $A_1, A_2, \ldots, A_n$ on the top of the braid α to the points $A_1', A_1', \ldots, A_n'$. The knot or link thus created is called the *closed braid* (or the *closure* of α). We assign to each string an orientation downward. The following theorem due to Alexander connects the braid theory to knot theory.

Theorem 3.1[A1] *Any knot (or link) is equivalent to some closed braid.*

Therefore, to each knot, we can associate an element of some braid group. However, since a knot can be represented as many different closed braids, an element of a braid assigned to a knot is not a knot invariant.

(5) **Braid index**

By Theorem 3.1, we see that any knot (or link) K is a closed n-braid for some $n \geq 1$. If K is a closed m-braid, but K cannot be represented by less number of strings, then m is called the *braid index* of K, denoted by $b(K)$. Obviously, $b(K)$ is an invariant of a knot (or link). It is a hard problem to decide the braid index, and there is no algorithm to determine $b(K)$. If L is an n-component link, then $b(L) \geq n$. For a link L, $b(L)$ depends on the orientation of each component. If L is a closed pure braid, then each component of L is unknotted, but the converse is not true.

4. Special families of knots (I): Torus knots

In the following two sections, we introduce two familiar classes of knots, tours knots and 2-bridge knots. These knots have been studied quite extensively, and, in particular, these knots are completely classified. Now, we begin with an introduction of a new concept.

(1) **Peripheral system**

Let T be a torus in $\mathbb{R}^3$. T may be considered as the boundary of a solid obtained by fatting a knot K in $\mathbb{R}^3$. (Such a solid is called a *tubular neighbourhood* of K.) See

Fig. 4.1 (a). On this torus T, there are two simple (oriented) closed curves, called a *meridian* m and a *longitude* ℓ, see Fig. 4.1. A meridian m is a simple closed curve on T that bounds a disk *inside* T, while ℓ is a simple closed curve on T that bounds an orientable surface *outside* T. (For an existence of a longitude, see Remark 7.1.) Both are oriented as are shown in Fig. 4.1 (a) and (b), where the orientation of ℓ is given so that two oriented simple closed curves ℓ and K are parallel. A pair $\{m, \ell\}$ is called a *peripheral system* of a knot K.

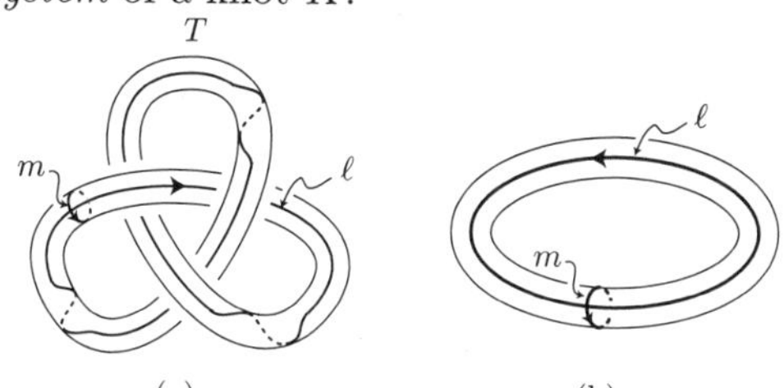

(a) (b)

Figure 4.1

Two curves m and ℓ represents two elements of the knot group $G(K)$ in a natural way.

A pair $\{m, \ell\}$ with the knot group $G(K)$ characterizes completely the knot type. More precisely, we have the following theorem due to Waldhausen:

Theorem 4.1 [Wal] *Two oriented knots K_1 and K_2 are equivalent if and only if there exists an isomorphism $f : G(K_1) \longrightarrow G(K_2)$ such that f maps a peripheral system of K_1 to a (conjugate of a) peripheral system of K_2.*

In fact, Dehn's proof of non-amphicheirality of the knot in Fig. 1.1 and also Trotter's proof of the non-invertibility of the knot in Fig. 1.6 are based upon Theorem 4.1.

(2) **Torus knots**

A torus knot is a knot embedded in a *standard torus* T in $\mathbb{R}^3$, where a standard torus means a surface generated by revolving a circle C (on a plane) about a line disjoint to C. Therefore, a standard torus can be considered as the boundary of a tubular neighbourhood of a trivial knot. Let $\{m, \ell\}$ be a peripheral system of a trivial knot K, see Fig. 4.1 (b). A torus knot (or link) of type (p, q), denoted by $T(p, q)$, is an oriented knot (or link) such that (i) $T(p, q)$ crosses a meridian m at exactly $|p|$ points from right to left if $p > 0$ (or from left to right if $p < 0$), (ii) $T(p, q)$ crosses a longitude ℓ at exactly $|q|$ points from right to left if $q > 0$ (or from left to right if $q < 0$). Any (non-trivial) knot on T can be deformed on T to $T(p, q)$ for some integers p and q, and $\gcd(p, q) = 1$. See Fig. 4.2.

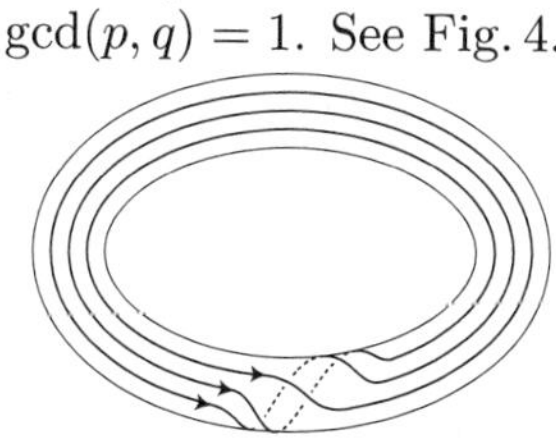

Figure 4.2: $T(3, -2)$

In particular, a meridian m may be considered as a torus knot of type $(0, -1)$ and ℓ as $T(1, 0)$. The number of components of $T(p, q)$ is given by $\gcd(p, q)$. Furthermore, if $p > 0$, then $T(p, q)$ is represented as a closed p-braid $(\sigma_1\sigma_2 \cdots \sigma_{p-1})^{-q}$. For $p < 0$, see Theorem 4.4. These torus knots have been completely classified.

Theorem 4.2 *A torus knot $T(p,q)$ is a trivial knot if and only if $pq = 0$, or p or q is ± 1.*

Theorem 4.3 *Any torus knot is invertible, and any non-trivial torus knot is not amphicheiral.*

A proof of the second statement will be outlined later.

Theorem 4.4 *Assume that none of p, q, p', q' is 0 or ± 1, and $gcd(p,q) = gcd(p',q') = 1$. Then, (i) $T(p,q) = T(q,p)$ (ii) $T(p,q) = T(p',q')$ if and only if $\{p,q\} = \{p',q'\}$.*

Proof. (i) $T(p,q)$ can be deformed into $T(q,p)$ in $\mathbb{R}^3$ (not on T). A proof of (ii) will be given later.

Theorem 4.5 *The braid index of a non-trivial torus knot $T(p,q)$ is equal to $min\{|p|,|q|\}$.*

5. Special families of knots (II) 2-bridge knots (or rational knots)

(1) **Bridge index**

Any knot (or link) has a diagram with only finitely many local maximal points. The minimal number of local maximal points among all diagrams K can have is called the *bridge index* of K, denoted by $br(K)$. $br(K)$ is an invariant of a knot (or link) K. The bridge index of an n-component link is at least n. Since a knot with the bridge index 1 is a trivial knot, the class of 2-bridge knots is the simplest class of non-trivial knots with respect to the bridge index. It is not easy to determine the bridge index for a knot or link, and there is no algorithm to decide the bridge index.

(2) **2-bridge knot (or link)**

A knot (or link) K is called a 2-*bridge* knot (or link) if K has a diagram D that has only 2 local maximal (and 2 local minimal) points. See Fig. 5.1.

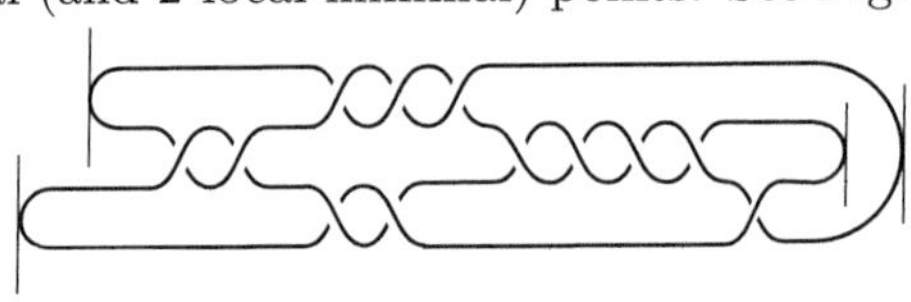

Figure 5.1

A 2-bridge knot (or link) K is characterized by two non-zero coprimes, α and β, where $-\alpha < \beta < \alpha$, or equivalently a rational number $-1 < \beta/\alpha < 1$. K is called a 2-bridge knot (or link) of *type* (α, β) and is denoted by $B(\alpha,\beta)$. K is a knot if and only if α is odd.

(3) **Diagram of** $B(\alpha,\beta)$

Given a rational number β/α, a 2-bridge knot (or link) $B(\alpha,\beta)$, $\gcd(\alpha,\beta) = 1$, is constructed in the following way. First express β/α as a continued fraction:

$$\frac{\beta}{\alpha} = \cfrac{1}{a_1 - \cfrac{1}{a_2 - \cfrac{1}{a_3 - \cfrac{1}{\ddots - \cfrac{1}{a_{m-1} - \cfrac{1}{a_m}}}}}}$$

$$= [a_1, a_2, \cdots, a_m],$$

where $a_i \neq 0$. This expression is not unique. To $[a_1, a_2, \dots, a_m]$ associate a 3-braid $\gamma = \sigma_2^{a_1}\sigma_1^{a_2}\sigma_2^{a_3}\sigma_1^{a_4}\dots\sigma_1^{a_m}$ if m is even, or $\gamma = \sigma_2^{a_1}\sigma_1^{a_2}\sigma_2^{a_3}\sigma_1^{a_4}\dots\sigma_2^{a_m}$, if m is odd.

Finally, join the top and bottom as is shown in Fig. 5.2.

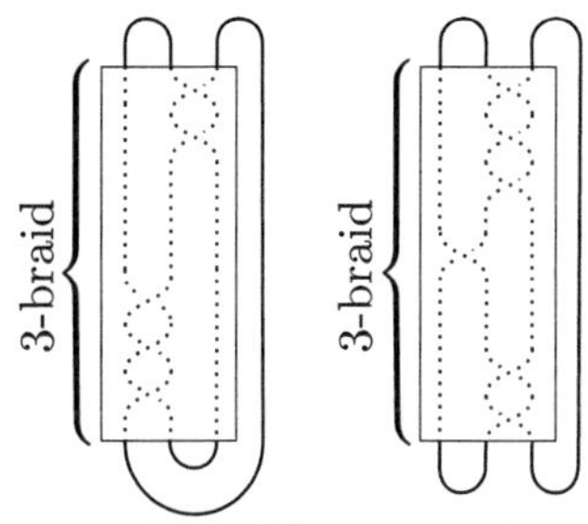

(a) $m \equiv 0 \pmod 2$ (b) $m \equiv 1 \pmod 2$

Figure 5.2

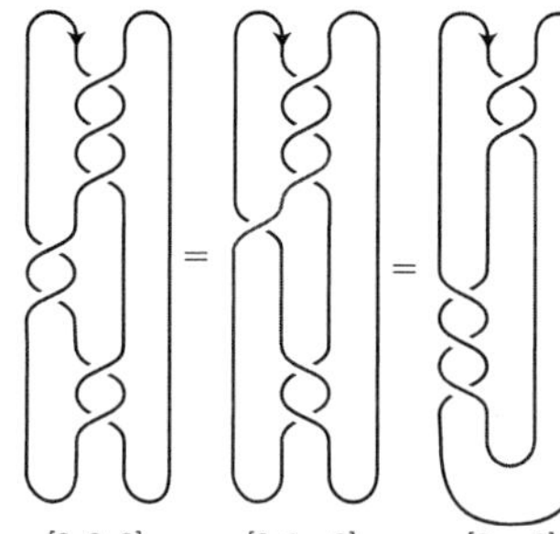

[3, 2, 2] [3, 1, −2] [2, −3]

Figure 5.3

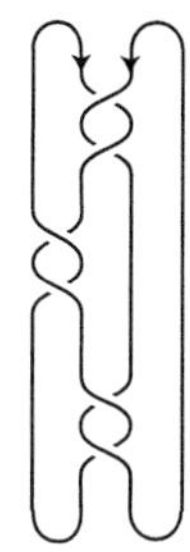

[2, −2, −2]

Figure 5.4

The diagram of a knot (or link) thus obtained is a diagram of $B(\alpha, \beta)$. It is shown that no matter what continued fractions of β/α is used, we obtain the same (i.e. equivalent) knot (or link).

Example 5.1 (1) Since $\frac{3}{7} = [3, 2, 2] = [3, 1, -2] = [2, -3]$, three diagrams in Fig. 5.3 represent the same 2-bridge knot $B(7, 3)$.

(2) Since $\frac{3}{8} = [2, -2, -2]$, Fig. 5.4 depicts a 2-bridge link $B(8, 3)$.

(4) **Classification**

The orientation of $B(\alpha, \beta)$ is given as follows. The second string is always oriented downward (see Fig. 5.3). If $B(\alpha, \beta)$ is a link, then the third string forms one component and is oriented downward (see Fig. 5.4). Thus, we obtain an oriented 2-bridge knot (or link) $B(\alpha, \beta)$. These knots (or links) are classified by the following theorem.

Theorem 5.1 [Sc] *For oriented 2-bridge knots (or links), $B(\alpha, \beta)$ is equivalent to $B(\alpha', \beta')$ if and only if*

(i) $\alpha = \alpha'$ and $\beta \equiv \beta'$, (mod α), or

(ii) $\alpha = \alpha'$ and $\beta\beta' \equiv 1$ (mod α).

Example 5.2 (1) $B(3, 1) = B(3, -2)$ (2) $B(7, 4) = B(7, -3) = B(7, -5)$.

Theorem 5.2 *A 2-bridge knot $B(\alpha, \beta)$ is invertible, and $B(\alpha, \beta)$ is amphicheiral if and only if $\beta^2 \equiv -1$ (mod α).*

6. Minkowski units

In the following three sections, we discuss three classical knot invariants. As was stated in Introduction, the Minkowski unit is one of the earliest numerical invariants in knot theory. Since its value is either $+1$ or -1, it cannot be applied on the classification problem, but it gives an elegant solution to the amphicheirality problem.

(1) **Goeritz matrix**

Let D be a diagram of a knot (not a link) K. Then D divides $\mathbb{R}^2$ into finitely many domains, $D_1, D_2, \dots, D_n$, one of which is unbounded, say D_n. We classify these domains into two classes, black and white in such a way that no domains of the same colour have edges in common. For convenience, we assume that D_n is a black domain. (This assumption is not a serious restriction, because any domain can be deformed into an unbounded domain by a suitable deformation of K.) Now we define an *index* ε of a crossing point v of D, where $\varepsilon(v) = \pm 1$ according to Fig. 6.1.

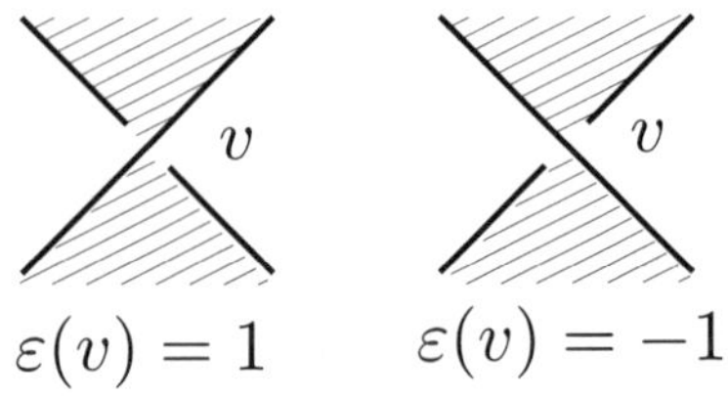

Figure 6.1

Here, a shaded area indicates a black domain and the orientation of K is irrelevant. Let $\{W_1, W_2, \ldots, W_{m+1}\}$ be the set of all white domains. Using these white domains, we define an $(m+1) \times (m+1)$ integer symmetric matrix $A = (a_{i,j})$, $1 \le i, j \le m+1$.
For $i \neq j$, $-a_{i,j} = \sum \varepsilon(v)$, where the summation runs over all crossing points of D that are common to W_i and W_j. For $i = 1, 2, \ldots, m+1$, $a_{i,i} = -\sum_{j=1, i\neq j}^{m+1} a_{i,j}$. This (integer symmetric) matrix A is called the *Goeritz matrix* of a knot K. Obviously, A is singular, so we eliminate one row, say the last row, and the last column from A to obtain an $m \times m$ matrix $\hat{A}$. Usually, $|\det \hat{A}|$ is called the *determinant* of a knot K. It is a knot invariant. However, the matrix $\hat{A}$ itself is not a knot invariant, since $\hat{A}$ depends on a diagram D. It is known that $|\det \hat{A}| \equiv 1 \pmod 2$.

(2) **Minkowski unit**
Take an odd prime p. If $p \nmid \det \hat{A}$, then we define the Minkowski unit $C_p(K)$ of K at p to be 1. So let $p | \det \hat{A}$. Suppose that there is an integer unimodular matrix B with which we can diagonalize $\hat{A}$ as follows, where $p \nmid b_j, j = 1, 2, \ldots, m$:

$$B^T \hat{A} B = \begin{bmatrix} b_1 & & & & & & 0 \\ & b_2 & & & & & \\ & & \ddots & & & & \\ & & & b_k & & & \\ & & & & pb_{k+1} & & \\ 0 & & & & & \ddots & \\ & & & & & & pb_m \end{bmatrix} \pmod{p^2},$$

where B^T denotes the transpose of B.

Then we define $C_p(K) = \left(\frac{(-1)^{[\frac{m-k}{2}]} b_{k+1} b_{k+2} \cdots b_m}{p}\right)$, where $\left(\frac{q}{p}\right)$ denotes Legendre symbol. The Minkowski unit $C_p(K)$ is an invariant of a knot K. For a link L, we can define the Goeritz matrix A in the same way, but $\det \hat{A}$ may be 0. Therefore, we need a different approach to define $C_p(L)$. See [Mu2].

Example 6.1 The Goeritz matrix of a knot K in Fig. 6.2 (a) is given by

$$A = \begin{pmatrix} 2 & -1 & -1 \\ -1 & 2 & -1 \\ -1 & -1 & 2 \end{pmatrix}, \text{ and hence } \det \hat{A} = 3.$$

Therefore, $C_p(K) = 1$ for $p \neq 3$. Let $p = 3$. Then $\hat{A} \sim \begin{pmatrix} 2 & 0 \\ 0 & -3 \end{pmatrix}$ (mod 9), and hence, $C_3(K) = \left(\frac{-1}{3}\right) = -1$. On the other hand, the Goeritz matrix of the mirror image K^* of K is $-A$. See Fig. 6.2 (b). Therefore, $C_3(K^*) = 1$. Since $C_3(K) \neq C_3(K^*)$, we conclude that K is not equivalent to K^*, and K is not amphicheiral.

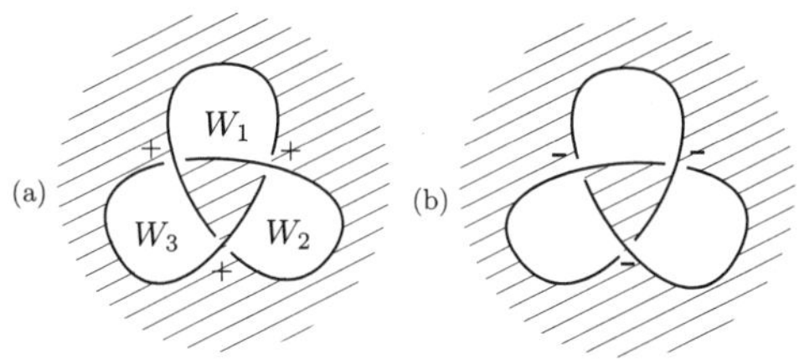

Figure 6.2

7. Seifert Matrix

In 1934, H.Seifert defined a new integer matrix N for a knot K. This matrix N is not necessarily symmetric and may be singular. However, using N, we can define two important invariants, called the *signature* and the *Alexander polynomial* of a knot K.

(1) **Seifert surface**

A connected orientable surface that bounds a given oriented knot (or link) K is called a *Seifert surface* of K. For an arbitrary knot (or link), such a surface always exists. In fact, Seifert gave an algorithm by which one can construct such a surface from a diagram D of K. Given a diagram D of K, first draw a small circle with one of the crossing point of D as its center. This circle intersects D at four points, say a, b, c and d. See Fig. 7.1 (a).

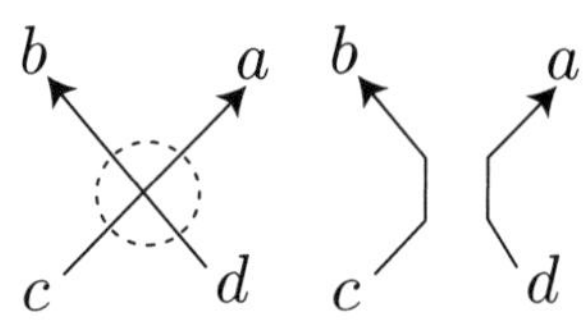

(a) (b)

Figure 7.1

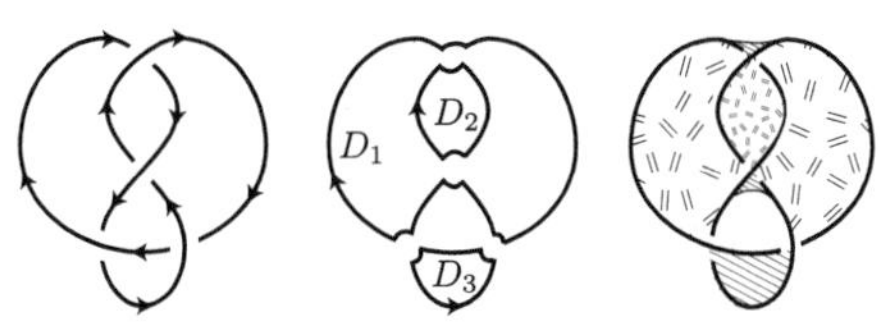

(a) (b) (c)

Figure 7.2

Then as is shown in Fig. 7.1 (b), splice this crossing point and connect a and d, and b and c. In this way we can remove the crossing point of D that lies within the circle. Now apply this process at every crossing point of D, and we remove all the crossing points from D. Then D becomes decomposed into several simple closed curves, and we span each simple closed curve by a disk. For the knot in Fig. 7.2 (a), there are three disks D_1, D_2, D_3. See Fig. 7.2 (b).

Finally, we attach a half-twisted band at the place of D that corresponds to a crossing point before they are removed. See Fig. 7.2 (c). Thus we obtain a connected, orientable surface S that bounds K. A shaded areas denotes the *face (or front)* of the surface S and a dotted area the *back* of S. (If K is a link, we alter

K, if necessary, in such a way that the projection of K is connected, then by the above method we can also obtain a connected orientable surface.)

By this simple process, we can guarantee the existence of at least one Seifert surface for any knot (or link). However, the surface obtained by this algorithm (called *Seifert algorithm*) may be unnecessarily complicated. For a Seifert surface S, we denote by $g(S)$ the genus of S. The genus of S is evaluated as $\frac{1}{2}(2-n-\chi(S))$, where n is the number of component of K and $\chi(S)$ denotes the Euler Characteristic of S. The minimum genus of all Seifert surface for a given knot (or link) K is an invariant of K. For an arbitrary knot (or link), there does exist an algorithm to actually calculate its genus, but it is exceedingly difficult to implement. In truth, to calculate its genus for an arbitrary knot (or link) is a difficult undertaking.

Remark 7.1 If a torus T is a boundary of a tubular neighbourhood of a knot K, then a longitude of T can be obtained as the intersections of a Seifert surface of K and T.

(2) **Linking number**

The *linking number*, denoted by $\text{lk}(\lambda, \mu)$, between two disjoint oriented simple closed curves λ and μ in $\mathbb{R}^3$ is the simplest, but important link invariant of a 2-component link $L = \lambda \cup \mu$. The linking number $\text{lk}(\lambda, \mu)$ is evaluated as follows. Consider a diagram D of a link L. See Fig. 7.3.

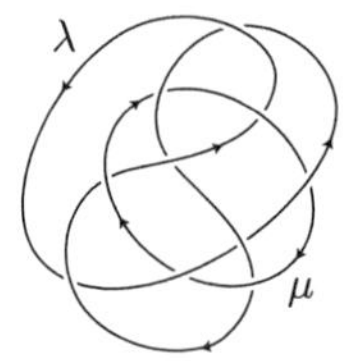

Figure 7.3

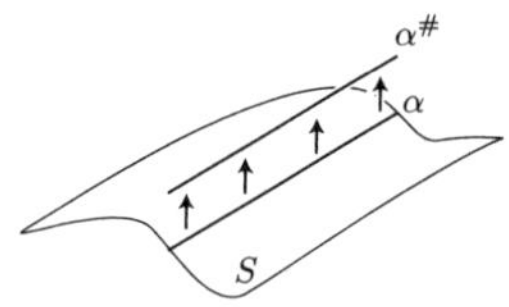

Figure 7.4

Let n_+ (or n_-) denote the number of times μ passes *under* λ from the right to left, (or the left to right). Then $\text{lk}(\lambda, \mu) = n_+ - n_-$. For the link L in Fig. 7.3, $\text{lk}(\lambda, \mu) = 1$. It is known that $\text{lk}(\lambda, \mu) = \text{lk}(\mu, \lambda)$.

Now we return to a Seifert surface.

(3) **Seifert Matrix**

Let S be a Seifert surface of a knot (or link) K and g the genus of S. Then the first homology group $H_1(S;\mathbb{Z})$ of S is a free abelian group of rank $2g+n-1$, where n is the number of components of K. For convenience, denote $m = 2g+n-1$. Now we can choose m oriented simple closed curves $\alpha_1, \alpha_2, \ldots, \alpha_m$ on the face of S in such a way that (a) For $i \neq j, \alpha_i$ and α_j intersect only at a finite number of points. (b) Homology classes $[\alpha_1], \ldots, [\alpha_m]$ form a basis for $H_1(S;\mathbb{Z})$. Using these curves, $\alpha_1, \ldots, \alpha_m$, we define an $m \times m$ integer matrix $N = (c_{i,j}), 1 \leq i, j \leq m$, where $c_{i,j} = \text{lk}(\alpha_i^{\#}, \alpha_j)$, Here $\alpha_i^{\#}$ is a simple closed curve in $\mathbb{R}^3$ that is obtained from α_i by lifting a bit in the positive normal direction, see Fig. 7.4. The orientation of $\alpha_i^{\#}$ is induced from that of α_i. Even if α_i and α_j may intersect, $\alpha_i^{\#}$ and α_j never meet, since $\alpha_i^{\#}$ is no longer on S. The matrix N is called the *Seifert matrix* of K. The matrix N is an integer matrix, but it may be singular, and in general it is not symmetric. Further, N itself is not an invariant of K, since N depends on a Seifert surface.

Example 7.1 Consider a Seifert surface S depicted in Fig. 7.5 (a). Choose α_1 and α_2 as is shown in Fig. 7.5 (a). Then the Seifert matrix $N = (c_{i,j})$ is given as follows. See Fig. 7.5 (b)-(e). $c_{1,1} = \text{lk}(\alpha_1^{\#}, \alpha_1) = -1$, $c_{1,2} = \text{lk}(\alpha_1^{\#}, \alpha_2) = 0$, $c_{2,1} = \text{lk}(\alpha_2^{\#}, \alpha_1) = 1$, $c_{2,2} = \text{lk}(\alpha_2^{\#}, \alpha_2) = 0$, and hence, $N = \begin{pmatrix} -1 & 0 \\ 1 & 0 \end{pmatrix}$

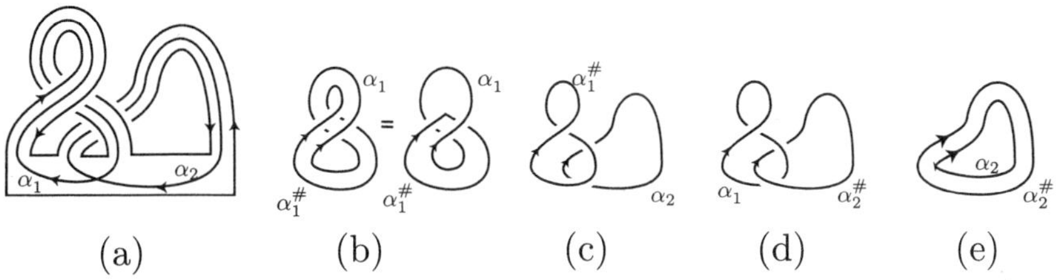

Figure 7.5

(4) **Signature**

Let N be an $m \times m$ Seifert matrix of an (oriented) knot (or link) K. Then, $N + N^T$ is a symmetric integer matrix. If K is a knot, it is not singular and $|\det(N + N^T)| = |\det \hat{A}|$, where$\hat{A}$ was defined in Section 6. Furthermore, it is shown [T2, Mu3] that the signature of $N + N^T$ is an invariant of a (oriented) knot (or link) K that is called the *signature* of K, and denoted by $\sigma(K)$. Here, the signature of $N + N^T$ is defined as follows: First, diagonalize $N + N^T$ over $\mathbb{Q}$ by means of some unimodular matrix B. Let $b_1, b_2, \ldots, b_m$ be diagonal elements. Then signature of $N + N^T$ is defined as the number of positive b_i's minus the number of negative b_j's.

Theorem 7.1 *Let K be an oriented knot and K^* the mirror image of K. Then $\sigma(K^*) = -\sigma(K)$.*

Therefore, if K is amphicheiral, then $\sigma(K) = 0$. The converse is not true. Let $\hat{K}$ be the knot obtained from K by reversing the orientation. Then, $\sigma(\hat{K}) = \sigma(K)$.

(5) **Alexander polynomial**

The Alexander polynomial $\Delta_K(t)$ of a knot K is redefined as, up to $\pm t^k$, $\Delta_K(t) = \det(N - tN^T)$. The original definition by Alexander was combinatorial [A2]. The Alexander polynomial of a knot is characterized in the following theorem due to Seifert.

Theorem 7.2 [Se] *If an integer polynomial $f(t)$ satisfies the following three conditions: (a) the degree of $f(t)$ is even, $2r$ say, (b) $f(t)$ is symmetric, i.e. $f(t) = t^r f(t^{-1})$, (c) $f(1) = 1$, then $f(t)$ is the Alexander polynomial of some knot in $\mathbb{R}^3$. The converse is also true.*

For a link L, the polynomial $\Delta_L(t) = \det(N - tN^T)$ is also an invariant of L, up to $\pm t^k$, where N is a Seifert matrix of L. This is called the *reduced* Alexander polynomial of L. From these definitions, we see

Theorem 7.3 *The degree of the (reduced) Alexander polynomial of a knot (or link) K is at most $m = 2g + n - 1$.*

In the next section, we will define the Alexander polynomial of a link, and also will discuss other formulations of the Alexander polynomial.

8. Alexander polynomials

Since Fox's free derivatives [F1] will be the most effective tool to define the Alexander polynomial for knots and links simultaneously, first we explain this operation briefly.

(1) **Free derivatives**

Let F be a free group generated freely by $x_1, x_2, \ldots, x_n$, and let $\mathbb{Z}F$ denote the integer group ring. With a word $w = x_{j_1}^{\epsilon_1} x_{j_2}^{\epsilon_2} \ldots x_{j_k}^{\epsilon_k}$ in F, we associate the *free derivatives* $\partial w/\partial x_1, \partial w/\partial x_2, \ldots, \partial w/\partial x_n$ defined as follows: $\partial w/\partial x_j = \epsilon_1 \delta_{j,j_1} x_{j_1}^{(\epsilon_1-1)/2} + \epsilon_2 \delta_{j,j_2} x_{j_1}^{\epsilon_1} x_{j_2}^{(\epsilon_2-1)/2} + \ldots + \epsilon_k \delta_{j,j_k} x_{j_1}^{\epsilon_1} \ldots x_{j(k-1)}^{\epsilon_{j(k-1)}} x_{j_k}^{(\epsilon_{j_k}-1)/2}$.

For example, if $w = x_1 x_2^{-1} x_1 x_1 x_2 x_1^{-1} x_2, \partial w/\partial x_1 = 1 + x_1 x_2^{-1} + x_1 x_2^{-1} x_1 - x_1 x_2^{-1} x_1 x_1 x_2 x_1^{-1}, \partial w/\partial x_2 = -x_1 x_2^{-1} + x_1 x_2^{-1} x_1 x_1 + x_1 x_2^{-1} x_1 x_1 x_2 x_1^{-1}$, where $1 = 1e$, e being the identity of F. Therefore, the free derivatives are mappings from F to $\mathbb{Z}F$, but they are obviously extended to mappings from $\mathbb{Z}F$ to $\mathbb{Z}F$. The following formulas are easily verified.

(a) $\partial 1/\partial x_j = 0, \partial x_j/\partial x_k = \delta_{j,k}$, and $\partial x_j^{-1}/\partial x_j = -x_j^{-1}$, for any $j = 1, 2, \ldots, n$,

(b) $\partial(uv)/\partial x_j = \partial u/\partial x_j + u\partial v/\partial x_j$, for any words u and v in F and $j = 1, 2, \ldots, n$,

(c) (Fundamental formula) $w - 1 = \sum_{j=1}^{n} \frac{\partial w}{\partial x_j}(x_j - 1)$, for any word w in F.

Note that since F is not commutative, $\frac{\partial w}{\partial x_j}(x_j - 1) \neq (x_j - 1)\frac{\partial w}{\partial x_j}$.

(2) **Alexander polynomials**

Let K be a link of r components, $r \geq 1$. Consider a Wirtinger presentation of the group $G(K)$ of K. $G(K) = \langle x_{i,j} | R_{i,j} = 1 \rangle, 1 \leq i \leq r, 1 \leq j \leq m_i$. See Section 2 (5). Compute $\partial R_{i,j}/\partial x_{k,\ell}$ for all i, j and k, ℓ, and obtain a square matrix $M = [\partial R_{i,j}/\partial x_{k,\ell}]$ of order $\sum_{i=1}^{r} m_i$, called the *Jacobian*. Now let F be the free group of rank $\sum_{i=1}^{r} m_i$ freely generated by $\{x_{i,j}\}, 1 \leq i \leq r, 1 \leq j \leq m_i$, and $\phi : F \to G$ be a natural homomorphism so that Kerϕ is the normal closure of $\{R_{i,j}\}$. Further, let $\psi : G \to G/G' = H$ be the abelianization of G, and hence, H is a free abelian group of rank r generated by $\{x_{i,1}\}, 1 \leq i \leq r$. For convenience, we use t_i for $x_{i,1}$, but if $r = 1$, we use t for x_1. Now evaluate M at $\phi\psi$, and hence we obtain a square matrix $M^{\phi\psi}$ of order $\sum m_i$, where an entry of $M^{\phi\psi}$ is an integer Laurent polynomial in $t_1, t_2, \ldots, t_r$. Note that $M^{\phi\psi}$ is singular.

Definition 8.1 The g.c.d. of all minors of $M^{\phi\psi}$ of order $\sum m_i - 1$ is called the *Alexander polynomial* of K, denoted by $\Delta(t_1, t_2, \ldots, t_r)$.

This is an invariant of K up to a unit of $\mathbb{Z}H$, i.e. $\pm t_1^{k_1} \ldots t_r^{k_r}$. If $r = 1$, this polynomial coincides with $\det(N - tN^T)$, where N is a Seifert matrix of K studied in section 7.

The Alexander polynomial may possibly be 0 for a link, but for a knot, it is never 0, since $|\Delta(1)| = 1$. Further, if $r > 1$, set $\widehat{\Delta}_K(t) = \Delta_K(t, t, \ldots, t)$ that is obtained from the Alexander polynomial of K by substituting t for every $t_i, 1 \leq i \leq r$. It is known that$(1 - t)\widehat{\Delta}_K(t) = \det(N - tN^T)$. Therefore $(1 - t)\widehat{\Delta}_K(t)$ is the reduced Alexander polynomial of a link K.

Remark 8.1 If $G(K)$ has a presentation such that the number of relations is less than $\sum m_i - 1$, then $\Delta_K(t_1, t_2, \ldots, t_r)$ is defined to be 0.

Example 2.2 (continued) From the presentation of $G(K)$, we obtain
$\partial R_1/\partial x_1 = 1, \partial R_1/\partial x_2 = -x_1 x_3 x_2^{-1}, \partial R_1/\partial x_3 = x_1 - x_1 x_3 x_2^{-1} x_3^{-1}$,
$\partial R_2/\partial x_1 = x_2 - x_2 x_1 x_3^{-1} x_1^{-1}, \partial R_2/\partial x_2 = 1, \partial R_2/\partial x_3 = -x_2 x_1 x_3^{-1}$.

Therefore, we have, $M^{\phi\psi} = \begin{bmatrix} 1 & -t & t-1 \\ t-1 & 1 & -t \\ * & * & * \end{bmatrix}$, and $\Delta_K(t) = 1 - t + t^2$.

(We do not need to evaluate $\partial R_3/\partial x_i$.)

Example 2.3 (continued) By Remark 8.1, $\Delta_L(t_1, \ldots, t_r) = 0$.

Example 2.4 (continued) Since $R_{1,1} = x_{1,1}x_{2,1}x_{1,1}^{-1}x_{2,1}^{-1}$, we have $\partial R_{1,1}/\partial x_{1,1} = 1 - x_{1,1}x_{2,1}x_{1,1}^{-1}$ and $\partial R_{1,1}/\partial x_{2,1} = x_{1,1} - x_{1,1}x_{2,1}x_{1,1}^{-1}x_{2,1}^{-1}$, and hence, $M^{\phi\psi} = \begin{bmatrix} 1-t_2 & t_1 - 1 \\ * & * \end{bmatrix}$. Therefore, $\Delta_L(t_1, t_2) = 1$.

In the rest of this section, we discuss the Alexander polynomials of torus knots and 2-bridge knots.

(3) **Torus knots**

With a Seifert matrix N of a torus knot $T(p,q)$, the calculation of $\det(N - tN^T)$ is cumbersome, but the final form is neat.

Theorem 8.1 *(1) If $pq = 0$, then $\Delta_{T(p,q)}(t) = 1$.*
(2) If $pq \neq 0$, then $\Delta_{T(p,q)}(t) = (t^{pq} - 1)(t-1)/(t^p - 1)(t^q - 1)$.

Using Theorem 8.1, we can prove Theorem 4.4 (ii).

(4) **2-bridge knots**

The calculation of the Alexander polynomial of a 2-bridge knot $B(\alpha, \beta)$, using a Seifert matrix, is also quite messy. However, a different approach makes it possible to write down the Alexander polynomial of $B(\alpha, \beta)$ in one form. See Theorem 9.5 in Section 9. Finally, we state a few properties of the Alexander polynomial of $B(\alpha, \beta)$. Since $\Delta_{B(\alpha,\beta)}(t) = \Delta_{B(\alpha,\beta)}(t) = \Delta_{B(\alpha,\alpha-\beta)}(t)$ and α is odd, we may assume that β is even. Consider the even continued fraction of a rational β/α:

$$\frac{\beta}{\alpha} = \cfrac{1}{2a_1 - \cfrac{1}{2a_2 - \cfrac{1}{\ddots - \cfrac{1}{2a_{m-1} - \cfrac{1}{2a_m}}}}}$$

$$= [2a_1, 2a_2, \cdots, 2a_m],$$

where $a_k \neq 0, 1 \leq k \leq m$. Then we have:

Theorem 8.2 [Kan] *(1) the degree of $\Delta_{B(\alpha,\beta)}(t)$ is exactly m,*
(2) $|\Delta_{B(\alpha,\beta)}(0)| = |a_1 a_2 \ldots a_m|$. Therefore, $\Delta_{B(\alpha,\beta)}(t)$ is monic if and only if $|a_1| = |a_2| = \cdots = |a_m| = 1$, namely, $\beta/\alpha = [\pm 2, \pm 2, \ldots, \pm 2]$.

9. Signature

In this section, we evaluate the signatures for torus knots and 2-bridge knots.

(1) **Torus knots**

The calculation of the signatures for torus knots is not easy. The earliest attempt was made by Hirzebruch in 1968, and he proved:

Theorem 9.1 [HizM] *Let $T(p,q)$ be a torus knot of type (p,q), where $p > 0$ and $q > 0$. Suppose that p and q both are odd. Then*

$$\sigma(T(p,q)) = 2\{\frac{(p-1)}{2}\frac{(q-1)}{2} + N_{p,q} + N_{q,p}\},$$

where $N_{p,q} = |\{x | 1 \leq x \leq (q-1)/2, ((px/q)) > 0\}|$, and

$$((y)) = \begin{cases} y - [y] - 1/2, & \text{if } y \in \mathbb{R} - \mathbb{Z}, \\ 0, & \text{if } y \in \mathbb{Z}. \end{cases}$$

Example 9.1 $\sigma(T(5,3)) = 2\{\frac{4}{2}\ \frac{2}{2} + N_{5,3} + N_{3,5}\} = 8$, since$N_{5,3} = N_{3,5} = 1$.

Hirzebruch also proves the following reduction formula.

Theorem 9.2 *Under the same assumptions of Theorem 9.1, the following formula holds:* $\sigma(T(p+2q,q)) = \sigma(T(p,q)) - (q^2-1)$.

In 1981, we proved the following complete reduction formula for $\sigma(T(p,q))$.

Theorem 9.3 [GLM] *Write* $\sigma(p,q)$ *for* $\sigma(T(p,q))$. *Assume* $p > 0$ *and* $q > 0$.

(1) $\sigma(p,q) = \sigma(q,p)$. *Therefore, we may assume further that* $0 < q < p$.

(2) Case $2q < p$.

(i) If $q \equiv 1 \pmod 2$, *then* $\sigma(p,q) = \sigma(p-2q,q) + q^2 - 1$.

(ii) If $q \equiv 0 \pmod 2$, *then* $\sigma(p,q) = \sigma(p-2q,q) + q^2$.

(3) Case $q < p < 2q$.

(i) If $q \equiv 1 \pmod 2$, *then* $\sigma(p,q) + \sigma(2q-p,q) = q^2 - 1$.

(ii) If $q \equiv 0 \pmod 2$, *then* $\sigma(p,q) + \sigma(2q-p,q) = q^2 - 2$.

(4) For $p > 0$, $\sigma(p,2) = p - 1$ *and* $\sigma(p,1) = 0$.

Example 9.2 For $p = 4$ and $q = 3$, we have from (3)(i) $\sigma(4,3) + \sigma(2,3) = 8$ and since $\sigma(2,3) = \sigma(3,2) = 2$, we see $\sigma(4,3) = 6$.

(2) **2-bridge knots**

In contrast to torus knots, an evaluation of the signature of a 2-bridge knot is straightforward. Let $B(\alpha,\beta)$ be a 2-bridge knot, where α is odd. Since $B(\alpha,-\beta)$ is the mirror image of $B(\alpha,\beta)$, it follows that $\sigma(B(\alpha,-\beta)) = -\sigma(B(\alpha,\beta))$, and hence, we may assume that $\alpha > \beta > 0$. Further, we may assume that β is odd. In fact, Theorem 5.1 yields $\sigma(B(\alpha,\beta)) = \sigma(B(\alpha,\beta-\alpha)) = -\sigma(B(\alpha,\alpha-\beta))$. Now consider the sequence $E = \{\beta, 2\beta, 3\beta, \ldots, (\alpha-1)\beta\}$. Take representatives $\overline{k\beta}$ of $k\beta \pmod{2\alpha}$ in $-\alpha < \overline{k\beta} < \alpha$, and obtain a new sequence $\bar{E} = \{\overline{\beta}, \overline{2\beta}, \overline{3\beta}, \ldots, \overline{(\alpha-1)\beta}\}$. Let ϵ_k be the sign of $\overline{k\beta}$, i.e. $\epsilon_k = \overline{k\beta}/|\overline{k\beta}|$.

Theorem 9.4 [Sh] *If* α *and* β *are odd and* $\alpha > \beta > 0$, *then* $\sigma(B(\alpha,\beta)) = \sum_{k=1}^{\alpha-1} \epsilon_k$.

Example 9.3 For $B(5,3)$, we have $E = \{3,6,9,12\}$ and $\bar{E} = \{3,-4,-1,2\}$ and therefore, $\sigma(B(5,3)) = 0$.

Finally, the Alexander polynomial of a 2-bridge knot $B(\alpha,\beta)$ can be calculated in the following simple form:

Theorem 9.5 [HiM1] *If* α *and* β *are both odd, and* $\alpha > \beta > 0$, *then:*

$$\Delta_{B(\alpha,\beta)}(t) = \sum_{k=0}^{\alpha-1} (-1)^k t^{q_k}, \text{where } q_k = \sum_{i=0}^{k} \epsilon_i, \text{and } \epsilon_0 = 1$$

.

10. Fibred knots

(1) **Commutator subgroup and Augmentation subgroup**

Almost the same time when Alexander defined his polynomial, Reidemeister also introduced the same polynomial from a different point of view.

Let K be a knot (not a link) and $G = G(K)$ the group of K. Let G' be the commutator subgroup of G. Then the infinite cyclic group $G/G' \cong Z = \langle t|-\rangle$ acts on G' by conjugation, and G'/G'' becomes a finitely generated $\mathbb{Z}[t^{\pm}]$-module with the relation matrix Λ. Then Reidemeister showed that $\det\Lambda$ is an invariant of K, up to $\pm t^k$, and in fact it is the Alexander polynomial of a knot K. For an (oriented) r-component link $L, r > 1$, the commutator subgroup G should be replaced by the kernel of $\phi\psi\tau$, where $\tau : H \to \langle t|-\rangle$, and $\tau(t_i) = t, 1 \leq i \leq r$. The subgroup $A = \ker \phi\psi\tau$ is called the *augmentation subgroup* of the group of L. Then the same argument above shows that the reduced Alexander polynomial $(1-t)\hat{\Delta}(t)$ is given by $\det \Lambda$. It is important to note that the augmentation subgroup A depends on the orientation of each component. If one of the components reverses its orientation, we have a different augmentation subgroup and hence the different (reduced) Alexander polynomial.

(2) **Fibred knots (or links)**

This reformulation suggests that the Alexander polynomial may provide us some information on the structure of G/G''. In fact, we have

Theorem 10.1 [Ra] *Let $\Delta_K(t)$ be the Alexander polynomial of a knot K. Then $\Delta_K(t)$ is monic if and only if G'/G'' is finitely generated, and then G'/G'' is a free abelian group of rank d, where d is the degree of $\Delta_K(t)$.*

On the other hand, L. Neuwirth studied the structure of the commutator subgroup G' of the group of a knot K, and, in particular, he proved

Theorem 10.2 [N] *If the commutator subgroup G' of the group of a knot K is finitely generated, then G' is a free group of rank d, the degree of the Alexander polynomial.*

Later, the property that G' be finitely generated was characterized topologically by J.Stallings.

Theorem 10.3 [St] *Let K be a knot. Suppose that the commutator subgroup of $G(K)$ is finitely generated. Then the complement of a small tubular neighbourhood of K in S^3 is the total space of a fibre space with base space a circle and with fibre a Seifert surface of K. The converse is also true. We call such a knot a fibred knot.*

Remark 10.1 For a link in S^3, analogous theorems to Theorems 10.2 and 10.3 hold if the augmentation subgroup A replaces the commutator subgroup. As an immediate consequence of Theorem 10.3, we obtain:

Theorem 10.4 *If K is a fibred knot, then the Alexander polynomial of K is monic. The converse is not true, however.*

The property that K be a fibred knot is a topological property, and hence, it is difficult to characterize this property by *algebraic* invariants. However, for 2-bridge knots or more generally for *alternating* knots, their Alexander polynomials characterize fibred knots. See Theorem 10.5 below. An alternating knot is a knot that has at least one *alternating* diagram. A diagram is called *alternating* if over-passing and under-passing alternate while moving along a diagram. Fig. 1.1 is an alternating diagram, while Fig. 2.1 is not an alternating diagram, but the knot itself is an alternating knot, since it has an alternating diagram. A 2-bridge knot is an alternating knot. Alternating knots form a large and special class in knot theory, and the classification problem is completely solved for these knots [MeT]. But, not all knots are alternating. A torus knot $T(p, q)$ is alternating if and only if $\min\{|p|, |q|\} \leq 2$. For alternating knots, the converse of Theorem 10.4 holds. Namely, we have:

Theorem 10.5 [Mu1] *Let K be an alternating knot. Then K is a fibred knot if and only if $\Delta_K(t)$ is monic.*

Recently, Gabai introduced some geometric methods by which one can decide whether many knots (or links) are fibred or not [G].

Remark 10.2 A trivial knot is a fibred knot, and all torus knots are fibred knots. A Hopf link is a fibred link.

11. Representation - Covering space

Let J_n denote a set of n elements, $\{1, 2, \ldots, n\}$, say, and $S(J_n)$ the group of all permutations of the set J_n.

(1) **Permutation Representation**

Let $G(K)$ be the group of a knot (or link) K in S^3. Let Φ be a transitive representation of $G(K)$ into $S(J_n)$ for some n. Corresponding to Φ is the covering space M of S^3 branched along K with the covering projection p (that we call simply the *branched covering space of* K in S^3). Two permutation representations Φ and Φ' are called *equivalent* if they differ by renumbering of J_n. Two covering spaces corresponding to Φ and Φ' are homeomorphic. Now if H is a subgroup of $G(K)$ with finite index n, then we obtain a transitive representation Φ of $G(K)$ into $S(J_n)$. (Any two such representations are equivalent.) This representation will be called a representation corresponding to the subgroup H. In fact, H is obtained as the stabilizer of some element in J_n under Φ.

(2) **Cyclic covering space - Homology invariants**

By taking various representations of $G(K)$, we obtain various compact 3-manifolds M as branched covering spaces of K. Since these manifolds are invariants of K, we can deduce many knot invariants from these manifolds. For example, the first homology group $H_1(M;\mathbb{Z})$ is one of many classical invariants that were already studied in 1930's. For example, the torsion number and Betti number of $H_1(M;\mathbb{Z})$ provide us handy and neat numerical invariants. In particular, the branched covering space of K associated with the cyclic representation $\Phi : G(K) \to \{(12\cdots n)\} \subset S(J_n)$ was the subject of investigation from the early stage of knot theory. In fact, we have

Theorem 11.1 *Let M_n be the n-fold cyclic branched covering space of a knot K. Let $\beta_1(K)$ and $\tau_1(K)$ denote respectively the first Betti number and torsion number of $H_1(M;\mathbb{Z})$. Then, if $\beta_1(K) = 0$, then $\tau_1(K) = \prod_{i=1}^{n-1} \Delta_K(\xi^i)$, where ξ is a primitive nth root of unity.*

Note that if $n = 2$, then $\beta_1(K) = 0$ for a knot K.

For a knot, any abelian covering is necessarily a cyclic covering. However, for a link L, there are infinitely many non-cyclic abelian coverings. But, if $\beta_1(L) = 0$, there is a formula that evaluates the torsion number of $H_1(M;\mathbb{Z})$ in terms of the Alexander polynomial of L and those of the various sublinks of L [MaM]. These formulas give us the torsion number of $H_1(M;\mathbb{Z})$, but do not give any information on the structure of $H_1(M;\mathbb{Z})$. The structure of $H_1(M;\mathbb{Z})$ is generally complex. A few classical results are known for the case of cyclic coverings of knots [Pl]. Later, we will discuss some results for the case of links.

(3) **Dihedral representations**

Besides abelian representations, representations of the knot group on the dihedral groups D_p of order $2p, p$ odd, have been studied extensively. The following theorem will easily be proved.

Theorem 11.2 *Let* $D_p = \langle a, b | a^2 = b^2 = (ab)^p = 1 \rangle$ *be the dihedral group of order* $2p$*, where* p *is odd. Then the knot group* $G(K)$ *has an (irregular) representation* $\Phi_p : G(K) \to D_p \subset S(J_p)$ *in such a way that* Φ_p *maps every meridian element of* K *to an element of order* 2 *in* D_p *if and only if* $\Delta_K(-1) \equiv 0 \pmod p$.

It is shown that if K is a 2-bridge knot, then the branched covering space corresponding to Φ_p is simply connected [F2], but in fact it is S^3 [Bu2]. One of the reasons to study (irregular) branched covering spaces of knots (or links) is hopefully to obtain a simply connected closed 3-manifold other than S^3. But, up to the present, nobody has found such a 3-manifold. Besides the dihedral coverings, various coverings have been studied, see [Ri], [HaM2].

(4) **Covering linkage invariants**

As was seen in the above subsection, the covering spaces themselves may not be significant invariants of a knot. However, even if the covering space M is simple, like S^3, we can deduce from M some important invariants of K. Let $\Phi : G(K) \to S(J_n)$ be a transitive representation of $G(K)$. Suppose that the branched covering space M corresponding to Φ is a homology 3-sphere, i.e. $H_1(M; \mathbb{Q}) = 0$ and, further, $p^{-1}(K)$, a lift of K in M is an r-component link $\tilde{K} = \tilde{K}_1 \cup \tilde{K}_2 \cup \cdots \cup \tilde{K}_r, r \geq 2$. Then the linking numbers $\mathrm{lk}(\tilde{K}_i, \tilde{K}_j) \pmod q, 1 \leq i \neq j \leq r$, are well-defined and they are invariants of K, where q is the order of $H_1(M; \mathbb{Z})$. They are called the *covering linkage invariants* associated to Φ. These invariants are sometimes very powerful tools to distinguish between two knots. In fact, two 2-bridge knots that have the same Alexander polynomials are distinguished by the covering linkage invariants associated to (irregular) dihedral representations Φ_p [BS]. The following theorem is one of many results about the covering linkage invariants.

Theorem 11.3 [Pe] *Suppose that the group of a 2-bridge knot* K *has an (irregular) representation on* $D_p \subset S(J_p)$*, where* p *is odd. Then the lift of* K *in the irregular dihedral covering space* M *of* K *is a link* $\tilde{K}$ *of* $(p+1)/2$ *components* $\tilde{K}_1, \tilde{K}_2, \ldots, \tilde{K}_{(p+1)/2}$*, and for* $i \neq j, \mathrm{lk}(\tilde{K}_i, \tilde{K}_j) \equiv 2 \pmod 4$.

Evaluation of these covering linkage invariants for a knot or link associated to an arbitrary finite representation Φ is completely formulated [Ha] and various applications can be found in [HaM1], [HaM2], [Mu5], [Lu].

12. Modular representations of 2-bridge knot groups

In this section, we will discuss an unimodular representation of the group of a 2-bridge knot $B(\alpha, \beta)$. The information we obtained in this section will be used in the next section.

(1) **Unimodular representations**

Suppose there is an unimodular representation $\rho : G(K) \to SL(2, \mathbb{Z})$ such that $\rho(m) = \begin{pmatrix} 1 & 1 \\ 0 & 1 \end{pmatrix}$, where m is one Wirtinger generator.

Such a representation ρ will be called a *canonical* representation. Since $PSL(2, \mathbb{Z}) \cong Z/(2) * Z/(3), \rho$ can be lifted to a homomorphism $\bar{\rho} : G(K) \to H = \langle a, b | a^2 = b^3 \rangle$ [HaM2]. Now H is the group of a torus knot of type $(3, 2)$ and $\Delta_{T(3,2)}(t) = 1 - t + t^2$, and hence, we have

Theorem 12.1 *If the group* $G(K)$ *of a knot* K *has a canonical unimodular representation* $\rho : G(K) \to SL(2, \mathbb{Z})$*, then* $\Delta_K(t) \equiv 0 \pmod{1 - t + t^2}$.

The converse of Theorem 12.1 is not true. For example, the Alexander polynomial of the 2-bridge knot $B(27,17)$ is $(1-t+t^2)(2-5t+2t^2)$, but its group does not have a canonical representation on $SL(2,\mathbb{Z})$. See Theorem 12.2.

(2) **Admissible form**

In this subsection, we characterize a 2-bridge knot $K = B(\alpha,\beta)$ that has a canonical representation on $SL(2,\mathbb{Z})$. First, we consider a continued fraction of β/α. Since the groups of 2-bridge knots $B(\alpha,\beta)$ and $B(\alpha,-\beta)$ are isomorphic, we may assume that $\alpha \equiv \beta \equiv 1 \pmod 2$ and $0 < \beta < \alpha$. Now express β/α as the continued fraction of the form:

$$\frac{\beta}{\alpha} = \cfrac{1}{2a_1 - \cfrac{1}{2a_2 - \cfrac{1}{\ddots - \cfrac{1}{2a_m - \cfrac{1}{c}}}}}$$

$$= [2a_1, 2a_2, \cdots, 2a_m, c],$$

where $a_i \neq 0$, c is odd and $|c| > 1$. This expression is called an *almost even continued fraction* of β/α, and this expression is unique. We say that β/α or $[2a_1, 2a_2, \ldots, 2a_m, c]$ is *admissible* if $G(B(\alpha,\beta))$ has a canonical representation on $SL(2,\mathbb{Z})$. Note that if $[2a_1, 2a_2, \ldots, 2a_m, c]$ is admissible, then $[-2a_1, -2a_2, \ldots, -2a_m, -c]$ is also admissible. The admissibility of an almost even continued fraction is completely determined by the following theorem.

Theorem 12.2 [HiM2] *Let $X = [2a_1, 2a_2, \ldots, 2a_m, c]$, c odd and $|c| > 1$, be the almost even continued fraction of β/α, where $\alpha > \beta > 0$ and $\alpha \equiv \beta \equiv 1 \pmod 2$. We call $m+1$ the length of X. (1) If X is admissible, then the length of X must be odd, i.e. m is even.*

(2) Suppose the length of X is 1. Then $[c]$ is admissible if and only if $c \equiv 3 \pmod 6$,

(3) Suppose the length of X is 3, i.e. $X = [2a_1, 2a_2, c]$. Then,

(i) $[6k, 2a_2, c], k \neq 0$, is admissible if and only if $[c]$ is admissible, i.e. $c \equiv 3 \pmod 6$.

(ii) $[6k+2, 2a_2, c], k \neq 0$, is admissible if and only if $2a_2 = -2$, and $[c-2]$ is admissible, i.e. $c \equiv 5 \pmod 6$.

(iii) $[6k+4, 2a_2, c], k \neq 0$, is admissible if and only if $2a_2 = 2$, and $[c-4]$ is admissible, i.e. $c \equiv 1 \pmod 6$.

(4) Suppose the length of $X \geq 5$.

(i) $[6k, 2a_2, \ldots, 2a_m, c], k \neq 0$, is admissible if and only if $[2a_3, 2a_4, \ldots, 2a_m, c]$ is admissible.

(ii) $[6k+2, 2a_2, \ldots, 2a_m, c], k \neq 0$, is admissible if and only if $2a_2 = -2$, and $[2a_3 - 2, 2a_4, \ldots, 2a_m, c]$ is admissible.

(iii) $[6k+4, 2a_2, \ldots, 2a_m, c], k \neq 0$, is admissible if and only if $2a_2 = 2$, and $[2a_3 - 4, 2a_4, \ldots, 2a_m, c]$ is admissible.

A proof needs a careful study of the modular diagram and a theorem due to Sakuma [Sa]. See also [GR] for a similar theorem to Theorem 12.2.

Example 12.1 Let K be a 2-bridge knot $B(27,17)$. Since $17/27 = [2,2,-2,3]$, it is not admissible, but $1-t+t^2$ divides $\Delta_K(t)$.

Example 12.2 Let $K = B(27,11)$. Since $11/27 = [2,-2,5]$, it is admissible.

13. Twisted Alexander polynomial

The twisted Alexander polynomial of a knot (or link) was defined by Wada [Wad] and Lin [Li] in early 1990's. This is a generalization of the Alexander polynomial, but it is a subtler invariant than the Alexander polynomial. In fact, it may distinguish between a knot and its mirror image. This is in general not possible by the Alexander polynomial.

(1) **Definition**

Originally, Wada defined this polynomial for a finitely presented group G with G/G' free. However, for simplicity, we consider only the group of a knot K. Let $\rho : G = G(K) \to GL(n, R)$ be a linear representation of the group of a knot K, where R is an unique factorization domain. Let $\alpha : G \to G/G' \cong \mathbb{Z} = \langle t|-\rangle$ be an abelianization. Then ρ and α can be extended to $\tilde{\rho} : \mathbb{Z}G \to M_n(R)$, and $\tilde{\alpha} : \mathbb{Z}G \to \mathbb{Z}[t^{\pm1}]$, where M_n denotes the ring of matrices of order n. Now, let $G = \langle x_1, x_2, \ldots, x_m | r_1 = 1, r_2 = 1, \ldots, r_{m-1} = 1\rangle$ be a Wirtinger presentation of $G(K)$. Let F be a free group freely generated by $\{x_1, x_2, \ldots, x_m\}$. Let $A = [\partial r_i/\partial x_j], 1 \leq i \leq m-1, 1 \leq j \leq m$, be an $(m-1) \times m$ matrix (over $\mathbb{Z}R$), where $\partial/\partial x_j$ denotes Fox's free derivatives. (See 8(1).) Now eliminate one column, say the last column to obtain $\hat{A}$. Let $\phi : F \to G$ be a natural homomorphism defined by $\phi(x_i) = x_i$ and $\phi : G \to M_n(R[t^{\pm1}])$ be a homomorphism defined by $\phi(x_i) = \rho(x_i)t$. Extend ϕ to a ring homomorphism from $\mathbb{Z}G$ to $M_n(Z[t^{\pm1}])$. Then $(\partial r_i/\partial x_j)^\phi \in M_n(R[t^{\pm1}])$, and $\hat{A}^\phi = ||(\partial r_i/\partial x_j)^\phi||$ is an $n(m-1) \times n(m-1)$ matrix over $R[t^{\pm1}]$. We define as in [Wad]:

$\Delta_{\rho,K}(t) = det\hat{A}^\phi / det(x_m^\phi - I)$, where I denotes the identity matrix of order n.

This is an invariant of a knot K up to $ut^{\pm k}$, where u is a unit of R.

Remark 13.1 [Wad] (1) If $\rho_o : G \to GL(n, \mathbb{Z})$ be a trivial representation, then $\Delta_{\rho_o,K}(t) = (\Delta_K(t)/(1-t))^n$. (2) $\Delta_{\rho,K}(t)$ is a rational function, but if there exists an element $w \in G'$ such that 1 is not an eigenvalue of $\rho(w)$, then $\Delta_{\rho,K}(t)$ is a Laurent polynomial with coefficients in the field of quotients R_0 of R. Therefore, $\Delta_{\rho,K}(t) \in R_0[t^{\pm1}]$.

Example 13.1 Let K be a knot in Fig. 13.1. Then a Wirtinger presentation of $G(K)$ is given by $\langle x, y | xyxy^{-1}x^{-1}y^{-1} = 1\rangle$. Then $\rho(x) = \begin{pmatrix} 1 & 1 \\ 0 & 1 \end{pmatrix}$ and $\rho(y) = \begin{pmatrix} 1 & 0 \\ -1 & 1 \end{pmatrix}$ define a linear representation. Then $\Delta_{\rho,K}(t) = (1-t)^2(1+t^2)/(1-t)^2 = 1 + t^2$.

(2) **Fibered knots**

As one of the general results, we mention the following theorem.

Theorem 13.1 [GKM] *If K is a fibred knot, then for any unimodular representation $\rho : G(K) \to SL(2n, F)$, $\Delta_{\rho,K}(t)$ is a rational function of monic polynomial, where F is a field.*

Remark 13.2 [Ki] If $F = \mathbb{C}$ or a finite field F_p, then $\Delta_{\rho,K}(t)$ is symmetric.

(3) **Twisted Alexander polynomial**

Now we consider the twisted Alexander polynomial $\Delta_{\rho,K}(t)$ of a 2-bridge knot $K = B(\alpha, \beta)$ associated to $\rho : G(K) \to SL(2, \mathbb{Z})$, where ρ is a canonical representation of $G(K)$. See Section 11. For convenience, we write a Wirtinger presentation of $G(K)$ as follows: $G(K) = \langle x, y | R = 1\rangle$, And assume that $\rho(x) = \begin{pmatrix} 1 & 1 \\ 0 & 1 \end{pmatrix}$ and $\rho(y) = \begin{pmatrix} 1 & 0 \\ -1 & 1 \end{pmatrix}$ Then xy^{-1} belongs to G', but none of the eigenvalues of $\rho(xy^{-1})$ is 1, and hence by Remark 13.1 (2), $\Delta_{\rho,K}(t)$ is a Laurent polynomial over $\mathbb{Q}$. Further, for any prime p, ρ induces a representation $\rho_p : G(K) \to SL(2, \mathbb{Z}/p)$, and $\Delta_{\rho_p,K}(t) \equiv \Delta_{\rho,K}(t) \pmod p$, and therefore, by Remark 13.2, $\Delta_{\rho,K}(t)$ is an integer symmetric polynomial. Noting that $\Delta_{\rho,T(3,2)}(t) = 1 + t^2$, (see Example 13.1,) we see that $\Delta_{\rho,K}(t)$ is divisible by $1 + t^2$. Now, set $\lambda(t) = \Delta_{\rho,K}(t)/(1 + t^2)$. Then, by Theorem 13.1, we see that if $K = B(\alpha, \beta)$ is fibred, then $\lambda(t)$ is monic. However, for a 2-bridge knot, the converse also holds. Namely, we have

Theorem 13.2 [HiM2] *Suppose that the group of a 2-bridge knot $B(\alpha, \beta)$ has a canonical representation ρ on $SL(2, \mathbb{Z})$. Then, $B(\alpha, \beta)$ is fibred if and only if $\Delta_{\rho,K}(t)$ is monic.*

A table of the twisted Alexander polynomials of about two hundred 2-bridge knots with canonical representation ρ on $SL(2, \mathbb{Z})$ by Stoimenow reveals the following facts. Since these facts are not proven yet, we leave them as conjectures.

Conjecture 13.1 *Let K be a 2-bridge knot with a canonical representation ρ on $SL(2, \mathbb{Z})$. Then*

(1) $\lambda(1) = 1$. Therefore, $\lambda(t)$ is the Alexander polynomial of some knot $\tilde{K}$ in $\mathbb{R}^3$. What is $\tilde{K}$?

(2) $\lambda(-1) = q^2$ for some integer q.

14. Milnor's Invariants (I)

Let G be a finitely generated group, and let $\{G^{(q)}, q \geq 1\}$ be the lower central series of G, i.e. $G^{(1)} = G$ and $G^{(q)} = [G, G^{(q-1)}]$ for $q \geq 2$. In 1957, J. Milnor defined a sequence of numerical invariants for an n-component link L, $n \geq 2$, using the specific presentation of the nilpotent group $G(L)/G(L)^{(q)}$. These invariants are called *Milnor's $\bar{\mu}$-invariants.* These invariants are trivial for a knot, since $G^{(2)} = G^{(3)} = \ldots$, for the group of a knot. In this section, we give a definition of Milnor's $\bar{\mu}$-invariants and state a few properties of these invariants. In the next section, we will give a topological interpretation of these invariants.

(1) **Chen-Milnor presentation**

Consider a Wirtinger presentation of the group of an n-component link $L = K_1 \cup K_2 \cup \cdots \cup K_n$. See Section 2.

The set of generators is $\{x_{1,1}, \ldots, x_{1,m_1}, x_{2,1}, \ldots, x_{2,m_2}, \ldots, x_{n,1}, \ldots, x_{n,m_n}\}$, and a set of relations is $\{R_{1,1} = 1, \ldots, R_{1,m_1} = 1, R_{2,1} = 1, \ldots, R_{2,m_2} = 1, \ldots, R_{n,1} = 1, \ldots, R_{n,m_n} = 1\}$. Although one of relations is irrelevant, we keep it in the set.

Each relation is of the form, for $1 \leq i \leq n, 1 \leq j \leq m_i$, $R_{i,j} = x_{i,j} a_{i,j} x_{i,j+1}^{-1} a_{i,j}^{-1}$, where $a_{i,j} = x_{k,\ell}^{\pm}$ for some $k, \ell, 1 \leq k \leq n, 1 \leq \ell \leq m_k$. Now replace $R_{i,2}$, by $R'_{i,2} = R_{i,1}(a_{i,1} R_{i,2} a_{i,1}^{-1}) = x_{i,1}(a_{i,1}, a_{j,2}) x_{i,3}^{-1}(a_{i,2}^{-1} a_{i,1}^{-1})$, and then replace $R_{i,3}$ by $R'_{i,3} = R'_{i,2}(a_{i,1} a_{i,2} R_{i,3} a_{i,2}^{-1} a_{i,1}^{-1}) = x_{i,1}(a_{i,1} a_{i,2} a_{i,3})\ x_{i,4}^{-1}(a_{i,3}^{-1} a_{i,2}^{-1} a_{i,1}^{-1})$. Inductively, the set $\{R_{i,1}, R_{i,2}, \ldots, R_{i,m_i}\}$ is replaced by $\{R_{i,1}, R'_{i,2}, \ldots, R'_{i,m_i}\}$, where $R'_{i,j} = x_{i,j} W_{i,j} x_{i,j+1}^{-1} W_{i,j}^{-1}$, for $1 \leq i \leq n, 1 \leq j \leq m_i$, and $W_{i,j} = a_{i,1} a_{i,2} \cdots a_{i,j}$. In particular, $R'_{i,m_i} = x_{i,1} W_{i,m_i} x_{i,1}^{-1} W_{i,m_i}^{-1}$. Write $a_{i,j} = x_{k_{i,j}\ell_{i,j}}^{\varepsilon_{i,j}}$ and let $\lambda_i = \sum \varepsilon_{i,j}$, where the summation runs over all $a_{i,j}$ in W_{i,m_i} such that $k_{i,j} = i$. Then for

$i = 1, 2, \ldots, n, \eta_i = W_{i,m_i} x_{i,1}^{-\lambda_i}$ represents a longitude of the ith component of K_i of L. Let F and F_n, respectively, be free groups that are freely generated, respectively, by $\{x_{i,j}, 1 \leq i \leq n, 1 \leq j \leq m_i\}$ and $\{x_1, x_2, \ldots, x_n\}$. Now we define the homomorphisms $\theta_q : F \to F_n$ inductively.

(14.1) For $1 \leq i \leq n, 1 \leq j \leq m_i$ and $q \geq 1$, $x_{i,j}^{\theta_1} = x_i$, and $x_{i,1}^{\theta_{q+1}} = x_i, x_{i,j+1}^{\theta_{q+1}} = (W_{i,j}^{-1} x_{i,1} W_{i,j})^{\theta_q}$.

Then the following theorem is proved by Milnor [Mi] [C]:

Theorem 14.1 *For $q \geq 1, G(L)/G(L)^{(q+1)}$ has a presentation:* $\langle x_1, x_2, \ldots, x_n | \eta_i^{\theta_q} x_i \eta_i^{-\theta_q} x_i^{-1} = 1, F_n^{(q+1)}, 1 \leq i \leq n \rangle$, *where $F_n^{(q+1)}$ denotes the $(q+1)$th member of the lower central series of F_n.*

(2) **Milnor's $\bar{\mu}$ - invariants**

Obviously, θ_q can be extended to the ring homomorphisms $\mathbb{Z}F$ to $\mathbb{Z}F_n$. The trivializer $o : \mathbb{Z}F_n \to \mathbb{Z}$ is a homomorphism defined by $\sum (a_i x_{i_1} \ldots x_{i_p})^o = \sum a_i$. Now following Milnor [Mi], we define an integer $\mu(i_1 i_2 \ldots i_p k)$ for a sequence of integers $\{i_1, i_2, \ldots, i_p, k\}, p \geq 1, 1 \leq i_1, i_2, \ldots, i_p, k \leq n$, as follows:

(14.2) $\mu(i_1 i_2 \ldots i_p k) = (\partial^p \eta_k^{\theta_p} / \partial x_{i_1} \partial x_{i_2} \cdots \partial x_{i_p})^o$.

Let $\Delta(i_1 i_2 \ldots i_r) = \gcd \mu(j_1 j_2 \ldots j_s)$, where $j_1 j_2 \ldots j_s, (2 \leq s \leq r)$ is to arrange over all sequences obtained by canceling at least one of the indices $i_1, i_2, \ldots, i_r$ and permuting the remaining indices cyclically. Then Milnor proved:

Theorem 14.2 [Mi] $\bar{\mu}(i_1 i_2 \cdots i_p k) \equiv \mu(i_1 i_2 \ldots i_p k) \pmod{\Delta(i_1 i_2 \cdots i_p k)}$ *is an invariant of L.*

This is called *Milnor's $\bar{\mu}$-invariant.*

Remark 14.1 (1) $\bar{\mu}(j_1 j_2 \cdots j_r)$ is in fact an isotopy invariant of L, where two n-component link L and L' are said to be *isotopic* if there is a continuous family of homoomorphismo $\{f_t, 0 \leq t \leq 1\}, f_t : L \to \mathbb{R}^3$ (or S^3) such that f_0 is an identity map and $f_1(L) = L'$.

(2) We define $\mu(i) = 0$ for any i.

(3) For $i \neq j, \mu(ij) = \mathrm{lk}(K_i, K_j)$, and $\mu(ii) = 0$. Due to these facts, Milnor's $\bar{\mu}$-invariants are sometimes called a *higher linking invariant.*

(4) $\bar{\mu}(\underbrace{ii \cdots i}_{k \text{ times}} j) \equiv \binom{\mu(ij)}{k} \mod \Delta(\underbrace{ii \ldots i}_{k \text{ times}} j)$.

(5) If a sequence $\{i_1, i_2, \ldots, i_r\}$ does not contain $j_1, j_2, \ldots, j_k$, then $\bar{\mu}(i_1 i_2 \cdots i_r)$ is equal to Milnor's invariant for the sublink L' that excludes k components, $K_{j_1}, K_{j_2}, \ldots, K_{j_k}$ from L.

(6) For a trivial n-component link, all Milnor's invariants vanish.

These invariants satisfy many relations. Some important relations are listed in the following theorem.

Theorem 14.3 [Mi] *(1) cyclic symmetry: $\bar{\mu}(i_1 i_2 \cdots i_r) = \bar{\mu}(i_2 i_3 \cdots i_r i_1)$*
(2) shuffle equality: If $i_1, \ldots, i_r$ and $j_1, \ldots, j_s$ are given sequences with $r, s \geq 1$, then $\sum \bar{\mu}(h_1 \cdots h_{r+s} k) \equiv 0 \mod \gcd \Delta(h_1 \cdots h_{r+s} k)$, where the summation runs over all proper shuffles of $i_1, \ldots, i_r$ and $j_1, \ldots, j_s$. (A proper shuffle $h_1, \ldots, h_{r+s}$ of $i_1, \ldots, i_r$ and $j_1, \ldots, j_s$ is one of the $\binom{r+s}{s}$ sequences obtained by intermeshing $i_1, \ldots, i_r$ with $j_1, \ldots, j_s$.)

Like 2-bridge links or closed pure n-braids, if the group of an n-component link L is generated by exactly n Wirtinger generators, then $G(L)$ has a presentation: $G(L) = \langle x_1, x_2, \ldots, x_n | \eta_i x_i \eta_i^{-1} x_i^{-1} = 1, 1 \leq i \leq n \rangle$. For this case, θ_q becomes an identity, and hence, $G(L)/G(L)^{(q+1)}$ has a presentation:

$\langle x_1, x_2, \ldots, x_n | \eta_i x_i \eta_i^{-1} x_i^{-1} = 1, F_n^{(q+1)}, 1 \le i \le n\rangle$. Therefore, the computation of $\bar{\mu}$-invariants is much simpler.

(3) **Examples**

Example 14.1 Let L be a link depicted in Fig. 14.1. For simplicity, we use x_i, y_j for $x_{1,i}, x_{2,j}$, and R_i, S_j for $R_{1,i}, R_{2,j}$. (In Fig. 14.1, we do not distinguish between an over-passing arc and a generator of $G(K)$.) Then $G(L) = \langle x_i, y_j, | R_i = 1, S_j = 1, i = 1,2,3, j = 1,2,\ldots,5\rangle$, where $R_1 = x_1 y_1 x_2^{-1} y_1^{-1}$, $R_2 = x_2 y_3 x_3^{-1} y_3^{-1}$, $R_3 = x_3 y_5 x_1^{-1} y_5^{-1}$, and $S_1 = y_1 y_5 y_2^{-1} y_5^{-1}$, $S_2 = y_2 x_1 y_3^{-1} x_1^{-1}$, $S_3 = y_3 y_5 y_4^{-1} y_5^{-1}$, $S_4 = y_4 x_1 y_5^{-1} x_1^{-1}$, $S_5 = y_5 x_1 y_1^{-1} x_1^{-1}$. The set of new relations $R_2', R_3', S_2', \ldots, S_5'$ are given below:

$R_2' = x_1(y_1 y_3) x_3^{-1}(y_3^{-1} y_1^{-1})$, $R_3' = x_1(y_1 y_3 y_5) x_1^{-1}(y_5^{-1} y_3^{-1} y_1^{-1})$, and hence, $\eta_1 = y_1 y_3 y_5$, and $S_2' = y_1(y_5 x_1) y_3^{-1}(x_1^{-1} y_5^{-1})$, $S_3' = y_1(y_5 x_1 y_5) y_4^{-1}(y_5^{-1} x_1^{-1} y_5^{-1})$, $S_4' = y_1(y_5 x_1 y_5 x_1)\ y_5^{-1}\ (x_1^{-1} y_5^{-1} x_1^{-1} y_5^{-1})$, $S_5' = y_1(y_5 x_1 y_5 x_1^2) y_1^{-1}(x_1^{-2} y_5^{-1} x_1^{-1} y_5^{-1})$, and hence, $\eta_2 = y_5 x_1 y_5 x_1^2 y_1^{-2}$. Now we evaluate some of $\bar{\mu}$-invariants. $\bar{\mu}(1) = \bar{\mu}(2) = 0$. To evaluate higher $\bar{\mu}$-invariants, first we compute θ_p. They are: $\theta_1(x_i) = x_1$ and $\theta_1(y_i) = y_1$, $\theta_2(x_1) = x_1$, $\theta_2(x_2) = y_1^{-1} x_1 y_1$, $\theta_2(x_3) = y_1^{-2} x_1 y_1^2$, and $\theta_2(y_1) = y_1, \theta_2(y_2) = y_1$, $\theta_2(y_3) = x_1^{-1} y_1 x_1$, $\theta_2(y_4) = (y_1^{-1} x_1^{-1}) y_1 (x_1 y_1)$, $\theta_2(y_5) = (x_1^{-1} y_1^{-1} x_1^{-1}) y_1 (x_1 y_1 x_1)$.

Using these formulas, we have $\eta_1^{\theta_1} = y_1^3$, $\eta_1^{\theta_2} = y_1 x_1^{-2} y_1 x_1 y_1 x_1$, and $\eta_2^{\theta_1} = y_1 x_1 y_1 x_1^2 y_1^{-1}$, $\eta_2^{\theta_2} = x_1^{-1} y_1^{-1} x_1^{-1} y_1 x_1 y_1 x_1 y_1^{-1} x_1^{-1} y_1 x_1 y_1 x_1^3 y_1^{-2}$. Now we are in position to evaluate $\bar{\mu}$-invariants. $\bar{\mu}(12) = (\partial \eta_2^{\theta_1}/\partial x_1)^o = 3$, $\bar{\mu}(21) = (\partial \eta_1^{\theta_1}/\partial y_1)^o = 3$, $\bar{\mu}(121) = (\partial^2 \eta_1^{\theta_2}/\partial x_1 \partial y_1)^o = 0$, since $\Delta(121) = 3$, $\bar{\mu}(112) = (\partial^2 \eta_2^{\theta_2}/\partial x_1^2)^o = 0$, since $\Delta(112) = 3$, $\bar{\mu}(122) = (\partial^2 \eta_2^{\theta_2}/\partial x_1 \partial y_1)^o = 0$, since $\Delta(122) = 3$. Similarly, we can show that $\bar{\mu}(1122) = 2$, since $\Delta(1122) = 3$. Further, by using Theorem 14.3, we have $\bar{\mu}(1212) = 2$. Therefore, $\Delta(j_1 j_2 j_3 j_4 j_5) = 1$, and $\bar{\mu}(i_1 i_2 \cdots i_k) = 0$ for $k \ge 5$.

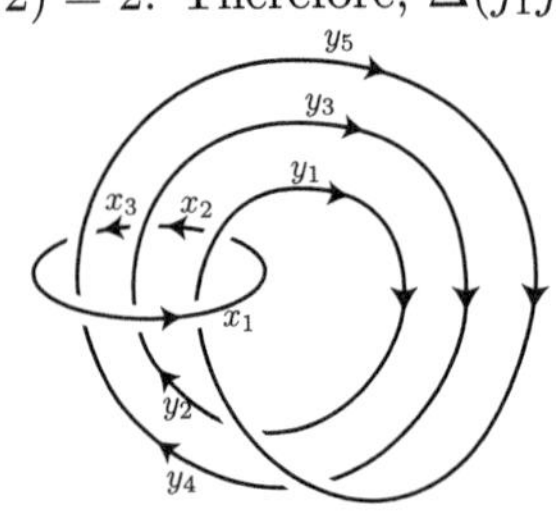

Figure 14.1

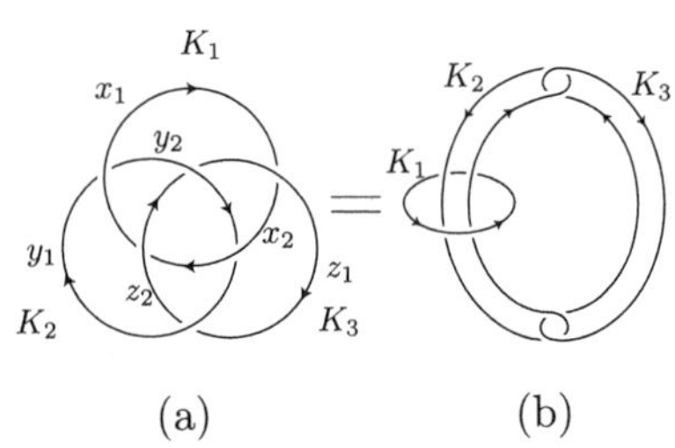

Figure 14.2

Example 14.2 Let L be a link depicted in Fig. 14.2 (a) or (b). This link is called the *Borromean rings*. (This link has a property that a removal of one arbitrary component results a trivial link.) As before, we use x_i, y_j, z_k for $x_{1,i}, x_{2,j}, x_{3,k}$, and R_i, S_j, U_k for relations. The group $G(L)$ has the following presentation:

$\langle x_1, x_2, y_1, y_2, z_1, z_2 | R_1 = 1, R_2 = 1, S_1 = 1, S_2 = 1, U_1 = 1, U_2 = 1\rangle$, where $R_1 = x_1 z_1^{-1} x_2^{-1} z_1, R_2 = x_2 z_2 x_1^{-1} z_2^{-1}, S_1 = y_1 x_1^{-1} y_2^{-1} x_1, S_2 = y_2 x_2 y_1^{-1} x_2^{-1}, U_1 = z_1 y_1^{-1} z_2^{-1} y_1, U_2 = z_2 y_2 z_1^{-1} y_2^{-1}$. And new relations are:
$R_2' = x_1(z_1^{-1} z_2) x_1^{-1}(z_2^{-1} z_1), S_2' = y_1(x_1^{-1} x_2) y_1^{-1}(x_2^{-1} x_1), U_2' = z_1(y_1^{-1} y_2) z_1^{-1}(y_2^{-1} y_1)$, and $\eta_1 = z_1^{-1} z_2, \eta_2 = x_1^{-1} x_2, \eta_3 = y_1^{-1} y_2$. Now, $\bar{\mu}(12) = \bar{\mu}(13) = \bar{\mu}(23) = 0$, and $\bar{\mu}(123) = 1, \bar{\mu}(132) = -1$. Therefore, $\Delta(j_1 j_2 j_3 j_4) = 1$, and $\bar{\mu}(i_1 i_2 \cdots i_k) = 0$ for $k \ge 4$.

(4) **Lower central quotients**

Let $G = G(L)$ be the group of an n-component link. If all $\bar{\mu}$-invariants vanish, then it is known that for any $q \geq 1, G/G^{(q)} \cong F/F_n^{(q)}$. However, if $\bar{\mu}(ij) = \pm 1$ for any i and $j, i \neq j$, then $\bar{\mu}(i_1 i_2 \dots i_k) = 0$ for $k \geq 3$. In this case, the following theorem is proved.

Theorem 14.4 *Let G be the group of an n-component link. Suppose that $\bar{\mu}(ij) = \pm 1$ for any i and $j, i \neq j$. Then for any $q \geq 2, G^{(q)}/G^{(q+1)} \cong F_{n-1}^{(q)}/F_{n-1}^{(q+1)}$, where F_{n-1} is a free group of rank $n-1$.*

This theorem was first proved by Kojima for the case $\bar{\mu}(ij) = 1$ for any i and $j, i \neq j$ [Ko]. Later, Maeda proved this theorem under weaker conditions [Ma], also see [MaT] [La]. In 2000, Morishita proved a number theoretical version of this theorem [Mo].

15. Milnor's invariants (II)

Milnor's $\bar{\mu}$-invariants are interpreted as covering linkage invariants in a certain covering space. Since $\bar{\mu}(i_1 i_2 \cdots i_r)$ is defined via a nilpotent group $G/G^{(q)}$, the covering will be most likely a nilpotent covering.

(1) **Nilpotent Representation**

To find an appropriate covering, first we define a free group action on $\mathbb{Z}^k$, the set of k-tuples $\{(a_1, a_2, \dots, a_k) | a_i \in \mathbb{Z}\}$. Let F be a free group of rank n freely generated by $x_1, x_2, \dots, x_n$. Given a sequence of integers $\xi = \{j_1, j_2, \dots, j_k\}$, $1 \leq j_1, j_2, \dots, j_k \leq n$, we define an action Φ of F on $\mathbb{Z}^k$: $\Phi_\xi(x_i)(a_1, a_2, \dots, a_k) = (b_1, b_2, \dots, b_k)$, where $b_1 = a_1 + \delta_{i,j_1}, b_2 = a_2 + \delta_{i,j_2} a_1, \cdots, b_k = a_k + \delta_{i,j_k} a_{k-1}$. For example, if $\xi = \{1, 2, 1, 1, 2\}$, then $\Phi_\xi(x_1)(a_1, a_2, \dots, a_5) = (a_1 + 1, a_2, a_3 + a_2, a_4 + a_3, a_5), \Phi_\xi(x_2)(a_1, a_2, \dots, a_5) = (a_1, a_2 + a_1, a_3, a_4, a_5 + a_4), \Phi_\xi(x_i) = \text{id.}$ for $i \geq 3$.

This action is written in the matrix form as follows:

$$\Phi_\xi(x_i) = \begin{bmatrix} 1 & \delta_{i,j_1} & & & 0 \\ & 1 & \delta_{i,j_2} & & \\ & & \ddots & \ddots & \\ & 0 & & 1 & \delta_{i,j_k} \\ & & & & 1 \end{bmatrix} \in GL(k+1, \mathbb{Z}).$$

Let Ω be the orbit of $(0, 0, \dots, 0)$ under Φ_ξ and $S(\Omega)$, the group of permutations on Ω. Then we have a homomorphism $\Phi_\xi : F \to S(\Omega)$. Using this (transitive) homomorphism, we can define a permutation representation of $G(L)$. Let $\hat{\xi} = \{j_1, j_2, \dots, j_k, j_{k+1}, j_{k+2}\}$ be a sequence of the length $k+2$.

Let $m = \Delta(j_1 j_2 \cdots j_k j_{k+1} j_{k+2})$. Now we consider a nilpotent representation $\hat{\Phi}_\xi : G(L) \to S((\mathbb{Z}/m)^k)$, where $S((\mathbb{Z}/m)^k)$ denotes the group of permutations on k-tuples $\{(a_1, a_2, \dots, a_k) | 0 \leq a_i \leq m-1\}$.

(2) **Covering linkage invariants**

Let L be an n-component link in S^3, and M_Φ the covering space corresponding to $\hat{\Phi}_\xi$. Let $\tilde{L}$ be the lift of L in M_Φ. Assume $m \neq 0$. Let $\{\mathcal{Q}_0^{(p)}, \mathcal{Q}_1^{(p)}, \dots, \mathcal{Q}_\lambda^{(p)}\}$ be the set of all orbits in Ω under the actions $\hat{\Phi}_\xi(x_{p,1})$ and $\hat{\Phi}_\xi(\eta_p), p = j_1, j_2, \dots, j_{k+2}$. Then each orbit $\mathcal{Q}_i^{(p)}$ corresponds to one and only one component of $\tilde{L}$, denoted by $\tilde{K}_{p,i}$.

Now to each orbit $\mathcal{Q}_i^{(j_{k+1})}$, we define $q_i = \sum \beta_k(b_1, \dots, b_k)$, where the summation runs over all $(b_1, \dots, b_k)$ in $\mathcal{Q}_i^{(j_{k+1})}$ and $\beta_k(b_1, \dots, b_k) = b_k$. Then set $\alpha = \sum_{i=1}^\lambda q_i \tilde{K}_{j_{k+1},i} \in H_1(\tilde{L}; \mathbb{Z})$.

It is known that α bounds a 2-chain B in M_Φ and we have:

Theorem 15.1 [Mu6] *Given a sequence* $\hat{\xi} = \{j_1, j_2, \dots, j_k, j_{k+1}, j_{k+2}\}$, *let* $\tilde{K}_{j_{k+2},0}$ *be the knot corresponding to the orbit* $\mathcal{Q}_0^{(j_{k+2})}$ *that contains* $(0, 0, \dots, 0)$. *Assume that* $m \neq 0$. *Then* $\bar{\mu}(j_1, j_2, \dots, j_{k+2}) \equiv \mathrm{int}\ (\tilde{K}_{j_{k+2},0}, B) \pmod m$, *where* int(,) *stands for the intersection number in* M_Φ. *Therefore, if* $\tilde{K}_{j_{k+2},0}$ *is homologous to* 0 *in* M_Φ, *then* $\bar{\mu}(j_1, j_2, \dots, j_{k+2}) \equiv \mathrm{lk}(\tilde{K}_{j_{k+2},0}, \alpha) \pmod m$.

Remark 15.1 (1) If $m = 0$, then consider a finite nilpotent representation $\hat{\Phi}_\xi : G(L) \to S((\mathbb{Z}/q)^k)$ for a sufficiently large $q > 0$. Then 'mod q' will be interpreted as 'equality'.
(2) By Cyclic Symmetry (Theorem 14.3), we may assume without loss of generality that $j_{k+1} \neq j_{k+2}$.

Example 15.1 For $\xi = \{i, j, k\}$, we have from Theorem 15.1 that $\bar{\mu}(ijk) \equiv$ *int*$(\tilde{K}_{k,0}, B) \pmod{\bar{\mu}(jk)}$, where $\partial B = \sum p\tilde{K}_{j,p} \in H_1(\tilde{L}, \mathbb{Z})$, and $\tilde{L}$ is a lift of L in the m-fold cyclic covering space of K_i.

(3) **Examples**

Example 14.1 (continued) First we see that $\bar{\mu}(12) = 3$. Since $\Delta(1jk) = 3$, M_Φ is the 3-fold cyclic covering space of K_1, and $M_\Phi \approx S^3$. Then the lift of L in M_Φ is depicted in Fig. 15.1.

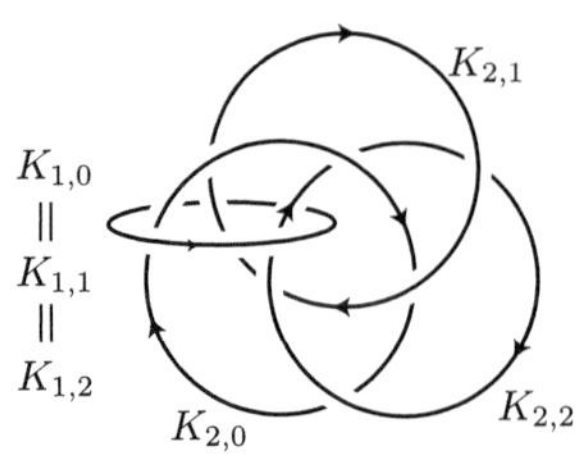

Figure 15.1

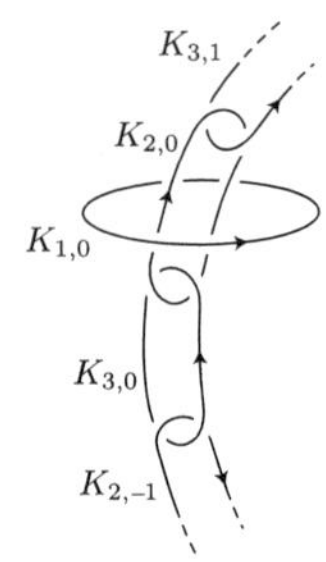

Figure 15.2

From this diagram, we can see easily that $\bar{\mu}(112) \equiv \mathrm{lk}(K_{2,0}, K_{1,1} + 2K_{1,2}) = 3 \equiv 0 \pmod 3$, $,\bar{\mu}(121) \equiv \mathrm{lk}(K_{1,0}, K_{2,1} + 2K_{2,2}) = 3 \equiv 0 \pmod 3, \bar{\mu}(122) \equiv \mathrm{lk}(K_{2,0}, K_{2,1} + 2K_{2,2}) = 3 \equiv 0 \pmod 3$.

Example 14.2 (continued) Since $\bar{\mu}(12) = \bar{\mu}(13) = \bar{\mu}(23) = 0$, we see $\Delta(ijk) = 0$. To see $\bar{\mu}(123)$, consider the lift $\tilde{L}$ of L in the infinite cyclic covering space of K_1 (or the m-fold cyclic covering space with a sufficiently large m). $\tilde{L}$ is depicted in Fig. 15.2. Let $\alpha = \sum_i K_{2,i}$ and $\beta = \sum_i K_{3,i}$. Then we see $\bar{\mu}(123) = \mathrm{lk}(K_{3,0}, \alpha) = \mathrm{lk}(K_{3,0}, -K_{2,-1}) = 1, \bar{\mu}(132) = \mathrm{lk}(K_{2,0}, \beta) = \mathrm{lk}(K_{2,0}, K_{3,-1}) = -1$.

16. Questions and Problems

In this last section, we discuss some problems in knot theory related to the number theory.

(1) **Structure of the homology group**

Let M_Φ be the double covering space of an n-component link $L = K_1 \cup K_2 \cup \cdots \cup K_n$. Suppose $|H_1(M_\Phi; \mathbb{Z})| < \infty$. Then the 2-sylow subgroup Γ of $H_1(M_\Phi; \mathbb{Z})$ is of the form: $\Gamma = \mathbb{Z}/2^{p_1} + \mathbb{Z}/2^{p_2} + \cdots + \mathbb{Z}/2^{p_{n-1}}, p_i \geq 1$. Define $e_k(L) = |\{p_i | p_i \geq k\}|$. Then we have $e_1(L) = n - 1$ [Sa1]. Further, $e_2(L)$ is determined as follows. Let Λ be an $n \times n$ matrix over $\mathbb{Z}/2$:

$$\Lambda = \begin{bmatrix} \mu_{11} & \mu(12) & \cdots & \mu(1n) \\ \mu(12) & \mu_{22} & \cdots & \mu(2n) \\ \vdots & \vdots & \ddots & \vdots \\ \mu(n1) & \mu(n2) & \cdots & \mu_{nn} \end{bmatrix} \pmod 2,$$

where $\mu(ij) = \mu(ji) = \mathrm{lk}(K_i, K_j), i \neq j$, and $\mu_{kk} = -\sum_{i=1, i\neq k}^{m} \mu(ki)$.

Then $e_2(L) = n - 1 - \mathrm{rank}\ \Lambda$.

Question 16.1 Can $e_k(L), k \geq 3$, be determined by Milnor's $\bar{\mu}$-invariants ?

Question 16.2 Given a sequence of positive integers, $\lambda_1 \geq \lambda_2 \geq \cdots \geq \lambda_m, (m \geq 1)$. Does there exist a link L in S^3 such that $e_1(L) = \lambda_1, e_2(L) = \lambda_2, \ldots, e_m(L) = \lambda_m$?

Remark 16.1 Question 16.1 has been solved completely by Hillmann, Matei and Morishita [HMM].

(2) **Galois group of the Alexander polynomials**

Let $\Delta_K(t)$ be the Alexander polynomial of a knot K.

Question 16.3 What is the Galois group of $\Delta_K(t)$?

Since $\Delta_K(t)$ is characterized by two conditions, see Theorem 7.2, this problem is to characterize the Galois group of the polynomials satisfying these conditions.

(3) **Hilbert class field**

(i) **Periodic knots**

We say that a knot K has *period* p if there exists an orientation preserving homeomorphism ϕ from S^3 to S^3 such that (a) $\phi(K) = K$, (b) $\phi^p = \mathrm{id}$, but $\phi^q \neq \mathrm{id}$. for $0 < q < p$, (c) $\mathrm{Fix}(\phi)$ and K are disjoint, where $\mathrm{Fix}(\phi)$ denotes the set of points in S^3 that are fixed under ϕ. (We know that $\mathrm{Fix}(\phi)$ is a trivial knot.)

Example 16.1 The knot K depicted in Fig. 16.1 has period 3. In fact, ϕ is a rotation about the z-axis through $\frac{2}{3}\pi$, and obviously $\mathrm{Fix}(\phi)$ is the z-axis. This knot K also has period 2. More generally, a 2-bridge knot has period 2, and a torus knot of type (p, q) has periods p and q. A non-trivial knot can have only finitely many periods. The set of periods that a knot can have is certainly a knot invariant, but there is no algorithm to determine all periods of a knot.

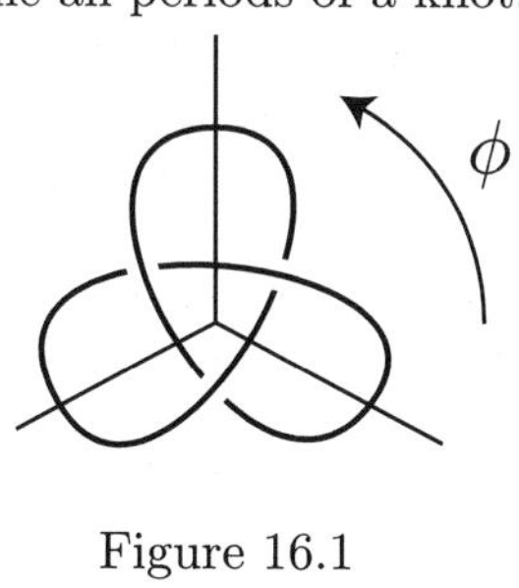

Figure 16.1

(ii) **Criteria on periodic knots**

The first result on the periodic knots was obtained by Trotter in 1960.

Theorem 16.1 [T1] *Suppose a knot K has period n. If (1)K is a fibred knot, and (2)$\Delta_K(t)$ has no repeated roots, then the splitting field F_Δ of $\Delta_K(t)$ over $\mathbb{Q}$ contains a primitive nth root of* 1.

Later, using the theory of the covering space, the following theorem was proved without any restriction.

Theorem 16.2 [Mu4] *Suppose a knot K has a period p, where p is a prime. Then*

$$\Delta_K(t) \equiv (1+t+t^2+\cdots+t^{\lambda-1})^{p-1} f(t)^p \pmod{p},$$

where $\lambda \geq 1$ and $f(t)$ is the Alexander polynomial of some knot.

The number λ and $f(t)$ are interpreted as follows. Suppose a knot K has period p, p a prime. Let $K_o = \mathrm{Fix}(\phi)$. Then S^3 is the p-fold cyclic covering space of a 3-manifold Σ branched along K_o, where $\Sigma \approx S^3$. Let $\psi : S^3 \to \Sigma$ be the projection. Then $\psi(K) = \hat{K}$ is a knot in Σ and it is shown that $f(t) = \Delta_{\hat{K}}(t)$, and further $\lambda = |\mathrm{lk}(K_o, \hat{K})| \neq 0$. (If $\lambda = 0$, then $\psi^{-1}(\hat{K})$ would be a link (not a knot). Often $\hat{K}$ is unknotted, so $f(t) = 1$.

(3) **Heilbronn's Conjecture**

By looking at above two theorems from the number theoretical points of view, Heilbronn suggested the following conjecture:

Conjecture *Let K be a knot. Suppose p is a prime. Then, if $\Delta_K(t) \equiv (1+t+t^2+\cdots+t^{\lambda-1})^{p-1} \pmod{p}$, $\lambda \geq 1$, then Hilbert class field $\hat{F}$ of F_Δ, the splitting field of $\Delta_K(t)$ over $\mathbb{Q}$, contains a primitive pth roof of* 1.

This conjecture is in general not true. The simplest example is the following

Example 16.2 Consider $\Delta_K(t) = 1 - 6t + 11t^2 - 6t^3 + t^4 (= (1-3t+t^2)^2) \equiv (1+t)^4 \pmod{5}$. Then $F_\Delta = \mathbb{Q}(\sqrt{5}) \in \mathbb{R}$, but $\mathbb{Q}(\sqrt{5})$ has class number 1 and hence $\hat{F} = F_\Delta$ does not contain a primitive 5th root of 1.

However, for some knot, the conjecture holds.

Example 16.3 Let $\Delta_K(t) = 4 - 7t + 4t^2 \equiv (1+t)^2 \pmod{3}$. Then $F_\Delta = \mathbb{Q}(\sqrt{-15})$ and $F_\Delta(\omega), \omega$ a primitive 3rd root of 1, is unramified over F_Δ, and hence $\hat{F}$ contains ω.

(3) **Revised Problem**

These two examples suggest that the conjecture should be revised, and Morishita now proposes the following question:

Question 16.4 (Morishita) Let p be a prime. Which Alexander polynomial $\Delta_K(t)$ of degree $p-1$ has the splitting field F_Δ over $\mathbb{Q}$ so that $F_\Delta(\xi), \xi$ a primitive pth root of 1, is unramified over F_Δ ?

Finally, we mention one of the recent result due to Morishita and Taguchi that gives a sufficient condition for $\Delta_K(t)$ to hold Heilbronn Conjecture.

Theorem 16.3 [MoT] *Let p be a prime $\neq 2$. Let $g(t)$ be an Eisenstein polynomial, i.e. $g(t) = t^{p-1} + a_1 t^{p-2} + \cdots + a_{p-1}$, where $p | a_i, 1 \leq i \leq p-1$, but $p^2 \nmid a_{p-1}$. Then, Hilbert class field of the splitting filed F of $g(t)$ contains a pth root of* 1.

Therefore, if $g(t+1) = \Delta_K(t)$ for some knot K, then $\Delta_K(t) \equiv (t+1)^{p-1} \pmod{p}$ and Hilbert class field of the splitting field of $\Delta_K(t)$ contains a primitive pth root of 1. For example, let $g(t) = t^4 + 5t^3 - 40t^2 + 70t - 35$. Then $g(t+1) = t^4 + 9t^3 - 19t^2 + 9t + 1 = \Delta_K(t)$ for some fibred knot K [Bu1], and $\Delta_K(t) \equiv (1+t)^4 \pmod{5}$. Since the splitting field F of $\Delta_K(t)$ does not contain a primitive 5th root of 1 (due to T. Komatsu), K does not have period 5 by Theorem 16.1, but by Theorem 16.3, Hilbert class field of F contains a primitive 5th root of 1.

Recently, using Theorem 16.1, T. Komatsu shows [Kom] that for every odd prime number, there are infinitely many knots satisfying Heilbronn's conjecture.

Acknowledgements The author wishes to thank M. Hirasawa, T. Komatsu and M. Morishita for his helpful suggestions and comments.

References

[A1] J. W. Alexander, *A lemma on systems of knotted curves*, Proc. Nat. Acad. Sci. USA 9 (1923), 93–95.

[A2] ———, *Topological invariants of knots and links*, Trans. Amer. Math. Soc. 30 (1928), 275–306.

[BS] C. Bankwitz and H.G. Schumann, *Über Viergeflechte*, Abh. Math. Sem. Univ. hamburg, 10 (1934) 263–284.

[Bu1] G. Burde, *Alexanderpolynome Neuwirthscher Knoten*, Topology 5 (1966), 321–330.

[Bu2] ———, *On branched coverings of* S^3, Canad. J. Math. 23 (1971), 84–89.

[BZ] G. Burde and H. Zieschang, Knots. Second edition, de Gruyter Studies in Mathematics, 5. Walter de Gruyter & Co., Berlin, 2003.

[C] K.T. Chen, *Isotopy invariants of links*, Ann. of Math. (2) 56 (1952), 343–353.

[F1] R. Fox, *Free differential calculus. I. Derivation in the free group ring*, Ann. of Math. (2) 57 (1953), 547–560.

[F2] ———, *Construction of simply connected 3-manifolds*, 1962 Topology of 3-manifolds and related topics (Proc. The Univ. of Georgia Institute, 1961) pp. 213–216 Prentice-Hall, Englewood Cliffs, N.J.

[G] D. Gabai, *Detecting fibred links in* S^3, Comment. Math. Helv. 61 (1986), 519–555.

[GKM] H. Goda, T. Kitano and T. Morifuji, *Reidemeister torsion, twisted Alexander polynomial and fibered knots*, preprint,

[GLM] C. Gordon, R. Litherland and K. Murasugi, *Signatures of covering links*, Canad. J. Math. 33 (1981), 381–394.

[GR] F. González-Acuña, and A. Ramírez, *Two-bridge knots with property Q*, Q. J. Math. 52 (2001), 447–454.

[Ha] R. Hartley, Ph. D. Thesis, Univ. of Toronto, 1975.

[HaM1] R. Hartley and K. Murasugi, *Covering linkage invariants*, Canad. J. Math. 29 (1977), 1312–1339.

[HaM2] ———, *Homology invariants*, Canad. J. Math. 30 (1978), 655–670.

[HMM] J. Hillman, D. Matei and M. Morishita, *Pro-p link groups and p-homology groups*, to appear in this volume

[HiM1] M. Hirasawa and K. Murasugi, *Fibered double torus knots which are band-sums of torus knots*, preprint (2002)

[HiM2] ———, *Twisted Alexander polynomials*, in preparation (2003)

[HizM] F. Hirzeburch and K.H. Mayer, $O(n)$-Manningfaltigkeiten, exotische Sphären und Singularitäten, Lect. Notes in Math. 57 (1968), Springer.

[Kan] T. Kanenobu, *Alexander polynomials of two-bridge links*, J. Austral. Math. Soc. Ser. A 36 (1984), no. 1, 59–68.

[Ki] T. Kitano, *Twisted Alexander polynomial and Reidemeister torsion*, Pacific J. Math. 174 (1996), 431–442.

[Kaw] A. Kawauchi, A Survey of Knot Theory, Birkhäser Verlag, Basel, 1996.

[Ko] S. Kojima, *Nilpotent completions and Lie rings associated to link groups*, Comment. Math. Helv. 58 (1983), 115–134.

[Kom] T. Komatsu, Note on Heilbronn's Conjecture (preprint) 2005.

[La] J.P. Labute, *The Lie algebra associated to the lower central series of a link group and Murasugi's conjecture*, Proc. Amer. Math. Soc. 109 (1990), 951–956.

[Li] X.S. Lin, *Representation of knot groups and twisted Alexander polynomials*, preprint (1990)

[Lu] U. Lüdicke, *Zyklische Knoten*, Arch. Math. (Basel) 32 (1979), no. 6, 588–599.

[Ma] T. Maeda, Ph.D. Thesis, Univ of Toronto, 1983

[MaM] J. P. Mayberry and K. Murasugi, *Torsion-groups of abelian coverings of links*, Trans. Amer. Math. Soc. 271 (1982), 143–173.

[MaT] W. S. Massey and L. Traldi, *On a conjecture of K. Murasugi*, Pacific J. Math. 124 (1986), 193–213.

[MeT] W. Menasco and M. Thistlethwaite, *The classification of alternating links*, Ann. of Math. (2) 138 (1993), 113–171.

[Mi] J. Milnor, *Isotopy of links. Algebraic geometry and topology*, A symposium in honor of S. Lefschetz, pp. 280–306. Princeton University Press, Princeton, N. J., 1957.

[Mo] M. Morishita, *On certain analogies between knots and primes*, J. Reine Angew. Math. 550 (2002), 141–167.
[MoT] M. Morishita and Yuichiro Taguchi, Private communication, March 2003.
[Mu1] K. Murasugi, *On a certain subgroup of the group of an alternating link*, Amer. J. Math. 85 (1963), 544–550.
[Mu2] ———, *On the Minkowski unit of slice links*, Trans. Amer. Math. Soc. 114 (1965), 377–383.
[Mu3] ———, *On a certain numerical invariant of link types*, Trans. Amer. Math. Soc. 117 (1965), 387–422.
[Mu4] ———, *On periodic knots*, Comment. Math. Helv. 46 (1971), 162–174.
[Mu5] ———, *On symmetry of knots*, Tsukuba J. Math. 4 (1980), no. 2, 331–347.
[Mu6] ———, *Nilpotent coverings of links and Milnor's invariant*, Low-dimensional topology (Chelwood Gate, 1982), 106–142, London Math. Soc. Lecture Note Ser., 95, Cambridge Univ. Press, Cambridge, 1985.
[N] L. Neuwirth, *The algebraic determination of the genus of knots*, Amer. J. Math. 82 (1960), 791–798.
[Pe] K. A. Perko Jr., *On dihedral covering spaces of knots*, Invent. Math. 34 (1976), no. 2, 77–84.
[Pl] A. Plans, *Contribution to the study of the homology groups of the cyclic ramified coverings corresponding to a knot*, Revista Acad. Ci. Madrid 47 (1953). 161–193.
[Ra] E. S. Rapaport, *On the commutator subgroup of a knot group*, Ann. of Math. (2) 71 (1960), 157–162.
[Ri] R. Riley, *Homomorphisms of knot groups on finite groups*, Math. Comp. 25 (1971), 603–619; addendum, ibid. 25 (1971), no. 115, loose microfiche suppl. A-B.
[Sa] M. Sakuma, *On regular coverings of links*, Math. Ann. 260 (1982), 303–315.
[Sa2] ———, *Variations of McShane's identity for the Riley slice and 2-bridge links*, Sūrikaisekikenkyūsho Kōkyūroku No. 1104 (1999), 103–108.
[Sc] H. Schubert, *Knoten mit zwei Brüken*, Math. Z. 65 (1956), 133–170.
[Se] H. Seifert, *Über das Geschecht von Knoten*, Math. Ann., 110 (1934), 571-592.
[Sh] Y. Shinohara, *On the signature of a link with two bridges*, Kwansei Gakuin Univ. Annual Stud. 25 (1976), 111–119.
[St] J. Stallings, *On fibering certain 3-manifolds*, 1962 Topology of 3-manifolds and related topics (Proc. The Univ. of Georgia Institute, 1961) pp. 95–100 Prentice-Hall, Englewood Cliffs, N.J.
[T1] H. F. Trotter, *Periodic automorphisms of groups and knots*, Duke Math. J. 28 (1961), 553–557.
[T2] ———, *Homology of group systems with applications to knot theory*, Ann. of Math. (2) 76 (1962), 464–498.
[T3] ———, *Non-invertible knots exist*, Topology 2 (1963), 275–280.
[Wad] M. Wada, *Twisted Alexander polynomial for finitely presentable groups*, Topology 33 (1994), 241–256.
[Wal] F. Waldhausen, *On irreducible 3-manifolds which are sufficiently large*, Ann. of Math. (2) 87 (1968), 56–88.

Department of Mathematics, University of Toronto, Toronto, Ontario M5S 2E4 Canada
E-mail address: murasugi@math.toronto.edu

Contemporary Mathematics
Volume **416**, 2006

Harmonic and equianharmonic equations in the Grothendieck-Teichmüller group, II

Hiroaki Nakamura and Hiroshi Tsunogai

Abstract. In [LS], Lochak and Schneps introduced the "harmonic parameter g" of the Grothendieck-Teichmüller group $\widehat{GT}$. We closely study the behavior of g on the absolute Galois group $G_{\mathbb{Q}}$ using a family of lemniscate elliptic curves. We obtain a relationship of the adelic beta function and the harmonic parameter specialized in the matrix group $\mathrm{SL}_2(\hat{\mathbb{Z}})$.

§1. Introduction.

Let $\widehat{GT}$ be the profinite Grothendieck-Teichmüller group introduced and studied by Drinfeld [Dr], Ihara [I1]. It contains the Galois group $G_{\mathbb{Q}} = \mathrm{Gal}(\overline{\mathbb{Q}}/\mathbb{Q})$ and has the standard parametrization $(\lambda, f) : \widehat{GT} \hookrightarrow \hat{\mathbb{Z}}^{\times} \times \hat{F}_2$ ($\hat{\mathbb{Z}}$, $\hat{F}_2$ are respectively the profinite completions of the integer ring $\mathbb{Z}$ and of F_2, the free group generated by non-commutative symbols x and y.) In [NT], we closely studied Galois behaviors of the auxiliary parameters g and $h : \widehat{GT} \to \hat{F}_2$ which had been introduced in Lochak-Schneps [LS] so as to decompose (uniquely) the main parameter $f : \widehat{GT} \to \hat{F}_2$ (actually f terminates into the commutator subgroup $[\hat{F}_2, \hat{F}_2]$ by definition) as follows:

$$f(x,y) = g(y,x)^{-1}g(x,y) = \begin{cases} y^{-\frac{\lambda-1}{2}} h(y,z)^{-1}h(x,y) & (\lambda \equiv 1 \mod 6), \\ y^{-\frac{\lambda-1}{2}} h(y,z)^{-1}y^{-1}h(x,y) & (\lambda \equiv -1 \mod 6), \end{cases}$$

where $z = (xy)^{-1} \in \hat{F}_2$. We showed in [NT] that the parameters g and h can be directly written by (λ, f) on the image of $G_{\mathbb{Q}}$ in $\widehat{GT}$, and presented several new-type equations satisfied by the Galois image.

In this paper, we continue our study with mainly concentrating on the parameter g. Our motivating problem to the present paper concerns with the matrix specialization of this parameter. In fact, in [N-I] Corollary 4.13, one of the authors explicitly computed the matrix $f_\sigma\left(\left(\begin{smallmatrix}1&2\\0&1\end{smallmatrix}\right), \left(\begin{smallmatrix}1&0\\-2&1\end{smallmatrix}\right)\right) \in \mathrm{SL}_2(\hat{\mathbb{Z}})$, which turned out later

---

1991 *Mathematics Subject Classification.* Primary 14H30; Secondary 11G05, 20E18, 20F36.
Key words and phrases. adelic beta functions, lemniscate elliptic curve.

in [NS] Remark 2.7 to be decomposed as the following "intriguing" form:

$$(*) \qquad f_\sigma\left(\begin{pmatrix}1&2\\0&1\end{pmatrix},\begin{pmatrix}1&0\\-2&1\end{pmatrix}\right)=(-1)^{\frac{\lambda_\sigma-1}{2}}\begin{pmatrix}1&0\\-8\rho_2(\sigma)&1\end{pmatrix}\begin{pmatrix}\lambda_\sigma^{-1}&0\\0&\lambda_\sigma\end{pmatrix}\begin{pmatrix}1&-8\rho_2(\sigma)\\0&1\end{pmatrix}.$$

Here, $\rho_2 : G_{\mathbb{Q}} \to \hat{\mathbb{Z}}$ designates the Kummer 1-cocycle along the positive roots of 2. This symmetric matrix should be decomposed into the form ${}^tG\cdot G$ for some G corresponding to the image $G = g_\sigma\left((\begin{smallmatrix}1&2\\0&1\end{smallmatrix}),(\begin{smallmatrix}1&0\\-2&1\end{smallmatrix})\right)$. However, unlike the decomposition $f(x,y) = g(y,x)^{-1}g(x,y)$ in $\hat{F}_2$, the determination of G in $\mathrm{SL}_2(\hat{\mathbb{Z}})$ does not follow from the decomposition property $f_\sigma((\begin{smallmatrix}1&2\\0&1\end{smallmatrix}),(\begin{smallmatrix}1&0\\-2&1\end{smallmatrix})) = {}^tG\cdot G$; namely, there remains apriori ambiguity modulo left multiplication by "rotation matrices" on G. In this paper, we shall determine the true choice of the matrix G representing $g_\sigma\left((\begin{smallmatrix}1&2\\0&1\end{smallmatrix}),(\begin{smallmatrix}1&0\\-2&1\end{smallmatrix})\right)$. Our answer is that the main factor is given by the special value of Anderson-Ihara's adelic beta function $B_\sigma(\frac{1}{4},\frac{1}{4})$. In terms of the measure version $\mathbb{B}_\sigma \in \hat{\mathbb{Z}}[[\hat{\mathbb{Z}}^2(1)]]$ of B_σ (see §2 below), we prove:

THEOREM D. (§5 (5.5)) *For each $\sigma \in G_{\mathbb{Q}}$, we have*

$$g_\sigma\left(\begin{pmatrix}1&2\\0&1\end{pmatrix},\begin{pmatrix}1&0\\-2&1\end{pmatrix}\right)=(-1)^{\frac{\lambda_\sigma-1}{2}}\mathbb{B}_\sigma\left(\begin{pmatrix}0&1\\-1&0\end{pmatrix},\begin{pmatrix}0&1\\-1&0\end{pmatrix}\right)\cdot\begin{pmatrix}\lambda_\sigma^{-1}&-8\rho_2(\sigma)\lambda_\sigma^{-1}\\0&(-1)^{\frac{\lambda_\sigma-1}{2}}\end{pmatrix}$$

in $\mathrm{GL}_2(\hat{\mathbb{Z}})$.

Once Theorem D is obtained, it is easy to recover $(*)$ by a basic property of $\mathbb{B}_\sigma$ (see (2.5.1)). The connection of the above two sides is, naively speaking, via the classical link between the harmonic ratio and the lemniscate elliptic curve $Y^2 = X^3 - X$. The right hand side reflects the adelic Galois version of the classical period integral via the beta function:

$$\int_0^1 \frac{dt}{\sqrt{1-t^4}} = \frac{1}{4\sqrt{2}}B(\frac{1}{4},\frac{1}{4}).$$

Since the lemniscate elliptic curve is dominated by the Fermat quartic, it is not surprising that the Galois action on the Tate module is given by the adelic beta values at the pair of 4-th roots of unity. However, we wish to take an explicit choice of a basis of this Tate module which allows well controls to be linked with the Grothendieck-Teichmüller parameters. Thereby we construct in §3 an explicit $G_{\mathbb{Q}}$-compatible immersion of the fundamental group of the (affine) lemniscate elliptic curve into the braid group with 4 strands. A close comparison of tangential base-points on the braid configuration space makes clear the appearance of the harmonic ratio $(0,\frac{1}{2},1,\infty)$ at an addressed object (§4). We settle the proof of Theorem D in §5.

In [NT] Theorem A, Proposition 6.1 (2), we proved the following equations in the braid group $\hat{B}_3 = \langle \tau_1,\tau_2 \mid \tau_1\tau_2\tau_1 = \tau_2\tau_1\tau_2 =: \eta\rangle$ for the image of $G_{\mathbb{Q}} \hookrightarrow \widehat{GT}$:

$$(\mathrm{GF}_0) \qquad g(\tau_1^2,\tau_2^2) = \eta^{2\rho_2-\rho_3}f(\tau_1,\eta)\tau_1^{-2\rho_2+3\rho_3},$$

$$(\mathrm{GF}_1) \qquad g(\tau_1^2,\tau_2^2) = f(\tau_1^2,\eta)\tau_1^{4\rho_2},$$

$$(\mathrm{GG}) \qquad g(\tau_1^2,\tau_2^2) = \eta^{2\rho_2-\rho_3}g(\tau_1,\tau_2)\tau_1^{-4\rho_2+3\rho_3}.$$

Here and henceforth, we sometimes omit σ from formulas for simplicity, when no confusion could occur from contexts. Combining Theorem D with the above (GF0), (GF1) and (GG) specialized with $\tau_1 \mapsto (\begin{smallmatrix}1&1\\0&1\end{smallmatrix})$, $\tau_2 \mapsto (\begin{smallmatrix}1&0\\-1&1\end{smallmatrix})$, we immediately obtain

COROLLARY E. *On the image of $G_{\mathbb{Q}}$ in $\widehat{GT}$, we have*

$$f\left(\left(\begin{smallmatrix}1&2\\0&1\end{smallmatrix}\right),\left(\begin{smallmatrix}0&1\\-1&0\end{smallmatrix}\right)\right)=(-1)^{\frac{\lambda-1}{2}}\mathbb{B}\left(\left(\begin{smallmatrix}0&1\\-1&0\end{smallmatrix}\right),\left(\begin{smallmatrix}0&1\\-1&0\end{smallmatrix}\right)\right)\cdot\begin{pmatrix}\lambda^{-1}&-12\rho_2\lambda^{-1}\\0&(-1)^{\frac{\lambda-1}{2}}\end{pmatrix}.$$

$$f\left(\left(\begin{smallmatrix}1&1\\0&1\end{smallmatrix}\right),\left(\begin{smallmatrix}0&1\\-1&0\end{smallmatrix}\right)\right)=(-1)^{\frac{\lambda-1}{2}-\rho_2}\left(\begin{smallmatrix}0&1\\-1&0\end{smallmatrix}\right)^{\rho_3}\mathbb{B}(\left(\begin{smallmatrix}0&1\\-1&0\end{smallmatrix}\right),\left(\begin{smallmatrix}0&1\\-1&0\end{smallmatrix}\right))\begin{pmatrix}\lambda^{-1}&(-6\rho_2-3\rho_3)\lambda^{-1}\\0&(-1)^{\frac{\lambda-1}{2}}\end{pmatrix},$$

$$g\left(\left(\begin{smallmatrix}1&1\\0&1\end{smallmatrix}\right),\left(\begin{smallmatrix}0&1\\-1&0\end{smallmatrix}\right)\right)=(-1)^{\frac{\lambda-1}{2}-\rho_2}\left(\begin{smallmatrix}0&1\\-1&0\end{smallmatrix}\right)^{\rho_3}\mathbb{B}(\left(\begin{smallmatrix}0&1\\-1&0\end{smallmatrix}\right),\left(\begin{smallmatrix}0&1\\-1&0\end{smallmatrix}\right))\begin{pmatrix}\lambda^{-1}&(-4\rho_2-3\rho_3)\lambda^{-1}\\0&(-1)^{\frac{\lambda-1}{2}}\end{pmatrix}.$$

One obvious remaining subject is to pursue the analogy of these results for the equianharmonic ratio $(0, e^{2\pi i/6}, 1, \infty)$. The lemniscate elliptic curve will then be replaced by the mordell elliptic curve $Y^2 = X^4 - X$ (*cf.* (3.4)). We hope to continue the investigation in a subsequent work. In the present paper, all algebraic varieties are treated as defined over $\mathbb{Q}$. We fix once and for all the embedding $\overline{\mathbb{Q}} \hookrightarrow \mathbb{C}$. The notation ζ_n always represents the specified primitive n-th root of unity $\exp(2\pi i/n) \in \mathbb{C}$.

§2. Short review of the adelic beta function.

(2.1) We shall start with the standard action of $\widehat{GT}$ on the free profinite group $\hat{F}_2$ generated by the symbols x and y. For each $\sigma \in \widehat{GT}$ associated are two parameters $\lambda = \lambda_\sigma \in \hat{\mathbb{Z}}^\times$ and $f = f_\sigma(x,y) \in [\hat{F}_2, \hat{F}_2]$ which are unique to satisfy $\sigma(x) = x^\lambda$, $\sigma(y) = f^{-1}y^\lambda f$. The abelianization $\hat{\mathbb{Z}}^2(1)$ of $\hat{F}_2$ is generated by the images $\bar{x}, \bar{y}$ of x, y respectively. The complete group algebra $\hat{\mathbb{Z}}[[\hat{F}_2]]$ and its abelianization $\hat{\mathbb{Z}}[[\hat{\mathbb{Z}}^2(1)]]$ play fundamental roles in Ihara's definition of the adelic beta function $\mathbb{B}_\sigma$. Recall that in the former ring $\hat{\mathbb{Z}}[[\hat{F}_2]]$, one has free differential operators $\partial_x := \frac{\partial}{\partial x}$, $\partial_y := \frac{\partial}{\partial y}$ as the profinite analog of Fox's operators in the discrete case (cf. [I1] Appendix). Using them, one may associates the Jacobian matrix

$$J_\sigma := \begin{pmatrix}\partial_x\sigma(x) & \partial_y\sigma(x)\\ \partial_x\sigma(y) & \partial_y\sigma(y)\end{pmatrix} \left(= \begin{pmatrix}\frac{x^\lambda-1}{x-1} & 0\\ (\partial_x f)(y^\lambda-1) & \frac{y^\lambda-1}{y-1}+(\partial_y f)(y^\lambda-1)\end{pmatrix}\right)$$

for each $\sigma \in \widehat{GT}$ whose abelianization image J_σ^{ab} belongs to $\mathrm{GL}_2(\hat{\mathbb{Z}}[[\hat{\mathbb{Z}}^2(1)]])$. Then,

DEFINITION. (Ihara [I2])

$$\mathbb{B}_\sigma(\bar{x},\bar{y}) := \frac{\bar{x}-1}{\bar{x}^\lambda-1}\frac{\bar{y}-1}{\bar{y}^\lambda-1}\det(J_\sigma^{\mathrm{ab}}) \left(= 1 + (\partial_y f)^{\mathrm{ab}}(\bar{y}-1)\right).$$

Since $\hat{\mathbb{Z}}[[\hat{\mathbb{Z}}^2(1)]]$ is identified with $\varprojlim_n \left(\hat{\mathbb{Z}}[\bar{x},\bar{y}]/(\bar{x}^n-1,\bar{y}^n-1)\right)$, one may speak about the specialization $\mathbb{B}_\sigma(\zeta,\zeta')$ where the variables $\bar{x}$, $\bar{y}$ are specialized to any roots of unity ζ, ζ'. The specialization is valued in $\hat{\mathbb{Z}} \otimes \mathbb{Z}(\zeta,\zeta')$. As an element of $\hat{\mathbb{Z}}[[\hat{\mathbb{Z}}^2(1)]]$, $\mathbb{B}_\sigma$ can be determined by the collection of all specialization values at pairs of roots of unity. G.Anderson [A] introduces the adelic beta function $B_\sigma : (\mathbb{Q}/\mathbb{Z})^2 \to \hat{\mathbb{Z}} \otimes \mathbb{Q}^{\mathrm{ab}}$ by setting $B_\sigma(\frac{a}{n},\frac{b}{n}) := \mathbb{B}_\sigma(\zeta_n^a,\zeta_n^b)$ which is a closer analogue of the classical beta function $B(\frac{a}{n},\frac{b}{n}) := \int_0^1 t^{\frac{a}{n}-1}(1-t)^{\frac{b}{n}-1}dt$. These special values control the complex multiplication theory of the quotients of Fermat Jacobians, especially the Galois actions on the torsion points of them. In this paper, however, we will establish the control directly for the lemniscate elliptic curve in an elementary way, so that we do not enter details of Anderson's deep device here. Also, we mainly use $\mathbb{B}_\sigma$ (the measure version of B_σ) for the usefulness of its Galois theoretic formations.

Some of the first basic properties of $\mathbb{B}_\sigma$ are:

(B0) $$\mathbb{B}_\sigma(\bar{x},\bar{y}) \in \hat{\mathbb{Z}}[[\hat{\mathbb{Z}}^2(1)]]^\times,$$

(B1) $$\mathbb{B}_\sigma(\bar{x},\bar{y}) = \mathbb{B}_\sigma(\bar{y},\bar{x}),$$

(B2) $$\mathbb{B}_\sigma(1,\bar{x}) = 1,$$

(B3) $$\mathbb{B}_\sigma(\bar{x},\bar{x}^{-1}) = \frac{1-\bar{x}}{1-\bar{x}^\lambda}\cdot \bar{x}^{\frac{\lambda-1}{2}}\cdot\lambda_\sigma,$$

$$\textit{in particular,}\ \mathbb{B}_\sigma(-1,-1) = (-1)^{\frac{\lambda-1}{2}}\lambda_\sigma.$$

(2.2) An alternative useful characterization of $\mathbb{B}_\sigma$ may be given as follows. Let the commutator subgroup $[\hat{F}_2,\hat{F}_2]$ be denoted $\hat{F}_2'$, and let the double commutator subgroup $[\hat{F}_2',\hat{F}_2']$ be denoted $\hat{F}_2''$. Then, the quotient $\hat{F}_2'/\hat{F}_2''$ is acted on by $\hat{\mathbb{Z}}[[\hat{\mathbb{Z}}^2(1)]]$ via (linear extension of) conjugation, and Ihara showed that $\hat{F}_2'/\hat{F}_2''$ is a free cyclic $\hat{\mathbb{Z}}[[\hat{\mathbb{Z}}^2(1)]]$-module generated by $[x,y]=xyx^{-1}y^{-1}$. We shall write this conjugate action by $*$. Then, there exists a unique $\mathbb{A}_\sigma(\bar{x},\bar{y})\in\hat{\mathbb{Z}}[[\hat{\mathbb{Z}}^2(1)]]$ such that

(FA) $$f_\sigma(x,y) \equiv \mathbb{A}_\sigma(\bar{x},\bar{y}) * [x,y] \bmod \hat{F}_2''.$$

Then, the theory of Blanchfield-Lyndon sequence relates this quantity with the adelic beta function as follows:

(BA) $$\mathbb{B}_\sigma(\bar{x},\bar{y}) = 1+\mathbb{A}_\sigma(\bar{x},\bar{y})\cdot(\bar{x}-1)(\bar{y}-1).$$

One remarkable use of $\mathbb{A}_\sigma$ is that if we define $\rho_N:\widehat{GT}\to\hat{\mathbb{Z}}$ by

$$\rho_N(\sigma) := \frac{1}{N}\sum_{c=0}^{N-1}(\zeta_N^c-1)\mathbb{A}_\sigma(\zeta_N^c,1),$$

then, restricted on $G_\mathbb{Q}\subset\widehat{GT}$, ρ_N gives the Kummer 1-cocycle on positive roots of N, i.e., $\sigma(\sqrt[k]{N}) = \zeta_N^{\rho_N(\sigma)}\sqrt[k]{N}$ for all $k=1,2,\ldots$ and $\sigma\in G_\mathbb{Q}$. From this, one can easily see that $\mathbb{A}_\sigma(-1,1) = -\rho_2(\sigma)$ (see Appendix (A4); *cf.* also [NS] Proposition 5.3).

(2.3) Let $\mathbb{W}_p$ be the ring of restricted Witt vectors of $\bar{\mathbb{F}}_p$, and let $\mathbb{W}=\prod_p\mathbb{W}_p$ be the direct product for all primes p which is topologized by the collection $\{n\mathbb{W}\}_{n\in\mathbb{N}}$ as the fundamental system of neighborhoods of 0. One can consider the enlarged ring $\mathbb{W}[[\hat{\mathbb{Z}}^2(1)]]$ naturally containing $\hat{\mathbb{Z}}[[\hat{\mathbb{Z}}^2(1)]]$. Ihara [I2] extends Anderson's hyperadelic Gamma function $\mathit{\Gamma}\!\!\Gamma_\sigma$ for all $\sigma\in\widehat{GT}$. In particular, he deduces that the 5-cyclic relation of $\widehat{GT}$ allows the "Γ-decomposition" of $\mathbb{B}_\sigma$:

(BΓ) $$\forall\sigma\in\widehat{GT},\ \exists\,\mathit{\Gamma}\!\!\Gamma_\sigma(\bar{x})\in\mathbb{W}[[\hat{\mathbb{Z}}]]\ \text{s.t.}\ \mathbb{B}_\sigma = \frac{\mathit{\Gamma}\!\!\Gamma_\sigma(\bar{x})\,\mathit{\Gamma}\!\!\Gamma_\sigma(\bar{y})}{\mathit{\Gamma}\!\!\Gamma_\sigma(\bar{x}\bar{y})}.$$

G.Anderson [A] proved that, for $\sigma\in G_\mathbb{Q}$, the hyperadelic Gamma function $\mathit{\Gamma}\!\!\Gamma_\sigma$ satisfies an analog of the Gauss n-multiplication formula:

(Γ_n) $$\prod_{nc=0}\frac{\mathit{\Gamma}\!\!\Gamma_\sigma(\zeta_n^c\bar{x})}{\mathit{\Gamma}\!\!\Gamma_\sigma(\zeta_n^c)}\cdot\frac{1}{\mathit{\Gamma}\!\!\Gamma_\sigma(\bar{x}^n)} = \bar{x}^{\,n\rho_n(\sigma)}.$$

It is unknown whether (Γ_n) holds on the total $\widehat{GT}$. Recently, Furusho [F] showed that the property (Γ_2) holds on a subgroup of $\widehat{GT}$ satisfying certain geometric relations (IV) and (K_2). The following lemma indicates that the two main parameters λ and f of $\widehat{GT}$ are "not totally independent."

(2.4) LEMMA. *If $\sigma \in \widehat{GT}$ satisifies (Γ_2) (e.g., $\sigma \in G_{\mathbb{Q}}$), then,*

$$(-1)^{\rho_2(\sigma)} = \frac{\sqrt{-1}^{\frac{\lambda_\sigma - 1}{2}}(1-\sqrt{-1}^{\lambda_\sigma})}{1-\sqrt{-1}}.$$

PROOF. For $\sigma \in G_{\mathbb{Q}}$, one can separately check that each side coincides with 1 when $\lambda_\sigma \equiv \pm 1 \mod 8$, and with -1 when $\lambda_\sigma \equiv \pm 5 \mod 8$ (Notice that $(-1)^{\rho_2(\sigma)}$ depends only on $\lambda \mod 8$ since $\sqrt{2} \in \mathbb{Q}(\zeta_8)$.) For general $\sigma \in \widehat{GT}$, the property (Γ_2) reads: $\frac{\mathit{\Gamma}_\sigma(\bar{x})\mathit{\Gamma}_\sigma(-\bar{x})}{\mathit{\Gamma}_\sigma(1)\mathit{\Gamma}_\sigma(-1)} \cdot \frac{1}{\mathit{\Gamma}_\sigma(\bar{x}^2)} = \bar{x}^{2\rho_2(\sigma)}$. Putting $\bar{x} = \sqrt{-1}$, we then obtain

$$(-1)^{\rho_2(\sigma)} = (-1)^{-\rho_2(\sigma)} = \frac{\mathbb{B}_\sigma(-1,-1)}{\mathbb{B}_\sigma(\sqrt{-1}, -\sqrt{-1})}.$$

The lemma follows from this and (B3). □

(2.5) We finally complement a few observations on the specialization $\mathbb{B}_\sigma((\begin{smallmatrix} 0 & 1 \\ -1 & 0 \end{smallmatrix}), (\begin{smallmatrix} 0 & 1 \\ -1 & 0 \end{smallmatrix}))$ in the 2 by 2 matrix algebra $M_2(\hat{\mathbb{Z}})$. Let R be the commutative subring consisting of matrices of the form $(\begin{smallmatrix} a & -b \\ b & a \end{smallmatrix})$ $(a, b \in \hat{\mathbb{Z}})$. The issued $\mathbb{B}_\sigma((\begin{smallmatrix} 0 & 1 \\ -1 & 0 \end{smallmatrix}), (\begin{smallmatrix} 0 & 1 \\ -1 & 0 \end{smallmatrix}))$ is defined as the specialization image of $\mathbb{B}_\sigma$ under the map $\hat{\mathbb{Z}}[\![\hat{\mathbb{Z}}^2(1)]\!] \to R$ $(x, y \mapsto (\begin{smallmatrix} 0 & 1 \\ -1 & 0 \end{smallmatrix}))$. Then, for all $\sigma \in \widehat{GT}$,

$$\text{(2.5.1)} \qquad {}^t\mathbb{B}_\sigma((\begin{smallmatrix} 0 & 1 \\ -1 & 0 \end{smallmatrix}), (\begin{smallmatrix} 0 & 1 \\ -1 & 0 \end{smallmatrix})) \cdot \mathbb{B}_\sigma((\begin{smallmatrix} 0 & 1 \\ -1 & 0 \end{smallmatrix}), (\begin{smallmatrix} 0 & 1 \\ -1 & 0 \end{smallmatrix})) = (-1)^{\frac{\lambda-1}{2}} \begin{pmatrix} \lambda & 0 \\ 0 & \lambda \end{pmatrix}$$

PROOF. Since the matrix transposition gives an algebra automorphism on R, the result amounts to the formula $\mathbb{B}_\sigma(\sqrt{-1}^{-1}, \sqrt{-1}^{-1})\mathbb{B}_\sigma(\sqrt{-1}, \sqrt{-1}) = (-1)^{\frac{\lambda-1}{2}}\lambda$. This is a simple result of (B2) (and $(\mathrm{B}\Gamma)$). □

Since the transposed inverse of $B = (\begin{smallmatrix} a & -b \\ b & a \end{smallmatrix})$ $(a, b \in \hat{\mathbb{Z}})$ is $\det(B)^{-1}B$, from (2.5.1) it is immediate to see

$$\text{(2.5.2)} \qquad \det \mathbb{B}_\sigma((\begin{smallmatrix} 0 & 1 \\ -1 & 0 \end{smallmatrix}), (\begin{smallmatrix} 0 & 1 \\ -1 & 0 \end{smallmatrix})) = (-1)^{\frac{\lambda-1}{2}}\lambda.$$

§3. Lemniscate curve.

(3.1) In this section, we shall consider the lemniscate elliptic curve

$$E^{\mathrm{lem}} : Y^2 = X^3 - X$$

as a cyclic branched cover over the projective t-line $\mathbf{P}^1_t$ ramified only over $t = 0, 1, \infty$. The fundamental group of $E^{\mathrm{lem}} - \{O\}$ is then recognized as a suitable subquotient of $\pi_1(\mathbf{P}^1_t - \{0, 1, \infty\}, \overrightarrow{01})$. We recall that the standard action of $\sigma = (\lambda, f) \in \widehat{GT}$ (where $\lambda \in \hat{\mathbb{Z}}^\times$, $f \in [\hat{F}_2, \hat{F}_2]$) on the latter group is given as follows:

$$\text{(3.1.1)} \qquad \begin{cases} x & \mapsto x^\lambda, \\ y & \mapsto f(y,x)y^\lambda f(x,y), \\ z & \mapsto x^{\frac{\lambda-1}{2}} f(z,x) z^\lambda f(x,z) x^{\frac{1-\lambda}{2}}, \end{cases}$$

where x, y, z are loops of $\pi_1(\mathbf{P}^1_t - \{0,1,\infty\}, \overrightarrow{01})$ with $xyz = 1$ chosen as in [NT] §1. The purpose of this section is to give explicit descriptions of the Galois representations in the Tate module of E^{lem} in terms of the adelic beta functions.

(3.2) The lemniscate elliptic curve $E^{\mathrm{lem}} : Y^2 = X^3 - X$ can be realized as a cyclic ramified cover of $\mathbf{P}^1_t$ of degree 4 by

$$X = \frac{1}{\sqrt{t}} \quad \text{and} \quad Y = \sqrt{\frac{1-t}{\sqrt{t}^3}}.$$

The geometric fundamental group of $E^{\mathrm{lem}} - \{O\}$ is naturally regarded as a subgroup of $\pi_1(\mathbf{P}^1_t - \{0,1,\infty\}; e_1|2, e_\infty|4) = \widehat{\Delta}(\infty, 2, 4)$ with free generators $x_1 := x^{-1}z$, $x_2 := xz^{-1}$. The commutator $[x_1, x_2] = x^{-4}$ generates an inertia subgroup over the point $O \in E^{\mathrm{lem}}$. Therefore, if the extra condition '$e_0|4$' is imposed on the above fundamental group of $\mathbf{P}^1_t - \{0,1,\infty\}$, then the finite-adelic Tate module $T_f(E^{\mathrm{lem}})$ appears as the corresponding subgroup fitting in the commutative diagram of exact sequences:

$$\begin{array}{ccccccccc} 1 & \longrightarrow & \pi_1(E^{\mathrm{lem}}_{\overline{\mathbb{Q}}} - \{O\}) & \longrightarrow & \widehat{\Delta}(\infty,2,4) & \longrightarrow & \mathbb{Z}/4\mathbb{Z} & \longrightarrow & 1 \\ & & \downarrow & & \downarrow & & \| & & \\ 1 & \longrightarrow & T_f(E^{\mathrm{lem}}) & \longrightarrow & \widehat{\Delta}(4,2,4) & \longrightarrow & \mathbb{Z}/4\mathbb{Z} & \longrightarrow & 1. \end{array}$$

Note here that $\Delta(4,2,4)$ denotes the triangle group of Euclidean type corresponding to a plane tiling by isosceles right triangles which refines the regular square tessellation. Since the quotient $\pi_1(\mathbf{P}^1_t - \{0,1,\infty\}; e_0|4, e_1|2, e_\infty|4)$ is preserved by the operation of $\widehat{GT}$, the $G_{\mathbb{Q}}$-action on the Tate module $T_f(E^{\mathrm{lem}})$ is naturally extended to the $\widehat{GT}$-action on it.

(3.3) Proposition. *Let $\bar{x}_1, \bar{x}_2$ be the basis of the Tate module $T_f(E^{\mathrm{lem}})$ which are the images of x_1, x_2 respectively, and define the action of the ring $\hat{\mathbb{Z}}[\sqrt{-1}]$ on $T_f(E^{\mathrm{lem}})$ by $\sqrt{-1} : \bar{x}_1 \mapsto -\bar{x}_2$, $\bar{x}_2 \mapsto \bar{x}_1$. Then, each element $\sigma = (\lambda, f) \in \widehat{GT}$ acts on $T_f(E^{\mathrm{lem}})$ by*

$$\begin{cases} \bar{x}_1 & \mapsto (-1)^{\frac{\lambda-1}{2} + \rho_2(\sigma)} \mathbb{B}_\sigma(\sqrt{-1}, \sqrt{-1}) \cdot \bar{x}_1, \\ \bar{x}_2 & \mapsto (-1)^{\rho_2(\sigma)} \mathbb{B}_\sigma(\sqrt{-1}, \sqrt{-1}) \cdot \bar{x}_2. \end{cases}$$

Proof. First, we compute the action of $\sigma = (\lambda, f)$ on $x_2 = xz^{-1}$. The formula in (3.1) implies

$$\begin{aligned} \sigma(x_2) &= x^\lambda x^{\frac{\lambda-1}{2}} f(z,x) z^{-\lambda} f(x,z) x^{\frac{1-\lambda}{2}} \\ &= \mathrm{Int}(x)^{\frac{\lambda-1}{2}} \left(\mathrm{Int}(x)^\lambda (f(x,z)^{-1}) \cdot x^\lambda z^{-\lambda} f(x,z) \right). \end{aligned}$$

We shall use the notation $\equiv$ to designate the congruence modulo the kernel of the projection of $\widehat{\Delta}(\infty,2,4)$ onto $\widehat{\Delta}(4,2,4)$. Since $[x,z] \equiv x_1^{-1}x_2$ in $T_f(E^{\mathrm{lem}})$, the formula (FA) of §2 implies that

$$f(x,z) \equiv (1 - \sqrt{-1}) \mathbb{A}_\sigma(\sqrt{-1}, \sqrt{-1}) \cdot \bar{x}_2.$$

Here, note that the inner (conjugate) actions $\mathrm{Int}(x)$ and $\mathrm{Int}(z)$ on $T_f(E^{\mathrm{lem}})$ are given by multiplication by $\sqrt{-1}$. We also have the congruence formula:

$$x^\lambda z^{-\lambda} \equiv \frac{1-\sqrt{-1}^\lambda}{1-\sqrt{-1}} \cdot \bar{x}_2 \qquad (\lambda \in \hat{\mathbb{Z}})$$

which can be easily proved by induction on $\lambda \in \mathbb{Z}_{>0}$ and by the standard continuity argument. Then, in total, it follows that

$$\sigma(x_2) \equiv \frac{\sqrt{-1}^{\frac{\lambda-1}{2}}(1-\sqrt{-1}^\lambda)}{1-\sqrt{-1}}\left(1+\mathbb{A}_\sigma(\sqrt{-1},\sqrt{-1})(1-\sqrt{-1})^2\right)\cdot \bar{x}_2.$$

Since the value $\sqrt{-1}^{\frac{\lambda-1}{2}}(1-\sqrt{-1}^\lambda)/(1-\sqrt{-1})$ is equal to $(-1)^{\rho_2(\sigma)}$ (*cf.* Lemma (2.4)), we finally obtain

$$\sigma(x_2) \equiv (-1)^{\rho_2(\sigma)}\mathbb{B}_\sigma(\sqrt{-1},\sqrt{-1})\cdot \bar{x}_2.$$

To compute $\sigma(\bar{x}_1)$, we apply the above formula to $\bar{x}_1 \equiv \mathrm{Int}(x)(x_2)$. Then,

$$\sigma(x_1) \equiv \mathrm{Int}(x^\lambda)\big((-1)^{\rho_2(\sigma)}\mathbb{B}_\sigma(\sqrt{-1},\sqrt{-1})\cdot\bar{x}_2\big) = \sqrt{-1}^{\lambda-1}(-1)^{\rho_2(\sigma)}\mathbb{B}_\sigma(\sqrt{-1},\sqrt{-1})\cdot\bar{x}_1.$$

This completes the proof. □

(3.4) Before closing this section, we shall discuss briefly the case of the *mordell elliptic curve* $E^{\mathrm{mor}} : Y^2 = X^4 - X$. This curve can be realized as a cyclic ramified cover of $\mathbf{P}^1_t$ of degree 6 by $t = X^3$: The function field of E^{mor} is generated by $X = \sqrt[3]{t}$ and $Y = \sqrt[6]{t}\sqrt{t-1}$. The geometric fundamental group of $E^{\mathrm{mor}} - \{O\}$ is naturally regarded as a subgroup of $\pi_1(\mathbf{P}^1_t - \{0,1,\infty\}; e_1|2, e_\infty|3) = \widehat{\Delta}(\infty,2,3)$ freely generated by $x_1' := x^{-2}z$, $x_2' := x^2z^{-1}$. An inertia group over the origin of E^{mor} is generated by the commutator $[x_1', x_2'] = x^{-6}$. Therefore, if the extra condition '$e_0|6$' is imposed on the fundamental group of $\mathbf{P}^1_t - \{0,1,\infty\}$, then the finite-adelic Tate module $T_f(E^{\mathrm{mor}})$ appears as the corresponding subgroup. This fits in the following commutative diagram of exact sequences:

$$\begin{array}{ccccccccc}
1 & \longrightarrow & \pi_1(E^{\mathrm{mor}}_{\overline{\mathbb{Q}}} - \{O\}) & \longrightarrow & \widehat{\Delta}(\infty,2,3) & \longrightarrow & \mathbb{Z}/6\mathbb{Z} & \longrightarrow & 1 \\
 & & \downarrow & & \downarrow & & \| & & \\
1 & \longrightarrow & T_f(E^{\mathrm{mor}}) & \longrightarrow & \widehat{\Delta}(6,2,3) & \longrightarrow & \mathbb{Z}/6\mathbb{Z} & \longrightarrow & 1.
\end{array}$$

The triangle group $\Delta(6,2,3)$ is of Euclidean type corresponding to a plane tiling by 30-60-90 triangles which refines the regular triangle tessellation. Since the quotient $\pi_1(\mathbf{P}^1_t - \{0,1,\infty\}; e_0|6, e_1|2, e_\infty|3)$ is preserved by the operation of $\widehat{GT}$, the $G_{\mathbb{Q}}$-action on the Tate module $T_f(E^{\mathrm{mor}})$ is naturally extended to the $\widehat{GT}$-action on it. By the similar argument to Proposition (3.3), we can show

(3.5) PROPOSITION. *Let $\bar{x}_1', \bar{x}_2'$ be the basis of the Tate module $T_f(E^{\mathrm{mor}})$ which are the images of x_1', x_2' respectively, and define the action of the ring $\hat{\mathbb{Z}}[\zeta_6]$ on $T_f(E^{\mathrm{mor}})$ by $\zeta_6 : \bar{x}_1' \mapsto (\bar{x}_1' - \bar{x}_2')$, $\bar{x}_2' \mapsto \bar{x}_1'$. Then, each element $\sigma = (\lambda, f) \in \widehat{GT}$ acts on $T_f(E^{\mathrm{mor}})$ by*

$$\begin{cases} \bar{x}_1' & \mapsto \zeta_6^{\lambda-1}\mathbb{B}_\sigma(\zeta_6,\zeta_6^2)\cdot\bar{x}_1', \\ \bar{x}_2' & \mapsto \mathbb{B}_\sigma(\zeta_6,\zeta_6^2)\cdot\bar{x}_2'. \end{cases}$$

□

We leave the proof as reader's exercise. This result may be regarded as the Galois correspondent of the classical integral:

$$\int_0^\infty \frac{dt}{\sqrt{t^3+1}} = \frac{1}{3}B(\frac{1}{6},\frac{1}{3}).$$

§4. Family of quartics and tangential base points.

(4.1) The purpose of this section is to give suitable embedding of $\mathbf{P}^1_t - \{0,1,\infty\}$ to the space of quartics which enables us to consider the fundamental group of $E^{\mathrm{lem}} \setminus \{O\}$ in the framework of braid groups. Let $\mathbf{A}^4_u$ be the affine 4-space (over $\mathbb{Q}$) where each point $u = (u_1,u_2,u_3,u_4)$ is regarded as a quartic $f_u(T) := T^4 + u_1T^3 + u_2T^2 + u_3T + u_4$ of one formal variable T. Let D be the discriminant locus on $\mathbf{A}^4_u$ corresponding to the quartics with duplicate roots. Over the affine variety $\mathbf{A}^4_u - D$, one has a Galois etale cover $\mathbf{A}^4_v - \Delta$ called *the space of solutions* whose points are the ordered tuples $v = (v_1,v_2,v_3,v_4)$ with distinct entries. (The divisor Δ denotes the hyper-diagonals of $\mathbf{A}^4_v$.) The morphism of $\mathbf{A}^4_v - \Delta$ to $\mathbf{A}^4_u - D$ is given by $v \mapsto f_u(T) := (T-v_1)(T-v_2)(T-v_3)(T-v_4)$. Over $\mathbf{A}^4_v - \Delta$, one has a standard tangential basepoint $\vec{b}$ given by $v = (0, t_1t_2t_3, t_2t_3, t_3)$ with values in the formal Puiseux power series ring $\overline{\mathbb{Q}}\{\{t_1,t_2,t_3\}\}$ (the maximal etale extension of $\overline{\mathbb{Q}}[\![t_1,t_2,t_3]\!][(t_1t_2t_3)^{-1}]$). This is so called the Ihara-Matsumoto tangential basepoint ([IM]). Writing the image of $\vec{b}$ on $\mathbf{A}^4_u - D$ by the same symbol $\vec{b}$, one can naturally identify $\pi_1(\mathbf{A}^4_u - D, \vec{b}) = G_{\mathbb{Q}} \ltimes \hat{B}_4$. The Galois action here can be extended to that of $\widehat{GT}$ in the form of the famous formula of Drinfeld ([Dr], see [IM]), i.e., each $\sigma = (\lambda, f) \in GT$ maps starndard generators as follows:

$$\text{(4.1.1)} \qquad \begin{cases} \tau_1 & \mapsto \tau_1^\lambda, \\ \tau_2 & \mapsto f(\tau_1^2,\tau_2^2)^{-1}\tau_2^\lambda f(\tau_1^2,\tau_2^2), \\ \tau_3 & \mapsto f(\eta^2,\tau_3^2)^{-1}\tau_3^\lambda f(\eta^2,\tau_3^2). \end{cases}$$

Here τ_1,τ_2,τ_3 are standard generators of the braid group B_4 ($\tau_1\tau_3 = \tau_3\tau_1$, $\tau_i\tau_{i+1}\tau_i = \tau_{i+1}\tau_i\tau_{i+1}$ $(i=1,2)$), and $\eta = \tau_1\tau_2\tau_1$.

(4.2) The space $\mathbf{A}^4_v - \Delta$ is an $\mathbf{A}^1$-torsor by parallel transformations $v \mapsto (v_1+a, v_2+a, v_3+a, v_4+a)$ $(a \in \mathbf{A}^1)$. This action does not affect the fundamental groupoid of $\mathbf{A}^4_v - \Delta$ as $\pi_1(\mathbf{A}^1_{\overline{\mathbb{Q}}}) = 1$. Two (tangential) basepoints b_1, b_2 on $\mathbf{A}^4_v - \Delta$ will be called equivalent (written $b_1 \sim b_2$) if they are transformed to each other by this $\mathbf{A}^1$-action. The multiplicative group $\mathbf{G}_m$ acts on $\mathbf{A}^4_v - \Delta$ by simultaneous multiplication on entries. The induced $\mathbf{G}_m$-action on $\mathbf{A}^4_u - D$ is given by $f_u(T) \mapsto a^4 f_u(a^{-1}T)$ $(a \in \mathbf{G}_m)$. We shall write those quotient spaces as

$$\mathcal{A}^4_v := (\mathbf{A}^4_v - \Delta)/\mathbf{G}_m, \quad \mathcal{A}^4_u := (\mathbf{A}^4_u - \Delta)/\mathbf{G}_m.$$

Their geometric fundamental groups are given by moding out by the center $\langle \omega_4 \rangle$ $(\omega_4 = (\tau_1\tau_2\tau_3)^4)$, i.e., $\pi_1(\mathcal{A}^4_u \otimes \overline{\mathbb{Q}})$ is isomorphic to $\hat{B}^*_4 := \hat{B}_4/\langle\omega_4\rangle$, and $\pi_1(\mathcal{A}^4_v \otimes \overline{\mathbb{Q}})$ is its pure part $\hat{P}^*_4$ which is by definition the kernel of the "string-permutation" homomorphism $\hat{B}^*_4 \to S_4$. Remarkable observation here is that we can find braids in $\hat{B}^*_4$ which behaves like the generators of the triangle group $\Delta(\infty,2,4)$: One can take

$$\text{(4.2.1)} \qquad \eta := \tau_1\tau_2\tau_1, \quad \eta_2 := \tau_1^{-1}\tau_2^{-1}\tau_3^{-1}\tau_1^{-1}\tau_2^{-1}\tau_1^{-1}, \quad \eta_4 := \tau_1\tau_2\tau_3$$

so that $\eta\eta_2\eta_4 = 1$, ${\eta_2}^2(=\omega_4^{-1}) = 1$, ${\eta_4}^4(=\omega_4) = 1$ in $\hat{B}_4^*$. Therefore, one has a homomorphism of $\widehat{\Delta}(\infty, 2, 4)$ into $\hat{B}_4^*$ with $x \mapsto \eta$, $y \mapsto \eta_2$, $z \mapsto \eta_4$. Under this mapping, the generators x_1, x_2 of the $\pi_1(E^* \setminus \{O\}) \subset \widehat{\Delta}(\infty, 2, 4)$ will be mapped into the braids ξ_1, ξ_2 of $\hat{B}_4^*$:

$$\text{(4.2.2)} \qquad \begin{cases} x_1 = x^{-1}z & \mapsto \xi_1 := \tau_3\tau_1^{-1}, \\ x_2 = xz^{-1} & \mapsto \xi_2 := \tau_2\tau_1\tau_3^{-1}\tau_2^{-1}\tau_3\tau_1^{-1} \end{cases}$$

so that $[x_1, x_2] \mapsto [\xi_1, \xi_2] = \omega_4\eta^{-4} = \eta^{-4}$. It is known that these two braids generate a free profinite subgroup in $\hat{B}_4^*$ (*cf.* §5 (5.2) below). From this one easily sees the injectivity of $\widehat{\Delta}(\infty, 2, 4) \hookrightarrow \hat{B}_4^*$.

(4.3) Now, let us introduce the embedding $f^{\text{lem}} : \mathbf{P}_t^1 - \{0, 1, \infty\} \to \mathbf{A}^4 \setminus D$ by giving the corresponding quartics as follows:

$$f_t^{\text{lem}}(T)(= f^{\text{lem}}(t)(T)) := T^4 - 2T^3 + \frac{3}{2} \cdot \frac{t}{t-1}T^2 - \frac{1}{2} \cdot \frac{t}{t-1}T + \frac{1}{16} \cdot \frac{t^2}{(t-1)^2}$$

The discriminant of $f_t^{\text{lem}}(T)$ is $\frac{4t^3}{(t-1)^6}$ so that the morphism f^{lem} is well defined. The injectivity is obvious from the coefficients in T. The tasks we are now going to work on are:

(a) to find and fix a suitable path from $\vec{b}$ to the image of $\overrightarrow{01}_t$ so that the image of loops in $\pi_1(\mathbf{P}_t^1 - \{0, 1, \infty\}, \overrightarrow{01}_t)$ can be uniquely identified with loops in $\mathcal{A}_u^4$;

(b) to check the coincidence of the image of x, y with η, η_2 respectively under the above identification of (a).

To approach (a), we shall solve the quartic equation $f_t^{\text{lem}}(T) = 0$ by the Cardano-Ferrari formula. The result is that the 4 zeros are given by

$$\text{(4.3.1)} \qquad T = \frac{1}{2} + \sqrt{\frac{1}{4(1-t)}} + \sqrt{\frac{1+\sqrt{t}}{4(1-t)}} + \sqrt{\frac{1-\sqrt{t}}{4(1-t)}}$$

where the (outer) three $\sqrt{*}$'s are taken so that the product comes to be $1/8(1-t)$. Near t being small > 0, after declaring that all $\sqrt{*}$ mean positive roots, one can specify 4 solutions as :

$$\text{(4.3.2)} \qquad T_1 \approx -\frac{1}{2}\sqrt{t}, \quad T_2 \approx 0, \quad T_3 \approx \frac{1}{2}\sqrt{t}, \quad T_4 \approx 2.$$

The corresponding lift of the tangential basepoint $f^{\text{lem}}(\overrightarrow{01}_t)$ on $\mathbf{A}_v^4 - \Delta$ is then estimated as :

$$(0, t_1t_2t_3, t_2t_3, t_3) = (-\frac{1}{2}\sqrt{t}, 0, \frac{1}{2}\sqrt{t}, 2) \sim (0, \frac{1}{2}\sqrt{t}, \sqrt{t}, 2 + \frac{1}{2}\sqrt{t}),$$

where $\sim$ means equivalence by $\mathbf{A}^1$-action ($cf.$(4.2)); namely $t_1 = 1/2$, $t_2 = \frac{\sqrt{t}}{2} = \sqrt{\frac{t}{4}}$ and $t_3 \risingdotseq 2$. Moding out $t_3 \in \mathbf{G}_m$, we can define a path r on $\mathcal{A}_v^4$ from (the image of) $\vec{b}$ to (that of) $f^{\text{lem}}(\frac{1}{4}\overrightarrow{01}_t)$ by letting t_1 move from $\overrightarrow{01}_{t_1}$ to $\frac{1}{2}$. Let ε denote the straight path from $\frac{1}{4}\overrightarrow{01}_t$ to $\overrightarrow{01}_t$ on $\mathbf{P}^1_t - \{0,1,\infty\}$.

$$\begin{array}{ccc} & \tfrac{1}{4}\overrightarrow{01}_t & \xrightarrow{\ \varepsilon\ } \overrightarrow{01}_t (\rightleftharpoons x, y) \\ & \Big\downarrow f^{\mathrm{lem}} & \\ \vec{b} \xrightarrow{\ r\ } & f^{\mathrm{lem}}(\tfrac{1}{4}\overrightarrow{01}_t) & \end{array}$$

By using (the images on $\mathcal{A}^4_u$ of) these r and ε, the loops x, y of $\pi_1(\mathbf{P}^1_t - \{0,1,\infty\}, \overrightarrow{01}_t)$ are mapped to the loops $rf^{\mathrm{lem}}(\varepsilon x \varepsilon^{-1})r^{-1}$, $rf^{\mathrm{lem}}(\varepsilon y \varepsilon^{-1})r^{-1}$ of $\pi_1(\mathcal{A}^4_u, \vec{b})$. For (b), it suffices to show that these correspond to the braids ξ_1 and ξ_2 of (4.2.2) respectively. The first one is not difficult: the image $rf^{\mathrm{lem}}(\varepsilon x \varepsilon^{-1})r^{-1}$ of x can be seen from (4.3.2) where t moves on a small circle around 0 in the anticlockwise way. The result is that T_1 and T_3 go around T_2 to be interchanged to each other with leaving T_2 (inside) and T_4 (outside) invariant. This is nothing but the braid $\eta = \tau_1\tau_2\tau_1$. When t is close to 1, the four roots T_1,...,T_4 behave approximately as follows:

$$T_1 \approx 1 - \sqrt{\frac{1}{1-t}}, \quad T_2 \approx \frac{1-\sqrt{2}}{2}\sqrt{\frac{1}{1-t}}, \quad T_3 \approx \frac{\sqrt{2}-1}{2}\sqrt{\frac{1}{1-t}}, \quad T_4 \approx 1 + \sqrt{\frac{1}{1-t}}.$$

From this one sees that, when t goes on a small circle around 1 in the anticlockwise way, the points T_1 and T_4 (resp. T_2 and T_3) move on a large circle clockwise with angle π to be switched to one another. This identifies $rf^{\mathrm{lem}}(\varepsilon y \varepsilon^{-1})r^{-1}$ with the braid $\eta_2 = \tau_1^{-1}\tau_2^{-1}\tau_3^{-1}\tau_1^{-1}\tau_2^{-1}\tau_1^{-1}$.

We summarize the above discussion as

(4.4) PROPOSITION. *Let r and ε be paths introduced as above. Then, each $\sigma \in G_{\mathbb{Q}}$ acts on them by*

$$\begin{cases} \sigma(r) &= g_\sigma(\tau_1^2, \tau_2^2)^{-1} \cdot r, \\ \sigma(\varepsilon) &= \varepsilon \cdot x^{-2\rho_2(\sigma)}. \end{cases}$$

The induced homomorphism $(*) \mapsto rf^{\mathrm{lem}}(\varepsilon(*)\varepsilon^{-1})r^{-1}$ of $\pi_1(\mathbf{P}^1_t - \{0,1,\infty\}, \overrightarrow{01}_t)$ into $\pi_1(\mathcal{A}^4_u)$ is injective, and maps $x \mapsto \eta$, $y \mapsto \eta_2$ and $z \mapsto \eta_4$. □

§5. Galois actions and proof of Theorem D.

(5.1) PROPOSITION. *Under the standard action of $G_{\mathbb{Q}}(\subset \widehat{GT})$ on $\hat{B}_4$ (4.1.1), the elements $\eta := \tau_1\tau_2\tau_1$, $\eta_2 := \tau_1^{-1}\tau_2^{-1}\tau_3^{-1}\tau_1^{-1}\tau_2^{-1}\tau_1^{-1}$, $\eta_4 := \tau_1\tau_2\tau_3$, $\xi_1 := \tau_3\tau_1^{-1}$ and $\xi_2 := \tau_2\tau_1\tau_3^{-1}\tau_2^{-1}\tau_3\tau_1^{-1}$ are explicitly transformed as follows:*

(1) $\qquad \eta \mapsto g(\tau_1^2,\tau_2^2)^{-1}\eta^\lambda g(\tau_1^2,\tau_2^2), = f(\tau_2^2,\tau_1^2)\eta^\lambda$

(2) $\qquad \eta_2 \mapsto g(\tau_1^2,\tau_2^2)^{-1}\eta^{-2\rho_2} f(\eta,\eta_2)\eta_2^\lambda f(\eta_2,\eta)\eta^{2\rho_2} g(\tau_1^2,\tau_2^2),$

(3) $\qquad \eta_4 \mapsto g(\tau_1^2,\tau_2^2)^{-1}\eta^{\frac{\lambda-1}{2}-2\rho_2} f(\eta_4,\eta)\eta_4^\lambda f(\eta,\eta_4)\eta^{-\frac{\lambda-1}{2}+2\rho_2} g(\tau_1^2,\tau_2^2);$

(4) $\qquad \xi_1 \mapsto \eta^{-8\rho_2} f(\xi_1, z_0)\xi_1^\lambda f(z_0,\xi_1)\eta^{8\rho_2},$

(5) $\qquad \xi_2 \mapsto f(\tau_2^2,\tau_1^2)\eta^{\lambda-1-8\rho_2} f(\xi_2^{-1}, z_0)\xi_2^\lambda f(z_0,\xi_2^{-1})\eta^{1-\lambda+8\rho_2} f(\tau_1^2,\tau_2^2).$

Here $z_0 = \eta^4\omega_4^{-1}$ is the braid satisfying $[\xi_1,\xi_2]z_0 = 1$.

PROOF. The result (1) dates back to [N-I] Proposition 4.12. The validity of the first three formulae (1)~(3) in $\hat{B}_4^*$ are consequences of Proposition (4.4) and the $G_{\mathbb{Q}}$-action on x, y, z described in (3.1). To lift them to $\hat{B}_4$, note that the kernel of $\hat{B}_4 \to \hat{B}_4^*$ is the center $\langle\omega_4\rangle$ of $\hat{B}_4$ and that the $G_{\mathbb{Q}}$-actions on η^4, $\eta_2^2 = \omega_4^{-1}$ and $\eta_4^4 = \omega_4$ are given by their λ powers. The abelianization map $\hat{B}_4 \to \hat{\mathbb{Z}}$ assures the coincidence of both sides of the issued lifts. (4) follows from the standard $G_{\mathbb{Q}}$-action on $\xi_1 = \tau_3\tau_1^{-1} \in \hat{B}_3$ according to Drinfeld's formula (4.1.1) together with the relation (IV) first proved in [N-I] Theorem 4.16:

$$\text{(IV)} \qquad f(\tau_3^2, \eta^2) = \eta^{-8\rho_2} f(\xi_1, \eta^4\omega_4^{-1})\tau_1^{4\rho_2}\tau_3^{-4\rho_2} = \eta^{-8\rho_2} f(\xi_1, \eta^4\omega_4^{-1})\tau_1^{4\rho_2}\xi_1^{-4\rho_2}$$

Note that $[\xi_1, \xi_2](\eta^4\omega_4^{-1}) = 1$. Applying (1),(4) to $\xi_2 = \eta\xi_1^{-1}\eta^{-1}$, we obtain (5). □

(5.2) Let $M_{1,2}$ be the moduli stack of smooth genus 1 curves with 2 ordered marked points and let $f_2 : M_{1,2} \to M_{1,1}$ be the morphism to the moduli stack of elliptic curves which forgets the second marked points. Each quartic $f_u(T)$ of $\mathbf{A}^4 \setminus D$ gives the elliptic curve $Y^2 = f_u(T)$ with two ordered points at infinity. This gives rise to an isomorphism of fundamental groups $\pi_1(\mathcal{A}_u^4) \xrightarrow{\sim} \pi_1(M_{1,2})$ (*cf.* [N-I] p.353, [N-II] §7.8). The homotopy exact sequence of this fibration is of the form:

$$\begin{array}{ccccccccc} & & & & \pi_1(\mathcal{A}_u^4) & & & & \\ & & & & \downarrow{\scriptstyle isom.} & & & & \\ 1 & \longrightarrow & \hat{F}_2 & \longrightarrow & \pi_1(M_{1,2}) & \xrightarrow{f_2} & \pi_1(M_{1,1}) & \longrightarrow & 1. \end{array}$$

The kernel part $\hat{F}_2$ corresponds to the geometric fundamental group of a fibre elliptic curve minus origin. In [N-I] §4.9, we saw that ξ_1, ξ_2 give a generator system of this $\hat{F}_2$. Thus we see: *ξ_1, ξ_2 generates a free profinite subgroup in $\pi_1(\mathcal{A}_u^4)$, hence also in $\hat{B}_4$.*

(5.3) REMARK. The above forgetful projection f_2 can be lifted to $\pi_1(\mathbf{A}^4 \setminus D_4) \to \pi_1(\mathbf{A}^3 \setminus D_3)$ coming from a certain morphism $\mathcal{F} : \mathbf{A}^4 \setminus D_4 \to \mathbf{A}^3 \setminus D_3$. This morphism $\mathcal{F}$, called the *Ferrari morphism* (cf. [N01]) associates to quartic polynomials their resolvent cubics, and will play basic roles in the theory of Weierstrass tangential base points. (See the forthcoming paper partly based on [N01]).

(5.4) In [N-I], we studied various actions on the free subgroup $\hat{F}_2 = \langle\xi_1, \xi_2\rangle$ of $\pi_1(M_{1,2}) \cong \hat{B}_4^*$ which embodies the fundamental group of punctured elliptic curves. Geometric monodromy on them can be summarized as the inner actions by τ_1, τ_2. This action was determined in the following simple way:

$$\text{(5.4.1)} \qquad \mathrm{Int}(\tau_1) : \begin{cases} \xi_1 \mapsto \xi_1, \\ \xi_2 \mapsto \xi_2\xi_1, \end{cases} \qquad \mathrm{Int}(\tau_2) : \begin{cases} \xi_1 \mapsto \xi_2\xi_1^{-1}, \\ \xi_2 \mapsto \xi_2. \end{cases}$$

Once given an elliptic curve with Weierstrass equation $Y^2 = 4X^3 - g_2X - g_3$, then we may associate in a standard way the Weierstrass tangential basepoint on it and the Galois action on $\hat{F}_2$ (*cf.* also [N98]). One important example is the case of the Tate elliptic curve

$$\mathrm{Tate}(q) : Y^2 = 4X^3 - g_2(q)X - g_3(q)$$

over the formal Laurant power series field $\mathbb{Q}((q))$, where $g_2(q)$, $g_3(q)$ are given as Eisenstein q-series of weight 4, 6 (constant terms are $1/12$, $-1/216$) respectively. From the corresponding Weierstrass tangential basepoint, the Galois group $G_{\mathbb{Q}}$ acts by

$$(5.4.2)\qquad \begin{cases} \xi_1 \mapsto f(\xi_1, z_0)\xi_1^{\lambda} f(z_0, \xi_1), \\ \xi_2 \mapsto f(\xi_1, z_0) x_1^{\frac{1-\lambda}{2}} f(\xi_2\xi_1^{-1}\xi_2^{-1}, \xi_1)\xi_2\xi_1^{\frac{\lambda-1}{2}} f(z_0, \xi_1), \\ z_0 \mapsto z_0^{\lambda}, \end{cases}$$

where $[\xi_1, \xi_2]z_0 = 1$. The difference between this action and the Drinfeld-Ihara-Matsumoto action has been determined in [N-I] §4. (In fact, the conjugation by $\eta^{-8\rho_2}\tau_1^{8\rho_2}$ makes (5.4.2) coincide with Proposition (5.1) (4)-(5).) Using this circle of ideas, we showed in [N-I] Corollary 4.13:

$$(5.4.3)\qquad f_\sigma\left(\begin{pmatrix} 1 & 2 \\ 0 & 1 \end{pmatrix}, \begin{pmatrix} 1 & 0 \\ -2 & 1 \end{pmatrix}\right) = (-1)^{\frac{\lambda_\sigma - 1}{2}} \begin{pmatrix} 1 & 0 \\ -8\rho_2(\sigma) & 1 \end{pmatrix} \begin{pmatrix} \lambda_\sigma^{-1} & 0 \\ 0 & \lambda_\sigma \end{pmatrix} \begin{pmatrix} 1 & -8\rho_2(\sigma) \\ 0 & 1 \end{pmatrix}.$$

Now, we are ready to prove the following

(5.5) THEOREM. *For $\sigma \in G_{\mathbb{Q}}$, we have the following equation:*

$$g_\sigma\left(\begin{pmatrix} 1 & 2 \\ 0 & 1 \end{pmatrix}, \begin{pmatrix} 1 & 0 \\ -2 & 1 \end{pmatrix}\right) = (-1)^{\frac{\lambda_\sigma - 1}{2}} \mathbb{B}_\sigma\left(\begin{pmatrix} 0 & 1 \\ -1 & 0 \end{pmatrix}, \begin{pmatrix} 0 & 1 \\ -1 & 0 \end{pmatrix}\right) \cdot \begin{pmatrix} \lambda_\sigma^{-1} & -8\rho_2(\sigma)\lambda_\sigma^{-1} \\ 0 & (-1)^{\frac{\lambda_\sigma - 1}{2}} \end{pmatrix}.$$

PROOF. Identify $T_f(E^{\mathrm{lem}})$ with $\hat{\mathbb{Z}}^2$ by $\bar{x}_1 \mapsto \binom{1}{0}$, $\bar{x}_2 \mapsto \binom{0}{1}$. By Proposition (3.3), the $G_{\mathbb{Q}}$-action is given as the right multiplication by the matrix

$$(\#)\qquad (-1)^{\rho_2(\sigma)} \mathbb{B}_\sigma\left(\begin{pmatrix} 0 & 1 \\ -1 & 0 \end{pmatrix}, \begin{pmatrix} 0 & 1 \\ -1 & 0 \end{pmatrix}\right) \cdot \begin{pmatrix} (-1)^{\frac{\lambda-1}{2}} & 0 \\ 0 & 1 \end{pmatrix}.$$

On the other hand, by Proposition (5.1) (4)-(5), the standard action from $\vec{b}$ on the abelianization of $\hat{F}_2 = \langle \xi_1, \xi_2 \rangle$ is given by

$$\left((-1)^{-4\rho_2}\begin{pmatrix} \lambda \\ 0 \end{pmatrix}, f_\sigma\left(\begin{pmatrix} 1 & 0 \\ -2 & 1 \end{pmatrix}, \begin{pmatrix} 1 & 2 \\ 0 & 1 \end{pmatrix}\right)(-1)^{\frac{\lambda-1}{2} - 4\rho_2}\begin{pmatrix} 0 \\ \lambda \end{pmatrix}\right) = \begin{pmatrix} \lambda & 8\rho_2(\sigma) \\ 0 & 1 \end{pmatrix},$$

Therefore, taking into consideration the factor coming from our path $rf^{\mathrm{lem}}(\varepsilon)$ from $\vec{b}$ to $f^{\mathrm{lem}}(\overrightarrow{01}_t)$, we see that the matrix (#) is equal to

$$(-1)^{\rho_2(\sigma)} g_\sigma\left(\begin{pmatrix} 1 & 2 \\ 0 & 1 \end{pmatrix}, \begin{pmatrix} 1 & 0 \\ -2 & 1 \end{pmatrix}\right)\begin{pmatrix} \lambda & 8\rho_2(\sigma) \\ 0 & 1 \end{pmatrix}.$$

This completes the proof. □

Appendix. A note on the cocycle $\Psi_n^{(0)}$.

In [NT] Remark (6.3), we mentioned about Ihara's 1-cocycle $\Psi_n^{(0)}$ ($n \in \mathbb{N}$) whose minus sign $-\Psi_n^{(0)}$ extends the Kummer 1-cocycle ρ_n on the positive roots of n. In fact, originated from a difference of path conventions between Ihara's papers [I1,I2,I3] and ours, the association manner $G_{\mathbb{Q}} \to \widehat{GT}$ delicately differs from each other so that the minus sign of "$-\Psi_n^{(0)}$" in that claim should be dropped if $\Psi_n^{(0)}$ is replaced by its italicized version "$\mathit{\Psi}_n^{(0)}$" introduced in our convention exactly in the same way as $\Psi_n^{(0)}$ in Ihara's paper [I1]. To normalize ambiguities buried in contexts of literatures, in this appendix we shall give an explicit exposition starting from the definition of our italicized $\Psi_n^{(0)}$ and then shall give a proof of the regularized statement:

(A1) PROPOSITION. *The 1-cocycle* $\Psi_n^{(0)} : \widehat{GT} \to \hat{\mathbb{Z}}(1)$ *extends the Kummer 1-cocycle* $\rho_n : G_{\mathbb{Q}} \to \hat{\mathbb{Z}}(1)$ *attached to the real positive roots of* n.

Let $\hat{F}_2$ be the free profinite group freely generated by the symbols x, y. For a positive integer n, let H_n be the kernel of the homomorphism $\hat{F}_2 \to \mu_n$ defined by $x \mapsto \zeta_n, y \mapsto 1$, which is a free profinite subgroup $\hat{F}_{n+1}$ freely generated by the $x^{-a}yx^a$ ($a \in \mathbb{Z}/n\mathbb{Z}$) and x^n. Since, for any $\sigma = (\lambda_\sigma, f_\sigma) \in \widehat{GT}$, f_σ belongs to $\hat{F}_2' \subset H_n$, there exists a unique element ($n+1$-term proword) $f_\sigma^{(n)} = f_\sigma^{(n)}(\xi, \eta_0, \eta_1, \dots, \eta_{n-1}) \in \hat{F}_{n+1}$ satisfying $f_\sigma^{(n)}(x^n, y, xyx^{-1}, \dots, x^{n-1}yx^{-(n-1)}) = f_\sigma(x, y)$. For $a \in \mathbb{Z}/n\mathbb{Z}$, consider the homomorphism $\psi_n^{(a)} : F_{n+1} \to \hat{\mathbb{Z}}$ given by $\eta_a \mapsto 1$ and $\xi, \eta_i \mapsto 0$ ($i \neq a$), and then:

(A2) DEFINITION. Define $\Psi_n^{(a)}(\sigma) := \psi_n^{(a)}(f_\sigma^{(n)})$ ($a \in \mathbb{Z}/n\mathbb{Z}$).

By the same argument as [I2] p.171, these $\Psi_n^{(a)}$ can also be characterized by the formula on $\widehat{GT}$:

$$\text{(A3)} \quad \left(\frac{\mathbb{B}_\sigma(\bar{x}, \bar{y}) - 1}{\bar{y} - 1}\right)_{\bar{y}=1} = (\bar{x} - 1)\mathbb{A}_\sigma(\bar{x}, 1) \equiv \sum_{a \in \mathbb{Z}/n\mathbb{Z}} \Psi_n^{(a)}(\sigma)\bar{x}^{-a} \mod (\bar{x}^n - 1).$$

In particular,

$$\text{(A4)} \qquad \Psi_n^{(0)}(\sigma) := \frac{1}{n}\sum_{c=0}^{n-1}(\zeta_n^c - 1)\mathbb{A}_\sigma(\zeta_n^c, 1).$$

To prove Proposition (A1), we shall compute $\Psi_n^{(0)}(\sigma)$ for $\sigma \in G_{\mathbb{Q}}$. Let $X_w = \mathbf{P}^1 \smallsetminus \{0, \mu_n, \infty\} \to X_z = \mathbf{P}^1 \smallsetminus \{0, 1, \infty\}$ be the cyclic cover given by $z = w^n$, and consider the tangential base points $\overrightarrow{01}_z$, $\overrightarrow{10}_z$ on the z-line and $\overrightarrow{01}_w$, $\overrightarrow{10}_w$, $n\overrightarrow{10}_w$ on the w-line, whose value fields are $\overline{\mathbb{Q}}\{\{z\}\}$, $\overline{\mathbb{Q}}\{\{z_1\}\}$ ($z_1 = 1 - z$), $\overline{\mathbb{Q}}\{\{w\}\}$, $\overline{\mathbb{Q}}\{\{w_1\}\}$ ($w_1 = 1 - w$), $\overline{\mathbb{Q}}\{\{nw_1\}\}$ respectively. Let $L_{\overrightarrow{01}_z}$ (resp. $L_{\overrightarrow{10}_z}$) be the maximal algebraic extention of $\mathbb{Q}(z)$ in $\overline{\mathbb{Q}}\{\{z\}\}$ (resp. $\overline{\mathbb{Q}}\{\{z_1\}\}$) unramified outside $0, 1, \infty$ (viz. over X_z), and let $L_{\overrightarrow{01}_w}$ (resp. $L_{\overrightarrow{10}_w}$, $L_{n\overrightarrow{10}_w}$) be the maximal algebraic extention of $\mathbb{Q}(w)$ in $\overline{\mathbb{Q}}\{\{w\}\}$ (resp. $\overline{\mathbb{Q}}\{\{w_1\}\}$, $\overline{\mathbb{Q}}\{\{nw_1\}\}$) unramified outside $0, \mu_n, \infty$ (viz. over X_w). We shall connect the above tangential base points by the standard paths along the real axis:

$$\text{(A5)} \qquad \begin{cases} p_z \in \pi_1(X_z; \overrightarrow{01}_z, \overrightarrow{10}_z) \simeq \mathrm{Isom}_{\mathbb{Q}(z)}(L_{\overrightarrow{10}_z}, L_{\overrightarrow{01}_z}), \\ p_w \in \pi_1(X_w; \overrightarrow{01}_w, \overrightarrow{10}_w) \simeq \mathrm{Isom}_{\mathbb{Q}(w)}(L_{\overrightarrow{10}_w}, L_{\overrightarrow{01}_w}), \\ \delta_n \in \pi_1(X_w; n\overrightarrow{10}_w, \overrightarrow{10}_w) \simeq \mathrm{Isom}_{\mathbb{Q}(w)}(L_{\overrightarrow{10}_w}, L_{n\overrightarrow{10}_w}). \end{cases}$$

The covering map $\beta_n : X_w \to X_z$ and the path δ_n give an embedding $\pi_1(X_w; \overrightarrow{01}_w) \hookrightarrow \pi_1(X_z; \overrightarrow{01}_z)$ such that $\xi \mapsto x^n, \eta_i \mapsto x^i y x^{-i}$ ($i = 0, 1, \dots, n-1$). Meanwhile, one has a natural open immersion $\iota_n : X_w \to X_z$ ($w \mapsto z$) which induces the surjection of $\pi_1(X_w; \overrightarrow{01}_w)$ onto $\pi_1(X_z; \overrightarrow{01}_z)$ such that $\xi \mapsto x$, $\eta_0 \mapsto y$, $\eta_i \mapsto 1$ ($i = 1, \dots, n-1$).

Writing the action of $\sigma \in G_{\mathbb{Q}}$ at a tangential base point b as σ_b, one may describe Galois effects on the above paths as follows:

$$\text{(A6)} \quad \begin{cases} \sigma(p_z) & = \sigma_{\overrightarrow{01}_z} p_z \sigma_{\overrightarrow{10}_z}^{-1} = f_\sigma(x,y)^{-1} p_z, \\ \sigma(p_w) & = \sigma_{\overrightarrow{01}_w} p_w \sigma_{\overrightarrow{10}_w}^{-1} =: F_\sigma(\xi, \eta_0, \eta_1, \dots, \eta_{n-1})^{-1} p_w, \\ \sigma(p_w \delta_n^{-1}) & = \sigma_{\overrightarrow{01}_w} p_w \delta_n^{-1} \sigma_{n\overrightarrow{10}_w}^{-1} = f_\sigma^{(n)}(\xi, \eta_0, \eta_1, \dots, \eta_{n-1})^{-1} p_w \delta_n^{-1}. \end{cases}$$

(A7) LEMMA. *(1)* $\sigma(\delta_n) = \sigma_{n\overrightarrow{10}_w} \delta_n \sigma_{\overrightarrow{10}_w}^{-1} = \delta_n \eta_0^{\rho_n(\sigma)}$.
(2) $\beta_n(n\overrightarrow{10}_w) = \overrightarrow{10}_z$ *(hence* $\beta_n(p_w \delta_n^{-1}) = p_z$*).*
(3) $f_\sigma^{(n)}(\xi, \eta_0, \eta_1, \dots, \eta_{n-1}) = \eta_0^{\rho_n(\sigma)} F_\sigma(\xi, \eta_0, \eta_1, \dots, \eta_{n-1})$.
(4) $\psi_n^{(0)}(F_\sigma(\xi, \eta_0, \eta_1, \dots, \eta_{n-1})) = 0$.

PROOF. Calculations:
(1) It suffices to check the effect of both sides on $w_1^{1/N} \in \mathbb{Q}\{\{w_1\}\}$:

$$\begin{aligned}\sigma(\delta_n)(w_1^{1/N}) &= \sigma_{n\overrightarrow{10}_w} \delta_n \sigma_{\overrightarrow{10}_w}^{-1}(w_1^{1/N}) = \sigma_{n\overrightarrow{10}_w} \delta_n(w_1^{1/N}) = \sigma_{n\overrightarrow{10}_w}(\sqrt[N]{n}^{-1}(nw_1)^{1/N}) \\ &= \zeta_N^{-\rho_n(\sigma)} \sqrt[N]{n}^{-1}(nw_1)^{1/N} = \delta_n(\zeta_N^{-\rho_n(\sigma)} w_1^{1/N}) = \delta_n \eta_0^{\rho_n(\sigma)}(w_1^{1/N}).\end{aligned}$$

Notice that $\eta_0(w_1^{1/N}) = \zeta_N^{-1} w_1^{1/N}$ in our convention.
(2) $\beta_n^*(z_1) = 1 - (1-w_1)^n = nw_1(1 - \frac{n-1}{2} w_1 + \cdots) \in \mathbb{Q}\{\{w_1\}\}$.
(3) Recalling the above definition of F_σ in (A6), one has

$$\begin{aligned}\sigma(p_w \delta_n^{-1}) &= F_\sigma(\xi, \eta_0, \eta_1, \dots, \eta_{n-1})^{-1} p_w (\delta_n \eta_0^{\rho_n(\sigma)})^{-1} \\ &= F_\sigma(\xi, \eta_0, \eta_1, \dots, \eta_{n-1})^{-1} \eta_0^{-\rho_n(\sigma)} p_w \delta_n^{-1} \\ &= (\eta_0^{\rho_n(\sigma)} F_\sigma(\xi, \eta_0, \eta_1, \dots, \eta_{n-1}))^{-1} p_w \delta_n^{-1}.\end{aligned}$$

Hence

$$\begin{aligned}f_\sigma(x,y) &= \beta_n(\eta_0^{\rho_n(\sigma)} F_\sigma(\xi, \eta_0, \eta_1, \dots, \eta_{n-1})) \\ &= y^{\rho_n(\sigma)} F_\sigma(x^n, y, xyx^{-1}, \dots, x^{n-1} y x^{-(n-1)}).\end{aligned}$$

(4) Since $\iota_n(p_w) = p_z$, it follows that

$$f_\sigma(x,y) = \iota_n(F_\sigma(\xi, \eta_0, \eta_1, \dots, \eta_{n-1})) = F_\sigma(x, y, 1, \dots, 1).$$

Thus, if we put $F_\sigma(\xi, \eta_0, \eta_1, \dots, \eta_{n-1}) \equiv \xi^a \eta_0^{b_0} \eta_1^{b_1} \cdots \eta_{n-1}^{b_{n-1}} \bmod \hat{F}'_{n+1}$, we have $a = b_0 = 0$ because $f_\sigma(x,y) \in \hat{F}'_2$. □

Proof of Proposition (A1): By Lemma (A2) and Lemma (A7) (3),(4), one sees

$$\begin{aligned}\Psi_n^{(0)}(\sigma) &= \psi_n^{(0)}(f_\sigma^{(n)}(\xi, \eta_0, \eta_1, \dots, \eta_{n-1})) \\ &= \psi_n^{(0)}(\eta_0^{\rho_n(\sigma)} F_\sigma(\xi, \eta_0, \eta_1, \dots, \eta_{n-1})) = \rho_n(\sigma). \quad \square\end{aligned}$$

References

[A] G.Anderson, *The hyperadelic gamma function,*, Invent. Math. **95** (1989), 63–131.

[Dr] V.G.Drinfeld, *On quasitriangular quasi-Hopf algebras and a group closely connected with $Gal(\bar{Q}/Q)$*, Leningrad Math. J. **2(4)** (1991), 829–860.

[F] H.Furusho, *Geometric and arithmetic subgroups of the Grothendieck-Teichmüller group*, Math. Res. Letters **10** (2003), 97–108.

[G] A.Grothendieck, *Esquisse d'un Programme, 1984*, Geometric Galois Actions I, P.Lochak, L.Schneps (eds.), vol. 242, London Math. Soc. Lect. Note Ser., 1997, pp. 5–48.

[I1] Y.Ihara, *On the embedding of $Gal(\bar{Q}/Q)$ into $GT^\wedge$*, The Grothendieck theory of Dessin's d'Enfants, London Math. Soc. Lect. Note Ser., vol. 200, Cambridge Univ. Press, 1994, pp. 289–306.

[I2] Y.Ihara, *On beta and gamma functions associated with the Grothendieck-Teichmüller modular group*, Aspects of Galois Theory, (H. Voelklein et.al (eds.)), London Math. Soc. Lect. Note Ser., vol. 256, 1999, pp. 144–179.

[I3] Y.Ihara, *On beta and gamma functions associated with the Grothendieck-Teichmüller modular group, II*, J. reine angew. Math. **527** (2000), 1–11.

[IM] Y.Ihara, M.Matsumoto, *On Galois actions on profinite completions of braid groups*, Recent developments in the inverse Galois problem (M.Fried et al. eds), vol. 186, Contemp. Math. (AMS), 1995, pp. 173–200.

[LS] P.Lochak, L.Schneps, *A cohomological interpretation of the Grothendieck-Teichmüller group*, Invent. Math. **127** (1997), 571–600.

[N-I] H.Nakamura, *Limits of Galois representations in fundamental groups along maximal degeneration of marked curves, I*, Amer. J. Math. **121** (1999), 315–358.

[N-II] ———, *Limits of Galois representations in fundamental groups along maximal degeneration of marked curves, II*, Proc. Symp. Pure Math. **70** (2002), 43–78.

[N98] ———, *Tangential base points and Eisenstein power series*, Aspects of Galois Theory, H.Voelklein et. al. (eds.), London Math. Soc. Lect. Note Ser., vol. 256, 1999, pp. 202–217.

[N01] ———, *Some arithmetic in fundamental groups of affine elliptic curves*, talk at Euresco Conference at Acquafredda Maratea (2001).

[NS] H.Nakamura, L.Schneps, *On a subgroup of the Grothendieck-Teichmüller group acting on the tower of profinite Teichmüller modular groups*, Invent. Math. **141** (2000), 503–560.

[NT] H.Nakamura, H.Tsunogai, *Harmonic and equianharmonic equations in the Grothendieck-Teichmüller group*, Forum Math. **15** (2003), 877–892.

Department of Mathematics, Okayama University, Okayama 700-8530, Japan
E-mail address: h-naka@math.okayama-u.ac.jp

Department of Mathematics, Sophia University, Tokyo 102-8544, Japan
E-mail address: tsuno@mm.sophia.ac.jp

Contemporary Mathematics
Volume **416**, 2006

ON p-ADIC PROPERTIES OF THE WITTEN-RESHETIKHIN-TURAEV INVARIANT

L. ROZANSKY

ABSTRACT. We prove the Lawrence conjecture about p-adic convergence of the series of Ohtsuki invariants of a rational homology sphere to its $SO(3)$ Witten-Reshetikhin-Turaev invariant. Our proof is based on the surgery formula for Ohtsuki series and on the properties of the expansion of the colored Jones polynomial of a knot in powers of $q-1$ and $q^{\alpha}-1$, α being the color of the knot.

1. INTRODUCTION

A quantum invariant of 3-manifolds (WRT invariant) discovered by E. Witten [31] and by N. Reshetikhin, V. Turaev [18], is an extension of the Jones polynomial from links in S^3 to links in any compact oriented 3-manifold M. For a link $\mathcal{L} \subset M$ with L components, the WRT invariant $Z_{\underline{\alpha}}(M, \mathcal{L}; K)$ is a complex number depending on positive integer parameters K and $\underline{\alpha} = (\alpha_1, \ldots, \alpha_L)$ such that $1 \leq \alpha_j \leq K-1$. The WRT invariant is quite effective in distinguishing between the topologically inequivalent links and 3-manifolds. However its relation to the 'classical' topology remains mysterious, especially if we use the mathematically rigorous definition of [18], which relies on the representation theory of quantum group $su_q(2)$ rather than the non-rigorous path integral definition of [31].

One could try to determine the topological content of the WRT invariant $Z(M; K)$ independently for every value of K. R. Kirby and P. Melvin have tried this approach in [5], but they succeeded in relating $Z(M; K)$ to classical invariants only for few small values of K. An alternative approach is to study how $Z(M; K)$ depends on K in a hope of identifying the topological invariants which parametrize this dependence. Two different methods of achieving this goal were developed – a number-theoretic and an analytic one. We showed in [25] that if M is a rational homology sphere ($\mathbb{Q}$HS) then, surprisingly, both methods produce the same sequence of invariants.

The number-theoretic method was advanced by H. Murakami and T. Ohtsuki. They worked with the modified WRT invariant $Z'(M; K)$ introduced by Kirby and Melvin [5]. Murakami proved [13], [14] (see also [12]) that for a $\mathbb{Q}$HS M and for an odd prime K, $Z'(M; K)$ belongs to the cyclotomic ring $\mathbb{Z}[q]$, $q = \exp(2\pi i/K)$.

2000 *Mathematics Subject Classification.* Primary 57M27, Secondary 11S99.
Key words and phrases. Witten-Reshetikhin-Turaev invariant, p-adic numbers.
The author was supported by NSF Grants DMS-9704893 and DMS-0509793.

Therefore one can write

$$(1.1)\quad Z'(M;K)=\sum_{n=0}^{K-2}a_n(M;K)\,(q-1)^n,\qquad \text{where } a_n(M;K)\in\mathbb{Z}.$$

The coefficients $a_n(M;K)$ still depend on K, however Ohtsuki observed [15], [16] that their remainders modulo K are almost K-independent. He showed that there exist K-independent invariants $\lambda_n(M)\in\mathbb{Q}$, $n\geq 0$ such that

$$(1.2)\quad a_n(M;K)\equiv\left(\frac{|H_1(M,\mathbb{Z})|}{K}\right)_{\mathrm{L}}\lambda_n(M)\pmod K\quad\text{for } n\leq \tfrac{K-3}{2}.$$

Here $|H_1(M,\mathbb{Z})|$ is the order of the first homology of M and $\left(\frac{|H_1(M,\mathbb{Z})|}{K}\right)_{\mathrm{L}}$ is the Legendre symbol. If $\lambda_n(M)$ is not integer, then it should be viewed in eq.(1.2) as a K-adic integer. Murakami showed that the first two invariants λ_0 and λ_1 are classical

$$(1.3)\quad \lambda_0(M)=1/|H_1(M,\mathbb{Z})|,\qquad \lambda_1(M)=3\lambda_{\mathrm{CW}}(M)/|H_1(M,\mathbb{Z})|,$$

where $\lambda_{\mathrm{CW}}(M)$ is the Casson-Walker invariant of M.

R. Lawrence [6] conjectured that $\lambda_n(M)$ have much more control over $Z'(M;K)$, than that displayed by eq.(1.2). She suggested that

$$(1.4)\qquad \lambda_n(M)\in\mathbb{Z}[1/2,1/|H_1(M,\mathbb{Z})|]$$

and that the series $\left(\frac{|H_1(M,\mathbb{Z})|}{K}\right)_{\mathrm{L}}\sum_{n=0}^{\infty}\lambda_n(M)\,(q-1)^n$ converges K-adically to $Z'(M;K)$ in the K-adic completion $\mathbb{Z}_K[q]$ of the cyclotomic ring $\mathbb{Z}[q]$, that is,

$$(1.5)\qquad Z'(M;K)=\left(\frac{|H_1(M,\mathbb{Z})|}{K}\right)_{\mathrm{L}}\sum_{n=0}^{\infty}\lambda_n(M)\,(q-1)^n.$$

Lawrence verified her conjecture for Seifert manifolds constructed by Dehn's surgery on $(2,m)$ torus knots.

The goal of this paper is to prove the Lawrence conjecture for a general $\mathbb{Q}$HS. Our proof is based on a surgery formula for the invariants $\lambda_n(M)$. This formula has first appeared [21] in the study of the asymptotic properties of $Z(M;K)$ at large values of K. This study led us to a formal power series

$$(1.6)\quad Z^{(\mathrm{tr})}(M;K)=|H_1(M,\mathbb{Z})|^{-1/2}\sum_{n=0}^{\infty}D_n(M)\,K^{-n},\qquad D_n(M)\in\mathbb{C},$$

defined by means of special surgery formulas [21]-[24]. In other words, if a $\mathbb{Q}$HS M is constructed by a surgery on a link $\mathcal{L}\subset S^3$, then the coefficients $D_m(M)$ are expressed in terms of the coefficients of the colored Jones polynomial of $\mathcal{L}$. We proved that for a given manifold M, the series $Z^{(\mathrm{tr})}(M;K)$ does not depend on the choice of a link $\mathcal{L}$, thus establishing it as a topological invariant of M. We called the series (1.6) *a trivial connection contribution* (TCC) to the full WRT invariant $Z(M;K)$ because of its conjectured relation to the asymptotic properties of the latter.

Our surgery formulas for $Z^{(\mathrm{tr})}(M;K)$ showed that, as we hoped, the first coefficients $D_n(M)$ in the sum of eq.(1.6) are classical topological invariants of M

$$(1.7)\quad D_0(M)=1/|H_1(M,\mathbb{Z})|,\qquad D_1(M)=6\pi i\lambda_{\mathrm{CW}}(M)/|H_1(M,\mathbb{Z})|.$$

There is an apparent similarity between eqs.(1.7) and (1.3). In [25] we showed that for $n \geq 2$ the invariants $D_n(M)$ and $\lambda_n(M)$ are also essentially equivalent. More precisely, we proved the equation

$$Z^{(\mathrm{tr})}(M;K) = \sum_{n=0}^{\infty} \lambda_n(M)\,(e^{2\pi i/K} - 1)^n, \tag{1.8}$$

which should be understood as a relation between two formal power series in K^{-1}.

Our proof of eq.(1.8) as well as an alternative proof of the Murakami and Ohtsuki results (1.1) and (1.3) which we provided in [25], was based on the propreties of the expansion of the colored Jones polynomial in powers of $(q-1)$ and colors (this expansion was first considered in [11]). We used a previously established bound [23] on the powers of colors versus the powers of $(q-1)$ which exists for algebraically split links, *i.e.* links whose linking numbers are zero (recently T. Le [8] has used the same idea in order to prove the analogs of eqs.(1.1) and (1.3) for $SU(N)$ WRT invariants). However, the bound itself did not allow us to relate $\lambda_n(M)$ to $Z'(M;K)$ beyond eq.(1.3).

In [27] we proved that the expansion of the colored Jones polynomial of a knot in S^3 in powers of $(q-1)$ and α (α being the color of the knot) can be rewritten as an expansion in powers of $(q-1)$ and $(q^\alpha - 1)$ (α being the color of the knot) with integer coefficients. The integrality of those coefficients allowed us [26] to give a simple proof of Lawrence's conjecture for a $\mathbb{Q}$HS, constructed by a surgery on a knot in S^3. In [29] we generalized the results of [27] to a knot in a $\mathbb{Q}$HS In this paper we will use the method of [26] in order to prove the Lawrence conjecture in the general case.

2. Notations, background and statement of results

First of all, let us introduce our multi-index, number-theoretic and topological notations.

We use multi-index notations for the colors of links. For a link $\mathcal{L}$ with L components we denote $\underline{x} = (x_1, \ldots, x_L)$ and

$$\underline{x}+\underline{y} = (x_1+y_1,\ldots,x_L+y_L),\qquad y\underline{x} = (yx_1,\ldots,yx_L),\qquad \underline{x}\underline{y} = (x_1y_1,\ldots,x_Ly_L),$$

$$x^{\underline{y}} = (x^{y_1},\ldots,x^{y_L}),\qquad \underline{x}^y = (x_1^y,\ldots,x_L^y),\qquad \underline{x}^{\underline{y}} = \textstyle\prod_{j=1}^L x_j^{y_j},$$

$$\{f(\underline{x})\} = \textstyle\prod_{j=1}^L f(x_j),\qquad |\underline{x}| = \textstyle\sum_{j=1}^L x_j.$$

When we consider two links $\mathcal{L}$ and $\mathcal{L}'$ simultaneously, we also use a multi-index notation

$$\underline{\underline{y}} = (y_1,\ldots,y_{L'}) \tag{2.1}$$

for the colors of the L'-component link $\mathcal{L}'$. We denote concatenation of multi-indices as

$$\underline{x},x - (x_1,\ldots,x_L,x),\qquad \underline{x},\underline{\underline{y}} - (x_1,\ldots,x_L,y_1,\ldots,y_{L'}). \tag{2.2}$$

Also $\underline{x} = x$ means $x_j = x$ for all $1 \leq j \leq L$.

We will use three variables K, q and h which are related by

$$q = e^{2\pi i/K},\qquad h = q-1. \tag{2.3}$$

A rational power of q is defined as

$$q^r = e^{2\pi i r/K}, \quad r \in \mathbb{Q}. \tag{2.4}$$

In many cases we will have to specialize K to a positive integer. In order to indicate this explicitly we use 'checked' symbols

$$\check{K} \in \mathbb{Z}_+, \qquad \check{q} = e^{2\pi i/\check{K}}, \qquad \check{h} = \check{q} - 1. \tag{2.5}$$

Also while α, β are variables, $\check{\alpha}$ and $\check{\beta}$ stand for positive integers. These elaborate notations will help us to avoid confusion between the formal parameters and their $\check{K}$-adic and cyclotomic versions. In most cases we will assume that $\check{K}$ is an odd prime, but we will always indicate this explicitly.

For $\check{K}$ being prime, $\mathbb{Z}_{\check{K}}$ denotes the ring of $\check{K}$-adic integers, $\mathbb{Z}[\check{q}]$ is the cyclotomic ring and $\mathbb{Z}_{\check{K}}[\check{q}]$ is its $\check{K}$-adic completion. Note that since we set $\check{q} = e^{2\pi i/\check{K}}$, then $\mathbb{Z}[\check{q}]$ for us is a particular subring of $\mathbb{C}$.

Denote by $\mathbb{Z}_{(\check{K})} \subset \mathbb{Q}$ the ring of rational numbers whose denominators are not divisible by $\check{K}$. There is a natural embedding

$$^\vee : \mathbb{Z}_{(\check{K})} \to \mathbb{Z}_{\check{K}}, \qquad \lambda \mapsto \lambda^\vee, \tag{2.6}$$

which maps fractions into power series in $\check{K}$. Whenever we use a notation $\lambda^\vee$, we assume that $\lambda \in \mathbb{Z}_{(\check{K})}$. The homomorphism (2.6) can be extended from $\mathbb{Z}_{(\check{K})}$ to formal power series with coefficients in $\mathbb{Z}_{(\check{K})}$

$$^\vee : \mathbb{Z}_{(\check{K})}[[h]] \to \mathbb{Z}_{\check{K}}[\check{q}] \tag{2.7}$$

by setting $h \mapsto \check{h}$. Indeed, since

$$(\check{q}^{\check{K}} - 1)/(\check{q} - 1) = 0, \tag{2.8}$$

then

$$\check{h}^{\check{K}-1} = \check{K}x, \qquad \text{for some } x \in \mathbb{Z}[\check{q}]. \tag{2.9}$$

As a result, any series of the form $\sum_{n=0}^{\infty} a_n \check{h}^n$, $a_n \in \mathbb{Z}_{\check{K}}$ has a $\check{K}$-adic limit in $\mathbb{Z}_{\check{K}}[\check{q}]$

$$\sum_{n=0}^{\infty} a_n \check{h}^n = A \in \mathbb{Z}_{\check{K}}[\check{q}], \tag{2.10}$$

which means by definition that for any $N_0 > 0$ there exists N_0' such that for any $N > N_0'$

$$A = \sum_{n=0}^{N} a_n \check{h}^n + \check{K}^{N_0+1}x, \qquad x \in \mathbb{Z}_{\check{K}}[\check{q}]. \tag{2.11}$$

Therefore the image of a formal power series $S(h) = \sum_{n=0}^{\infty} a_n h^n$, $a_n \in \mathbb{Z}_{(\check{K})}$ under the homomorphism (2.7) can be defined as the $\check{K}$-adic limit of the corresponding series $\sum_{n=0}^{\infty} a_n^\vee \check{h}^n$

$$^\vee : S(h) \mapsto S^\vee = \sum_{n=0}^{\infty} a_n^\vee \check{h}^n \in \mathbb{Z}_{\check{K}}[\check{q}]. \tag{2.12}$$

We will need a few more number-theoretic notations. Let $\check{K}$ be an odd prime. $\left(\frac{x}{\check{K}}\right)_{\mathrm{L}}$ denotes a Legendre symbol of x. If x is an integer not divisible by $\check{K}$, then

$\left(\frac{x}{\check{K}}\right)_{\mathrm{L}} = 1$ if there exists $y \in \mathbb{Z}$ such that $x \equiv y^2 \pmod{\check{K}}$, and $\left(\frac{x}{\check{K}}\right)_{\mathrm{L}} = -1$ otherwise. For an integer n not divisible by $\check{K}$ let $n\sigma$ denote any integer such that

$$n n\sigma \equiv 1 \pmod{\check{K}}. \tag{2.13}$$

Also denote

$$\kappa = \begin{cases} 1 & \text{if } \check{K} \equiv 1 \pmod 4 \\ -1 & \text{if } \check{K} \equiv -1 \pmod 4. \end{cases} \tag{2.14}$$

Finally, for $x \in \mathbb{Z}_{\check{K}}$, we denote by $[x]_N \in \mathbb{Z}$ the remainder of x modulo $\check{K}^{N+1}$

$$[x]_N \equiv x \pmod{\check{K}^{N+1}}. \tag{2.15}$$

Now let us fix some standard topological notations. Let $\mathcal{K}$ be a knot in a 3-manifold M. A *meridian* of $\mathcal{K}$ is a simple cycle on the boundary of its tubular neighborhood, which is contractible through that neighborhood. A *parallel* is a cycle on the same boundary which has a unit intersection number with the meridian. A knot is called *framed* if we made a choice of its parallel. A link is framed if all of its components are framed. We denote framed knots and links by putting a hat on top of their symbols: $\hat{\mathcal{K}}, \hat{\mathcal{L}}$.

Suppose that a framed knot $\hat{\mathcal{K}}$ is of finite order as an element of $H_1(M, \mathbb{Z})$ (we will denote the order of $\hat{\mathcal{K}}$ as o). Then its *self-linking number* p is the linking number between the knot and its parallel, $p \in \mathbb{Z}/o$. *Dehn's surgery* on a framed knot $\hat{\mathcal{K}} \subset M$ is a transformation of cutting out a tubular neighborhood of $\hat{\mathcal{K}}$ and gluing it back in such a way that the meridian of the tubular neighborhood matches the parallel left on the boundary of the knot complement. A result of this procedure is a new 3-manifold M'.

In order to shorten our formulas, we will denote the order of the first homology of a $\mathbb{Q}$HS M as $h_1(M)$:

$$h_1(M) = |H_1(M, \mathbb{Z})|. \tag{2.16}$$

Let us recall the quantum invariants appearing in 3-dimensional topology. The colored Jones polynomial

$$J_{\underline{\check{\alpha}}}(\hat{\mathcal{L}}; q) \in \mathbb{Z}[q^{1/2}, q^{-1/2}], \tag{2.17}$$

$$J_{\underline{\check{\alpha}}}(\hat{\mathcal{L}}; q) \in \mathbb{Z}[q, q^{-1}] \quad \text{if } \underline{\check{\alpha}} \text{ are odd} \tag{2.18}$$

is an invariant of a framed L-component link $\hat{\mathcal{L}} \subset S^3$. The positive integers $\underline{\check{\alpha}}$ are called 'colors', they are assigned to components of $\hat{\mathcal{L}}$. The WRT invariant is an invariant of a framed link $\hat{\mathcal{L}}$ in a 3-manifold M. It depends on a positive integer $\check{K}$, and we denote it as $Z_{\underline{\check{\alpha}}}(M, \hat{\mathcal{L}}; \check{K})$. N. Reshetikhin and V. Turaev defined the WRT invariant by a surgery formula and verified that $Z_{\underline{\check{\alpha}}}(M, \hat{\mathcal{L}}; \check{K})$ does not depend on the choice of a surgery link.

Definition 2.1 (N. Reshetikhin, V. Turaev [18]). If $M = S^3$, then

$$Z_{\underline{\check{\alpha}}}(S^3, \hat{\mathcal{L}}; \check{K}) = J_{\underline{\check{\alpha}}}(\hat{\mathcal{L}}; \check{q}), \tag{2.19}$$

If M is constructed by Dehn's surgery on a framed L'-component link $\hat{\mathcal{L}}' \subset S^3$, then

$$Z_{\underline{\check{\alpha}}}(M, \hat{\mathcal{L}}; \check{K}) = i^{-L'} (2\check{K})^{-L'/2} e^{-(3\pi i/4)\,\mathrm{sign}(\hat{\mathcal{L}}')} \check{q}^{(3/4)\,\mathrm{sign}(\hat{\mathcal{L}}')} \times \sum_{0<\check{\beta}<\check{K}} \{\check{q}^{\underline{\underline{\check{\beta}}}/2} - \check{q}^{-\underline{\underline{\check{\beta}}}/2}\} J_{\underline{\check{\alpha}},\underline{\underline{\check{\beta}}}}(\hat{\mathcal{L}} \cup \hat{\mathcal{L}}'; \check{q}), \tag{2.20}$$

where $\mathrm{sign}(\hat{\mathcal{L}}')$ is the signature of the linking matrix of $\hat{\mathcal{L}}'$.

For an empty link $\mathcal{L}$ we denote $Z_{\underline{\check{\alpha}}}(M, \hat{\mathcal{L}}; \check{K})$ simply as $Z(M; \check{K})$. Note that we use the normalization in which

$$Z(S^3; \check{K}) = 1. \tag{2.21}$$

The WRT invariant (and all other quantum invariants of links that we will consider in this paper) has a simple dependence on the choice of framing of $\hat{\mathcal{L}}$. As a result, if a component $\hat{\mathcal{K}}$ of a link $\hat{\mathcal{L}} \cup \hat{\mathcal{K}}$ has a finite order as an element of $H_1(M, \mathbb{Z})$ so that it has a well-defined self-linking number p, then we can introduce an invariant which does not depend on the framing of $\hat{\mathcal{K}}$

$$Z_{\underline{\check{\alpha}},\check{\beta}}(M, \hat{\mathcal{L}} \cup \mathcal{K}; \check{K}) = q^{-(1/4)p(\check{\beta}^2-1)} Z_{\underline{\check{\alpha}},\check{\beta}}(M, \hat{\mathcal{L}} \cup \hat{\mathcal{K}}; \check{K}). \tag{2.22}$$

If $\hat{\mathcal{K}}$ is homologically trivial, then $\mathcal{K}$ can be interpreted as a knot with zero self-linking number, otherwise $\mathcal{K}$ is just a symbolic notation for the framing-independent invariant $Z_{\underline{\check{\alpha}},\check{\beta}}(M, \hat{\mathcal{L}} \cup \mathcal{K}; \check{K})$.

For an odd number $\check{K}$, R. Kirby and P. Melvin introduced the $SO(3)$ WRT invariant $Z'(M; \check{K})$ which has a simple relation to $Z(M; \check{K})$

$$\begin{aligned} Z(M; \check{K}) &= Z(M;3)\, Z'(M; \check{K}) \quad \text{if } \check{K} \equiv -1 \pmod 4, \\ Z(M; \check{K}) &= \overline{Z(M;3)}\, Z'(M; \check{K}) \quad \text{if } \check{K} \equiv 1 \pmod 4. \end{aligned} \tag{2.23}$$

H. Murakami [13],[14] and T. Ohtsuki [15],[16] proved that $Z'(M; \check{K})$ satisfies the following properties.

Theorem 2.2. *If $\check{K}$ is an odd prime and M is a $\mathbb{Q}HS$ such that the order of the first homology $h_1(M)$ is not divisible by $\check{K}$, then*

(1) *[Murakami]*

$$Z'(M; \check{K}) \in \mathbb{Z}[\check{q}], \tag{2.24}$$

so that

$$Z'(M; \check{K}) = \textstyle\sum_{n=0}^{\check{K}-2} a_n(M; \check{K})\, \check{h}^n, \qquad \text{where } a_n(M; \check{K}) \in \mathbb{Z}; \tag{2.25}$$

(2) *[Ohtsuki] There exists a sequence of invariants $\lambda_n(M) \in \mathbb{Q}$, $n \geq 0$ such that if $n \leq \frac{\check{K}-3}{2}$ then $\lambda_n(M) \in \mathbb{Z}_{(\check{K})}$ and*

$$a_n(M; \check{K}) \equiv \left(\frac{h_1(M)}{\check{K}}\right)_{\mathrm{L}} \lambda_n^{\vee}(M) \pmod{\check{K}}, \tag{2.26}$$

where $\left(\frac{\cdot}{\check{K}}\right)_{\mathrm{L}}$ is the Legendre symbol;

(3) *[Murakami]*

$$\lambda_0(M) = 1/h_1(M), \qquad \lambda_1(M) = 3\lambda_{\mathrm{CW}}(M)/h_1(M), \tag{2.27}$$

where $\lambda_{\mathrm{CW}}(M)$ is the Casson-Walker invariant of M.

The goal of this paper is to prove the following conjecture made by R. Lawrence.

Conjecture 2.3 (R. Lawrence [6]). *For a* $\mathbb{Q}$*HS* M*,*

$$\lambda_n(M) \in \begin{cases} \mathbb{Z} & \text{if } h_1(M) = 1 \\ \mathbb{Z}[1/2, 1/h_1(M)] & \text{if } h_1(M) > 1 \end{cases} \tag{2.28}$$

and if $\check{K}$ *is an odd prime which does not divide* $h_1(M)$*, then the series* $\sum_{n=0}^{\infty} \lambda_n(M)\, \check{h}^n$ *converges* $\check{K}$*-adicly to the modified WRT invariant* $Z'(M; \check{K})$*:*

$$Z'(M;\check{K}) = \left(\frac{h_1(M)}{\check{K}}\right)_{\mathrm{L}} \sum_{n=0}^{\infty} \lambda_n^{\vee}(M)\, \check{h}^n \quad \text{in } \mathbb{Z}_{\check{K}}[\check{q}]. \tag{2.29}$$

The proof of the Lawrence conjecture will be based on the properties of the invariant of $\mathbb{Q}$HS which we call *the trivial connection contribution* to the WRT invariant, or simply the TCC invariant. We introduced it and studied its properties in [21]-[24]. The TCC invariant of a $\mathbb{Q}$HS M is a formal power series in powers of K^{-1}

$$Z_{\underline{\alpha}}^{(\mathrm{tr})}(M, \hat{\mathcal{L}}; K) = h_1^{-1/2}(M) \sum_{\substack{\underline{m}, n \geq 0 \\ |\underline{m}| \leq n}} D_{\underline{m};n}(M, \hat{\mathcal{L}})\, \underline{\alpha}^{2\underline{m}+1} K^{-n}, \tag{2.30}$$

whose coefficients $D_{\underline{m},n}(M, \hat{\mathcal{L}})$ are invariants of M and $\hat{\mathcal{L}}$. The invariant (2.30) can be defined by various equivalent surgery formulas. Here we will use the definition which is provided by the following

Theorem 2.4 ([24]). *There exists a unique invariant* $Z_{\underline{\alpha}}^{(\mathrm{tr})}(M, \hat{\mathcal{L}}; K)$ *of a framed link* $\hat{\mathcal{L}}$ *in a* $\mathbb{Q}$*HS* M *(called the TCC invariant) which has the form (2.30) and satisfies the following three properties:*

1. *If* $M = S^3$*, then eq. (2.30) coincides with the Melvin-Morton expansion* [11] *of the colored Jones polynomial in powers of* K^{-1} *at fixed values of colors* $\underline{\check{\alpha}}$*, that is*

$$Z_{\underline{\check{\alpha}}}^{(\mathrm{tr})}(S^3, \hat{\mathcal{L}}; K) = J_{\underline{\check{\alpha}}}(\hat{\mathcal{L}}; \check{q}). \tag{2.31}$$

2. *Let* $\hat{\mathcal{L}}$ *be a framed* L*-component link and* $\hat{\mathcal{K}}$ *be a framed knot with self-linking number* p *in a* $\mathbb{Q}$*HS* M*. Then the expression*

$$Z_{\underline{\alpha},\beta}^{(\mathrm{tr})}(M, \hat{\mathcal{L}} \cup \mathcal{K}; K) = q^{-p(\beta^2-1)/4}\, Z_{\underline{\alpha},\beta}^{(\mathrm{tr})}(M, \hat{\mathcal{L}} \cup \hat{\mathcal{K}}; K), \tag{2.32}$$

does not depend on the choice of the framing of $\hat{\mathcal{K}}$ *(that is,* $Z_{\underline{\alpha},\beta}^{(\mathrm{tr})}(M, \hat{\mathcal{L}} \cup \mathcal{K}; K)$ *is well-defined for unframed knots). Moreover, if* $m > n/2$*, then* $D_{\underline{m},m,n}(M, \hat{\mathcal{L}} \cup \mathcal{K}) = 0$ *or, in other words,*

$$\begin{aligned} Z_{\underline{\alpha},\beta}^{(\mathrm{tr})}&(M, \hat{\mathcal{L}} \cup \mathcal{K}; K) \\ &= h_1^{-1/2}(M) \sum_{\substack{\underline{m}, m, n \geq 0 \\ |\underline{m}| + m \leq n \\ m \leq n/2}} D_{\underline{m},m;n}(M, \hat{\mathcal{L}} \cup \mathcal{K})\, \underline{\alpha}^{2\underline{m}+1} \beta^{2m+1} K^{-n} \end{aligned} \tag{2.33}$$

(3) *If a $\mathbb{Q}HS$ M' is constructed by Dehn's surgery on a knot $\hat{\mathcal{K}}$ with self-linking number p in a $\mathbb{Q}HS$ M, then*

$$Z^{(\mathrm{tr})}_{\underline{\alpha}}(M',\hat{\mathcal{L}};K) = -i\,(8K)^{-1/2}\,e^{-(3\pi i/4)\,\mathrm{sign}(p)}\,q^{(3/4)\,\mathrm{sign}(p)} \times \int_{[\beta=0]} Z^{(\mathrm{tr})}_{\underline{\alpha},\beta}(M,\hat{\mathcal{L}}\cup\hat{\mathcal{K}};K)\,(q^{\beta/2}-q^{-\beta/2})\,d\beta \tag{2.34}$$

(here $\int_{[\beta=0]}$ denotes a stationary phase contribution of the point $\beta=0$), or more precisely,

$$\begin{aligned} Z^{(\mathrm{tr})}_{\underline{\alpha}}(M',\hat{\mathcal{L}};K) &= -i\,(8K)^{-1/2}\,e^{-(3\pi i/4)\,\mathrm{sign}(p)}\,q^{(3/4)\,\mathrm{sign}(p)}h_1^{-1/2}(M) \\ &\times \sum_{\substack{\underline{m},m,n\geq 0\\ |\underline{m}|+m\leq n\\ m\leq n/2}} \Bigg(D_{\underline{m},m;n}(M,\hat{\mathcal{L}}\cup\mathcal{K})\,\underline{\alpha}^{2\underline{m}+1}\,K^{-n} \\ &\times \int_{-\infty}^{+\infty} e^{(i\pi/2K)p(\beta^2-1)}\,\beta^{2m+1}\,(e^{(i\pi/K)\beta}-e^{-(i\pi/K)\beta})\,d\beta\Bigg). \end{aligned} \tag{2.35}$$

Equation (2.35) is a well-defined relation between formal power series. Indeed, since

$$K^{-1/2}\int_{-\infty}^{+\infty} e^{(i\pi/2K)p(\beta^2-1)}\,\beta^{2m+1}\,(e^{(i\pi/K)\beta}-e^{-(i\pi/K)\beta})\,d\beta = O(K^m)\quad\text{as } K\to\infty, \tag{2.36}$$

then only the coefficients $D_{\underline{m},m,n}(M,\hat{\mathcal{L}}\cup\mathcal{K})$ with $n-m\leq n'$ participate in the expression for $D_{\underline{m},n'}(M',\hat{\mathcal{L}})$. In view of the bound $m\leq n/2$ in the sum of eq.(2.33), this means that the number of such coefficients $D_{\underline{m},m,n}(M,\hat{\mathcal{L}}\cup\mathcal{K})$ is finite for any fixed n'.

We proved Theorem 2.4 in [24] by showing that eq.(2.33) (which indicates that the removal of the self-linking factor $q^{p(\beta^2-1)/4}$ reduces the power of β versus the power of K^{-1} in the expansion) is consistent with the surgery formula (2.34) and that the invariant $Z^{(\mathrm{tr})}_{\underline{\alpha}}(M,\hat{\mathcal{L}};K)$ does not depend on the choice of the sequence of knot surgeries which leads from S^3 to M. We also proved in [29] (Theorem 1.15 and Remark 1.17) that if we switch in eq.(2.30) from powers of K^{-1} to powers of $h=e^{2\pi i/K}-1$ by writing

$$Z^{(\mathrm{tr})}_{\underline{\alpha}}(M,\hat{\mathcal{L}};K) = h_1^{-1/2}(M)\sum_{n=0}^{\infty} P_{\underline{\alpha};n}(M,\hat{\mathcal{L}})\,h^n, \tag{2.37}$$

$$P_{\underline{\alpha};n}(M,\hat{\mathcal{L}}) = \sum_{\substack{\underline{m}\geq 0\\ |\underline{m}|\leq n}} d_{\underline{m};n}(M,\hat{\mathcal{L}})\,\underline{\alpha}^{2\underline{m}+1}, \tag{2.38}$$

then for fixed odd $\underline{\breve{\alpha}}$

$$P_{\underline{\breve{\alpha}};n}(M,\hat{\mathcal{L}}) \in \begin{cases} \mathbb{Z} & \text{if } h_1(M)=1 \\ \mathbb{Z}[1/2,1/h_1(M)] & \text{if } h_1(M)>1 \end{cases} \tag{2.39}$$

Remark 2.5. In fact, we proved relation (2.39) in [29] for a 0-framed link $\mathcal{L}$. However, since

$$Z^{(\mathrm{tr})}_{\underline{\alpha}}(M,\hat{\mathcal{L}};K) = q^{\frac{1}{4}\sum_{j=1}^{L} l_{jj}(\alpha_j^2-1)}\,Z^{(\mathrm{tr})}_{\underline{\alpha}}(M,\mathcal{L};K), \tag{2.40}$$

l_{jj} being self-linking numbers of the link components $\hat{\mathcal{L}}_j$, and since for odd $\underline{\breve{\alpha}}$

$$q^{\frac{1}{4}\sum_{j=1}^{L} l_{jj}(\breve{\alpha}_j^2-1)} \in \mathbb{Z}[1/h_1(M)]\,[[h]], \tag{2.41}$$

then relations (2.39) are also true for framed links $\hat{\mathcal{L}} \subset M$.

We proved in [21] and [23] that for an empty link $\hat{\mathcal{L}}$,

$$P_0(M) = 1/h_1(M), \qquad P_1(M) = 3\lambda_{\mathrm{CW}}(M)/h_1(M). \tag{2.42}$$

Then we showed in [25] that an apparent similarity between eqs. (2.27) and (2.42) can be extended.

Theorem 2.6. *For an empty link $\hat{\mathcal{L}}$, the coefficients $P_n(M)$ of eq. (2.37) coincide with Ohtsuki's invariants $\lambda_n(M)$, or in other words,*

$$\sum_{n=0}^{\infty} \lambda_n(M)\, h^n = h_1^{1/2}(M)\, Z^{(\mathrm{tr})}(M;K). \tag{2.43}$$

In this paper we will prove the Lawrence Conjecture 2.3 extended to the $SO(3)$ WRT invariant Z' of links in $\mathbb{Q}$HS The definition of this invariant is given by the following

Theorem 2.7 (*cf.* [5]). *Let $\hat{\mathcal{L}}$ be a framed L-component link in a 3-manifold M. Suppose that M is constructed by Dehn's surgery on a framed L'-component link $\hat{\mathcal{L}}' \subset S^3$. Then for an odd integer $\breve{K}$ and for a set of odd colors $\underline{\breve{\alpha}}$, the following expression*

$$\begin{aligned} Z'_{\underline{\breve{\alpha}}}(M,\hat{\mathcal{L}},\breve{K}) \;-\; & i^{-L'}\,\breve{K}^{-L'/2}\, e^{(i\pi/4)(\kappa+1)\,\mathrm{sign}(\hat{\mathcal{L}}')}\, \breve{q}^{3\,4\sigma\,\mathrm{sign}(\hat{\mathcal{L}}')} \\ & \times \sum_{\substack{0<\underline{\underline{\breve{\beta}}}<K \\ \underline{\underline{\breve{\beta}}}-\mathrm{odd}}} \{q^{\underline{\underline{\breve{\beta}}}/2} - q^{-\underline{\underline{\breve{\beta}}}/2}\}\, J_{\underline{\breve{\alpha}},\underline{\underline{\breve{\beta}}}}(\hat{\mathcal{L}}\cup\hat{\mathcal{L}}';\breve{q}) \end{aligned} \tag{2.44}$$

does not depend on the choice of the surgery link $\hat{\mathcal{L}}'$ which is used to construct M. We call $Z'_{\underline{\breve{\alpha}}}(M,\hat{\mathcal{L}};\breve{K})$ the $SO(3)$ invariant of $\hat{\mathcal{L}}$ and M. It satisfies the analog of eq. (2.23)

$$\begin{aligned} Z_{\underline{\breve{\alpha}}}(M,\hat{\mathcal{L}};\breve{K}) &= Z(M;3)\, Z'_{\underline{\breve{\alpha}}}(M,\hat{\mathcal{L}};\breve{K}) \quad \text{if } \breve{K} \equiv -1 \pmod 4 \\ Z_{\underline{\breve{\alpha}}}(M,\hat{\mathcal{L}};\breve{K}) &= \overline{Z(M;3)}\, Z'_{\underline{\breve{\alpha}}}(M,\hat{\mathcal{L}};\breve{K}) \quad \text{if } \breve{K} \equiv 1 \pmod 4 \end{aligned} \tag{2.45}$$

Proof. The independence of $Z'_{\underline{\alpha}}(M,\hat{\mathcal{L}};K)$ on the choice of a surgery link $\hat{\mathcal{L}}'$ can be proved in exactly the same way as the invariance of $Z'(M;K)$ in [5]. We will prove eq. (2.45) in Section 5. □

Theorem 2.8 (Main Theorem). *For a $\mathbb{Q}$HS M let $\breve{K}$ be an odd prime which does not divide $h_1(M)$. If $\hat{\mathcal{L}}$ is a framed link in M, then*

$$Z'_{\underline{\breve{\alpha}}}(M,\hat{\mathcal{L}};\breve{K}) \in \mathbb{Z}[\breve{q}] \tag{2.46}$$

and the series (2.37) converges $\breve{K}$-adicly to the colored $SO(3)$ WRT invariant $Z'_{\underline{\breve{\alpha}}}(M,\hat{\mathcal{L}};\breve{K})$ of a framed link $\hat{\mathcal{L}} \subset M$:

$$Z'_{\underline{\breve{\alpha}}}(M,\hat{\mathcal{L}};\breve{K}) = \left(\frac{h_1(M)}{\breve{K}}\right)_{\mathrm{L}} \sum_{n=0}^{\infty} P^{\vee}_{\underline{\breve{\alpha}};n}(M,\hat{\mathcal{L}})\,\breve{h}^n, \tag{2.47}$$

or equivalently,

$$Z'_{\check{\underline{\alpha}}}(M,\hat{\mathcal{L}};\check{K}) = \left(\frac{h_1(M)}{\check{K}}\right)_{\mathrm{L}} \left(h_1^{1/2}(M)\, Z^{(\mathrm{tr})}_{\check{\underline{\alpha}}}(M,\hat{\mathcal{L}};K)\right)^{\vee}. \tag{2.48}$$

Here $P_{\underline{\alpha};n}(M,\hat{\mathcal{L}})$ are the coefficients in the expansion (2.37) of $Z^{(\mathrm{tr})}_{\underline{\alpha}}(M,\hat{\mathcal{L}};K)$.

Remark 2.9. Since eqs. (2.29) and (2.47) imply that $\lambda_n(M) = P_n(M)$, then relation (2.28) follows from (2.39). In fact, (2.39) is a generalization of (2.28) for the case of a link $\hat{\mathcal{L}} \subset M$ (see also eqs. (1.87), (1.90) and (1.93) in Theorem 1.8 of [29] for a slightly stronger statement about the denominators of $P_{\underline{\alpha};n}(M,\hat{\mathcal{L}})$).

Sketch of the proof of Main Theorem. Suppose that a QHS M is constructed by a surgery on a framed algebraically split link $\hat{\mathcal{L}}' \subset S^3$ (*i.e.* all linking numbers between the components of $\mathcal{L}'$ are zero). We prove the theorem by induction in the number of components of $\hat{\mathcal{L}}'$. An easy Proposition 5.3 establishes Main Theorem for $M = S^3$. It remains to prove that if Main Theorem is true for a QHS M, then it is also true for a QHS M' constructed by a surgery on a homologically trivial framed knot $\hat{\mathcal{K}} \subset M$. The key to the proof is Corollary 4.2 which describes the structure of $Z^{(\mathrm{tr})}_{\underline{\alpha},\beta}(M,\hat{\mathcal{L}}\cup\hat{\mathcal{K}};K)$ and Corollary 4.6 which establishes the similarity between the gaussian sum of the surgery formula (2.44) and the gaussian integral of the surgery formula (2.34) as they appear after the substitution (4.6).

Our proof of Main Theorem relies exclusively on the properties of the TCC invariant. We use neither the Murakami-Ohtsuki Theorem 2.2, nor Theorem 2.6. In fact, Main Theorem is stronger than these theorems (except eq. (2.27) which we will prove in Appendix).

3. Path integral interpretation

An intriguing aspect of the $\check{K}$-adic covergence relations (2.29) and (2.48) is their connection to the asymptotic properties of the WRT invariants $Z(M;K)$ and $Z_{\check{\underline{\alpha}}}(M,\hat{\mathcal{L}};\check{K})$. At the 'physical' level of rigor, the asymptotic properties of $Z(M;K)$ follow from Witten's formula [31] which presents this quantum invariant as a path integral over $SU(2)$ connections in the (trivial) $SU(2)$ bundle over M. At large values of K the integral can be evaluated in the stationary phase approximation, the stationary phase points being the flat $SU(2)$ connections. Therefore Witten conjectured that $Z(M;K)$ splits into a sum of contributions coming from connected components $\mathcal{M}^{(c)}$ of the moduli space $\mathcal{M}$ of flat $SU(2)$ connections on M

$$Z(M;K) \simeq \sum_c Z^{(c)}(M;K), \tag{3.1}$$

where $\simeq$ denotes asymptotic convergence. Each contribution $Z^{(c)}(M;K)$ is proportional to a 'classical exponential'

$$Z^{(c)}(M;K) = e^{(iK/2\pi)S^{(c)}_{\mathrm{CS}}}X^{(c)}(M;K). \tag{3.2}$$

Here $S^{(c)}_{\mathrm{CS}}$ is the Chern-Simons invariant of connections in the component $\mathcal{M}^{(c)} \subset \mathcal{M}$ while $X^{(c)}(M;K)$ is an asymptotic series in powers of K^{-1} whose coefficients are invariants of M. These coefficients are independent of K and therefore they can be expected to be related to classical topology. If M is a QHS, then the trivial $SU(2)$ connection is an isolated point in the moduli space $\mathcal{M}$. Therefore according to path integral predictions, it should yield a distinct contribution to $Z(M;K)$ of the form of a power series in K^{-1}. In [21] and [22] we used path integral arguments in support of the conjecture that the trivial connection contribution is, in fact, the series (1.6). In [19] and [20] we verified this conjecture in case when M is a Seifert rational homology sphere.

If the conjecture about the series $Z^{(\mathrm{tr})}(M;K)$ representing the contribution of the trivial connection to the asymptotic expansion of the WRT invariant is true, then the $\check{K}$-adic convergence relation (2.48) implies that the whole WRT invariant of a $\mathbb{Q}$HS is determined just by the contribution of the trivial connection.

Similarly to the TCC invariant $Z^{(\mathrm{tr})}(M;K)$, the definition of $Z^{(\mathrm{tr})}_{\underline{\alpha}}(M,\hat{\mathcal{L}};K)$ was also motivated by an asymptotic approach to the study of the WRT invariant $Z_{\underline{\check{\alpha}}}(M,\hat{\mathcal{L}};\check{K})$. Namely, path integral arguments suggest that the series (2.37) represents the contribution of the trivial connection to the stationary phase asymptotic expansion of the Witten-Chern-Simons path integral formula for $Z_{\underline{\check{\alpha}}}(M,\hat{\mathcal{L}};\check{K})$.

4. Preliminary results

All our proofs are based on a particular case of Theorem 1.8 of [29].

Theorem 4.1. *Let $\hat{\mathcal{L}}$ be a framed link and let $\hat{\mathcal{K}}$ be a framed knot with self-linking number p in a $\mathbb{Q}HS$ M. Let o be the order of $\mathcal{K}$ as an element of $H_1(M,\mathbb{Z})$. For odd colors $\underline{\check{\alpha}}$ of $\hat{\mathcal{L}}$, there exist the polynomials*

$$P_{\underline{\check{\alpha}};n}(M,\hat{\mathcal{L}},\mathcal{K};t)\in\mathbb{Z}[t^{1/o},t^{-1/o},1/2,1/h_1(M)] \tag{4.1}$$

such that

$$P_{\underline{\check{\alpha}};n}(M,\hat{\mathcal{L}},\mathcal{K};t^{-1})=P_{\underline{\check{\alpha}};n}(M,\hat{\mathcal{L}},\mathcal{K};t) \tag{4.2}$$

and

$$\begin{aligned} &Z^{(\mathrm{tr})}_{\underline{\check{\alpha}},\beta}(M,\hat{\mathcal{L}}\cup\hat{\mathcal{K}};K) \\ &\quad= h_1^{-1/2}(M)\,q^{(1/4)p(\beta^2-1)}\,\frac{q^{\beta/2o}-q^{-\beta/2o}}{q^{1/2}-q^{-1/2}}\sum_{n=0}^{\infty}\frac{P_{\underline{\check{\alpha}};n}(M,\hat{\mathcal{L}},\mathcal{K};q^{\beta})}{\Delta_{\mathrm{A}}^{2n+1}(M,\mathcal{K};q^{\beta})}\,h^n. \end{aligned} \tag{4.3}$$

Here $\Delta_{\mathrm{A}}(M,\mathcal{K};t)$ is the Alexander polynomial of $\mathcal{K}$ normalized in such a way that

$$\Delta_{\mathrm{A}}(M,\mathcal{K};t^{-1})=\Delta_{\mathrm{A}}(M,\mathcal{K};t), \tag{4.4}$$

$$\Delta_{\mathrm{A}}(M,\mathcal{K};t)\big|_{t=1}=h_1(M)/o. \tag{4.5}$$

Proof. For the case of 0-framed link $\mathcal{L}$ and knot $\mathcal{K}$, this theorem is a special case of Theorem 1.8 and Remark 1.10 of [29]. Remark 2.5 allows us to extend it to framed links and knots.

Corollary 4.2. *If $\hat{\mathcal{K}}$ is a framed homologically trivial knot with self-linking number p in a $\mathbb{Q}HS$ M, then for odd colors $\underline{\check{\alpha}}$*

$$\begin{aligned} Z^{(\mathrm{tr})}_{\underline{\check{\alpha}},\beta}(M,\hat{\mathcal{L}}\cup\hat{\mathcal{K}};K) &= h_1^{-1/2}(M)\,q^{1/2}\,h^{-1}\,q^{(1/4)p(\beta^2-1)} \\ &\quad\times\sum_{m,n\geq 0}d_{\underline{\check{\alpha}},m;n}(M,\hat{\mathcal{L}},\mathcal{K})\,(q^{\beta/2}-q^{-\beta/2})^{2m+1}\,h^n, \end{aligned} \tag{4.6}$$

where

$$d_{\underline{\check{\alpha}},m;n}(M,\hat{\mathcal{L}},\mathcal{K})\in\mathbb{Z}[1/2,1/h_1(M)] \tag{4.7}$$

Proof. Since $\mathcal{K}$ is homologically trivial, then $o = 1$ and also

$$\Delta_{\mathrm{A}}(M, \mathcal{K}; t) \in \mathbb{Z}[t, t^{-1}]. \tag{4.8}$$

Therefore relations (4.2), (4.4) and (4.5) imply that $\Delta_{\mathrm{A}}(M, \mathcal{K}; t)$ and $P_{\underline{\check{\alpha}};n}(M, \hat{\mathcal{L}}, \mathcal{K}; t)$ are polynomials of $(t^{1/2} - t^{-1/2})^2$

$$\Delta_{\mathrm{A}}(M, \mathcal{K}; t) = h_1(M) + \sum_{m \geq 1} a_m \, (t^{1/2} - t^{-1/2})^{2m}, \qquad a_m \in \mathbb{Z}, \tag{4.9}$$

$$P_{\underline{\check{\alpha}};n}(M, \hat{\mathcal{L}}, \mathcal{K}; t) = \sum_{m \geq 0} b_m \, (t^{1/2} - t^{-1/2})^{2m}, \qquad b_m \in \mathbb{Z}[1/2, 1/h_1(M)]. \tag{4.10}$$

The expansion of denominators in the sum of eq.(4.3) in powers of $(q^{\beta/2} - q^{-\beta/2})^2$ leads to eq.(4.6). □

To prove Main Theorem, we will also need some simple facts.

Lemma 4.3 (H. Murakami [13]). *If $f(q) \in \mathbb{Z}[q, q^{-1}]$ and its expansion in powers of $h = q - 1$ is*

$$f(q) = \sum_{n=0}^{\infty} a_n \, h^n, \tag{4.11}$$

then for any $N > 0$

$$f(q) = \sum_{n=0}^{N} a_n \, h^n + h^{N+1} x, \qquad \text{where } x \in \mathbb{Z}[q, q^{-1}]. \tag{4.12}$$

Lemma 4.4. *If $\check{K}$ is prime and $m, n \in \mathbb{Z}$, n not divisible by $\check{K}$, then there is an equlity in $\mathbb{Z}_{\check{K}}[\check{q}]$*

$$\check{q}^{mn\sigma} = (q^{m/n})^{\vee}, \tag{4.13}$$

where $q^{m/n}$ is understood as a power series in h

$$q^{m/n} = (1 + h)^{m/n} = \textstyle\sum_{k=0}^{\infty} \binom{m/n}{k} h^k \tag{4.14}$$

$$\binom{m/n}{k} = \frac{\frac{m}{n}\left(\frac{m}{n} - 1\right) \cdots \left(\frac{m}{n} - k + 1\right)}{k!} = \frac{m(m-n) \cdots (m - n(k-1))}{n^k \, k!}. \tag{4.15}$$

Proof. It is easy to see that if n is not divisible by $\check{K}$, then

$$\binom{m/n}{k} \in \mathbb{Z}_{(\check{K})} \tag{4.16}$$

and as a result

$$\binom{m/n}{k}^{\vee} = \binom{m\,(1/n)^{\vee}}{k} \in \mathbb{Z}_{\check{K}}. \tag{4.17}$$

Therefore according to eq.(4.14), the *r.h.s.* of eq.(4.13) can be presented in $\mathbb{Z}_{\check{K}}[\check{q}]$ as a convergent power series in $\check{h}$

$$(q^{m/n})^{\vee} = \sum_{k=0}^{\infty} \binom{m/n}{k}^{\vee} \check{h}^k = \sum_{k=0}^{\infty} \binom{m\,(1/n)^{\vee}}{k} \check{h}^k. \tag{4.18}$$

Since $\check{q}^{mn\sigma}$ does not depend on the choice of $n\sigma$ which satisfies eq.(2.13), then according to definition (2.15), for any positive integer N we can set $n\sigma = [(1/n)^{\vee}]_N$

and present the *l.h.s.* of eq. (4.13) in $\mathbb{Z}_{\check{K}}[\check{q}]$ also as a convergent power series in h

$$(4.19)\;\check{q}^{mn\sigma} = \check{q}^{m\,[(1/n)^{\vee}]_N} = (1+\check{h})^{m\,[(1/n)^{\vee}]_N} = \sum_{k=0}^{\infty} \binom{m\,[(1/n)^{\vee}]_N}{k} \check{h}^k.$$

Now eq. (4.13) follows from the fact that each term of the series (4.19) converges $\check{K}$-adicly to the corresponding term of the series (4.18) as $N \longrightarrow \infty$. □

Our proof of the Lawrence conjecture is based on a simple relation between gaussian sums and gaussian integrals. We introduce the following notation: for a function $f(\check{\beta})$,

$$(4.20)\qquad \sum_{\substack{0<\check{\beta}<\check{K}\\ \check{\beta}-\mathrm{odd}}}{}' f(\check{\beta}) = \sum_{\substack{0<\check{\beta}<\check{K}\\ \check{\beta}-\mathrm{odd}}} f(\check{\beta}) + \frac{1}{2} f(\check{K}).$$

Lemma 4.5. *For $\check{K}$ being an odd prime and for $p, m \in \mathbb{Z}$, p not divisible by $\check{K}$, consider the expression*

$$(4.21)\qquad X_{\mathrm{cycl.}} = \check{K}^{-1/2}\, e^{(i\pi/4)(\kappa-1)\,\mathrm{sign}(p)} \sum_{\substack{0<\check{\beta}<\check{K}\\ \check{\beta}-\mathrm{odd}}}{}' \check{q}^{4\sigma p \check{\beta}^2} (\check{q}^{m\check{\beta}} + \check{q}^{-m\check{\beta}}) \in \mathbb{C}$$

and a function of q

$$(4.22)\; X_{\mathrm{asympt.}} = (8K)^{-1/2}\, |p|^{1/2}\, e^{-(i\pi/4)\,\mathrm{sign}(p)} \int_{-\infty}^{+\infty} q^{(1/4)p\beta^2} (q^{m\beta} + q^{-m\beta})\, d\beta.$$

which is well-defined for $0 < |q| < 1$. Then

$$(4.23)\qquad X_{\mathrm{cycl.}} = \left(\frac{|p|}{\check{K}}\right)_{\mathrm{L}} \check{q}^{-m^2 p\sigma} \in \mathbb{Z}[\check{q}] \subset \mathbb{C},$$

while (4.22) can be extended analyticly to the vicinity of $q = 1$ and expanded in power series of $h = q - 1$

$$(4.24)\qquad X_{\mathrm{asympt.}} = q^{-m^2/p} \in \mathbb{Z}[1/p]\,[[h]].$$

Finally,

$$(4.25)\qquad X_{\mathrm{cycl.}} = \left(\frac{|p|}{\check{K}}\right)_{\mathrm{L}} (X_{\mathrm{asympt.}})^{\vee}.$$

Proof. We calculate the sum

$$\sum_{\substack{0<\check{\beta}<\check{K}\\ \check{\beta}-\mathrm{odd}}}{}' \check{q}^{4\sigma p \check{\beta}^2} (\check{q}^{m\check{\beta}} + \check{q}^{-m\check{\beta}}) = \sum_{\substack{-\check{K}<\check{\beta}\le\check{K}\\ \check{\beta}-\mathrm{odd}}} \check{q}^{4\sigma p \check{\beta}^2 + m\check{\beta}} = \sum_{\substack{-\check{K}<\check{\beta}\le\check{K}\\ \check{\beta}-\mathrm{odd}}} \check{q}^{4\sigma p\,(\check{\beta}+2mp\sigma)^2 - m^2 p\sigma}$$

$$(4.26)\qquad = \check{q}^{-m^2 p\sigma} \sum_{\substack{-\check{K}<\check{\beta}\le\check{K}\\ \check{\beta}-\mathrm{odd}}} \check{q}^{4\sigma p \check{\beta}^2}$$

$$= \check{q}^{-m^2 p\sigma} \sum_{-(\check{K}+1)/2<\check{\gamma}\le(\check{K}-1)/2} \check{q}^{4\sigma p\,(2\check{\gamma}+1)^2}$$

$$(4.27)\qquad = \check{q}^{-m^2 p\sigma} \sum_{-(\check{K}+1)/2<\check{\gamma}\le(\check{K}-1)/2} \check{q}^{p\,(\check{\gamma}+2\sigma)^2}$$

$$= \check{q}^{-m^2 p\sigma} \sum_{\check{\gamma}=1}^{\check{K}} \check{q}^{p\check{\gamma}^2}$$

$$(4.28)\qquad = \left(\frac{|p|}{\check{K}}\right)_{\mathrm{L}} e^{(i\pi/4)(1-\kappa)\,\mathrm{sign}(p)}\, \check{K}^{1/2}\, \check{q}^{-m^2 p\sigma}$$

In the last line we used the formula for the gaussian sum which can be found, for example, in [3], Chapter 6:

$$\sum_{\check\gamma=1}^{\check K} \check q^{p\check\gamma^2} = \left(\frac{|p|}{\check K}\right)_{\mathrm{L}} e^{(i\pi/4)(1-\kappa)\,\mathrm{sign}(p)}\,\check K^{1/2}. \tag{4.29}$$

Equation (4.23) follows from eq.(4.28).

Next we calculate the gaussian integral

$$\begin{aligned}\int_{-\infty}^{+\infty} q^{(1/4)p\beta^2}(q^{m\beta}+q^{-m\beta})\,d\beta &= 2q^{-m^2/p}\int_{-\infty}^{+\infty} q^{(1/4)\,p\,(\beta+2m/p)^2}\,d\beta\\ &= 2q^{-m^2/p}\int_{-\infty}^{+\infty} e^{(i\pi/2K)p\beta^2}\,d\beta\\ &= (8K)^{1/2}\,e^{(i\pi/4)\,\mathrm{sign}(p)}\,|p|^{-1/2}\,q^{-m^2/p}.\end{aligned} \tag{4.30}$$

Relation (4.24) follows from this equation, because $q^{-m^2/p} = (1+h)^{-m^2/p} \in \mathbb{Z}[1/p]\,[[h]]$. Equation (4.25) follows from eqs.(4.23), (4.24) and from Lemma 4.4, because the *r.h.s.* of eq.(4.25) is equal to $\left(\frac{|p|}{\check K}\right)_{\mathrm{L}}\left(q^{-m^2/p}\right)^{\vee}$. □

Corollary 4.6. *For $\check K$ being an odd prime, $m,p\in\mathbb{Z}$, p not divisible by $\check K$, consider the expression*

$$\begin{aligned}&Y_{\mathrm{cycl.}}(p,m,\check q)\\ &\quad= \check h^{-1}\,\check K^{-1/2}\,e^{(i\pi/4)(\kappa-1)\,\mathrm{sign}(p)} \sum_{\substack{0<\check\beta<\check K\\ \check\beta-\mathrm{odd}}} \check q^{4\sigma p\check\beta^2}(\check q^{\check\beta/2}-\check q^{-\check\beta/2})^{2m+2}\in\mathbb{C}\end{aligned} \tag{4.31}$$

and a function

$$\begin{aligned}Y_{\mathrm{asympt.}}(p,m,q) &= h^{-1}\,(8K)^{-1/2}\,|p|^{1/2}\,e^{-(i\pi/4)\,\mathrm{sign}(p)}\\ &\quad\times\int_{-\infty}^{+\infty} q^{(1/4)p\beta^2}(q^{\beta/2}-q^{-\beta/2})^{2m+2}\,d\beta.\end{aligned} \tag{4.32}$$

which is well-defined for $0<|q|<1$. Then $Y_{\mathrm{cycl.}}(p,m,\check q)\in\mathbb{Z}[\check q]\subset\mathbb{C}$ and it is $\check K$-adicly small

$$Y_{\mathrm{cycl.}}(p,m,\check q) = \check h^m\,x,\qquad x\in\mathbb{Z}[\check q]. \tag{4.33}$$

$Y_{\mathrm{asympt.}}(p,m,q)$ can be extended analytically to the vicinity of $q=1$ and expanded there in powers of $h=q-1$, so that $Y_{\mathrm{asympt.}}(p,m,q)\in\mathbb{Z}[1/p]\,[[h]]$ and it is asymptotically small

$$Y_{\mathrm{asympt.}}(p,m,q) = h^m\,x',\qquad x'\in\mathbb{Z}[1/p]\,[[h]]. \tag{4.34}$$

Finally,

$$Y_{\mathrm{cycl.}}(p,m,\check q) = \left(\frac{|p|}{\check K}\right)_{\mathrm{L}} (Y_{\mathrm{asympt.}}(p,m,q))^{\vee} \tag{4.35}$$

Proof. Obviously,

$$\check q^{\check K/2}-\check q^{-\check K/2}=0, \tag{4.36}$$

so we can replace the sum $\sum_{\substack{0<\check\beta<\check K\\ \check\beta-\text{odd}}}$ in the expression (4.31) for $Y_{\text{cycl.}}$ by the extended sum $\sum'_{\substack{0<\check\beta<\check K\\ \check\beta-\text{odd}}}$. Since for $m \geq 0$

$$(4.37)\qquad (q^{\beta/2} - q^{-\beta/2})^{2m+2} = \textstyle\sum_{n=0}^{m+1} a_{m,n}\,(q^{n\beta} + q^{-n\beta}), \qquad a_{m,n} \in \mathbb{Z},$$

then we conclude from eq.(4.23) and (4.25) that

$$(4.38)\quad \check h\, Y_{\text{cycl.}} \in \mathbb{Z}[\check q], \quad h\, Y_{\text{asympt.}} \in \mathbb{Z}[1/p]\,[[h]], \quad \check h\, Y_{\text{cycl.}} = \left(\frac{|p|}{\check K}\right)_{\mathrm{L}} (h\, Y_{\text{asympt.}})^{\vee}$$

Since

$$(4.39)\quad K^{-1/2}\int_{-\infty}^{+\infty} q^{(1/4)p\beta^2}(q^{\beta/2} - q^{-\beta/2})^{2m+2}\, d\beta = O(K^{-m-1}) \quad \text{as } K \to \infty,$$

then the expansion of $h\, Y_{\text{asympt.}}$ in powers of h starts at h^{m+1}. Therefore $Y_{\text{asympt.}}$ is divisible by h^m in $\mathbb{Z}[1/p]\,[[h]]$, $Y_{\text{cycl.}}$ is divisible by $\check h^m$ in $\mathbb{Z}[\check q]$ and the relations (4.33)–(4.35) follow from (4.38). □

5. Proof of Main Theorem

We start with Symmetry Principle formulated by R. Kirby and P. Melvin in [5].

Theorem 5.1 (R. Kirby, P. Melvin [5]). *Let $\hat{\mathcal{L}}$ be a framed L-component link in S^3. For a fixed integer j, $1 \leq j \leq L$, consider two sets of colors $\underline{\check\alpha}$ and $\underline{\check\alpha}'$ such that*

$$(5.1)\qquad \check\alpha'_k = \check\alpha_k \quad \text{if } k \neq j, \qquad \check\alpha'_j = \check K - \check\alpha_j.$$

Then

$$(5.2)\qquad J_{\underline{\check\alpha}'}(\hat{\mathcal{L}};\check q) = i^{\kappa l_{jj}}\,(-1)^{l_{jj}\check\alpha_j + \sum_{\substack{1\leq k\leq L\\ k\neq j}} l_{jk}(\check\alpha_k - 1)}\, J_{\underline{\check\alpha}}(\hat{\mathcal{L}};\check q),$$

here l_{jk} are the linking numbers of $\hat{\mathcal{L}}$ while κ is defined by eq.(2.14).

Proof of eq.(2.45) of Theorem 2.7. Our proof is a slightly shortened version of the proof of eq.(2.23) in [5]. The Symmetry Principle applied to the colors $\underline{\underline{\check\beta}}$ in the surgery formula (2.20) suggests that the sum over $\underline{\underline{\check\beta}}$ can be split into a sum over the elements of the symmetry group generated by the transformations (5.1) acting on $\underline{\underline{\check\beta}}$, and over the orbits of that action. Since according to our assumptions, $\check K$ is odd, then each orbit contains a set of colors $\underline{\underline{\check\beta}}$ which are all odd. Therefore for odd colors $\underline{\check\alpha}$ we can rewrite eq.(2.20) as

$$(5.3)\quad Z_{\underline{\check\alpha}}(M,\hat{\mathcal{L}};\check K) = i^{-L'}\,(2\check K)^{-L'/2}\,e^{-(3\pi i/4)\,\mathrm{sign}(\hat{\mathcal{L}}')}\,\check q^{(3/4)\,\mathrm{sign}(\hat{\mathcal{L}}')} \times X(\hat{\mathcal{L}}';\kappa) \sum_{\substack{0<\underline{\underline{\check\beta}}<K\\ \underline{\underline{\check\beta}}-\text{odd}}} \{\check q^{\underline{\underline{\check\beta}}/2} - \check q^{-\underline{\underline{\check\beta}}/2}\}\, J_{\underline{\check\alpha},\underline{\underline{\check\beta}}}(\hat{\mathcal{L}}\cup\hat{\mathcal{L}}';\check q),$$

where

$$(5.4)\qquad X(\hat{\mathcal{L}}';\kappa) = \sum_{J\subset\{1,\ldots,L'\}} i^{\kappa\sum_{j\in J} l'_{jj}}\,(-1)^{\sum_{j,k\in J,\, j\leq k} l'_{j,k}}$$

and l'_{ij} are the linking numbers of $\hat{\mathcal{L}}'$. Note that

$$(5.5)\qquad X(\hat{\mathcal{L}}';1) = \overline{X(\hat{\mathcal{L}}';-1)}$$

and the sum $X(\hat{\mathcal{L}}';\kappa)$ depends on K only through κ. Consider the invariant $Z(M;K)$ at $K=3$. Since $J_{\underline{\breve{\alpha}}}(\hat{\mathcal{L}};\breve{q})\Big|_{\underline{\breve{\alpha}}=1}=1$, then according to eq.(5.3),

$$Z(M;3)=2^{-L'/2}\,e^{-(i\pi/4)\,\mathrm{sign}(\hat{\mathcal{L}}')}\,X(\hat{\mathcal{L}}';-1). \tag{5.6}$$

A combination of eqs.(5.3), (5.5) and (5.6) shows that eq.(2.45) is satisfied if the invariant $Z'_{\underline{\breve{\alpha}}}(M,\hat{\mathcal{L}};\breve{K})$ was given by the expression

$$\begin{aligned} Z'_{\underline{\breve{\alpha}}}(M,\hat{\mathcal{L}};\breve{K}) &= i^{-L'}\,\breve{K}^{-L'/2}\,e^{-(i\pi/4)(\kappa+3)\,\mathrm{sign}(\hat{\mathcal{L}}')}\,\breve{q}^{(3/4)\,\mathrm{sign}(\hat{\mathcal{L}}')} \\ &\quad\times\sum_{\substack{0<\underline{\underline{\breve{\beta}}}<K\\ \underline{\underline{\breve{\beta}}}-\mathrm{odd}}}\{\breve{q}^{\underline{\underline{\breve{\beta}}}/2}-\breve{q}^{-\underline{\underline{\breve{\beta}}}/2}\}J_{\underline{\breve{\alpha}},\underline{\underline{\breve{\beta}}}}(\hat{\mathcal{L}}\cup\hat{\mathcal{L}}';\breve{q}). \end{aligned} \tag{5.7}$$

In fact, since

$$\frac{1-\kappa\breve{K}}{4}\equiv 4\sigma\pmod{\breve{K}}, \tag{5.8}$$

then it is easy to see that

$$e^{-(i\pi/4)(\kappa+3)\,\mathrm{sign}(\hat{\mathcal{L}}')}\,\breve{q}^{(3/4)\,\mathrm{sign}(\hat{\mathcal{L}}')}=e^{(i\pi/4)(\kappa+1)\,\mathrm{sign}(\hat{\mathcal{L}}')}\,\breve{q}^{3\,4\sigma\,\mathrm{sign}(\hat{\mathcal{L}}')} \tag{5.9}$$

and eq.(5.7) is equivalent to eq.(2.44). □

Proposition 5.2. *Suppose that $\breve{K}$ is an odd integer. Let $\hat{\mathcal{K}}\subset M$ be a framed knot with self-linking number $p\neq 0$. If M' is constructed by Dehn's surgery on $\hat{\mathcal{K}}$, then for odd colors $\underline{\breve{\alpha}}$*

$$\begin{aligned} Z'_{\underline{\breve{\alpha}}}(M',\hat{\mathcal{L}};\breve{K}) &= \breve{K}^{-1/2}\,\mathrm{sign}(p)\,e^{(i\pi/4)(\kappa-1)\,\mathrm{sign}(p)}\,\breve{q}^{3\,4\sigma\,\mathrm{sign}(p)} \\ &\quad\times\sum_{\substack{0<\breve{\beta}<\breve{K}\\ \breve{\beta}-\mathrm{odd}}}(q^{\breve{\beta}/2}-q^{-\breve{\beta}/2})\,Z'_{\underline{\breve{\alpha}},\breve{\beta}}(M,\hat{\mathcal{L}}\cup\hat{\mathcal{K}};\breve{K}). \end{aligned} \tag{5.10}$$

Proof. Suppose that M is constructed by a surgery on a framed link $\hat{\mathcal{L}}'\subset S^3$. Then M' can be constructed by a surgery on $\hat{\mathcal{L}}'\cup\hat{\mathcal{K}}$ with $\hat{\mathcal{K}}\subset S^3$ having a self-linking number

$$p_0=p+\sum_{j,k=1}^{L'}l'_{0j}\,l'_{0k}\,(l')^{-1}_{jk}, \tag{5.11}$$

here l'_{0j} are the linking numbers between $\mathcal{K}$ and components of $\hat{\mathcal{L}}'$ while $(l')^{-1}_{jk}$ is the inverse linking matrix of $\hat{\mathcal{L}}'$. Since

$$\mathrm{sign}(\hat{\mathcal{L}}'\cup\hat{\mathcal{K}})=\mathrm{sign}(p)+\mathrm{sign}(\hat{\mathcal{L}}') \tag{5.12}$$

and for $p\neq 0$

$$i^{-1}\,e^{(i\pi/4)(\kappa+1)\,\mathrm{sign}(p)}=\mathrm{sign}(p)\,e^{(i\pi/4)(\kappa-1)\,\mathrm{sign}(p)}, \tag{5.13}$$

then comparing eq.(2.44) for the surgeries on $\hat{\mathcal{L}}'$ and on $\hat{\mathcal{L}}'\cup\hat{\mathcal{K}}$ we arrive at eq.(5.10). □

The following two propositions will allow us to prove Main Theorem by induction.

Proposition 5.3. *The Main Theorem is true for $M=S^3$.*

Proof. According to eq.(2.44) applied in the trivial case of an empty surgery link,

$$Z'_{\underline{\check{\alpha}}}(S^3, \hat{\mathcal{L}}; \check{K}) = J_{\underline{\check{\alpha}}}(\hat{\mathcal{L}}; \check{q}). \tag{5.14}$$

Therefore relation (2.46) follows from (2.18). Equation (2.31) indicates that the TCC invariant $Z^{(\mathrm{tr})}_{\underline{\check{\alpha}}}(S^3, \hat{\mathcal{L}}; K)$ is the expansion of $J_{\underline{\check{\alpha}}}(\hat{\mathcal{L}}; q)$ in powers of h, so eq.(2.48) follows from Lemma 4.3 together with eq.(2.9) and the definition of $\check{K}$-adic convergence (2.11). □

Proposition 5.4. *Suppose that a $\mathbb{Q}$HS M' is constructed by a surgery on a homologically trivial framed knot $\hat{\mathcal{K}}$ in another $\mathbb{Q}$HS M. If the Main Theorem is true for M, then it is true for M'.*

Proof. Suppose that $h_1(M')$ is not divisible by an odd prime $\check{K}$. Since

$$h_1(M') = |p|\, h_1(M), \tag{5.15}$$

where p is the self-linking number of $\hat{\mathcal{K}}$, then neither p nor $h_1(M)$ is divisible by $\check{K}$. We assume that eq.(2.47) is true for the link $\hat{\mathcal{L}} \cup \hat{\mathcal{K}} \subset M$

$$Z'_{\underline{\check{\alpha}},\check{\beta}}(M, \hat{\mathcal{L}} \cup \hat{\mathcal{K}}; \check{K}) = \left(\frac{h_1(M)}{\check{K}}\right)_{\mathrm{L}} \sum_{n=0}^{\infty} P^{\vee}_{\underline{\check{\alpha}},\check{\beta};n}(M, \hat{\mathcal{L}} \cup \hat{\mathcal{K}})\, \check{h}^n. \tag{5.16}$$

On the other hand, if $\underline{\check{\alpha}}$ and $\check{\beta}$ are odd, then in view of eqs. (2.37), (4.6) and Lemma 4.3, for any $N > 0$

$$\begin{aligned} &\sum_{n=0}^{N} P_{\underline{\check{\alpha}},\check{\beta};n}(M, \hat{\mathcal{L}} \cup \hat{\mathcal{K}})\, h^n \\ &= \check{q}^{1/2}\, \check{h}^{-1}\, \check{q}^{(1/4)p\,(\check{\beta}^2-1)} d_{\underline{\check{\alpha}},m;n}(M, \hat{\mathcal{L}}, \mathcal{K}) \sum_{\substack{m,n\geq 0\\ m+n\leq N}} (\check{q}^{\check{\beta}/2} - \check{q}^{-\check{\beta}/2})^{2m+1}\, \check{h}^n + \check{h}^{N+1} x_1, \end{aligned} \tag{5.17}$$

where

$$d_{\underline{\check{\alpha}},m;n}(M, \hat{\mathcal{L}}, \mathcal{K}) \in \mathbb{Z}[1/2, 1/h_1(M)], \qquad x_1 \in \mathbb{Z}[q, q^{-1}, 1/2, 1/h_1(M)]. \tag{5.18}$$

Since we assumed that $\check{K}$ does not divide $h_1(M)$, then the first relation of (5.18) implies that $d_{\underline{\check{\alpha}},m;n}(M, \hat{\mathcal{L}}, \mathcal{K}) \in \mathbb{Z}_{(\check{K})}$. Therefore $d^{\vee}_{\underline{\check{\alpha}},m;n}(M, \hat{\mathcal{L}}, \mathcal{K})$ is well-defined and we can also use $\left[d^{\vee}_{\underline{\check{\alpha}},m;n}(M, \hat{\mathcal{L}}, \mathcal{K})\right]_N$ as its $\check{K}$-adic approximation. Thus equations (5.16) and (5.17) together with the relations (2.9) and

$$\check{q}^{(1/4)p\,(\check{\beta}^2-1)} = \check{q}^{4\sigma p\,(\check{\beta}^2-1)} \text{ if } \check{\beta} \text{ is odd}, \qquad \check{q}^{1/2} = -\check{q}^{2\sigma} \tag{5.19}$$

imply that for any N_0 there exists N_0' such that for any $N > N_0'$

$$\begin{aligned} Z'_{\underline{\check{\alpha}},\check{\beta}}(M, \hat{\mathcal{L}} \cup \hat{\mathcal{K}}; \check{K}) &= -\left(\frac{h_1(M)}{\check{K}}\right)_{\mathrm{L}} \check{q}^{2\sigma}\, \check{h}^{-1}\, \check{q}^{4\sigma p\,(\check{\beta}^2-1)} \\ &\times \sum_{\substack{m,n\geq 0\\ m+n\leq N}} \left[d^{\vee}_{\underline{\check{\alpha}},m;n}(M, \hat{\mathcal{L}}, \mathcal{K})\right]_N (\check{q}^{\check{\beta}/2} - \check{q}^{-\check{\beta}/2})^{2m+1}\, \check{h}^n + \check{K}^{N_0+2} x_2, \qquad x_2 \in \mathbb{Z}[\check{q}]. \end{aligned} \tag{5.20}$$

We substitute eq.(5.20) into eq.(5.10). Since

$$\check{K}^{1/2}\, e^{(i\pi/4)(\kappa-1)} = \sum_{\check{\gamma}=1}^{\check{K}} \check{q}^{\check{\gamma}^2} \in \mathbb{Z}[\check{q}] \tag{5.21}$$

and for odd $\check{\beta}$

(5.22)
$$\check{q}^{\check{\beta}/2} - \check{q}^{-\check{\beta}/2} = \check{q}^{1/2}(\check{q}^{(\check{\beta}-1)/2} - \check{q}^{-(\check{\beta}+1)/2}) = -\check{q}^{2\sigma}(\check{q}^{(\check{\beta}-1)/2} - \check{q}^{-(\check{\beta}+1)/2}) \in \mathbb{Z}[\check{q}],$$

then it is easy to see that the contribution of the term $\check{K}^{N_0+2}\,x_2$ to $Z'_{\check{\underline{\alpha}}}(M',\hat{\mathcal{L}};\check{K})$ is equal to

$$\check{K}^{-1/2}\,\mathrm{sign}(p)\,e^{(i\pi/4)(\kappa-1)\,\mathrm{sign}(p)}\,\check{q}^{3\,4\sigma\,\mathrm{sign}(p)} \sum_{\substack{0<\check{\beta}<\check{K}\\ \check{\beta}-\mathrm{odd}}} (\check{q}^{\check{\beta}/2} - \check{q}^{-\check{\beta}/2})\,\check{K}^{N_0+2}\,x_2 = \check{K}^{N_0+1}\,x_3, \qquad x_3 \in \mathbb{Z}[\check{q}]. \tag{5.23}$$

Thus we come to the following statement: for any $N_0 > 0$ there exists N such that

$$Z'_{\check{\underline{\alpha}}}(M',\hat{\mathcal{L}};\check{K}) = -\left(\frac{h_1(M')}{\check{K}}\right)_{\mathrm{L}} \mathrm{sign}(p)\,\check{q}^{4\sigma(2-p+3\,\mathrm{sign}(p))} \times \sum_{\substack{m,n\geq 0\\ m+n\leq N}} \left[d^{\vee}_{\check{\underline{\alpha}},m;n}(M,\hat{\mathcal{L}},\mathcal{K})\right]_N Y_{\mathrm{cycl.}}(p,m,\check{q})\,\check{h}^n \quad + \check{K}^{N_0+1}\,x_3, \qquad x_3 \in \mathbb{Z}[\check{q}], \tag{5.24}$$

where $Y_{\mathrm{cycl.}}(p,m,\check{q})$ is defined by eq.(4.31). Since eq.(4.33) implies that

$$Y_{\mathrm{cycl.}}(p,m,\check{q}) \in \mathbb{Z}[\check{q}],$$

then the claim (2.46) of Main Theorem for the manifold M' follows easily from eq.(5.24).

It remains to prove eq.(2.48) for M'. If we substitute the expression for the TCC invariant $Z^{(\mathrm{tr})}_{\underline{\alpha},\beta}(M,\hat{\mathcal{L}}\cup\hat{\mathcal{K}};K)$ from eq.(4.6) into the surgery formula (2.34) and use the relations (5.15) and

$$i\,e^{-(i\pi/2)\,\mathrm{sign}(p)} = \mathrm{sign}(p) \quad \text{if } p \neq 0, \tag{5.25}$$

then we find that

$$h_1^{1/2}(M')\,Z^{(\mathrm{tr})}_{\check{\underline{\alpha}}}(M',\hat{\mathcal{L}};K) = -\,\mathrm{sign}(p)\,q^{(2-p+3\,\mathrm{sign}(p))/4} \sum_{\substack{m,n\geq 0\\ m+n\leq N}} d_{\check{\underline{\alpha}},m;n}(M,\hat{\mathcal{L}},\mathcal{K})\,Y_{\mathrm{asympt.}}(p,m,q)\,h^n + h^{N+1}\,x_4, \qquad x_4 \in \mathbb{Z}[1/2, 1/h_1(M')][[h]], \tag{5.26}$$

where $Y_{\mathrm{asympt.}}(p,m,q)$ is defined by eq. (4.32). The remainder $h^{N+1}x_4$ in this formula comes from the terms $d_{\check{\underline{\alpha}},m;n}(M,\hat{\mathcal{L}},\mathcal{K})\,Y_{\mathrm{asympt.}}(p,m,q)$ with $m+n > N$ which were not included in the sum of eq. (5.26). Their contribution is estimated with the help of relation (4.34).

Now if we compare eqs.(5.24) and (5.26), then eq.(2.48) for M' follows easily from eq.(4.35) if we recall eq.(2.15) which relates $d^{\vee}_{\check{\underline{\alpha}},m;n}(M,\hat{\mathcal{L}},\mathcal{K})$ and $\left[d^{\vee}_{\check{\underline{\alpha}},m;n}(M,\hat{\mathcal{L}},\mathcal{K})\right]_N$ and if we use the relation

$$\left(q^{(2-p+3\,\mathrm{sign}(p))/4}\right)^{\vee} = \check{q}^{4\sigma(2-p+3\,\mathrm{sign}(p))} \tag{5.27}$$

which follows from Lemma 4.4. □

Corollary 5.5. *Main Theorem is true for a* $\mathbb{Q}$*HS* M *which is constructed by Dehn's surgery on an algebraically split framed link* $\hat{\mathcal{L}}' \subset S^3$.

Proof. The proof is by induction in the number L' of components of $\hat{\mathcal{L}}'$. Proposition 5.3 demonstrates that the claim is true for $L' = 0$, that is, when $\hat{\mathcal{L}}'$ is an empty link and as a result $M = S^3$. Suppose that the claim is true for L'-component links. Consider an $(L'+1)$-component algebraically split link $\hat{\mathcal{L}}'$. The surgery on the first L' components of $\hat{\mathcal{L}}'$ produces a $\mathbb{Q}$HS M', and the $(L'+1)$-st component of $\hat{\mathcal{L}}'$ is homologically trivial inside M'. Therefore Main Theorem is true for the surgery on the whole link $\hat{\mathcal{L}}'$ because of Proposition 5.4. □

Lemma 5.6. *Suppose that a $\mathbb{Q}$HS $M' = M\#L_{p,1}$ is a connected sum of a lens space $L_{p,1}$ and of a $\mathbb{Q}$HS M which contains a framed link $\hat{\mathcal{L}}$. Suppose that Main Theorem holds for M'. If an odd prime K divides neither p, nor $h_1(M)$, then the claim of Main Theorem is true for M.*

Proof. The lens space $L_{p,1}$ can be constructed by Dehn's surgery on a framed unknot in S^3 with self-linking number $-p$. Since

$$J_{\breve{\beta}}(\text{unknot}; q) = \breve{q}^{-p\,(\beta^2-1)/4}\,\frac{q^{\breve{\beta}/2} - q^{-\breve{\beta}/2}}{q^{1/2} - q^{-1/2}}, \tag{5.28}$$

then according to eq. (2.34)

$$\begin{aligned} Z^{(\mathrm{tr})}(L_{p,1};K) &= -i\,(8K)^{-1/2}\,e^{(3\pi i/4)\,\mathrm{sign}(p)}\,q^{-(3/4)\,\mathrm{sign}(p)}\,q^{p/4}\,(q^{1/2}-q^{-1/2})^{-1} \\ &\quad\times \int_{-\infty}^{+\infty} e^{-(i\pi/2K)p\beta^2}\,(e^{(i\pi/K)\beta} - e^{-(i\pi/K)\beta})^2\,d\beta \\ &= \mathrm{sign}(p)\,|p|^{-1/2}\,q^{(p+2/p-3\,\mathrm{sign}(p))/4}\,\frac{q^{1/2p}-q^{-1/2p}}{q^{1/2}-q^{-1/2}} \end{aligned} \tag{5.29}$$

(this formula was first derived by L. Jeffrey in [4]). Corollary 5.5 indicates that Main Theorem is true for $L_{p,1}$, so since $h_1(L_{p,1}) = |p|$ then

$$\begin{aligned} Z'(L_{p,1};\breve{K}) &= \left(\frac{|p|}{\breve{K}}\right)_{\mathrm{L}} \left(|p|^{1/2}\,Z^{(\mathrm{tr})}(L_{p,1};K)\right)^{\vee} \\ &= \left(\frac{|p|}{\breve{K}}\right)_{\mathrm{L}} \mathrm{sign}(p)\,\breve{q}^{4\sigma(p+2p\sigma-3\,\mathrm{sign}(p))}\,\frac{\breve{q}^{2\sigma p\sigma}-\breve{q}^{-2\sigma p\sigma}}{\breve{q}^{2\sigma}-\breve{q}^{-2\sigma}}. \end{aligned} \tag{5.30}$$

Note that since $\breve{q}^{2\sigma} = (\breve{q}^{2\sigma p\sigma})^p$, then

$$\frac{\breve{q}^{2\sigma}-\breve{q}^{-2\sigma}}{\breve{q}^{2\sigma p\sigma}-\breve{q}^{-2\sigma p\sigma}} \in \mathbb{Z}[\breve{q}] \tag{5.31}$$

and as a result

$$\frac{1}{Z'(L_{p,1};\breve{K})} \in \mathbb{Z}[\breve{q}]. \tag{5.32}$$

Suppose that an odd prime K divides neither p, nor $h_1(M)$. Since $h_1(L_{p,1}) = |p|$, then in view of multiplicativity of $h_1(M)$ under the operation of connected sum,

$$h_1(M') = h_1(L_{p,1})\,h_1(M) = |p|\,h_1(M) \tag{5.33}$$

and, as a result, K does not divide $h_1(M')$. Therefore, according to our assumptions, the claims of Main Theorem are true for M'

$$Z'_{\underline{\breve{\alpha}}}(M',\hat{\mathcal{L}};\breve{K}) \in \mathbb{Z}[\breve{q}], \tag{5.34}$$

$$Z'_{\underline{\breve{\alpha}}}(M',\hat{\mathcal{L}};\breve{K}) = \left(\tfrac{h_1(M')}{\breve{K}}\right)_{\mathrm{L}} \left(h_1(M')^{1/2}\,Z^{(\mathrm{tr})}_{\underline{\breve{\alpha}}}(M',\hat{\mathcal{L}};K)\right)^{\vee}. \tag{5.35}$$

Both quantum invariants $Z'_{\underline{\alpha}}(M,\hat{\mathcal{L}};K)$ and $Z^{(\mathrm{tr})}_{\underline{\alpha}}(M,\hat{\mathcal{L}};K)$ are also multiplicative under the operation of connected sum, so

$$Z'_{\underline{\check{\alpha}}}(M,\hat{\mathcal{L}};\check{K}) = Z'_{\underline{\check{\alpha}}}(M',\hat{\mathcal{L}};\check{K})/Z'(L_{p,1};\check{K}), \tag{5.36}$$

$$Z^{(\mathrm{tr})}_{\underline{\alpha}}(M,\hat{\mathcal{L}};K) = Z^{(\mathrm{tr})}_{\underline{\alpha}}(M',\hat{\mathcal{L}};K)/Z^{(\mathrm{tr})}(L_{p,1};K), \tag{5.37}$$

Therefore relation (2.46) follows from eq.(5.36) and from relations (5.34) and (5.32), while eq.(2.48) follows from eqs.(5.37), (5.35) and (5.30). □

The last statement that we need in order to prove Main Theorem is the lemma that H. Murakami introduced in [14] at the suggestion of T. Ohtsuki.

Lemma 5.7 (H. Murakami [14]). *For a* $\mathbb{Q}$*HS* M *and an odd prime* $\check{K}$ *there exists an algebraically split framed link* $\hat{\mathcal{L}}' \subset S^3$ *such that Dehn's surgery on* $\hat{\mathcal{L}}'$ *produces a* $\mathbb{Q}$*HS* M' *which is a connected sum of* M *and of a finite number of lens spaces* $L_{p_j,1}$, $1\leq j\leq N$,

$$M' = M\#L_{p_1,1}\#\cdots\#L_{p_N,1}, \tag{5.38}$$

such that neither of p_j *is divisible by* $\check{K}$.

Proof of Main Theorem. For a $\mathbb{Q}$HS M let $\check{K}$ be an odd prime number which does not divide $h_1(M)$. Consider the $\mathbb{Q}$HS M' of Lemma 5.7 which can be constructed by Dehn's surgery on a framed algebraically split link $\hat{\mathcal{L}}' \subset S^3$. Corollary 5.5 says that Main Theorem is true for M' and then Lemma 5.6 implies that it is also true for M. □

6. Habiro's cyclotomic expansion

Our proof of the Lawrence conjecture is a byproduct of the study of the properties of the $U(1)$-reducible connection contribution (4.3) to the WRT invariant of links in rational homology spheres. Although the series (4.3) is interesting in virtue of its manifest relation to classical topology (the Alexander polynomial appears in the denominator of each term of expansion), nevertheless, it might not present the best instrument for the study of number-theoretic properties of the WRT invariant.

Kazuo Habiro [2] suggested an alternative and potentially more powerful approach. Let $\mathcal{K}$ be a knot in S^3. Instead of expansions (4.3) and (4.6) for its colored Jones polynomial, Habiro suggests a different ('cyclotomic') expansion:

$$J_\alpha(\mathcal{K};q) = \sum_{n=0}^{\infty} C_{n,\mathcal{K}}(q)\, f_n(q,q^\alpha), \tag{6.1}$$

where

$$f_n(q,t) = \prod_{-n\leq\beta\leq n} \left((q^\beta t)^{1/2} - (q^\beta t)^{-1/2}\right) \tag{6.2}$$

while $C_{n,\mathcal{K}}(q) \in \mathbb{Z}[q^{\pm1}]$ and $C_{0,\mathcal{K}}(q,t) = 1$. The convenience of this presentation stems from an obvious fact that $f_n(q,q^\alpha)\equiv 0$ for $\alpha\in\mathbb{Z}_+$, $n\geq\alpha$. This means that only the first α terms in the sum (6.1) may be non-zero. Hence, if M is an integer homology sphere, constructed by a 1-surgery on $\mathcal{K}$, then one can replace the Jones polynomial $J_\alpha(\mathcal{K};q)$ in the Reshetikhin-Turaev surgery formula

$$Z(M;\check{K}) = e^{\frac{3}{4}i\pi}\,\check{q}^{\frac{1}{2}}\,(2\check{K})^{-\frac{1}{2}} \sum_{0<\alpha<\check{K}} \check{q}^{\frac{1}{4}\alpha^2}\,(q^{\frac{\alpha}{2}} - q^{-\frac{\alpha}{2}})\,J_\alpha(\mathcal{K};q) \tag{6.3}$$

by a truncated series $\sum_{n=0}^{N} C_{n,\mathcal{K}}(q)\, f_n(q, q^\alpha)$ with any $N \geq \check{K} - 2$. Then the sum in eq.(6.3) becomes purely Gaussian and it can be easily calculated with the help of eq.(4.28).

Pursuing this approach Habiro showed that if M is an integer homology sphere, then its TCC invariant has a presentation similar to (6.3):

$$Z^{(\mathrm{tr})}(M; K) = \sum_{n=0}^{\infty} C_{n,M}(q)\, f_n(q), \tag{6.4}$$

where

$$f_n(q) = \prod_{m=1}^{n} (q^m - 1), \qquad f_0(q) = 1, \tag{6.5}$$

while $C_{n,M}(q) \in \mathbb{Z}[q^{\pm 1}]$. Obviously, $f_n(\check{q}) = 0$ for $n \geq \check{K}$, so if we replace q with $\check{q}$ in the *r.h.s.* of eq.(6.4), then only the first $\check{K}$ terms will survive. Habiro showed that

$$Z(M; \check{K}) = \sum_{n=0}^{\check{K}-1} C_{n,M}(q)\, f_n(q). \tag{6.6}$$

This equation together with eq.(6.4) implies the Lawrence p-adic conjecture, yet the presentation (6.5) is more powerful than the mere statement of p-adic convergence.

K. Habiro and T. Le extended the cyclotomic presentations to WRT invariants based on other simple Lie algebras. For details see [9], references therein and their forthcoming papers.

Acknowledgements. I am thankful to R. Lawrence and A. Vaintrob for their comments and suggestions. I am greatly indebted to A. Vaintrob for his persistence and patience in helping me to improve the exposition of this paper. This work was supported by NSF Grants DMS-9704893 and DMS-0509793.

Appendix

Appendix A. Casson's invariant as the second coefficient of the TCC invariant

In [21] we used path integral arguments in the framework of the Chern-Simons quantum field theory of [31] in order to derive a much weaker version of Theorem 4.1. Considering that result as a conjecture, we used it to derive relations (2.42). Now that we have Theorem 4.1 at our disposal, we repeat the arguments of [21] in order to present a rigorous proof of (2.42) which in conjunction with Theorem 2.6 constitutes an alternative proof of Murakami's result (2.27).

The first of eqs.(2.42) was already established in [29], so it remains to show that for a $\mathbb{Q}$HS M

$$\frac{1}{3}\, h_1(M)\, P_1(M) = \lambda_{\mathrm{CW}}(M). \tag{A.1}$$

We have to establish some preliminary facts. We proved in [29] that for a knot $\mathcal{K} \subset M$

$$P_0(M, \mathcal{K}; t) = 1, \tag{A.2}$$

here $P_n(M,\mathcal{K};t)=1$ are the polynomials of eq. (4.3) in the case of an empty link $\hat{\mathcal{L}}$. A simple corollary of eqs. (A.2) and (4.5) is that if $\mathcal{K}$ is homologically trivial in M, then

$$d_{0;0}(M,\mathcal{K}) = 1/h_1(M), \tag{A.3}$$

$$d_{1;0}(M,\mathcal{K}) = \tfrac{1}{2}\, h_1^{-2}(M)\, \Delta''_{\mathrm{A}}(M,\mathcal{K}). \tag{A.4}$$

Here

$$\Delta''_{\mathrm{A}}(M,\mathcal{K}) = \frac{d^2}{dt^2}\, \Delta_{\mathrm{A}}(M,\mathcal{K};t)\big|_{t=1} \tag{A.5}$$

and the coefficients $d_{m,n}(M,\mathcal{K})$ come from eq. (4.6) written for an empty link $\hat{\mathcal{L}}$.

Suppose that a $\mathbb{Q}$HS M' is constructed by a surgery on a framed homologically trivial knot $\hat{\mathcal{K}}\subset M$ with self-linking number p. Then according to eq. (2.34), (2.37) and (4.6)

$$\begin{aligned} P_0(M') + h\,P_1(M') = {} & -i\,(8K)^{-1/2}\, e^{-(3\pi i/4)\,\mathrm{sign}(p)}\, q^{(3/4)\,\mathrm{sign}(p)}\, |p|^{1/2}\, (q^{1/2}-q^{-1/2})^{-1} \\ & \times \int_{-\infty}^{+\infty} q^{(1/4)p\,(\beta^2-1)} \Big(d_{0;0}(M,\mathcal{K})\,(q^{\beta/2}-q^{-\beta/2})^2 + d_{1;0}(M,\mathcal{K})\,(q^{\beta/2}-q^{-\beta/2})^4 \\ & + h\, d_{0;1}(M,\mathcal{K})\,(q^{\beta/2}-q^{-\beta/2})^2 \Big)\, d\beta + O(h^2) \end{aligned} \tag{A.6}$$

and as a result,

$$P_1(M') = \frac{1}{|p|}\left[\left(\frac{3}{4}\,\mathrm{sign}(p) - \frac{p}{4} - \frac{1}{2p} \right) d_{0;0}(M,\mathcal{K}) + d_{0;1}(M,\mathcal{K}) - \frac{6}{p}\, d_{1;0}(M,\mathcal{K}) \right], \tag{A.7}$$

or in view of eqs. (A.3), (A.4)

$$P_1(M') = \frac{1}{h_1(M')} \left(\frac{3}{p}\,\mathrm{sign}(p) - \frac{p}{4} - \frac{1}{2p} - \frac{3}{p}\,\frac{\Delta''_{\mathrm{A}}(M,\mathcal{K})}{h_1(M)} + h_1(M)\, d_{0;1}(M,\mathcal{K}) \right). \tag{A.8}$$

Proposition A.1. *Equation (A.1) is true for a $\mathbb{Q}$HS M constructed by a surgery on a framed algebraically split link $\hat{\mathcal{L}}'\subset S^3$.*

Proof. (*cf.* the proof of Corollary 5.5) We prove the proposition by induction in the number of components L' of $\hat{\mathcal{L}}'$.

If $L'=0$, then $M=S^3$. Since

$$\lambda_{\mathrm{CW}}(S^3) = 0, \tag{A.9}$$

then eq. (A.1) is true in view of eqs. (2.21) and (2.31).

Suppose that eq. (A.1) is true for all $\mathbb{Q}$HS constructed by surgeries on L'-component links in S^3. Let us prove that it is also true for manifolds constructed by surgeries on $(L'+1)$ component links. It is enough to prove the following: if eq. (A.1) is true for a $\mathbb{Q}$HS M then it is also true for a $\mathbb{Q}$HS M' constructed by Dehn's surgery on a framed homologically trivial knot $\hat{\mathcal{K}}\subset M$.

Equations (2.37), (2.38) imply that

$$Z^{(\mathrm{tr})}_{\beta}(M,\hat{\mathcal{K}};K) = h_1^{-1/2}(M) \sum_{\substack{m,n\geq 0 \\ |m|\leq n/2}} d_{m;n}(M,\mathcal{K})\, \beta^{2m+1}\, h^n. \tag{A.10}$$

If we substitute $\beta=1$ into this equation and take into account the relation

$$Z^{(\mathrm{tr})}_{\beta}(M,\hat{\mathcal{K}};K)\Big|_{\beta=1} = Z^{(\mathrm{tr})}(M;K), \tag{A.11}$$

then we conclude that

$$d_{0;1}(M,\mathcal{K}) = P_1(M). \tag{A.12}$$

Assume that eq. (A.1) is true for M. Then in view of eq. (A.12)

$$d_{0;1}(M,\mathcal{K}) = 3\lambda_{\mathrm{CW}}(M)/h_1(M), \tag{A.13}$$

and therefore according to eq. (A.8),

$$h_1(M')\,P_1(M') = \frac{1}{4}\,\mathrm{sign}(p) - \frac{p}{12} - \frac{1}{6p} - \frac{1}{p}\,\frac{\Delta''_{\mathrm{A}}(M,\mathcal{K})}{h_1(M)} + \lambda_{\mathrm{CW}}(M). \tag{A.14}$$

According to [30], the *r.h.s.* of this equation is equal to $\lambda_{\mathrm{CW}}(M')$, hence we proved eq. (A.1) for M'. □

Since a lens space $L_{p,1}$ is constructed by Dehn's surgery on an unknot in S^3, then Proposition A.1 implies that eq. (A.1) is true for $L_{p,1}$. Suppose that eq. (A.1) holds for a manifold $M' = M\#L_{p,1}$, with M being a $\mathbb{Q}$HS Since both $\lambda_{\mathrm{CW}}(M)$ and $(1/3)\,h_1(M)\,P_1(M')$ are additive under the operation of connected sum (the latter additivity follows from the multiplicativity of $Z^{(\mathrm{tr})}(M;K)$), then eq. (A.1) should also be true for M. In view of Lemma 5.7 and Proposition A.1, this proves eq. (A.1) for any $\mathbb{Q}$HS (*cf.* Proof of Main Theorem).

References

1. D. Bar-Natan, S. Garoufalidis, L. Rozansky, D. Thurston, *The Aarhus invariant of rational homology spheres II: Invariance and universality* Selecta Math. N.S. **8** (2002), 341-371.
2. K. Habiro, *On the quantum sl_2 invariants of knots and integral homology spheres,* In: 'Invariants of knots and 3-manifolds' (Kyoto, 2001), 55-68 (electronic), Geom. Topol. Monogr., 4, Geom. Topol. Publ., Coventry, 2002.
3. K. Ireland, M. Rosen, *A classical introduction to modern number theory,* Second Edition, Springer-Verlag, 1990.
4. L. Jeffrey, *Chern-Simons-Witten Invariants of lens spaces and torus bundles, and the semi-classical approximation* Commun. Math. Phys. **147** (1992), 563-604.
5. R. Kirby, P Melvin, *The 3-manifold invariants of Witten and Reshetikhin-Turaev for $sl(2,\mathbb{C})$* Invent. Math. **105** (1991), 473-545.
6. R. Lawrence, *Asymptotic expansions of Witten-Reshetikhin-Turaev invariants for some simple 3-manifolds* Journ. Math. Phys. **36** (1995), 6106-6129.
7. R. Lawrence, L. Rozansky, *Witten-Reshetikhin-Turaev invariants of Seifert manifolds,* preprint, to appear in Commun. Math. Phys.
8. T. Le, *On the perturbative $PSU(n)$ invariants of rational homology 3-spheres* Topology **39** (2000), 813-849.
9. T. Le, *Finite type invariants of 3-manifolds,* preprint math.GT/0507145.
10. T. Le, J. Murakami, T. Ohtsuki, *On a universal perturbative invariant of 3-manifolds* Topology **37** (1998), 539-574.
11. P. Melvin, H. Morton, *The coloured Jones function* Commun. Math. Phys. **169** (1995), 501-520.
12. G. Masbaum, J. D. Roberts, *A simple proof of integrality of quantum invariants at prime roots of unity* Math. Proc. Cam. Phil. Soc. **121** (1997), 443-454.
13. H. Murakami, *Quantum $SU(2)$ invariants dominate Casson's $SU(2)$ invariant* Math. Proc. Camb. Phil. Soc. **115** (1993), 253-281.
14. H. Murakami, *Quantum $SO(3)$-invariants dominate the $SU(2)$-invariant of Casson and Walker* Math. Proc. Camb. Phil. Soc. **117** (1995), 237-249.
15. T. Ohtsuki, *A polynomial invariant of integral homology 3-spheres* Math. Proc. Camb. Phil. Soc. **117** (1995), 83-112.
16. T. Ohtsuki, *A polynomial invariant of rational homology 3-spheres* Invent. Math. **123** (1996), 241-257.
17. T. Ohtsuki, *Finite type invariants of integral homology 3-spheres* Journ. of Knot Theory and its Ramifications **5** (1996), 101-115.

18. N. Reshetikhin, V. Turaev, *Invariants of 3-manifolds via link polynomials and quantum groups* Invent. Math. **103** (1991), 547-597.
19. L. Rozansky, *A large k asymptotics of Witten's invariant of Seifert manifolds* Commun. Math. Phys. **171** (1995), 279.
20. L. Rozansky, *Residue formulas for the large k asymptotics of Witten's invariants of Seifert manifolds. The Case of* $SU(2)$ Commun. Math. Phys. **178** (1996), 27-60.
21. L. Rozansky, *A contribution of the trivial connection to the Jones polynomial and Witten's Invariant of 3d manifolds I* Commun. Math. Phys. **175** (1996), 275-296.
22. L. Rozansky, *A contribution of the trivial connection to the Jones polynomial and Witten's Invariant of 3d manifolds II* Commun. Math. Phys. **175** (1996), 297-318.
23. L. Rozansky, *The trivial connection contribution to Witten's invariant and finite type invariants of rational homology spheres* Commun. Math. Phys. **183** (1997), 23-54.
24. L. Rozansky, *On finite type invariants of links and rational homology spheres derived from the Jones polynomial and Witten-Reshetikhin-Turaev invariant,* Proc. of the Aarhus Conf. Geometry and physics, (J. E. Andersen, J. Dupont, H. Pedersen, and A. Swann, eds.), Lecture notes in pure and applied mathematics 184 (1997) 379–397, Marcel Dekker, New-York.
25. L. Rozansky, *Witten's invariant of rational homology spheres at prime values of* K *and trivial connection contribution* Commun. Math. Phys. **180** (1996), 297-324.
26. L. Rozansky, *On p-adic convergence of perturbative invariants of some rational homology spheres* Duke Math. Journ. **91** (1998), 353-380.
27. L. Rozansky, *The universal R-matrix, Burau representation and the Melvin-Morton expansion of the colored Jones polynomial* Adv. in Math. **134** (1998), 1-31.
28. L. Rozansky, *A contribution of a* $U(1)$*-reducible connection to quantum invariants of links I: R-matrix and Burau representation*, preprint math.QA/9806004.
29. L. Rozansky, *A contribution of a* $U(1)$*-reducible connection to quantum invariants of links II: Links in rational homology spheres*, preprint math.QA/9806066.
30. K. Walker, *An extension of Casson's invariant,* Annals of Mathematics Studies, 126, Princeton University Press, Princeton, 1992.
31. E. Witten, *Quantum field theory and the Jones polynomial* Commun. Math. Phys. **121** (1989), 351-399.

Department of Mathematics, CB # 3250, Phillips Hall Chapel Hill, NC 27599
E-mail address: rozansky@math.unc.edu

Contemporary Mathematics
Volume **416**, 2006

Seiberg-Witten Integrable Systems and Periods of Rational Elliptic Surfaces

Yuji Shimizu

Dedicated to Professor Tetsuji Shioda on his 60-th birthday

ABSTRACT. We consider a rational elliptic surface with a section and a non-singular fiber. A marking for such a triple means a usual homological marking for the mixed Hodge structure on the associated open surface as well as a specification of a holomorphic 1-form on the chosen fiber.

We describe a symplectic structure on the moduli space of marked rational elliptic surfaces with solution to a differential equation. This last equation, which we call Seiberg-Witten differential equation, is the equation for the Seiberg-Witten differential which governs the Seiberg-Witten integrable system. Our integrable system is related with E-strings.

1. Introduction

In their celebrated work [**SW**] determining exact solution of $4DN = 2SU(2)$ supersymmetric Yang-Mills theory, Seiberg and Witten introduced the following elliptic curve:

$$y^2 = (x^2 - 1)(x - u) \quad (u \neq \pm 1, \infty)$$

The parameter u, which is introduced as a vacuum expectation value of (the square of) the scalar field in the $N = 2$ supermultiplet, describes the modulus of this elliptic curve. In the description of the exact solution, the next meromorphic differential introduced by them plays an important role.

$$\lambda = \frac{ydx}{1 - x^2} = u\frac{dx}{y} - \frac{xdx}{y}$$

The quantity

$$a = \int_\alpha \lambda, \; a_D = \int_\beta \lambda$$

based on this Seiberg-Witten differential $\lambda = \lambda_{SW}$ gives rise to an algebraically integrable system (cf.[**Do**]). (Here α, β are a symplectic basis of first homology

1991 *Mathematics Subject Classification.* Primary 14J15 14H70 ; Secondary 14J26 14J27.
Key words and phrases. Algebraic geometry, Integrable systems.

of the above elliptic curve.) The exact solution of the supersymmetric Yang-Mills (SYM) theory can be described by the prepotential $F = F(u)$ defined as

$$\frac{\partial F}{\partial a} = a_D$$

for a, a_D. For example, the modulus of this elliptic curve is given as

$$\tau = \frac{da_D}{du} \Big/ \frac{da}{du} = \frac{da_D}{da}.$$

The above story so far is for SYM theory without matter, but the one with matter can be treated as well. In that case, the Seiberg-Witten differential contains a term of a differential of the 3rd kind and the residue at hte pole (of order 1) corresponds to the mass.

The work of Seiberg and Witten was immediately extended to other gauge groups of A, D type other than $SU(2)$. In such an extension, the role of elliptic curves is played by the spectral curves. But the extension to the groups of E type was not straightforward and Minahan-Nemeschansky [**MN**] found the Seiberg-Witten differential on the elliptic curves. Lerche-Warner [**LW**] proposed the senario of relating the type IIB superstring theory compactified on K3 fibered Calabi-Yau 3-folds and the spectral curves in the E_6 case. Noguchi-Terashima-Yang [**NTY**] examined the F theory compactified on elliptic K3 surfaces by D3-brane probe and related D3-branes and elliptic curves of [**MN**]. Actually they constructed SYM theory with matter of E type from rational elliptic surfaces $\pi : X \to \mathbf{P}^1$ and their sections instead of K3 surfaces. There appears the Seiberg-Witten differential as a relative 1-form on each fiber of π. Their rational elliptic surface had the characteristic that its fiber at infinity is of type II in the Kodaira classification of singular fibers.

In this article, we extend the work of Noguchi-Terashima-Yang to the case where the fiber at infinity is non-singular and describes the moduli and periods of rational elliptic surfaces in order to construct SYM theory with matter of E type. We also formulate the notion of Seiberg-Witten differential and construct an integrable system for E series.

The content is as follows. In 2.1 we introduce the notion of marking necessary for the construction of the moduli which we explain using the deformation theory in 2.2. In 2.3, we describe the period domain of mixed Hodge structure corresponding to the pairs of a rational elliptic surface and its non-singular fiber. The main result of the section 2 is the infinitesimal Torelli theorem for the period mapping (2.4.1) and the local isomorphy (2.44.2), (2.4.3).

In 3.1, we introduce the Gauss-Manin connection for the elliptic fibration (π). In 3.2, we formulate the notion of Seiberg-Witten differential and consider its expansion at inifnity of the base $\mathbf{P}^1$. In 3.3, we construct an integrable system using the moduli space in section 2.

The content of this talk[1] is a part of joint work with A.Tsuchiya, H.Awata (Nagoya), A.Kato, Y.Saito (Tokyo). The author thanks them for their collaboration. He also thanks Profs. Masaki Morishita, Toshitake Kohno, and Jack Morava,

[1]It was also given at the conference "Approaches to open Calabi-Yau manifolds from algebraic geometry and string theory" at Hokkaido University (December 11–14, 2001).

the organizers of the conference "Primes and Knots" (March 7 - 16, 2003) at JAMI, as well as Japan-U.S. Mathematics Institute (JAMI) for support.

2. Moduli and periods of rational elliptic surfaces

A rational elliptic surface (abbreviated as RES) is an elliptic surface X which is birational to $\mathbf{P}^2$. X has a structure of an elliptic fibration $\pi : X \to B$ whose base space is $B \simeq \mathbf{P}^1$. Note that the elliptic fibration may be given as the anti-canonical map. We only consider a complex algebraic variety and/or a complex manifold throughout this article.

In the following, we consider only a rational elliptic surface with a section. For a section $\sigma : B \to X, \pi \cdot \sigma = id_B$, we put $S := \sigma(B)$. We say that two rational elliptic surfaces with a section $(X, S), (X', S')$ are isomorphic if there exists an isomorphism $f : X \to X'$ with $f(S) = S'$.

Denote the set of critical values of the elliptic fibration π by $\Sigma(\subset B)$. We fix a non-singular fiber $Y = \pi^{-1}(b)$ $(b \in B \setminus \Sigma)$ of π and consider the triple (X, S, Y).

REMARK 2.0.1. *For a Milnor fiber X° of an E_8 simply elliptic singular point, there exists a triple (X, S, Y) such that $X^\circ = X \setminus (Y \cup S)$.*

2.1. Marked rational elliptic surfaces. In order to consider the moduli space of triples (X, S, Y) consisting of a rational elliptic surface with a section and a non-singular fiber, let us introduce a marking, which kills a non-trivial automorphism. First let us examine the cohomology.

PROPOSITION 2.1.1. *The Betti numbers and the Hodge numbers are given as follows:*

$$b_0 = b_4 = 1, b_1 = b_3 = 0, b_2 = h^{1,1} = 10, h^{2,0} = h^{0,2} = 0$$

The long exact sequence for the pair (X, Y) is as follows:

$$0 \to H^0(Y, \mathbf{Z})(-1) \to H^2(X, \mathbf{Z}) \to H^2(X - Y, \mathbf{Z}) \to H^1(Y, \mathbf{Z})(-1) \to 0$$

Here (-1) means the Tate twist. The cohomology of $X - Y$ is an extension of $H^1(Y, \mathbf{Z})(-1)$ by $H^2(X, \mathbf{Z})/\mathbf{Z} \cdot [Y]$ ($[Y]$ denotes the cohomology class of Y).

Note that $H^1(Y, \mathbf{Z})$ and $H^2(X, \mathbf{Z})$ have an alternating form and a symmetric form respectively (by the cup product).

REMARK 2.1.2. *The above exact sequence is an exact sequence of mixed Hodge structures. The weight 2 part W_2 of $H^2(X - Y, \mathbf{Z})$ equals $H^2(X, \mathbf{Z})/\mathbf{Z} \cdot [Y]$ and the weight 3 part $Gr_3^W = W_3/W_2$ is $H^1(Y, \mathbf{Z})(-1)$.*

Similarly the mixed Hodge structure of $H^2(X - (Y \cup S), \mathbf{Z})$ has $H^2(X, \mathbf{Z})/(\mathbf{Z} \cdot [Y] \oplus \mathbf{Z} \cdot [S])$ as the weight 2 part W_2 and $H^1(Y, \mathbf{Z})(-1)$ as the weight 3 part $Gr_3^W = W_3/W_2$.

In order to define a marking, we need to fix an object, but we sparc it and fix a triple (X_0, S_0, Y_0) for reference once and for all.

DEFINITION 2.1.3 (marking). *1) A homological marking of the triple (X, S, Y) is a triple $\varphi_2, \varphi, \varphi_3$ of isomorphisms of $\mathbf{Z}$-modules which make the following diagram commutative*

$$\begin{array}{ccccccccc}
0 \to & H^0(Y, \mathbf{Z})(-1) & \to & H^2(X, \mathbf{Z}) & \to & H^2(X - Y, \mathbf{Z}) & \to & H^1(Y, \mathbf{Z})(-1) & \to 0 \\
 & \downarrow\simeq & & \varphi_2 \downarrow\simeq & & \varphi \downarrow\simeq & & \varphi_3 \downarrow\simeq & \\
0 \to & H^0(Y_0, \mathbf{Z})(-1) & \to & H^2(X_0, \mathbf{Z}) & \to & H^2(X_0 - Y_0, \mathbf{Z}) & \to & H^1(Y_0, \mathbf{Z})(-1) & \to 0
\end{array}$$

and such that φ_2 *(resp.* φ_3 *) preserves the symmetric form on* $H^2(X, \mathbf{Z})$ *(resp. the alternating form on* $H^1(Y, \mathbf{Z})$*) and that* $\varphi_2([Y]) = [Y_0]$ *and* $\varphi_2([S]) = [S_0]$ *holds.*

2) An analytic marking of the triple (X, S, Y) *is a non-zero form* $\omega \in H^0(Y, \Omega^1_Y)$.

3) We call such a datum $(X, S, Y, \varphi, \omega)$ *a marked rational elliptic surface.*

REMARK 2.1.4. *The Poincaré residue gives the isomorphism* $(\mathbf{C} \simeq) H^0(Y, \Omega^1_Y) \simeq H^0(X, \Omega^2_X(\log Y))$. *We denote the class on* X *corresponding to* ω *under this isomorphism by* Ω.

2.2. Moduli of marked rational elliptic surfaces. We construct analytically a moduli space of marked rational elliptic surfaces $(X, S, Y, \varphi, \omega)$. Miranda [**Mi**] constructed algebraically the moduli space of elliptic surfaces over $\mathbf{P}^1$ with only semistable singular fibers using Geometric Invariant Theory. However, we would like to have a space for those with all possible singular fibers and construct it using the Kuranishi space in the deformation theory of complex structure.

Now let us consider the pairs (X, Y) and $(X, Y + S)$. Since Y is a smooth divisor of X and $Y + S$ is a divisor with normal crossings on X, their Kuranishhi spaces $Def_{(X,Y)}, Def_{(X,Y+S)}$ are available by [**Kw**].

PROPOSITION 2.2.1. *Let* $(X, S, Y, \varphi, \omega)$ *be a marked rational elliptic surface. Then we have the following:*

$$h^i(X, T_X(-\log Y)) = \begin{cases} 0 & i = 0, 2 \\ 10 & i = 1 \end{cases}$$

$$h^i(X, T_X(-\log Y + S)) = \begin{cases} 0 & i = 0, 2 \\ 10 & i = 1 \end{cases}$$

Here $T_X(-\log Y)), T_X(-\log Y + S))$ *denote the sheaf of holomorphic vector fields which vanish along the divisors with normal crossings.*

According to this proposition, the Kuranishi spaces $Def_{(X,Y)}, Def_{(X,Y+S)}$ are smooth and their automophism groups are discrete. The automorphism of the pair (X, Y) (resp. $(X, Y + S)$) which preserves the homological marking φ is only the identity. Therefore we can glue the local deformation spaces of the triple (X, Y, φ) (resp. $(X, Y + S, \varphi)$) as well as the corresponding families over them.

DEFINITION 2.2.2. *We denote the resulting families by*

$$f_0 : (\mathcal{X}, \mathcal{Y}) \to \mathcal{N}_0 \qquad (resp.\ f : (\mathcal{X}, \mathcal{Y} + \mathcal{S}) \to \mathcal{N}).$$

Note that this construction entirely parallels that of the analytic moduli space of (homologically) marked K3 surfaces, cf. [**BPV**].

We denote by $\widetilde{\mathcal{N}_0}$ *(resp.* $\widetilde{\mathcal{N}}$*) the total space with zero-section deleted of the line bundles over* $\mathcal{N}_0$ *(resp.* $\mathcal{N}$*) corresponding to the sheaf* $f_{0*}(\Omega^1_{\mathcal{Y}/\mathcal{N}_0})$ *(resp.* $f_*(\Omega^1_{\mathcal{Y}/\mathcal{N}})$*).*

REMARK 2.2.3. *1) Pairs* (X, Y) *are called by Saito et al.* [**ST**] *as Okamoto-Painlevé pairs of elliptic type.*

2) For a general point $(X, Y, \varphi) \in N_0$, X *is obtained as the nine points blown-up of* $\mathbf{P}^2$ *and* $h^0(-K_X) = 1$ *holds.*

3) The forgetful map $u : \mathcal{N} \to \mathcal{N}_0$ *which maps* $(X, Y + S, \varphi)$ *to* (X, Y, φ) *is a local isomorphism.*

DEFINITION 2.2.4. *Consider the set* $\mathcal{M}_0 := \{(X, Y, \varphi) \in \mathcal{N}_0 \mid h^0(-K_X) = 2\}$ *(resp.* $\mathcal{M} := \{(X, Y + S, \varphi) \in \mathcal{N} \mid h^0(-K_X) = 2\}$*), which is an analytic closed set of* $\mathcal{N}_0$ *(resp.* $\mathcal{N}$*) by upper semi-continuity of the dimension of the cohomology of a coherent sheaf. We also put*

$$\widetilde{\mathcal{M}_0} := \widetilde{\mathcal{N}_0} \times_{\mathcal{N}_0} \mathcal{M}_0 \qquad (resp. \widetilde{\mathcal{M}} := \widetilde{\mathcal{N}} \times_{\mathcal{N}} \mathcal{M}).$$

REMARK 2.2.5. *We can analyze the space* $\mathcal{M}$ *by applying the deformation theory of holomorphic maps* [**Ho**] *to the elliptic fibration* π. *This approach is useful in 3.3.*

DEFINITION 2.2.6. *Let us denote the moduli space of (isomorphism classes of) marked elliptic curves* (Y, φ_1) *by* $\mathcal{E}\ell\ell$, *where* φ_1 *is a homological marking. Let us also denote the moduli space of (isomorphism classes of) triples* (Y, φ_1, ω) *by* $\widetilde{\mathcal{E}\ell\ell}$, *where* $\omega(\neq 0) \in H^0(Y, \Omega^1_Y)$ *is an analytic marking.*

Then we can consider a natural map $\mathcal{M} \to \mathcal{E}\ell\ell$ *(resp.* $\widetilde{\mathcal{M}} \to \widetilde{\mathcal{E}\ell\ell}$*) which associates* $(Y, \varphi_3(1))$ *(resp.* $(Y, \varphi_3(1), \omega)$*) to a marked rational elliptic surface* (X, S, Y, φ) *(resp.* $(X, S, Y, \varphi, \omega)$*).*

2.3. Periods of marked rational elliptic surfaces. We fix a triple (X_0, S_0, Y_0) consisting of a rational elliptic surface, its secton and a non-singular fiber.Let us consider the period domain of mixed Hodge structure for the cohmology $H^2(X_0 - Y_0, \mathbf{Z})$ (resp. $H^2(X_0 - (Y_0 \cup S_0), \mathbf{Z})$). We have seen their mixed Hodge structure in 2.1.2. Letting W be the weight filtration as usual, there exists a symmetric form ψ_2 on $Gr_2^W = W_2$ (resp. an alternating form ψ_3 on Gr_3^W) by cup product. These forms define a graded polarization $\psi = (\psi_2, \psi_3)$ on the mixed Hodge structure.

For th period domain, we refer the reader to [**SSU**] and the references therein.

DEFINITION 2.3.1. *Denote the period domain of mixed Hodge structure on* $(H^2(X_0 - Y_0, \mathbf{Z}), W, \psi)$ *(resp.* $(H^2(X_0 - (Y_0 \cup S_0), \mathbf{Z}), W, \psi)$*) with Hodge numbers* $(h^{1,1}, h^{2,1}, h^{1,2}) = (9, 1, 1)$ *(resp.* $(h^{1,1}, h^{2,1}, h^{1,2}) = (8, 1, 1)$*) by* $\mathcal{P}_0 = \mathcal{P}_{(X_0, Y_0)}$ *(resp.* $\mathcal{P} = \mathcal{P}_{(X_0, Y_0 + S_0)}$*).*

By definition, $\mathcal{P}_0$ (resp. $\mathcal{P}$) is the space of 2 step filters $F = \{F^p\}$ on $H^2(X_0 - Y_0, \mathbf{C})$ (resp. $H^2(X_0 - (Y_0 \cup S_0), \mathbf{C})$):

$$F^0 = F^1 \supset F^2 \supset F^3 = 0 \qquad \dim_{\mathbf{C}} F^1/F^2 = h^{1,1}.$$

In fact, the polarized Hodge structure on the weight 2 part Gr_2^W is unique, since it is isomorphic to a direct sum of Tate's Hodge structure $\mathbf{Z}(-1)$. On the other hand, the polarized Hodge structure on the weight 3 part Gr_3^W corresponds to the moduli of an elliptic curve by associating $Gr_3^W(1)$ of weight 1.

The coorespondence $F \mapsto F$ on $Gr_3^W(1)$ defines a map onto the upper half plane $\varpi_0 : \mathcal{P}_0 \to \mathbf{H}$ (resp. $\varpi : \mathcal{P} \to \mathbf{H}$). Its fiber is noting but the space $\mathrm{Ext}^1_{MHS}(Gr_3^W, W_2)$ of isomorphism classes of extensions of Gr_3^W by W_2. Due to Carlson's theorem [**Ca**], we have the following:

PROPOSITION 2.3.2.

$$\mathrm{Ext}^1_{MHS}(Gr_3^W, W_2) \simeq (W_2)_{\mathbf{Z}} \otimes_{\mathbf{Z}} J(Gr_3^W(1), F)$$

Here $J(Gr_3^W(1), F)$ *denotes the Jacobian variety* $Gr_3^W(1)_{\mathbf{C}}/(F(1)^1 + Gr_3^W(1)_{\mathbf{Z}})$ *for the polarized Hodge structure of weight 1* $(Gr_3^W(1), F, \psi_3(1))$.

Applying this proposition to the mixed Hodge structure $H^2(X - Y, \mathbf{Z})$ and $H^2(X - (Y \cup S), \mathbf{Z})$ for an arbitrary marked rational elliptic surface $(X, S, Y, \varphi, \omega)$, we obtain $J(Gr_3^W(1), F) \simeq Y$.

Recall a well-known fact on the periods of ellptic curves.

LEMMA 2.3.3. *The period map from the moduli space of marked elliptic curves to the period domain of weight 1 is an isomorphism:*

$$\mathcal{E}\ell\ell \xrightarrow{\sim} \mathbf{H}, \qquad \widetilde{\mathcal{E}\ell\ell} \xrightarrow{\sim} \widetilde{\mathbf{H}} := \{(\tau_1, \tau_2) \in \mathbf{C}^2 \mid \operatorname{Im}(\tau_2/\tau_1) > 0\}$$

DEFINITION 2.3.4. *Denote the period map which associates to a homologically marked rational elliptic surface (X, Y, φ) (resp. (X, S, Y, φ)) the mixed Hodge structure $H^2(X - Y, \mathbf{Z})$ (resp. $H^2(X - (Y \cup S), \mathbf{Z})$) by*

$$p_0 : \mathcal{N}_0 \to \mathcal{P}_0 \qquad (\text{resp. } p : \mathcal{N} \to \mathcal{P}).$$

2.4. Torelli type theorem for marked rational elliptic surfaces.

THEOREM 2.4.1 (Infinitesimal Torelli theorem). *For a homologically marked rational elliptic surface (X, Y, φ), the tangent map*

$$(dp_0)_{(X,Y,\varphi)} : T_{(X,Y,\varphi)}\mathcal{N}_0 \to T_{p_0(X,Y,\varphi)}\mathcal{P}_0$$

is an isomorphism.

COROLLARY 2.4.2. *p_0 is a local isomorphism.*

We fix (once and for all) a splitting of a natural homomorphism $H^2(X, \mathbf{Z})/\mathbf{Z} \cdot [Y] \to H^2(X, \mathbf{Z})/(\mathbf{Z} \cdot [Y] \oplus \mathbf{Z} \cdot [S])$ on the reference space. Then it induces a natural map $\mathcal{P} \hookrightarrow \mathcal{P}_0$. One can then show that the commutative square

$$\begin{array}{ccc} \mathcal{N} & \xrightarrow{p_0} & \mathcal{P}_0 \\ \cup & & \cup \\ \mathcal{M} & \xrightarrow{p} & \mathcal{P} \end{array}$$

is Cartesian. Then we can prove the next theorem.

THEOREM 2.4.3. *The period map $p : \mathcal{M} \to \mathcal{P}$ is a local isomorphism. Thus $\mathcal{M}$ is smooth.*

3. Seiberg-Witten integrable system of type E

In 4D $N = 2$ SYM theory, a rational 1-form $\lambda = \lambda_{SW}$ on an elliptic curve (or a spectral curve), called "Seiberg-Witten differential," played an important role. It depends meromorphically on a parameter u in the family of elliptic curves (or spectral curves). One of its characteristic is the relation

$$\frac{d}{du}\lambda = \omega$$

where ω is a holomorphic 1-form on the curve (at least in the case of elliptic curves), cf. Introduction. We formulate this notion in 3.1 using the Gauss-Manin connection of the ellptic fibration. Since

$$\frac{d}{du}\lambda = 0$$

is the equation for flat sections of the Gauss-Manin connection, a Seiberg-Witten differential is a solution to an inhomogeneous equation of linear ordinary differential equation. We will study its behviour at infinity in 3.2.

Seiberg-Witten integrable system is an algebraically integrable Hamiltonian system, cf. [**Do**]. There is a universal feature that the Liouville 1-form defining the Hamiltonian structure is obtained as the 1-form on the total space of parameters involved and curves. In 3.3 we construct such a Hamiltonian system in the case of E type using the consideration in 3.2.

3.1. Gauss-Manin connection on the elliptic fibration. Let us fix a rational elliptic fibration $\pi : X \to B$. Denote the set of critical values of π by $\Sigma(\subset B)$. The totality of the cohomologies $H^1(X_b, \mathbf{C})$ of non-singular fibers $X_b := \pi^{-1}(b)$ $(b \in B \setminus \Sigma)$ of π forms the de Rham cohomology sheaf

$$\mathcal{H}^1_{DR}(X/B) = R^1\pi_*(\Omega^{\cdot}_{X/B}).$$

In order to deal with mass in the SYM theory, we should introduce the points on the fibers which correspond to the poles of a Seiberg-Witten differential. In [**NTY**], they assume the poles being situated at the intersection of sections and fibers.

DEFINITION 3.1.1. *For a (rational) elliptic surface $\pi : X \to B$ with a section, denote the set of its sections by $MW(X/B)$. One can introduce a positive definite pairing on $MW(X/B)$ which we call the Mordell-Weil lattice, cf.* [**S**]

In what folllows, let us consider the case where the Mordell-Weil lattice $MW(X/B)$ is the E_8 root lattice, for simplicity. Denote the union of all sections corresponding to the 240 roots by E and put $E_b := X_b \cap E$. We may assume that E is a divisor with normal crossings over $B \setminus \Sigma$ by enlarging Σ if necessary.

DEFINITION 3.1.2. *Put $\mathcal{H}^1_{DR}(X - E/B) = R^1\pi_*(\Omega^{\cdot}_{X/B}(\log E))|_{B-\Sigma}$.
This underlies a variation of mixed Hodge structure and carries the Gauss-Manin connection*

$$\nabla : \mathcal{H}^1_{DR}(X - E/B) \longrightarrow \Omega^1_{B-\Sigma} \otimes \mathcal{H}^1_{DR}(X - E/B)$$

the flat sections of which form the local system $R^1\pi_*\mathbf{C}_{X-E}$ of rank 241.

Put $Y = X_{b_0}$ the chosen non-singular fiber for $b_0 \in B$. From the exact sequence over $B - \Sigma$

$$0 \to \pi^*\Omega^1_B(b_0) \to \Omega^1_X(\log E + Y) \to \Omega^1_{X/B}(\log E) \to 0$$

we immediately obtain the following:

LEMMA 3.1.3. $\Omega^2_X(\log E + Y) \simeq \pi^*\Omega^1_B(b_0) \otimes \Omega^1_{X/B}(\log E)$.
Therefore, one has $\pi_*\Omega^2_X(\log E + Y) \simeq \Omega^1_B(b_0) \otimes \pi_*\Omega^1_{X/B}(\log E)$.

Let $\Omega \in H^0(X, \Omega^2_X(\log Y))$ be the rational 2-form corresponding to the analytic marking $\omega \in H^0(Y, \Omega^1_Y)$ (2.1.4).

DEFINITION 3.1.4. *We call the equation*

$$\nabla(\lambda) = \Omega$$

for $\lambda \in \mathcal{H}^1_{DR}(X - E/B)$ the Seiberg-Witten (differential) equation. Here ∇ is the Gauss-Manin connection and Ω is regarded as a section of $\Omega^1_{B-\Sigma} \otimes \mathcal{H}^1_{DR}(X - E/B)$ via the composition of the next homomorphisms and by 3.1.3.

$$H^0(X, \Omega_X^2(\log Y)) \hookrightarrow H^0(X, \Omega_X^2(\log E + Y)) \xrightarrow{\sim} H^0(B, \pi_*\Omega_X^2(\log E + Y))$$

Δ being the set of roots of E_8, let us denote the section corresponding to $\alpha \in \Delta$ by $P_\alpha \in MW(X/B)$ so that we have $E = \sum_{\alpha\in\Delta} P_\alpha$.

LEMMA 3.1.5. *Let λ be a solution to the Seiberg-Witten equation. The the residue $m_\alpha := \mathrm{Res}_{P_\alpha}\lambda$ is flat.*

There is a freedom of adding flat sections to a solution λ of the Seiberg-Witten (differential) equation. We further make a condition of the linearity of residues of a solution λ by the requirement from the physics side.

DEFINITION 3.1.6. *A solution of the Seiberg-Witten equation is said to have linearity if there exists an $m \in \mathrm{Hom}_{\mathbf{Z}}(Q(E_8), \mathbf{C})$ such that $m_\alpha = m(\alpha)$ holds for $\forall\alpha \in \Delta$. Here $Q(E_8)$ is the E_8 root lattice.*

3.2. Expansion of solutions of the Seiberg-Witten differential equation. Given an analytically marked rational elliptic surface (X, S, Y, ω), we can talk about the Seiberg-Witten equation 3.1.4. Let us examine expansion of a solution at the point $b_0 = \pi(Y) \in B$.

Denote the inhomogeneous coordinate at b_0 by v and put $u = 1/v$. We make an identification $B \simeq \mathbf{P}^1$ by this choice. Let $\pi : X \to B$ correspond to the Weierstrass model $\pi' : X' \to B$. (If the singular fibers of π are all irreducible, then we have $\pi = \pi'$.)

Here again we consider the case where the Mordell-Weil lattice $MW(X/B)$ is the E_8 root lattice, for simplicity. Denote the union of all sections corresponding to 240 roots and put $E_b := X_b \cap E$. We assume that E is a divisor with normal crossings over $B \setminus \Sigma$ enlarging Σ if necessary.

Then we have the following:

LEMMA 3.2.1. *If we express the analytic marking as $\omega = \frac{dx}{y}$, then the 2-form on X (2.1.4) is expressed as*

$$\Omega = \frac{dv}{v} \wedge \frac{dx}{y}.$$

REMARK 3.2.2. *In the Weierstrass model of a rational elliptic surface, we glue two coordinate systems $(x, y), (X, Y)$ over the coordinates v and u of $\mathbf{P}^1$ as follows:*

$$u = \frac{1}{v}, \ x = v^2 X, \ y = v^3 Y$$

Then we have $\Omega = -du \wedge \frac{dX}{Y}$.

Let us denote the section corresponding to an E_8 root $\alpha \in \Delta$ by $P_\alpha \in MW(X/B)$ and the coordinates of P_α by (x_α, y_α).

We now express the equation for flat sections

$$\nabla\lambda = 0$$

of the Gauss-Manin connection ∇ of $\mathcal{H}^1_{DR}(X/B)$ in terms of coordinates. A section of $\mathcal{H}^1_{DR}(X/B)$ can be expressed around $v = 0$ as

$$\lambda = c_1(v)\frac{dx}{y} + c_2(v)\frac{xdx}{y}$$

Then the equation for flat sections becomes a system of linear ordinary differential equations of first order:

$$\frac{d}{dv}\begin{pmatrix} c_1 \\ c_2 \end{pmatrix} = \begin{pmatrix} p & q \\ r & s \end{pmatrix}\begin{pmatrix} c_1 \\ c_2 \end{pmatrix}$$

This translates into a single equation of second order concerning c_1:

$$\ddot{c_1} - (p + s + \frac{\dot{q}}{q})\dot{c_1} + (ps - qr + \frac{p\dot{q} - \dot{p}q}{q})c_1 = 0$$

Denote the second order operator in the left hand side by L ($\dot{f} = \frac{df}{dv}$).

REMARK 3.2.3. *If we express p, q, r, s in terms of the polynomials in v – the coefficients of powers of x – in the Weierstrass model of X, we find the discriminant of the cubic polynomial in the denominator. Since b_0 corresponds to a non-singular fiber, the matrix elements of the differential equation is regular.*

Now examine the expression of the Seiberg-Witten equation $\nabla\lambda = \Omega$ for a section λ of $\mathcal{H}^1_{DR}(X - E/B)$ in terms of coordinates. Since λ allows a pole, we have an expansion:

$$\lambda = c_1(v)\frac{dx}{y} + c_2(v)\frac{xdx}{y} + c_3(v)\sum_{\alpha\in\Delta}\frac{m_\alpha y_\alpha}{x - x_\alpha}\frac{dx}{y}.$$

It turns out that, from the Seiberg-Witten equation, c_3 is a constant and that the corresponding system of ordinary linear differential equations of first order has the form:

$$\frac{d}{dv}\begin{pmatrix} c_1 \\ c_2 \end{pmatrix} = \begin{pmatrix} p & q \\ r & s \end{pmatrix}\begin{pmatrix} c_1 \\ c_2 \end{pmatrix} + \begin{pmatrix} k \\ l \end{pmatrix}$$

Here p, q, r, s, l is holomorphic in a small neibourhood of b_0 but k has a singularity:

$$k = k(v) = \frac{1}{v} + (\text{holomophic at } v = 0).$$

If we translate it into a linear ordinary differential equation of order 2 for c_1, it becomes

$$Lc_1 + (s + \frac{\dot{q}}{q})k - ql - \dot{k} = 0.$$

PROPOSITION 3.2.4. *Given an analytically marked rational elliptic surface (X, S, Y, ω), the Seiberg-Witten equation around b_0 has the form*

$$Lc_1 = c_0.$$

Here L is the ordinary linear differential operator of second order corresponding to the Gauss-Manin connection of $\mathcal{H}^1_{DR}(X/B)$ and c_0 has such a singularity as

$$c_0(v) = -\frac{1}{v^2} + \frac{1}{v}c_0^{(-1)}(v) + c_0^{(0)}(v).$$

$c_0^{(-1)}(v), c_0^{(0)}(v)$ is regular around $v = 0$.

Moreover, $c_0^{(0)}(v)$ is additive with respect to m_α.

LEMMA 3.2.5. *Let L, c_0 be as in 3.2.4. Denote the fundamental solutions of the equation $Lc_1 = 0$ at the point b_0 by φ_1, φ_2. Then the equation $Lc_1 = c_0$ has*

$$\varphi_0 := \varphi_2(v)\int_{v_0}^{v}\varphi_1(t)\frac{c_0(t)}{W(t)}dt - \varphi_1(v)\int_{v_0}^{v}\varphi_2(t)\frac{c_0(t)}{W(t)}dt$$

as its special solution. ($W(t) = \varphi_1(t)\varphi_2'(t) - \varphi_1'(t)\varphi_2(t)$ *is the Wronskian for* φ_1, φ_2 *in* $v_0 \neq 0, |v_0| << 1$.) *It diverges at* $v = 0$ *of order* v^{-1} *and has also multivaluedness of* log.

The general solutions of the equation $Lc_1 = c_0$ *are given as* $\varphi_0 + d_1\varphi_1 + d_2\varphi_2 (d_1, d_2 \in \mathbf{C})$.

REMARK 3.2.6. *For any* $m_\alpha (\alpha \in \Delta)$, *the Seiberg-Witten equation* $Lc_1 = c_0$ *has a solution.*

Taking the multivaluedness of log *into account, the totality of the solutions is bijective to*

$$\varphi_0 + \mathbf{C}\varphi_1 + \mathbf{C}\varphi_2 / (\mathbf{Z}\varphi_1 + \mathbf{Z}\varphi_2)$$

using the notations of the above Lemma.

3.3. Seiberg-Witten integrable systems. Consider now the space of the Seiberg-Witten differentials with linearity, namely the local solutions at infinity of the Seiberg-Witten differential equation 3.1.4. It has a structure of an integrable system, which we think is the Seiberg-Witten integrable system of E type.

We introduced the natural projection $\mathcal{M} \to \mathcal{E}\ell\ell$ (resp. $\widetilde{\mathcal{M}} \to \widetilde{\mathcal{E}\ell\ell}$) from the moduli space of marked rational elliptic surfaces onto that of marked elliptic curves (2.2.6). Denote the relative cotangent bundle by $T^*(\mathcal{M}/\mathcal{E}\ell\ell)$.

PROPOSITION 3.3.1. *For* $(X, Y, S, \varphi) \in \mathcal{M}$, *we have the following natural isomorphism:*

$$T^*_{(X,Y,S,\varphi)}(\mathcal{M}/\mathcal{E}\ell\ell) \simeq \mathrm{Hom}_{\mathbf{Z}}(Q(E_8), \mathbf{C})$$

DEFINITION 3.3.2. *Let* $\mathcal{M}'$ *be the (open) subset of* $\mathcal{M}$ *of marked rational elliptic surfaces whose Mordell-Weil lattice is isomorphic to* $Q(E_8)$:

$$\mathcal{M}' := \{(X, Y, S, \varphi) \in \mathcal{M} \mid MW(X/B) \simeq Q(E_8)\ \}$$

Put also $T^*(\mathcal{M}'/\mathcal{E}\ell\ell) := T^*(\mathcal{M}/\mathcal{E}\ell\ell) \times_{\mathcal{M}} \mathcal{M}'$.

For each $m \in \mathrm{Hom}_{\mathbf{Z}}(Q(E_8), \mathbf{C})$, $m_\alpha = m(\alpha)(\alpha \in \Delta)$ is determined. Accordingly there is a two dimensional family of local solutions around b_0 of Seiberg-Witten differential equation (3.2.6) and also that of Seiberg-Witten differentials λ_m.

DEFINITION 3.3.3. *Let us denote by* SW' *the set of all 6-tuples* $(X, Y, S, \varphi, m, \lambda)$ *consisting of a point* (X, Y, S, φ, m) *of the relative cotangent bundle* $T^*(\mathcal{M}'/\mathcal{E}\ell\ell)$ *and Seiberg-Witten differential (around* b_0*)* λ *whose residues along the sections is* m.

There is a natural surjection $\mu : SW' \to T^*(\mathcal{M}'/\mathcal{E}\ell\ell)$.

LEMMA 3.3.4. $\mu : SW' \to T^*(\mathcal{M}'/\mathcal{E}\ell\ell)$ *is a fiber space whose fibers are isomorphic to* $\mathbf{C}^2/\mathbf{Z}^2 \simeq (\mathbf{C}^*)^2$. *Thus* SW' *is a complex manifold.*

THEOREM 3.3.5. *Put* $\widetilde{SW}' := SW \times_{\mathcal{M}} \widetilde{\mathcal{M}}$. *The natural projection* $\widetilde{SW}' \to \mathcal{E}\ell\ell$ *is a fiber space whose fibers are (holomorphic) symplectic manifolds.*

Moreover, the relative symplectic 2-form on $\widetilde{SW}'$ is conjectured to be somehow encoded in the exterior derivative of the tautological 1-form (determined by the Seiberg-Witten differential) on $\widetilde{SW}'$.

References

[AKSST] H. Awata, K Kato, Y. Saito, Y. Shimizu and A. Tsuchiya, Rational elliptic surfaces and the period domain, in preparation.

[BPV] W. Barth, C. Peters and A. Van de Ven, *Compact complex surfaces*, Springer-Verlag (1980).

[Ca] James A. Carlson, Extensions of mixed Hodge structures, in *Journées de Géometrie Algébrique d'Angers, Juillet 1979/Algebraic Geometry, Angers*, Sijthoff & Noordhoff, Alphen aan den Rijn—Germantown, Md. (1980) 107–127.

[Do] R.Donagi, Seiberg-Witten integrable systems, *Algebraic geometry—Santa Cruz 1995*, Proc. Sympos. Pure Math., 62, Part 2, Amer. Math. Soc., Providence, RI, (1997) 3–43.

[Ho] E. Horikawa, On deformations of holomorphic maps I, II, III, I. J. Math. Soc. Japan 25 (1973), 372–396; II. J. Math. Soc. Japan 26 (1974), 647–667; III. Math. Ann. 222 (1976), no. 3., 275–282.

[Ks] A. Kas, Weierstrass normal form and invariants of elliptic surfaces,Trans. Amer. math. Soc. **225** (1977) 259–266.

[Kw] A. Kawamata, On deformations of compactifiable complex manifolds, Math. Ann. **235** (1978) 247–265.

[LW] W. Lerche, N. P. Warner, Exceptional SW geometry from ALE fibrations, Phys. Lett. B 423 (1998), no. 1-2, 79–86.

[Lo] E. Looijenga, On the semi-universal deformation of a simple-elliptic hypersurface singularity. II. The discriminant. Topology 17 (1978), no. 1, 23–40.

[MN] J. A. Minahan, D. Nemeschansky, An $N = 2$ superconformal fixed point with E_6 global symmetry, Nuclear Phys. B 482 (1996), no. 1-2, 142–152 ; Superconformal fixed points with E_n global symmetry, Nuclear Phys. B 489 (1997), no. 1-2, 24–46.

[Mi] R.Miranda, The moduli of Weierstrass fibrations over $\mathbf{P}^1$, Math. Ann. 255 (1981), no. 3, 379–394.

[NTY] Masayuki Noguchi, Seiji Terashima and Sung-Kil Yang, $N = 2$ superconformal field theory with ADE global symmetry on a D3-brane probe, Nuclear Phys. B 556 (1999), no. 1-2, 115–151.

[Pv] V.P. Palamodov, Deformations of complex spaces, Russian Math.Surveys **31** (1976) 129–197.

[Sa] Hidetaka Sakai, Rational surfaces associated with affine root systems and geometry of the Painlevé equations, Dept. of Math. Kyoto Univ. Preprint series, 1999, No. 10.

[SSU] M-H. Saito, Y. Shimizu and S. Usui, Variation of mixed Hodge structure and the Torelli problem, Adv. Stud. in Pure Math. **10** (1987) 649–693 .

[ST] Masa-Hiko Saito and Taro Takebe, Classification of Okamoto–Painlevé pairs, preprint, 2000, Feb., math/AG 0006028.

[S] T.Shioda, *The theory of Mordell-Weil lattices and their applications*, Graduate School of Mathematical Science, Tokyo Univ. Lecture Note No.1 (1993).

[SW] N.Seiberg, E.Witten, Electric-magnetic duality, monopole condensation, and confinement in $N = 2$ supersymmetric Yang-Mills theory, Nuclear Phys. B 426 (1994), no. 1, 19–52; Monopoles, duality and chiral symmetry breaking in $N = 2$ supersymmetric QCD, Nuclear Phys. B 431 (1994), no. 3, 484–550.

Department of Mathematics, International Christian University, 10-2, Osawa 3-chome, Mitaka-shi, Tokyo 181-8585, JAPAN

E-mail address: shimizu@icu.ac.jp

Contemporary Mathematics
Volume **416**, 2006

On the finiteness of various Galois representations

Yuichiro Taguchi

Introduction. This is a written version of my talk under the same title at the Conference, except for the last section whose contents I did not mention in the talk. The first three sections reproduce rather closely the actual talk. This part is a survey on the finiteness of the number of isomorphism classes of various Galois representations. For mod p representations, we have already a survey article [**14**], but our emphasis is on the analogy with knot theory. Thus I tried to keep the article readable (at least in Section 1) to those with less background in number theory. The last section explains a certain relation between finiteness conjectures in the mod p and p-adic cases.

I thank JAMI and the organizers of the Conference, most notably Professors Jack Morava and Masanori Morishita, for inviting me to this exciting conference and for their hospitality during my stay at JAMI.

Let us begin with the following vague

Problem. Classify Galois representations

$$\rho : G_{K,S} \to \mathrm{GL}_d(E).$$

Here and elsewhere, we denote by $G_{K,S}$ the Galois group of a field K (= our "base field") unramified outside a finite set S of places of K (resp. the fundamental group of a 3-manifold K minus a link S). E is a coefficient field. It is often a topological field, and if so, we always consider continuous representations ρ from the Krull topology (resp. discrete toporolgy) of $G_{K,S}$. In this broad sense, such a problem is prevalent in both sides of Knots and Primes (cf. other articles in this volume).

This "Problem" in general is too hard (or too "soft") for me, so let me consider the following modest problem:

Finiteness Problem. In particular, do there exist only finitely many isomorphism classes of $\rho : G_{K,S} \to \mathrm{GL}_d(E)$ such that $***$?

The condition $***$ depends on the context. The adjective "various" in my title means that we have various cases (K, E and the conditions $***$) to consider, as follows: As the base field K, we may consider a local field [1] or a global field [2]. It will also be interesting to consider the case where K is an infinite algebraic extension

2000 *Mathematics Subject Classification.* 11G99, 11R32, 11S20.

[1] A finite extension of either the p-adic field $\mathbf{Q}_p$ or the power series field $\mathbf{F}_p((t))$ over the finite field $\mathbf{F}_p$ of p elements.

[2] A finite extension of the rational number field $\mathbf{Q}$ or the rational function field $\mathbf{F}_p(t)$ in one variable over $\mathbf{F}_p$.

of one of these fields. In this paper, however, we shall mainly be concerned with the case of algebraic number fields of finite degree over $\mathbf{Q}$. As the coefficient field E, we may consider $\mathbf{C}$, $\overline{\mathbf{Q}}_p$, $\overline{\mathbf{F}}_p$, $\overline{\mathbf{F}_p((T))}$, etc. Here, the bar $^-$ means an algebraic closure. We remark that considering representations with coefficients in, say $\overline{\mathbf{Q}}_p$, means in effect considering representations with coefficients in various subfields of $\overline{\mathbf{Q}}_p$ of finite degree over $\mathbf{Q}_p$.

If the answer to the finiteness question is Yes, then the next problem will be to *enumerate* them. It will be interesting, for example, to consider the generating function (or the zeta function) for them. I understand that the work of Sink [**28**] has been done from such a point of view, in search of a version of the Casson invariant of knots. As Morishita points out, it would be interesting to define a Casson invariant for algebraic number fields.

Now let me mention some motivation for these problems form the side of Primes. First, we have:

Serre's Conjecture ([**26**]). Any odd and irreducible $\rho : G_{\mathbf{Q}} \to \mathrm{GL}_2(\overline{\mathbf{F}}_p)$ is modular of type $(N(\rho), k(\rho))$.

Here, a representaion ρ of $G_{\mathbf{Q}}$ is said to be *odd* if $\det \rho(\text{complex conjugation}) = -1$. The positive integers $N(\rho)$ and $k(\rho)$, called respectively the Artin conductor outside p and the Serre weight, will be defined in Section 2. Note that we have $k(\rho) \leq p^2-1$.

Some consequences of this conjecture are:

- Any such ρ lifts to characteristic 0. (In this direction, R. Ramakrishna proved the liftablity in many cases ([**19**], [**20**]).)
- Up to isomorphisms, there exist only finitely many such ρ's with $N(\rho)$ dividing a fixed integer N. (This is because the space of cusp forms of a given weight and level is of finite dimension.)
- For $p = 2, 3, 5, 7$, there exists no such ρ unramified outside p. (This is because, after twisting ρ by a power of the cyclotomic character, one can normalize the Serre weight $k(\rho)$ to be at most $p+1$, and there is no cusp form of level 1 and weight less than 12.)

Next, we have:

Fontaine-Mazur's Conjecture ([**6**]). Let K be an algebraic number field of finite degree and S a finite set of finite places of K. Then there exist only finitely many isomorphism classes of irreducible geometric p-adic representations with bounded inertial level and fixed Hodge-Tate type.

The precise meaning of this statement will be explained in Section 3. Morally, it asserts that there exist only finitely many irreducible p-adic representations which look like the ones coming from a certain class of smooth projective varieites over K with good reduction outside S. Behind this conjecture is a recognition that there should exist only finitely many varieties over K (up to K-isomorphisms) in a certain class. As a typical example, we have a theorem of Faltings [**5**] (conjectured by Shafarevich [**27**]) that there exist only finitely many isomorphism classes of principally polarized abelian varieties over K of a given dimension with good reduction outside S.

Throughout the paper, K denotes an algebraic number field of finite degree over $\mathbf{Q}$, S a finite set of places of K, and p a fixed prime number. A representation $\rho : G_{K,S} \to \mathrm{GL}_d(E)$ is often thought of as an E-vector space V of dimension d

equipped with a continuous $G_{K,S}$-action, so that we may say "an E-representation V of $G_{K,S}$" in place of ρ.

1. Classical Case ($E = \mathbf{C}$). Let $G_K = \mathrm{Gal}(\overline{K}/K)$ be the absolute Galois group, and let $\rho : G_K \to \mathrm{GL}_{\mathbf{C}}(V)$ be a continuous representation, where V is a finite-dimensional $\mathbf{C}$-vector space. Here, we consider the coefficient field $\mathbf{C}$ with the discrete topology, so that the continuity means that ρ has finite image, since the Krull topology of G_K is totally disconnected. Note that ρ is unramified outside a finite set S of primes of K, so that ρ factors through the quotient $G_{K,S}$ of G_K. We define the *Artin conductor* $N(\rho)$ of ρ, which is an integral ideal of K, to be the product

$$N(\rho) \ := \ \prod_{\mathfrak{q}} \mathfrak{q}^{n_{\mathfrak{q}}(\rho)}$$

over all prime ideals of K, with

$$n_{\mathfrak{q}}(\rho) \ := \ \sum_{i=0}^{\infty} \frac{1}{(G_{\mathfrak{q},0} : G_{\mathfrak{q},i})} \dim_{\mathbf{C}}(V/V^{G_{\mathfrak{q},i}}),$$

where $V^{G_{\mathfrak{q},i}}$ denotes the fixed subspace of V by the action of the ith ramification subgroup $G_{\mathfrak{q},i}$ at $\mathfrak{q}$ of $G := \mathrm{Im}(\rho)$ (for the definition of ramification groups, see e.g. [**21**]). Although not obvious from the definition, $n_{\mathfrak{q}}(\rho)$ is an integer. The Artin conductor $N(\rho)$ measures how deeply ρ is ramified. In particular, we have $n_{\mathfrak{q}}(\rho) = 0$ if and only if ρ is unramified at $\mathfrak{q}$ (so $n_{\mathfrak{q}}(\rho) = 0$ if $\mathfrak{q} \notin S$). Thus the product in the definition of $N(\rho)$ is in fact a finite product, and the set S as above contains all the prime factors of $N(\rho)$.

One can easily prove the following proposition:

Proposition 1.1 (cf. [**1**]). *For any K, d, N, there exist only finitely many isomorphism classes of $\rho : G_K \to \mathrm{GL}_d(\mathbf{C})$ with $N(\rho)$ dividing N.*

Before giving a sketch of the proof, let us consider some possible link analogues of these things. Let $G = \pi_1(M \setminus L)$, where M is a 3-manifold and $L = K_1 \cup \cdots \cup K_r$ is a link in M. Let $m_i \in G$ be a meridian at K_i. Then the subgroup $\langle m_i \rangle$ generated by m_i will be the link analogue of the inertia subgroup $G_{\mathfrak{q},0}$ at $\mathfrak{q}$ (or perhaps its tame quotient $G_{\mathfrak{q},0}/G_{\mathfrak{q},1}$, since there is no "wild ramification" in characteristic 0). For a continuous representation $\rho : G \to \mathrm{GL}_{\mathbf{C}}(V)$, define $N(\rho)$ to be the formal sum $\sum_{i=1}^{r} n_i K_i$, where

$$n_i \ = \ \dim_{\mathbf{C}}(V/V^{\langle m_i \rangle}).$$

Or, one may simply think of $N(\rho)$ as an r-tuple $(n_1, ..., n_r)$. Then:

Question. Given an $N = (n_1, ..., n_r)$, how many $\rho : G \to \mathrm{GL}_{\mathbf{C}}(V)$ are there (up to isomorphisms) with $N(\rho) = N$?

Similarly, one may consider as follows: for $N = (n_1, ..., n_r)$, let $G(N)$ be the subgroup of G *normally* generated by $m_i^{n_i}$ for $r = 1, ..., r$. Then what is the structure of the quotient group $G/G(N)$? Is it finite or infinite? What (how many) representations $\rho : G/G(N) \to \mathrm{GL}_{\mathbf{C}}(V)$ are possible? Concerning these questions, the Referee communicated to me the following remark due to J. Hillman, for which I am grateful to both Hillman and the Referee.

Remark. The quotient group $G/G(N)$ is known as the orbifold fundamental group associated to a weighted link in M. In the most important special case where $r = 1$ and $M = S^3$, the group $G/G(N)$ maps onto $\mathbf{Z}/n\mathbf{Z}$ and the kernel is the orinary fundamental group of the n-fold branched cyclic cover of S^3, branched over the knot. In general such groups are usually infinite.

Now we turn to the proof of the above Proposition. By Jordan's Theorem, there exists an integer J depending only on d such that, for any finite subgroup G of $\mathrm{GL}_d(\mathbf{C})$, there exists an abelian normal subgroup H of G with index at most J. We apply this with $G = \mathrm{Im}(\rho)$. Let L be the extension of K corresponding to the kernel of ρ, and K_1 the fixed subfield by the subgroup H as in Jordan's theorem. Then the lower half K_1/K is bounded in degree, and the upper half L/K_1 is abelian. Both have bounded conductor. For the lower half, the Hermite-Minkowski Theorem, recalled below, implies that there exist only finitely many such extensions K_1/K. For the upper half, class field theory (or the finiteness of the ray class groups) implies that there exist only finitely many such extensions L/K_1.

Remark. We can prove similar statements for function fileds over finite fields or local fields, assuming suitable conditions such as the boundedness of the extension degree of the constant (or residue) fields.

Here we recall:

Theorem 1.2 (Hermite-Minkowski). *For any K, S and a natural number n, there exist only finitely many extensions of K of degree n which are unramified outside S.*

Motivated by the above discussion, let us agree to say[3] that a topologcal group G is *of type (HM)* if it has, for each $n \geq 1$, only finitely many closed subgroups of index n. A finitely generated group is of type (HM) (here and elsewhere, "finitely generated" means "topologically finitely generated"). The Galois group $G_{K,S}$ and the fundamental group $\pi_1(M \smallsetminus L)$ are of type (HM), but it is not known whether $G_{K,S}$ is finitely generated or not. There are infinitely generated profinite groups of type (HM), such as the direct product of ${\mathbf{Z}_p}^{\oplus p}$ for infinitely many different primes p (cf. [**23**], Chap. III, §4.1, Exercice 2; this example was communicated to me by M. Yamagishi, to whom I am grateful). What are other infinitely generated profinite groups of type (HM)? Can one find a nice condition for a profinite group of type (HM) to be finitely generated?

Also, to have the analogue for links of the finiteness of ideal class groups, we want $H_1(M', \mathbf{Z})$ to be finite (i.e. M' be a $\mathbf{Q}$-homology sphere) for any finite covering M' of M involved in the above discussion. Suppose M is a $\mathbf{Q}$-homology sphere and $M' \to M$ a finite covering of M ramified along a link in M. When is M' again a $\mathbf{Q}$-homology sphere?

2. Mod p Case ($E = \overline{\mathbf{F}}_p$). In this section, we consider similar problems as above for representations with coefficients in $\overline{\mathbf{F}}_p$. This case is interesting not only in its own right but also as the "residual case" of the p-adic case, because problems in the p-adic case (as in Sect. 3) are often solved by first considering their reduction modulo p and then lifting (or deforming) to characteristic 0. This kind of relations between the mod p and p-adic cases will be discussed in Section 4.

[3]I thank A. Tamagawa who informed me that such a profinite group was said to be of type (F) in [**23**], Chap. III, §4.1, and that, recently, the terminology "small" seemed standard for such a group. Thus I use the terminology "type (HM)" here only to preserve the atmosphere of the Conference.

Let G_K denote the absolute Galois groups of K. For a continuous representation $\rho : G_K \to \mathrm{GL}_d(\overline{\mathbf{F}}_p)$, its *Artin conductor outside* p is defined as follows:

$$N(\rho) := \prod_{\mathfrak{q} \nmid p} \mathfrak{q}^{n_{\mathfrak{q}}(\rho)}.$$

Here, the exponent $n_{\mathfrak{q}}(\rho)$ is defined by the same formula as in the classical case (Sect. 1), but the product is only over prime ideals $\mathfrak{q}$ which do not divide p. As before, the ρ in fact factors through the quotient $G_{K,S}$ of G_K if S contains the primes of K which divide $N(\rho)$.

We are concerned with:

Conjecture ($\mathbb{F}$). For any K, d, p, and any integral ideal N of K, there exist only finitely many isomorphism classes of semisimple representations $\rho : G_K \to \mathrm{GL}_d(\overline{\mathbf{F}}_p)$ with $N(\rho)$ dividing N.

Such a statement may have been considered vaguely since before, but as far as I know, Khare [**7**] and Moon [**12**] (independently) were the first to jot down this statement (as a "Conjecture" in the former, and as a "Problem" in the latter). First results in the case $d > 2$ were given in [**12**]. Some quantitative and effective results on the number of monomial mod p representations are given in [**13**].

Remarks. (1) We need to assume the semisimplicity. (Otherwise, for a fixed finite Galois extension L/K, there may exist infinitely many mutually unisomorphic embeddings $\mathrm{Gal}(L/K) \hookrightarrow \mathrm{GL}_d(\overline{\mathbf{F}}_p)$.)
(2) We need to bound the conductor $N(\rho)$.
(3) We do *not* need to bound the ramification of ρ at p. (This is because the ramification at p is automatically bounded because of the group structure of $\mathrm{GL}_d(\overline{\mathbf{F}}_p)$.)
(4) If we put a finite field $\mathbf{F}_q$ instead of $\overline{\mathbf{F}}_p$ in the Conjecture, then the finiteness follows from the Hermite-Minkowski Theorem.
(5) The case of $d = 1$ is OK by class field theory (or rather the finiteness of the ray class groups).
(6) So one may say that this statement generalizes both the Hermite-Minkowski Theorem and the finiteness of ideal class groups.
(7) The case of $d = 2$ is related to Serre's conjecture.
(8) The case of $d \geq 3$ is also related to a conjecture of Ash et al. ([**3**], [**2**]), which is a higher dimensional generalization of Serre's conjecture.
(9) If K is a local field, the finiteness holds true if we impose suitable conditions on the representations (such as the boundedness of the degree of the residual extensions).

Our first result in this section is:

Theorem 2.1 ([**15**]). *For any K, d, p, N, there exist only finitely many isomorphism classes of semisimple representations $\rho : G_K \to \mathrm{GL}_d(\overline{\mathbf{F}}_p)$ with $N(\rho)$ dividing N and $\mathrm{Im}(\rho)$ solvable.*

A similar statement is also true for function fields over finite fields under suitable conditions.

For the proof, we use a theorem of Larsen and Pink [**8**] instead of Jordan's. It states that there exist two constants J_1, J_2 depending only on d such that, for any finite subgroup G in $\mathrm{GL}_d(\overline{\mathbf{F}}_p)$ (in fact, the field $\overline{\mathbf{F}}_p$ here may be any field of characteristic $p \geq 0$) there exist normal subgoups G_i of G with $G \supset G_1 \supset G_2 \supset G_3$

such that
(1) G_1 has index at most J_1;
(2) G_1/G_2 is a product of a finite number of finite simple groups of Lie type in characteristic p, and the number of components is at most J_2;
(3) G_2/G_3 is an abelian group of order prime to p; and
(4) G_3 is a p-group.

We apply this to $G = \mathrm{Im}(\rho)$. If G is solvable, then the factor G_1/G_2 is trivial. So the extension L/K corresponding to $\mathrm{Ker}(\rho)$ is a successive extension $K \subset K_1 \subset K_3 \subset L$, with K_i corresponding to G_i. The bottom part K/K_1 is taken care of by Hermite-Minkowski, and the middle part K_3/K_1 by class field theory. The top part L/K_3 is a successive extension of elementary p-extensions, and the number of steps is bounded in terms of d, so this is also done by class field theory.

Next let us look at the non-existence or finiteness in the case where $K = \mathbf{Q}$, d and p are small, and $N = 1$.

In 1973, Tate [**30**] showed that there exists no irreducible representation $\rho : G_{\mathbf{Q}} \to \mathrm{GL}_2(\overline{\mathbf{F}}_2)$ unramified outside 2. Soon after this, Serre pointed out that the same method applies to $p = 3$.

In 1999, Brueggeman [**4**] proved the same for $p = 5$, assuming the GRH (= Generalized Riemann Hypothesis).

In 2000, Moon proved the following:

Theorem 2.2 ([**12**]). *Besides the results cited above, the finiteness* ($\mathbb{F}$) *for* $K = \mathbf{Q}$ *and* $N = 1$ *holds true in the following cases:*
(1) ($d \leq 2$; $p = 5$; totally real),
    ($d \leq 2$; $p = 7, 11, 13$; totally real and GRH).
(2) ($d \leq 4$; $p = 2$; totally real),
    ($d \leq 4$; $p = 2$; GRH),
    ($d \leq 4$; $p = 3$; totally real and GRH).
(3) ($d \leq 8$; $p = 2$; totally real and GRH).

Here, the condition "totally real" (resp. "GRH") in the last component of each of the triples describing the cases means that the finiteness holds true if we restrict to those ρ's such that the field L corresponding to $\mathrm{Ker}(\rho)$ is totally real (resp. if we assume the GRH).

These results are proved by comparing two different inequalities, called the Tate and Odlyzko bounds, for the discriminant $d_{L/\mathbf{Q}}$ (or the different $\mathcal{D}_{L/\mathbf{Q}}$) of $L/\mathbf{Q}$. The Tate bound is an upper bound and is proved algebraically. The Odlyzko bound is a lower bound and is proved analytically. One shows that these two inequalities contradict when the degree of L goes to infinity. Then one knows that the degree is bounded, and hence the set of extensions $L/\mathbf{Q}$ is finite by Hermite-Minkowski. So the set of isomorphism classes of semisimple representations is also finite.

The Tate bound is the following: If $d = 2$, then ([**30**]):

$$v_p(\mathcal{D}_{L/\mathbf{Q}}) \;\leq\; 2 + \frac{1}{p} - \frac{1}{(p-1)p^{m-1}}.$$

For a general d, if d is at most 2^r, then ([**12**]):

$$v_p(\mathcal{D}_{L/\mathbf{Q}}) \;\leq\; r + 1 + \frac{r}{p-1}.$$

There are several versions of the Odlyzko bound, one of which is the following asymptotic inequality ([**24**]):

$$\log |d_{L/\mathbf{Q}}| \;>\; r_1 \log(8\pi e^{\gamma+\pi/2}) + 2r_2 \log(8\pi e^{\gamma}) + o(n)$$

as $n = [L : \mathbf{Q}] \to \infty$. Effective versions of the Odlyzko bound are found, e.g., in [**17**].

To go further (by the same method) beyond the above theorem, we need to improve these inequalities, but it seems almost impossible to improve the Odlyzko bound. So we try to refine the Tate bound in terms of the Serre weight. For a mod p representation $\rho : G_{\mathbf{Q}_p} \to \mathrm{GL}_2(\overline{\mathbf{F}}_p)$ of the Galois group of the p-adic field $\mathbf{Q}_p$, Serre defined the Serre weight $k(\rho)$, which is an integer taking values between 1 and $p^2 - 1$. We define the *reduced Serre weight* $\tilde{k}(\rho)$ to be the minimum of the Serre weights of the twists of ρ by powers of the mod p cyclotomic character χ;

$$\tilde{k}(\rho) \;:=\; \min_{\alpha \in \mathbf{Z}} k(\chi^{-\alpha} \otimes \rho).$$

Note that χ, taking values in $\mathbf{F}_p^\times$, has order $p-1$.

If ρ is *wildly ramified*, then ρ has the following form when restricted to the inertia group I_p at p:

$$\rho|_{I_p} \sim \begin{pmatrix} \chi^{\beta} & * \\ & \chi^{\alpha} \end{pmatrix} = \chi^{\alpha} \begin{pmatrix} \chi^{k-1} & * \\ & 1 \end{pmatrix},$$

with $* \neq 0$ and some integers α, β, k; we normalize k by $2 \leq k \leq p$. Then we have

$$\tilde{k}(\rho) \;=\; \begin{cases} p+1, & \text{if } k = 2 \text{ and } \chi^{-\alpha} \otimes \rho \text{ is not finite,} \\ k, & \text{otherwise.} \end{cases}$$

Here, a representation of $G_{\mathbf{Q}_p}$ over $\overline{\mathbf{F}}_p$ is said to be *finite* if it comes from a finite flat groups scheme over $\mathbf{Z}_p$.

Proposition 2.3 ([**16**]). *If $\rho : G_{\mathbf{Q}_p} \to \mathrm{GL}_2(\overline{\mathbf{F}}_p)$ is wildly ramified, then the valuation of the different $\mathcal{D}_{L/\mathbf{Q}_p}$ of the extension $L/\mathbf{Q}_p$ corresponding to the kernel of ρ is:*

$$v_p(\mathcal{D}_{L/\mathbf{Q}_p}) \;=\; \begin{cases} 1 + \frac{k-1}{p-1} - \frac{k-1+\delta}{(p-1)p^m}, & \text{if } 2 \leq k \leq p, \\ 2 + \frac{2}{(p-1)p} - \frac{2}{(p-1)p^m}, & \text{if } k = p+1, \end{cases}$$

where the valuation v_p of L is normalized by $v_p(p) = 1$, $k = \tilde{k}(\rho)$ and $\delta =$ $\gcd(\alpha, \beta, p-1) = \gcd(\alpha, k-1, p-1)$.

To prove this, we consider the subextensions $L/K_1/K_0/\mathbf{Q}_p$ such that $K_0/\mathbf{Q}_p$ is unramified, K_1/K_0 is tamely ramified, and L/K_1 is wildly ramified. The tame part is well-known. The point is the calculation of the different of the wild part, which is done by the Führerdiskriminantenproduktformel:

$$d_{L/K_1} \;=\; \prod_{\psi} \mathfrak{f}_{\psi}.$$

Since $\mathrm{Gal}(K_1/K_0)$ acts on $\left(\begin{smallmatrix} 1 & * \\ & 1 \end{smallmatrix}\right)$ by $\chi^{\beta-\alpha} = \chi^{k-1}$, one knows which quotient of the unit group of K_1 the characters ψ factor through, and can calculate the conductor $\mathfrak{f}_{\psi}$.

Using this, we can prove the following theorem:

Theorem 2.4 ([**16**]). *When $K = \mathbf{Q}$, $d = 2$, $N = 1$, $2 \le p \le 31$, and $k \le p + 1$, the non-existence/finiteness of irreducible $\rho : G_{\mathbf{Q}} \to \mathrm{GL}_2(\overline{\mathbf{F}}_p)$ is summarized as follows:*

$k \backslash p$	2	3	5	7	11	13	17	19	23	29	31
2	$\times$	$\times$	$\times$	$\times$	$\times$	$\times$	$\times$	$\times$	$\times_{\mathrm{R}}$	$\times_{\mathrm{R}}$	$\times_{\mathrm{R}}$
3	$\times$	$\times$	$\times$	$\times$	$\times$	$\times$	$\times$	$\times$	f	f	f
4		$\times$	$\times$	$\times$	$\times_{\mathrm{R}}$	$\times_{\mathrm{R}}$	$\times_{\mathrm{R}}$	$\times_{\mathrm{R}}$	$\times_{\mathrm{R}}$	f_{R}	f_{R}
5			$\times$	$\times$	$\times$	$\times$	$\times$	$\times$	f	f	f
6			$\times_{\mathrm{R}}$	$\times_{\mathrm{R}}$	$\times_{\mathrm{R}}$	$\times_{\mathrm{R}}$	f_{R}	f_{R}	?	?	?
7				$\times$	$\times$	$\times$	$\times$	$\times$	f	f	f_{R}
8				?	?	?	?	?	?	?	?
9					$\times_{\mathrm{R}}$	$\times_{\mathrm{R}}$	$\times_{\mathrm{R}}$	$\times_{\mathrm{R}}$	f_{R}	f_{R}	f_{R}
10					?	?	?	?	?	?	?
11					$\times_{\mathrm{R}}$	$\times_{\mathrm{R}}$	$\times_{\mathrm{R}}$	$\times_{\mathrm{R}}$	f_{R}	f_{R}	f_{R}
12					$\exists$	$\exists$	$\exists$	$\exists$	?	$\exists$	$\exists$
13						f_{R}	$\times_{\mathrm{R}}$	$\times_{\mathrm{R}}$	f_{R}	f_{R}	f_{R}
14						?	?	?	?	?	?
15							f_{R}	f_{R}	f_{R}	f_{R}	f_{R}
16							$\exists$	$\exists$	$\exists$	$\exists$	?
17							?	?	?	f_{R}	f_{R}
18							$\exists$	$\exists$	$\exists$	$\exists$	$\exists$
19								?	?	?	?
20								$\exists$	$\exists$	$\exists$	$\exists$

Here, the meaning of the notations are as follows: There exists no such ρ which is unramified outside p and of reduced Serre weight k in the cases marked with $\times$, and the same is true if we assume the GRH in the cases marked with $\times_{\mathrm{R}}$; An f (resp. f_{R}) means that, unconditionally (resp. under GRH), there exist only finitely many such ρ in that case, and an $\exists$ means that there does exist an irreducible representation in that case. A ? means that the non-existence/finiteness is unknown (at present) in that case. For $k > 20$, the results are either $\exists$ or ?.

3. p-adic case ($E = \overline{\mathbf{Q}}_p$). Let us first recall some terminologies used in the conjecture of Fontaine and Mazur. Let E be a subfield of $\overline{\mathbf{Q}}_p$. A representation $\rho : G_K \to \mathrm{GL}_d(E)$ is said to be *geometric* if it is
(1) unramified outside a finite number of places, and
(2) potentially semistable at all places v dividing p.
(The condition "potentially semistable" is equivalent to being de Rham according to a recent work of L. Berger.)

Let $\rho : G_{K,S} \to \mathrm{GL}_E(V)$ be a geometric representation. The *inertial level* $\mathcal{L}(\rho)$ of ρ is a family $(\mathcal{L}_v(\rho))_v$, where, for each finite place v of K, $\mathcal{L}_v(\rho)$ is the maximal open normal subgroup of an inertia group I_v at v such that ρ becomes semistable when restricted there. One has $\mathcal{L}_v(\rho) = I_v$ is $v \notin S$. An *inertial level* $\mathcal{L} = (\mathcal{L}_v)_v$ for (K, S) is a family of open normal subgroups $\mathcal{L}_v$ of I_v for all finite places v of K such that $\mathcal{L}_v = I_v$ if $v \notin S$. The *E-Hodge-Tate type* $h(\rho)$ of ρ is a family $(h_v(\rho))_{v|p}$ of $h_v(\rho)$'s for all primes v dividing p, where $h_v(\rho)$ is the isomorphism class of the graded $(E \otimes_{\mathbf{Q}_p} K_v)$-module $\mathrm{Hom}_{\mathbf{Q}_p[G_{K_v}]}(V, \oplus_{r \in \mathbf{Z}} \mathbf{C}_p(r))$. An *$E$-Hodge-Tate type* for K at p is a family $(h_v)_{v|p}$, for all primes v dividing p, of isomorphism classes of graded free $(E \otimes_{\mathbf{Q}_p} K_v)$-modules h_v, all having a same rank (called the *degree* of h).

We denote by $\mathrm{Geom}(K, S, h; E)$ (resp. $\mathrm{Geom}(K, S, \mathcal{L}, h; E)$) the set of geometric irreducible E-representations of $G_{K,S}$ with E-Hodge-Tate type h (resp. with inertial level $\mathcal{L}$ and E-Hodge-Tate type h). Then the Finiteness Conjecture of Fontaine and Mazur ([**6**], Conj. 2) is:

Conjecture. For any $K, S, p, \mathcal{L}, h$, we have:
FM(a). The set $\mathrm{Geom}(K, S, \mathcal{L}, h; \overline{\mathbf{Q}}_p)$ is finite.
Further, if E is a *finite* extension of $\mathbf{Q}_p$, we have:
FM(b). The set $\mathrm{Geom}(K, S, \mathcal{L}, h; E)$ is finite.
FM(c). The set $\mathrm{Geom}(K, S, h; E)$ is finite.

We have the relations: FM(a) $\Rightarrow$ FM(b) $\Leftrightarrow$ FM(c). Here, the only nontrivial one is (b) $\Rightarrow$ (c), and this is proved by showing that, if E is of finite degree over $\mathbf{Q}_p$, then the inertial level of a geometric E-representation is automatically bounded. Our result is that FM(a) is true if we restrict ourselves to *potentially abelian* ρ's, i.e. those which become abelian after a finite extension of the base field K:

Theorem 3.1 ([**29**]). *There exist only finitely many isomorphism classes of irreducible potentially abelian geometric representations $\rho : G_{K,S} \to \mathrm{GL}_d(\overline{\mathbf{Q}}_p)$ with bounded inertial level and given Hodge-Tate type.*

This is proved by combining Serre's theory of locally algebraic representations ([**22**]) and the following lemma:

Lemma 3.2. *All such ρ become abelian after a* fixed *finite extension of K.*

This lemma, in turn, follows from a Jordan type theorem on "potentially toric" algebraic subgroups of the general linear group GL_d, together with the Hermite-Minkowski Theorem.

4. Relation between the mod p and p-adic cases.

There seems to be a substantial difference between Conjectures FM(a) and FM(b) of the last section. We expect that, in some sense, Conjecture $(\mathbb{F})$ of Section 2 fills this gap, i.e., that FM(b) + $(\mathbb{F})$ $\Rightarrow$ FM(a). At the moment of writing this paper, however, some technicality concerning the universal deformation of Galois representations has not been fully checked, so we give here a proof of a variant (Prop. 4.3 below) of the expected relation in a restricted situation, leaving the details to a future publication. Let E be an algebraic extension of $\mathbf{Q}_p$, with ring of integers $\mathcal{O}_E$, valuation ideal $\mathfrak{m}_E$, and residue field $k_E = \mathcal{O}_E/\mathfrak{m}_E$. We shall restrict the objective of the finiteness conjecture of Fontaine and Mazur to residually absolutely irreducible representations. Here we say that a continuous E-representation V of $G_{K,S}$ is *residually absolutely irreducible* if there is a $G_{K,S}$-stable $\mathcal{O}_E$-lattice T of V of which the reduction $\overline{V} = T/\mathfrak{m}_E T$ modulo $\mathfrak{m}_E$ is absolutely irreducible as a k_E-representation of $G_{K,S}$. Then the isomorphism class of $\overline{V}$ does not depend on the choice of the lattice T.

For our purpose, it is convenient to employ the language of deformation theory of Galois representations ([**10**], [**11**]). So let k be a finite field of characteristic p, and let $\overline{\rho} : G_{K,S} \to \mathrm{GL}_d(k)$ be an absolutely irreducible representation. (The condition "absolutely irreducible" can be weakned to that "$\mathrm{End}(\overline{\rho}) \simeq k$", but for simplicity, we content ourselves with this stronger condition.) Let $W = W(k)$ be the ring of Witt vectors over k. Let $\mathcal{D}$ be a "deformation data", i.e., a full subcategory of the category $\underline{\mathrm{Rep}}^{\mathrm{f}}_W(G_{K,S})$ of continuous $W[G_{K,S}]$-modules of finite length which is closed under taking subobjects, quotients and direct sums. Suppose that our $\overline{\rho}$ is in $\mathcal{D}$. Then by Ramakrishna ([**18**] or [**11**], Sect. 25), there exist a universal deformation ring $R_{\mathcal{D}}(\overline{\rho})$ of $\overline{\rho}$ of type $\mathcal{D}$ in the category $\mathcal{C}_W$ of complete noetherian local W-algebras with residue field equal to k, and a universal deformation $\rho_{\mathcal{D}}$: $G_{K,S} \to \mathrm{GL}_d(R_{\mathcal{D}}(\overline{\rho}))$ of $\overline{\rho}$ of type $\mathcal{D}$. The universality refers to the property that there exists a canonical bijection

$$\{\text{type-}\mathcal{D} \text{ deformations of } \overline{\rho} \text{ to } A\} \simeq \mathrm{Hom}_{W\text{-alg}}(R_{\mathcal{D}}(\overline{\rho}), A),$$

functorially in $A \in \mathcal{C}_W$.

Fontaine and Mazur formulated a deformation-theoretic version of their finiteness conjecture. To state this, for an inertial level $\mathcal{L}$ for (K, S) and two integers $a \leq b$, let us take, as our deformation data $\mathcal{D}$, the full subcategory $\underline{\mathrm{Rep}}^{\mathrm{f}}_W(G_{K,S})_{\mathrm{st},\mathcal{L},[a,b]}$ of $\underline{\mathrm{Rep}}^{\mathrm{f}}_W(G_{K,S})$ consisting of objects T such that, for each $v \in S$,

(1) if $v \nmid p$, then $\mathcal{L}_v$ acts trivially on T; and

(2) if $v \mid p$, there exists a semistable p-adic representation V of $\mathcal{L}_v$ such that $(\mathbf{C}_v(r) \otimes_{\mathbf{Q}_p} V)^{\mathcal{L}_v} = 0$ if $r \notin [a, b]$ and that T is isomorphic to a subquotient of V as a $\mathbf{Z}_p[\mathcal{L}_v]$-module.

Let $\overline{\rho} \in \mathcal{D}$ be an absolutely irreducible representation. Then the deformation-theoretic version of the finiteness conjecture of Fontaine and Mazur ([**6**], Conj. 5) is:

Conjecture FM(D). The universal deformation ring $R_{\mathcal{D}}(\overline{\rho})$ is finite as a W-algebra.

A priori, this $R_{\mathcal{D}}(\overline{\rho})$ is not known to be flat over W (although it seems quite possible that such a deformation ring is always flat). Since we are interested in E-representations (rather than torsion $\mathcal{O}_E$-representations), let us consider the image $R_{\mathcal{D}}(\overline{\rho})_0$ of the natural map $R_{\mathcal{D}}(\overline{\rho}) \to R_{\mathcal{D}}(\overline{\rho}) \otimes_W F$, where F is the fraction field of W. Then FM(D) implies:

Conjecture FM($\mathbf{D}_0$). The W-algebra $R_{\mathcal{D}}(\overline{\rho})_0$ is finite.

In order to see the relation of FM(D_0) with FM(b), it is convenient to introduce a slightly wider (a priori) class of representations than geometric ones. Let $E/\mathbf{Q}_p$ be a *finite* extension, $\mathcal{L}$ an inertial level for (K, S), and $a \leq b$ two integers. We say that an E-representation V of $G_{K,S}$ is *piecewise geometric* with inertial level bounded by $\mathcal{L}$ and Hodge-Tate weights in $[a, b]$ if it admits a $G_{K,S}$-stable $\mathcal{O}_E$-lattice of which all quotients of finite length lie in $\mathcal{D} = \underline{\mathrm{Rep}}^{\mathrm{f}}_W(G_{K,S})_{\mathrm{st},\mathcal{L},[a,b]}$. A typical example of such a representation is one which arises from the type-$\mathcal{D}$ universal deformation of some residual representation $\overline{\rho}$ by way of a continuous W-algebra homomorphism $\phi : R_{\mathcal{D}}(\overline{\rho}) \to E$.

Let $\mathrm{Geom}'_d(K, S, \mathcal{L}, [a, b]; E)$ denote the set of isomorphism classes of d-dimensional irreducible piecewise geometric representations with inertial level bounded by $\mathcal{L}$ and Hodge-Tate weights in $[a, b]$. We also denote by $\mathrm{Geom}'_d(K, S, \mathcal{L}, [a, b]; \overline{\mathbf{Q}}_p)$ the inductive limit of these sets when E moves through the set of finite extensions of $\mathbf{Q}_p$ contained in $\overline{\mathbf{Q}}_p$. A geometric representation is trivially piecewise geometric, so that $\mathrm{Geom}(K, S, \mathcal{L}, h; E) \subset \mathrm{Geom}'_d(K, S, \mathcal{L}, [a, b]; E)$ if d is the degree of the Hodge-Tate type h and $[a, b]$ contains all the Hodge-Tate weights of h. Conversely, Fontaine and Mazur point out ([**6**], Rem. (a) after Conj. 5) that it should not be very hard to prove that:

Conjecture FM(P). A piecewise geometric representation with bounded inertial level and Hodge-Tate weights is geometric.

At this moment, I do not know to what extent this statement is proved.

On the contrary, a piecewise geometric representation with *unbounded* inertial level or Hodge-Tate weights may not be geometric (i.e. an E-representation may not be geometric no matter how its subquotients of finite length come from geometric representations, if their inertial levels are unbounded or their Hodge-Tate weights do not stay in a fixed interval $[a, b]$).

Now following Fontaine and Mazur, we may state:

Conjecture. For any $K, S, \mathcal{L}, a \leq b$, we have:
FM′(a). The set $\mathrm{Geom}'_d(K, S, \mathcal{L}, [a, b]; \overline{\mathbf{Q}}_p)$ is finite.
FM′(b). The set $\mathrm{Geom}'_d(K, S, \mathcal{L}, [a, b]; E)$ is finite for any finite extension E of $\mathbf{Q}_p$.

We have:

$$\begin{array}{ccc} \mathrm{FM}'(\mathrm{a}) & \Rightarrow & \mathrm{FM}(\mathrm{a}) \\ \Downarrow & & \Downarrow \\ \mathrm{FM}'(\mathrm{b}) & \Rightarrow & \mathrm{FM}(\mathrm{b}) \\ & & \Updownarrow \\ & & \mathrm{FM}(\mathrm{c}). \end{array}$$

In the rest of this paper, we concentrate on residually absolutely irreduible representations. For a set (or a conjecture) X, we denote by X_{rair} the corresponding object which concerns only with such representations; thus $\mathrm{Geom}'_d(K, S, \mathcal{L}, [a, b]; \overline{\mathbf{Q}}_p)_{\mathrm{rair}}$ is the subset of isomorphism classes of residulally absolutely irreducible representations in $\mathrm{Geom}'_d(K, S, \mathcal{L}, [a, b]; \overline{\mathbf{Q}}_p)$, and the conjecture $\mathrm{FM}'(\mathrm{a})_{\mathrm{rair}}$ states that this set is finite.

After Conj. 5 in [**6**], it is explained how FM(D) [4] implies FM(b), but what is actually implied is a little bit weaker (the assertion "the fact that V is irreducible implies that one can find an $\mathcal{O}_E$-lattice T of V stable under $G_{K,S}$ and such that, if $\overline{V} = k \otimes_{\mathcal{O}_E} T$, the map $k \to \mathrm{End}_{k[G_{K,S}]}(\overline{V})$ is an isomorphism" is not true in general, so FM(D) implies nothing about those V which do not have the above mentioned property). However, the proof there does show that FM(D) (or even $\mathrm{FM}(\mathrm{D}_0)$) implies $\mathrm{FM}'(\mathrm{b})_{\mathrm{rair}}$. In fact, the converse is also true:

Proposition 4.1. $\mathrm{FM}'(\mathrm{b})_{\mathrm{rair}} \iff \mathrm{FM}(\mathrm{D}_0)$.

Proof. Suppose $R_{\mathcal{D}}(\overline{\rho})_0$ is not finite over the Witt ring W. Then by factoring by a minimal prime, we would have an integral quotient R of $R_{\mathcal{D}}(\overline{\rho})_0$ which contains W and is not finite over W. By the structure theorem on complete noetherian local integral ring (e.g. [**9**], Sect. 29), such an R is finite over a subring R_0 which is isomorphic to a power series ring $W[\![X_1, ..., X_m]\!]$ over W in $m \geq 1$ variables. Recall that F denotes the fraction field of W. We claim that:

Lemma 4.2. *For any finite extension E_0/F, there exists a finite extension E/E_0 such that any W-algebra homomorphism $\phi_0 : R_0 \to \mathcal{O}_{E_0}$ extends to a W-algebra homomorphism $\phi : R \to \mathcal{O}_E$.*

It will follow from this lemma that the set $\mathrm{Hom}_{W\text{-alg}}(R, \mathcal{O}_E)$ is infinite since, R_0 being a power series ring over W in $m \geq 1$ variables, there are infinitely many $\phi_0 : R_0 \to \mathcal{O}_{E_0}$. This proves that $\mathrm{FM}'(\mathrm{b})_{\mathrm{rair}} \Rightarrow \mathrm{FM}(\mathrm{D}_0)$.

---

[4]Exactly speaking, a more general version of FM(D) for $\overline{\rho}$ with $\mathrm{End}(\overline{\rho}) \simeq k$.

Now we prove the lemma. Since R is finite over R_0, it can be obtained from R_0 by a successive extension

$$R_0 \subset R_1 \subset \cdots \subset R_n = R$$

with

$$R_j = R_{j-1}[Y_j]/I_j,$$

where I_j is an ideal of $R_{j-1}[Y_j]$. For simplicity, assume that I_j is principal and generated by a monic polynomial $f_j(Y_j) \in R_{j-1}[Y_j]$ (see a future publication for the general case). Let $d_j := \deg(f_j)$. Since there exist only finitely many extensions of E_0 (contained in $\overline{\mathbf{Q}}_p$) of a given degree ([**25**]), we can find inductively a series of finite extensions $E_0 \subset E_1 \subset \cdots \subset E_n = E$ such that E_j contains all finite extensions of E_{j-1} of degree $\leq d_j$. Let $\phi_0 : R_0 \to \mathcal{O}_{E_0}$ be any W-algebra homomorphism. By definition of E_1, the image $\phi_0 f_1 \in \mathcal{O}_{E_0}[Y_1]$ of the monic polynomial $f_1 \in R_0[Y_1]$ has a root in $\mathcal{O}_{E_1}$. Thus ϕ_0 extends to a W-algebra homomorphism $\phi_1 : R_1 = R_0[Y_1]/(f_1) \to \mathcal{O}_{E_1}$. Proceeding inductively, we obtain a W-algebra homomorphism $\phi : R \to \mathcal{O}_E$ which extends ϕ_0.

Finally, we prove:

Proposition 4.3. $\mathrm{FM}'(\mathrm{b})_{\mathrm{rair}}+ (\mathbb{F}) \Longrightarrow \mathrm{FM}'(\mathrm{a})_{\mathrm{rair}}$.

Proof. By $(\mathbb{F})$, there appear only finitely many isomorphism classes of $\overline{\rho} : G_{K,S} \to \mathrm{GL}_d(\overline{\mathbf{F}}_p)$ as the reduction of elements of $\mathrm{Geom}'_d(K, S, \mathcal{L}, [a, b]; \overline{\mathbf{Q}}_p)_{\mathrm{rair}}$. So it is enough to show that, for each absolutely irreducible representation $\overline{\rho} : G_{K,S} \to \mathrm{GL}_d(\overline{\mathbf{F}}_p)$, there are only finitely many elements ρ of $\mathrm{Geom}'_d(K, S, \mathcal{L}, [a, b]; \overline{\mathbf{Q}}_p)_{\mathrm{rair}}$ reducing to the class of $\overline{\rho}$. The $\overline{\mathbf{F}}_p$-representation $\overline{\rho}$ is in fact defined over some finite field k. Then the $\overline{\mathbf{Q}}_p$-representation ρ, being residually absolutely irreducible, is in fact defined over a ring in the category $\mathcal{C}_W$, with $W = W(k)$. Indeed, if ρ is defined over a finite extension E of $\mathbf{Q}_p$, then by a lemma of Carayol and Serre (cf. e.g. Prop. in Sect. 6 of [**11**]), it is actually defined over the subring generated over W by the traces $\mathrm{Tr}\, \rho(g)$ for all $g \in G_{K,S}$. But since $\mathrm{Tr}\, \rho(g) \pmod{\mathfrak{m}_E}$ are all in k, this ring is contained in the inverse image $\mathcal{O}_{E,k}$ of k by the reduction map $\mathcal{O}_E \to k_E$. The ring $\mathcal{O}_{E,k}$ is in $\mathcal{C}_W$. Thus the ρ comes from some W-algebra homomorphism $\phi : R_{\mathcal{D}}(\overline{\rho})_0 \to \mathcal{O}_{E,k} \hookrightarrow \overline{\mathbf{Q}}_p$. By $\mathrm{FM}(\mathrm{D}_0)$, which is equivalent to $\mathrm{FM}'(\mathrm{b})_{\mathrm{rair}}$, there are only finitely many such ϕ's.

References

[1] G. Anderson, D. Blasius, R. Coleman and G. Zettler, *On representations of the Weil group with bounded conductor,* Forum Math. **6** (1994), 537–545

[2] A. Ash, D. Doud and D Pollack, *Galois representations with conjectural connections to arithmetic cohomology,* Duke Math. J. **112** (2002), 521–579

[3] A. Ash and W. Sinnott, *An analogue of Serre's conjecture for Galois representations and Hecke eigenclasses in the mod-p cohomology of* $\mathrm{GL}(n, \mathbf{Z})$, Duke Math. J. **105** (2000), 1–24

[4] S. Brueggeman, *The nonexistence of certain Galois extensions unramified outside 5,* J. Number Theory **75** (1999), 47–52

[5] G. Faltings, *Endlichkeitssätze für abelsche Varietäten über Zahlkörpern,* Invent. Math. **73** (1983), 349–366

[6] J.-M. Fontaine and B. Mazur, *Geometric Galois representations,* in: Elliptic Curves, Modular Forms, and Fermat's Last Theorem, J. Coates and S.-T. Yau (eds.), Internat. Press, Cambridge, MA, 1995 pp. 41–78 in the first ed. (pp. 190–227 in the 2nd ed.)

[7] Ch. Khare, *Conjectures on finiteness of mod p Galois representations,* J. Ramanujan Math. Soc. **15** (2000), 23–42

[8] M. J. Larsen and R. Pink, *Finite subgroups of algebraic groups,* preprint (1998)

[9] H. Matsumura, *Commutative Ring Theory,* 2nd ed., Cambridge University Press, Cambridge, 1989

[10] B. Mazur, *Deforming Galois representations,* in: Galois Groups over **Q**, Y. Ihara, K. Ribet and J.-P. Serre (eds.), Springer-Verlag, New York, 1989, pp. 385–437

[11] B. Mazur, *An introduction to the deformation theory of Galois representations,* in: Modular Forms and Fermat's Last Theorem, G. Cornell, J.H. Silverman and G. Stevens (eds.), Springer-Verlag, New York, 1997, pp. 243–311

[12] H. Moon, *Finiteness results on certain mod p Galois representations,* J. Number Theory, **84** (2000), 156–165

[13] H. Moon, *The number of monomial mod p Galois representations with bounded conductor,* Tôhoku Math. J. **55**(2003), 89–98

[14] H. Moon, *On the non-existence of certain Galois extensions,* in: Galois Theory and Modular Forms, K. Hashimoto, K. Miyake and H. Nakamura (eds.), Kluwer Academic Publishers, 2003, pp. 221–231

[15] H. Moon and Y. Taguchi, *Mod p Galois representations of solvable image,* Proc. Amer. Math. Soc. **129** (2001), 2529–2534

[16] H. Moon and Y. Taguchi, *Refinement of Tate's discriminant bound and non-existence theorems for mod p Galois representations,* Documenta Math. Extra Volume: Kazuya Kato's Fiftieth Birthday (2003), 641–654

[17] G. Poitou, *Sur les petits discriminants,* Sém. Delange-Pisot-Poitou **18** (1976/77),

[18] R. Ramakrishna, *On a variation of Mazur's deformation functor,* Compositio Math. **87** (1993), 269–286

[19] R. Ramakrishna, *Lifting Galois representations,* Invent. Math. **138** (1999), 537–562

[20] R. Ramakrishna, *Deforming Galois representations and the conjectures of Serre and Fontaine-Mazur,* Ann. of Math. **156** (2002), 115–154

[21] J.-P. Serre, *Corps Locaux,* Hermann, Paris, 1968

[22] J.-P. Serre, *Abelian ℓ-adic representations and elliptic curves,* W. A. Benjamin, Inc., New York-Amsterdam, 1968

[23] J.-P. Serre, *Cohomologie Galoisienne,* 4th ed., Springer-Verlag, 1973

[24] J.-P. Serre, *Minorations de discriminants,* 1975, in: Œuvres III, Springer-Verlag, 1986, pp. 240–243

[25] J.-P. Serre, *Une "formule de masse" pour les extensions totalement ramifiees de degre donne d'un corps local,* C. R. Acad. Sci. Paris, Ser. A-B **286** (1978), A1031–A1036

[26] J.-P. Serre, *Sur les représentations modulaires de degré 2 de* $\mathrm{Gal}(\overline{\mathbf{Q}}/\mathbf{Q})$, Duke Math. J. **54** (1987), 179–230

[27] I. Shafarevich, *Algebraic number fields,* Proc. Internat. Congr. Mathematicians (Stockholm, 1962), Inst. Mittag-Leffler, Djursholm, 1963, pp. 163–176; also in: Collected Mathematical Papers, Springer-Verlag, 1984, pp. 283–294

[28] J.M. Sink, *A zeta-function for a knot using* $\mathrm{SL}_2(\mathbb{F}_{p^s})$ *representations,* in: Knots in Hellas '98 (Delphi), C. McA. Gordon, V.F.R. Jones, L.H. Kauffman, S. Lambropoulou and J.H. Przytycki (eds.), Ser. Knots Everything **24**, World Sci. Publishing, River Edge, NJ, 2000, pp. 452–470

[29] Y. Taguchi, *On potentially abelian geometric representations,* The Ramanujan Journal **7** (2003), 477-483

[30] J. Tate, *The non-existence of certain Galois extensions of* **Q** *unramified outside 2,* Contemp. Math. **174** (1994), 153–156

Graduate School of Mathematics, Kyushu University 33, Fukuoka 812-8581, Japan
E-mail address: taguchi@math.kyushu-u.ac.jp

Contemporary Mathematics
Volume **416**, 2006

Some new-type equations in the Grothendieck-Teichmüller group arising from geometry of $\mathcal{M}_{0,5}$

Hiroshi Tsunogai

ABSTRACT. It is a fascinating problem to determine the image of the absolute Galois group $G_{\boldsymbol{Q}}$ of the rationals in the Grothendieck-Teichmüller group $\widehat{\mathrm{GT}}$. In this article we give some new-type equations satisfied by the Galois images in $\widehat{\mathrm{GT}}$ arising from various embeddings of a punctured projective line into the moduli space of projective lines with five marked points.

0. Introduction

0.1. The *Grothendieck-Teichmüller group* $\widehat{\mathrm{GT}}$ is the profinite group defined by
(0.1.1)

$$\widehat{\mathrm{GT}} = \left\{ \sigma \in \mathrm{Aut}\hat{F}_2 \,\middle|\, \begin{array}{c} \sigma(x) = x^{\lambda}, \sigma(y) = f^{-1}y^{\lambda}f, \ (\lambda \in \hat{\boldsymbol{Z}}^{\times}, f = f(x,y) \in \hat{F}_2') \\ \text{s.t. (I),(II),(III)} \end{array} \right\}$$

$$\text{(I)} \quad f(x,y)f(y,x) = 1,$$

$$\text{(II)} \quad f(z,x)z^{\frac{\lambda-1}{2}}f(y,z)y^{\frac{\lambda-1}{2}}f(x,y)x^{\frac{\lambda-1}{2}} = 1 \quad \text{if } xyz = 1,$$

$$\text{(III)} \quad f(x_{12},x_{23})f(x_{34},x_{45})f(x_{51},x_{12})f(x_{23},x_{34})f(x_{45},x_{51}) = 1 \quad \text{in } \hat{P}_5^*,$$

where $\hat{F}_2 = \langle x, y \rangle$ is the free profinite group of rank two and $\hat{F}_2'$ is the topological commutator subgroup of $\hat{F}_2$ (for $\hat{P}_5^*$ and its elements x_{ij}, see §1). It admits an injection $G_{\boldsymbol{Q}} \hookrightarrow \widehat{\mathrm{GT}}$ from the absolute Galois group $G_{\boldsymbol{Q}}$ of the rational number field $\boldsymbol{Q}$, defined by the action on the fundamental group $\pi_1(\mathbf{P}^1 \smallsetminus \{0,1,\infty\}; \overrightarrow{01})$ (Belyĭ[**Be**], Drinfel'd [**Dr**], Ihara [**I1, I2**]). It is a basic and fascinating open problem to determine the image of $G_{\boldsymbol{Q}}$ in $\widehat{\mathrm{GT}}$.

0.2. Recall how we assign to $\sigma \in G_{\boldsymbol{Q}}$ an element $(\lambda_\sigma, f_\sigma) \in \widehat{\mathrm{GT}}$. Let χ be the cyclotomic character $G_{\boldsymbol{Q}} \to \hat{\boldsymbol{Z}}^{\times}$. Then $\lambda_\sigma = \chi(\sigma)$. To obtain f_σ, we fix an embedding of $\bar{\boldsymbol{Q}}$ into $\boldsymbol{C}$. This determines a pair of free generators x, y of

2000 *Mathematics Subject Classification.* Primary 14H30; Secondary 11R32,20F34,20F36.
Key words and phrases. Galois representation, Grothendieck-Teichmüller group .
This research was partly supported by MEXT Grant-in-Aid for Young Scientists (A) 09740041 and 12740026.

$\pi_1(\mathbf{P}^1 \setminus \{0,1,\infty\};\overrightarrow{01})$ and an element $p \in \pi_1(\mathbf{P}^1 \setminus \{0,1,\infty\};\overrightarrow{01},\overrightarrow{10})$ as in Figure 1 appearing later. Since the Galois group $G_{\boldsymbol{Q}}$ acts on the set $\pi_1(\mathbf{P}^1\setminus\{0,1,\infty\};\overrightarrow{01},\overrightarrow{10})$, there exists a unique element $f_\sigma \in \pi_1(\mathbf{P}^1 \setminus \{0,1,\infty\};\overrightarrow{01})$ satisfying

$$\sigma(p) = f_\sigma^{-1}p. \tag{0.2.1}$$

By the identification between $\pi_1(\mathbf{P}^1 \setminus \{0,1,\infty\};\overrightarrow{01})$ and $\hat{F}_2$ determined by x, y, we regard f_σ as an element of $\hat{F}_2$. We know that f_σ belongs to $\hat{F}_2'$ and that $(\lambda_\sigma, f_\sigma)$ belongs to $\widehat{\mathrm{GT}}$.

0.3. Now return to our current problem. By considering the action of $G_{\boldsymbol{Q}}$ on mapping class groups of higher genera, we know that any Galois image satisfies the following equations (Nakamura [**N1**], Lochak-Nakamura-Schneps [**LNS1**]):

$$\text{(III}'\text{)}\quad g(x_{45},x_{51})f(x_{12},x_{23})f(x_{34},x_{45}) = f(\tau_1\tau_3,\tau_2^2) \quad \text{in } \hat{B}_5^*,$$

$$\text{(IV)}\quad f(\tau_1,\tau_2^4) = \tau_2^{8\rho_2}f(\tau_1^2,\tau_2^2)\tau_1^{4\rho_2}(\tau_1\tau_2)^{-6\rho_2} \quad \text{in } \hat{B}_3,$$

where $g(x,y)$ is the unique element in $\hat{F}_2$ satisfying $f(x,y) = g(y,x)^{-1}g(x,y)$ (the existence and uniqueness of such $g(x,y)$ for an element of $\widehat{\mathrm{GT}}$ is proved in Lochak-Schneps [**LS**].) and $\rho_2 = \rho_2(\sigma) \in \hat{\boldsymbol{Z}}$ is defined by $g(x,y) \equiv (xy)^{\rho_2} \bmod \hat{F}_2'$. This 1-cocycle $\rho_2 : \widehat{\mathrm{GT}} \to \hat{\boldsymbol{Z}}(1)$ extends the Kummer 1-cocycle $\sigma \mapsto (\sqrt[N]{2})^{\sigma-1}$ on $G_{\boldsymbol{Q}}$ associated to the real positive roots $\sqrt[N]{2} \in \boldsymbol{R}_{>0}$ of 2, whose importance was firstly pointed out in [**N2**].

0.4. Although these new-type equations came up from the consideration of the moduli of curves of higher gerena, they live in braid groups on the sphere. The first motivation of this article is to deduce these equations within the "genus 0 world". For the equation (IV), Nakamura-Schneps [**NS**] obtained a refined version of (IV):

$$\text{(IV}'\text{)}\quad f(\tau_1,\tau_2^2) = \tau_2^{4\rho_2}f(\tau_1^2,\tau_2^2)\tau_1^{2\rho_2}(\tau_1\tau_2)^{-2\rho_2} \quad \text{in } \hat{B}_3$$

by considering suitable double coverings of $\mathbf{P}^1 \setminus \{0,1,\infty\}$. Furusho [**F**] combined his close study on these double coverings with Ihara's investigation on Kummer coverings [**I4**], and show that another equation (K_2) coming from a double covering together with (I), (II), and (IV) implies the equation (A_2^0) of different arithmetic nature arising from the multiplication formula of the adelic Beta function [**A, I3, I4**].

0.5. In their joint work of Nakamura and the author, they considered the "Legendre-Jacobi" $\mathfrak{S}_3$-covering of $\mathbf{P}^1\setminus\{0,1,\infty\}$ and its subcoverings besides double coverings so that they obtained a number of new-type equations including the "harmonic equation"

$$\text{(I}'\text{)}\quad f(\tau_1^2,\tau_2^2) = \tau_2^{-4\rho_2}f(\tau_2^2,\tau_1\tau_2\tau_1)^{-1}f(\tau_1^2,\tau_1\tau_2\tau_1)\tau_1^{4\rho_2},$$

and the "equianharmonic equation"

$$\text{(II}'\text{)}\quad f(\tau_1^2,\tau_2^2) = \tau_2^{-3\rho_3-\frac{\lambda-1}{2}}f(\tau_2^2,\tau_1\tau_2)^{-1}(\tau_1\tau_2)^{\frac{\lambda-1}{2}}f(\tau_1^2,\tau_1\tau_2)\tau_1^{3\rho_3-\frac{\lambda-1}{2}},$$

both of which hold in $\hat{B}_3$.

0.6. In this article, after the preparatory two sections, in Section 3 we reprove the equation (III′) by considering a certain embedding of $\mathbf{P}^1 \setminus \{0,1,\infty\}$ into the moduli of the projective lines of (partially) unordered five marked points. To do this we actually construct an embedding $\mathbf{P}^1 \setminus \{0,\pm 1,\infty\} \to \mathcal{M}_{0,5}$ such that the signature change $w \mapsto -w$ induces a permutation on the image.

0.7. In section 4 we try to proceed our method to obtain more new-type equations. The key point of the above method is to construct the following commutative diagram:

$$\begin{array}{ccc} \mathbf{P}^1 \setminus \beta^{-1}(\{0,1,\infty\}) & \xrightarrow{\varphi} & \mathcal{M}_{0,5} \\ {\scriptstyle\beta}\downarrow & & \downarrow \\ \mathbf{P}^1 \setminus \{0,1,\infty\} & \xrightarrow{\bar{\varphi}} & \mathcal{M}_{0,5}/G, \end{array} \tag{0.7.1}$$

where G is a subgroup of $\mathfrak{S}_5$, $\beta : \mathbf{P}^1 \to \mathbf{P}^1$ is a branched Galois covering with Galois group G unramified outside $\{0,1,\infty\}$, and φ is equivariant with respect to G-actions on both sides. In Proposition 4.4, we give the list of the diagrams such that φ, β and the G-actions are defined over $\boldsymbol{Q}$. Using these diagrams, one can obtain many new-type equations, which are enumerated in the latter part of this section. Similarly to be mentioned in other papers on this subject, we must say that we do not know whether these new-type equations can be deduced from the defining equations (I), (II) and (III) of $\widehat{\mathrm{GT}}$ nor whether they can characterize the image of $G_{\boldsymbol{Q}}$.

0.8. Recently Schneps obtained some of what we found with clear geometric interpretation from the view of Grothendieck-Teichmüller theory [**S1, S2**]. She studied special loci in the moduli of curves, which is the set of the fixed points of an automorphism of finite order acting on the moduli, in other words, the moduli of curves with special automorphims. In Section 5 we give some comments relevant to her work.

Acknowledgement: The author would like to thank Professors J. Morava and M.Morishita, the organizers of the conference "Knots and Primes" in JAMI (2003), for giving him the opportunity to join this exciting conference. Professor Y.Ihara kindly recommended strongly to the author to improve his old version of this article, which included only a few example in the list in Propsition 4.4. Without his comment, this work could not be completed. The author would like to express his graditude to Professor Y.Ihara. The author is sincerely grateful to Professor H.Nakamura for his many useful suggestion in their joint works and other discussions and for his constant encouragement. The author is grateful also for Dr.H.Furusho who kindly told him on the recent works of Schneps deeply relevant to this subject. Finally, the author thanks Professor L.Schneps, who kindly showed him the latest status of her work in her communication with him.

1. Tangential base points on $\mathbf{P}^1 \setminus \{0,1,\infty\}$

In these two sections we explain the basic setup mainly based on Ihara [**I2**]. In this section, we review the formulation of tangential base points $\overrightarrow{01}, \overrightarrow{10}$ on a punctured projective line $\mathbf{P}^1 \setminus \{0,1,\infty\}$ via Puiseux series expansion and the definition of $f_\sigma(x,y) \in \hat{F}_2$ for $\sigma \in G_{\boldsymbol{Q}}$, which yields an inclusion $G_{\boldsymbol{Q}} \hookrightarrow \widehat{\mathrm{GT}}$. Ihara

[**I2**] (and preceding works of Anderson-Ihara [**AI**]) gave a formulation of them using Puiseux series. Matsumoto [**Ma**] formulated a notion of *tangential morphisms* more generally. Here we do not mention general formulations of them.

1.1. Let $X_4 := \mathcal{M}_{0,4}$ be the moduli space of projective lines with ordered four marked points and $\bar{X}_4 = X_4 \times_{\boldsymbol{Q}} \overline{\boldsymbol{Q}}$. We identify X_4 with $\mathbf{P}^1_{\boldsymbol{Q}} \setminus \{0,1,\infty\} = \mathrm{Spec}(\boldsymbol{Q}[t,t^{-1},(1-t)^{-1}])$. Denote the maximal algebraic extention of the function field $\overline{\boldsymbol{Q}}(t)$ of $\bar{X}_4$ unramified outside $t = 0,1,\infty$ by L. Then we have

(1.1.1)

$$\begin{array}{ccccccccc} 1 & \longrightarrow & \pi_1(\bar{X}_4) & \longrightarrow & \pi_1(X_4) & \longrightarrow & G_{\boldsymbol{Q}} & \longrightarrow & 1 \\ & & \wr\downarrow & & \wr\downarrow & & \wr\downarrow & & \\ 1 & \longrightarrow & \mathrm{Gal}(L/\overline{\boldsymbol{Q}}(t)) & \longrightarrow & \mathrm{Gal}(L/\boldsymbol{Q}(t)) & \longrightarrow & \mathrm{Gal}(\overline{\boldsymbol{Q}}(t)/\boldsymbol{Q}(t)) & \longrightarrow & 1. \end{array}$$

A $\boldsymbol{Q}$-ratinal point $\mathrm{Spec}\boldsymbol{Q} \to X_4$ of X_4 gives a section $G_{\boldsymbol{Q}} \to \pi_1(X_4) \simeq \mathrm{Gal}(L/\boldsymbol{Q}(t))$, that is, an extension of the coeffientwise $G_{\boldsymbol{Q}}$-action on $\overline{\boldsymbol{Q}}(t)$ to the action on L. A tangential base point is not a $\boldsymbol{Q}$-rational point of X_4 in a usual sense, but gives a section $G_{\boldsymbol{Q}} \to \mathrm{Gal}(L/\boldsymbol{Q}(t))$ and can be considered as a base point of a fundamental group.

1.2. A tangential base point of X_4 written by $\overrightarrow{01}$ is defined as follows. Denote the maximal algebraic extention of the function field $\overline{\boldsymbol{Q}}(t)$ of $\bar{X}_4$ unramified outside $t = 0,1,\infty$ in the Puiseux series field $\overline{\boldsymbol{Q}}\{\{t\}\} = \bigcup_{N\geq 1} \overline{\boldsymbol{Q}}((t^{1/N}))$ by $L_{\overrightarrow{01}}$ ($\simeq L$). Any element $f \in L_{\overrightarrow{01}}$ is expanded in the form $f(t) = \sum_{i\gg -\infty} a_i t^{i/N}$ ($\exists N \geq 1, a_i \in \overline{\boldsymbol{Q}}$). We define the action of $\sigma \in G_{\boldsymbol{Q}}$ on $L_{\overrightarrow{01}}$ by the coefficientwise action in this expansion:

$$(\sigma f)(t) = \sum_{i \gg -\infty} \sigma(a_i) t^{i/N}. \tag{1.2.1}$$

Similarly, considering the maximal algebraic extention $L_{\overrightarrow{10}}$ of $\overline{\boldsymbol{Q}}(t)$ unramified outside $t = 0,1,\infty$ in $\overline{\boldsymbol{Q}}\{\{1-t\}\}$, we define another tangential base point $\overrightarrow{10}$. These are a natural generalization of an usual $\boldsymbol{Q}$-rational point $t = a$ ($a \in \boldsymbol{Q} \setminus \{0,1\}$), which gives a section $G_{\boldsymbol{Q}} \to \mathrm{Gal}(L/\boldsymbol{Q}(t))$ by the Laurent series expansion in $\overline{\boldsymbol{Q}}((t-a))$.

For a small positive real number $0 < \varepsilon < 1$, the real segment $(0,\varepsilon)$ is contained in $X_4(\boldsymbol{R})$. Any element of $L_{\overrightarrow{01}}$ can be regarded as an analytic function on $(0,\varepsilon)$ by taking the values of $t^{1/N}$ to be real positive values on $(0,\varepsilon)$. Note that it has a positive convergence radius because it is algebraic over $\boldsymbol{Q}(t)$.

Define $x, y \in \pi_1(\bar{X}_4; \overrightarrow{01}) \simeq \mathrm{Gal}(L_{\overrightarrow{01}}/\overline{\boldsymbol{Q}}(t))$ and $p \in \pi_1(\bar{X}; \overrightarrow{01}, \overrightarrow{10}) \simeq \mathrm{Isom}_{\overline{\boldsymbol{Q}}(t)}(L_{\overrightarrow{10}}, L_{\overrightarrow{01}})$ by analytic continuation backward along the paths in $X_4(\boldsymbol{C})$ presenting in Figure 1. For example, we have $x(t^{1/N}) = \zeta_N^{-1} t^{1/N}$ (where $\zeta_N := \exp(2\pi i/N)$).

REMARK 1.3. In this article, employing the convention of path composition used in [**N1, NT1**], we write composition from left to right in braid chasing. This convention is referred as the "τ-convention" in [**LNS2**]. We must notice, for avoiding the readers' confusion, that in the author's old article [**T**] (in Japanese) he used the contrary "σ-convention".

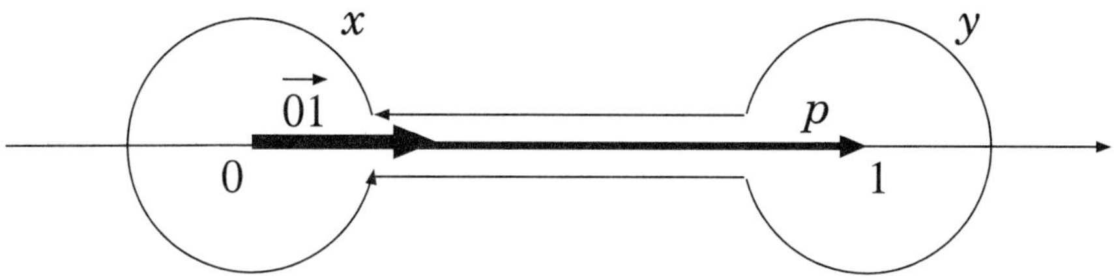

FIGURE 1. $\mathbf{P}^1 \setminus \{0, 1, \infty\}$

1.4. For $\sigma \in G_{\boldsymbol{Q}}$, let $\sigma_{\overrightarrow{01}}$ (resp. $\sigma_{\overrightarrow{10}}$) be the extension of σ to $L_{\overrightarrow{01}}$ (resp. $L_{\overrightarrow{10}}$) corresponding to $\overrightarrow{01}$ (resp. $\overrightarrow{10}$). We define $f_\sigma = f_\sigma(x, y)$ as follows:

$$f_\sigma := p\sigma_{\overrightarrow{10}} p^{-1} \sigma_{\overrightarrow{01}}^{-1} = p\sigma(p)^{-1}. \tag{1.4.1}$$

Then we have $G_{\boldsymbol{Q}} \hookrightarrow \widehat{\mathrm{GT}}$ by $\sigma \mapsto (\chi(\sigma), f_\sigma)$, where $\chi : G_{\boldsymbol{Q}} \to \hat{\boldsymbol{Z}}^\times$ is the cyclotomic character [**I2**].

The geometric interpretation of the cocycle $g_\sigma = g_\sigma(x, y)$ introduced by Lochak-Schneps is as follows [**LS, LNS1**]: Let $r \in \pi_1(\bar{X}_4; \overrightarrow{01}, \frac{1}{2})$ be the path from $\overrightarrow{01}$ to $\frac{1}{2}$ along the real axis, then $g_\sigma = r\sigma(r)^{-1}$.

1.5. When we consider the $G_{\boldsymbol{Q}}$-action, sometimes we must distinguish tangential base points corresponding with the same (homotopic) simply connected set in $X_4(\boldsymbol{C})$ [**N1, N2, Ma**]. For a positive rational number $a \in \boldsymbol{Q}_{>0}$, denote a tangential base point defined via Puiseux series expansion in $\overline{\boldsymbol{Q}}\{\{at\}\}$ by $a\overrightarrow{01}$. Then, though the corresponding simply connected sets in $X_4(\boldsymbol{C})$ are the same, the $G_{\boldsymbol{Q}}$-actions corresponding to $\overrightarrow{01}$ and $a\overrightarrow{01}$ are different. Indeed, denoting the positive N-th root of a by $\sqrt[N]{a}$, a natural isomorphism $\overline{\boldsymbol{Q}}\{\{t\}\} \to \overline{\boldsymbol{Q}}\{\{at\}\}; t^{1/N} \mapsto \sqrt[N]{a}^{-1}(at)^{1/N}$ is *not* $G_{\boldsymbol{Q}}$-equivariant. This isomorphism should be regarded as "analytic continuation from $\overrightarrow{01}$ to $a\overrightarrow{01}$ along the real axis". Let $\delta_a \in \pi_1(\bar{X}_4; a\overrightarrow{01}, \overrightarrow{01}) \simeq \mathrm{Isom}_{\overline{\boldsymbol{Q}}(t)}(L_{\overrightarrow{01}}, L_{a\overrightarrow{01}})$ be the element of the fundamental groupoid defined by the above isomorphism. Then we have $\sigma(\delta_a) := \sigma_{a\overrightarrow{01}} \delta_a \sigma_{\overrightarrow{01}}^{-1} = \delta_a x^{\rho_a(\sigma)}$, where $\rho_a : G_{\boldsymbol{Q}} \to \hat{\boldsymbol{Z}}(1)$ is the Kummer 1-cocycle attached to $\sqrt[N]{a}$; $\sigma(\sqrt[N]{a}) = \zeta_N^{\rho_a(\sigma)} \sqrt[N]{a}$. We need this kind of modification of "scaling" to obtain the correct formulae. This is the reason why Kummer 1-cocycles appear in the resulted equations.

2. The base pentagon

In this section we first make a brief review of the *base pentagon in* $\mathcal{M}_{0,5}$ studied by Ihara [**I2**], and then introduce some additional settings used in the following section.

2.1. Let $X_5 := \mathcal{M}_{0,5}$ be the moduli space of projective lines with ordered five marked points:

$$X_5 = \{(x_i)_{i=1}^5 \in (\mathbf{P}^1)^5 | x_i \neq x_j \ (i \neq j)\}/\mathrm{PGL}(2). \tag{2.1.1}$$

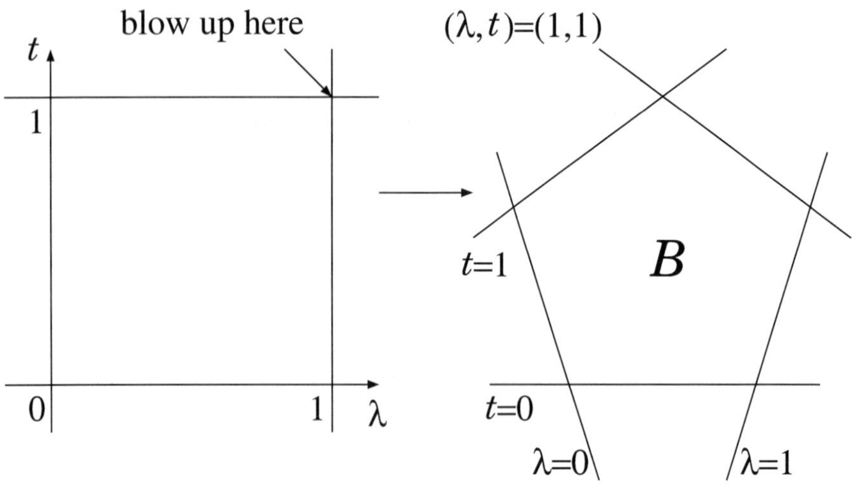

FIGURE 2. The base pentagon $\mathcal{B}$

We denote the class of $(x_1,\dots,x_5)$ modulo PGL(2) by $[x_1,\dots,x_5]$. We identify X_5 with an affine open subvariety of $(\mathbf{P}^1)^2$ by

$$(2.1.2)\qquad \begin{aligned} X_5 &= \{(\lambda,t)\in(\mathbf{P}^1)^2 \mid \lambda,t\neq 0,1,\infty,\ \lambda t\neq 1\}\\ &= \mathrm{Spec}(\boldsymbol{Q}[\lambda,t,\lambda^{-1},t^{-1},(1-\lambda)^{-1},(1-t)^{-1},(1-\lambda t)^{-1}]) \end{aligned}$$

$$[0,\lambda,1,t^{-1},\infty] \leftrightsquigarrow (\lambda,t).$$

Let X_5^* be the smooth compactification of X_5 by adding the stable points. We can construct X_5^* by blowing up $\mathbf{P}^1_\lambda\times\mathbf{P}^1_t$ at the three points $(\lambda,t)=(0,\infty),(1,1),(\infty,0)$. The irreducible components of $X_5^*\setminus X_5$ are the divisors corresponding to $x_i = x_j$ $(1\le i,j\le 5, i\neq j)$.

2.2. Define a subset $\mathcal{B}$ of $X_5(\boldsymbol{C})$ (in fact, of $X_5(\boldsymbol{R})$) by

$$(2.2.1)\qquad \mathcal{B}=\{(\lambda,t)\in\boldsymbol{R}\mid 0<\lambda,t<1\}.$$

In terms of five point configuration, an element of $\mathcal{B}$ corresponds to $(x_i)_{i=1}^5$ with $x_1,\dots,x_5$ lying on $\boldsymbol{R}\cup\{\infty\}$ in this order (permitting the case of passing through ∞). Since $\mathcal{B}$ is simply connected, we can consider the (topological) fundamental group $\pi_1(X_5(\boldsymbol{C});\mathcal{B})$ based at $\mathcal{B}$. This is isomorphic to the group P_5^* defined as follows. Let P_5 be the pure Artin braid group on 5 strings, generated by x_{ij} $(i,j=1,\dots,5)$ in Figure 3. The group P_5^* is defined to be the quotient group of P_5 by the relations $x_{i1}x_{i2}x_{i3}x_{i4}x_{i5}=1$ $(i=1,\dots,5; x_{ii}=1)$ (sphere relation) and $x_{12}x_{13}x_{14}x_{15}x_{23}x_{24}x_{25}x_{34}x_{35}x_{45}=1$. (Again notice that we write composition from left to right in braid chasing.)

2.3. On X_5 and X_5^* the symmetric group $\mathfrak{S}_5$ of degree 5 acts from the left by permutation of coordinates; $s((x_i)_{i=1}^5)=(x_{s^{-1}(i)})_{i=1}^5$ for any $s\in\mathfrak{S}_5$. The stabilizer $S(\mathcal{B})$ of $\mathcal{B}$ in $\mathfrak{S}_5$ is generated by $\rho=(1\ 4\ 2\ 5\ 3)$ and $\theta=(1\ 3)(4\ 5)$, isomorphic to the dihedral group D_5 of order 10. As automorphisms of X_5, they are described in

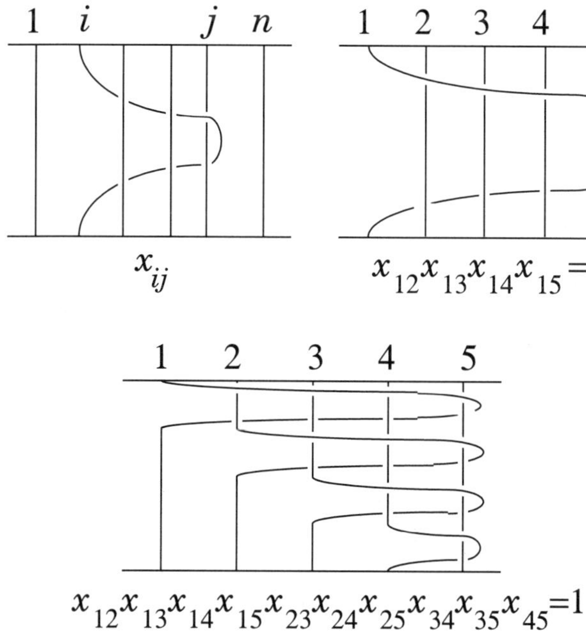

FIGURE 3. Generators and relators of P_5^*

terms of (λ, t)-coorinate by

$$\rho : \begin{cases} \lambda \longmapsto \dfrac{1-t}{1-\lambda t}, \\ t \longmapsto \lambda \end{cases} \qquad \theta : \begin{cases} \lambda \longmapsto \dfrac{1-\lambda}{1-\lambda t}. \\ t \longmapsto t \end{cases} \tag{2.3.1}$$

2.4. Let $V_0 := \{P_i | i = 1, \ldots, 5\}$ be the set consisting of the vertices of the pentagon $\mathcal{B}$, and for each point $P \in V_0$ let $\mathcal{B}_P$ be a simply connected open subset in $\mathcal{B}$ as in Figure 4. (Notice that this is different from the numbering in [**I2**]. Our convention of numbering is compatible with $S(\mathcal{B})$-action; $s(P_i) = P_{s(i)}$ for any $s \in S(\mathcal{B})$.) We have the topological fundamental groupoid $\pi_1(X_5(\boldsymbol{C}); V) = \{\pi_1(X_5(\boldsymbol{C}); \mathcal{B}_P, \mathcal{B}_{P'})\}_{P,P' \in V}$. By the inclusion $\mathcal{B}_P \hookrightarrow \mathcal{B}$, we identify $\pi_1(X_5(\boldsymbol{C}), \mathcal{B}_P)$ with $\pi_1(X_5(\boldsymbol{C}), \mathcal{B})$. We write the element of $\pi_1(X_5(\boldsymbol{C}), \mathcal{B}_P)$ corresponding to $x \in \pi_1(X_5(\boldsymbol{C}), \mathcal{B})$ by x_P if we want to distinguish them. For any two points $P, P' \in V$, let $\overrightarrow{PP'} \in \pi_1(X_5(\boldsymbol{C}); \mathcal{B}_P, \mathcal{B}_{P'})$ be the element determined by any path from $\mathcal{B}_P$ to $\mathcal{B}_{P'}$ inside $\mathcal{B}$. Then, for an element $x \in \pi_1(X_5(\boldsymbol{C}); \mathcal{B}_P, \mathcal{B}_{P'})$, we have $x_P \overrightarrow{PP'} = \overrightarrow{PP'} x_{P'}$.

2.5. We can regard $\mathcal{B}_P$'s as tangential base points. For simplicity, we also write P for $\mathcal{B}_P$. Let L_{P_3} be the maximal extension of the function field $\overline{\boldsymbol{Q}}(\lambda, t)$ of X_5 unramified over X_5 in $\overline{\boldsymbol{Q}}\{\{\lambda, t\}\} := \bigcup_{N \geq 1} \overline{\boldsymbol{Q}}((\lambda^{1/N}, t^{1/N}))$. A tangential base point P_3 is defined to be attached to the coefficientwise action of $\sigma \in G_{\boldsymbol{Q}}$ on L_{P_3} extending that on $\overline{\boldsymbol{Q}}(\lambda, t)$. We define P_i for $i = 1, \ldots, 5$ to be $s(P_i) = P_{s(i)}$ for $s \in S(\mathcal{B})$. For each one of these tangential base points Q, the fundamental group

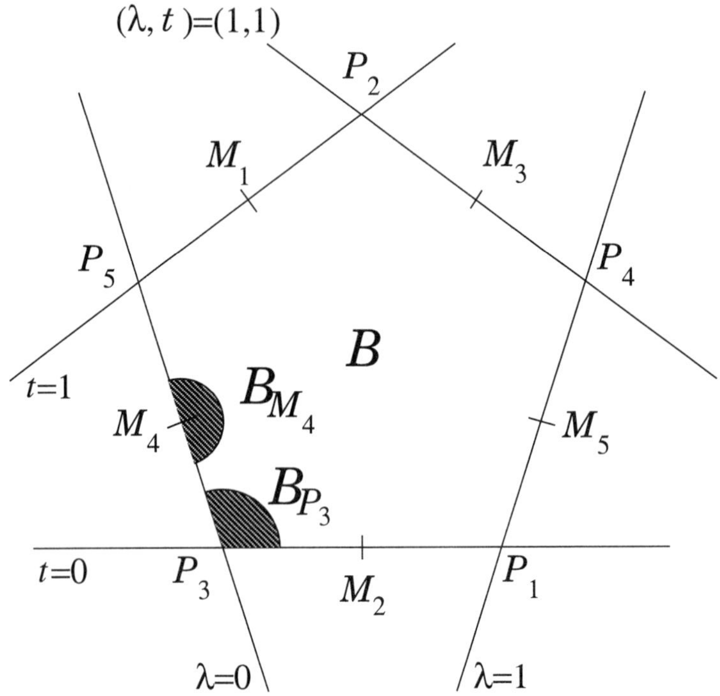

FIGURE 4. The sets V_0, V and the base points $\mathcal{B}_P$'s

$\pi_1(X_5; P)$ of X_5 with a base point P is defined. Its geometric part $\pi_1(\bar{X}_5; Q)$ is isomorphic to $\hat{P_5^*}$, the profinite completion of P_5^*.

The algebraic (profinite) fundamental groupoid

$$\pi_1(X_5; V_0) = \{\pi_1(X_5; \mathcal{B}_P, \mathcal{B}_{P'})\}_{P,P' \in V_0} \tag{2.5.1}$$

based at such tangential base points is defined by

$$\pi_1(X_5; \mathcal{B}_P, \mathcal{B}_{P'}) := \mathrm{Isom}_{\overline{\boldsymbol{Q}(\lambda,t)}}(L_{P'}, L_P). \tag{2.5.2}$$

The path $\overrightarrow{PP'}$ gives the element of this set by analytic continuation backward along it. This groupoid is canonically isomorphic to the profinite completion of the topological fundamental groupoid $\pi_1(X_5(\boldsymbol{C}); V)$, and has $G_{\boldsymbol{Q}}$-action. Since any tangential base point $P \in V$ is $\boldsymbol{Q}$-rational, an element $\sigma \in G_{\boldsymbol{Q}}$ stabilizes each $\pi_1(X_5; P, P')$. For any $\gamma \in \pi_1(X_5; P, P')$ and $\sigma \in G_{\boldsymbol{Q}}$, there exists a unique element $x = x_P \in \pi_1(X_5; P)$ satisfying $\sigma(\gamma) = x^{-1}\gamma$, because the group $\pi_1(X_5; P)$ acts on $\pi_1(X_5; P, P')$ simply transitively on the left. The following proposition is basic and important to yield the 5-cycle relation (III).

PROPOSITION 2.6 (Ihara [**I2**] Lemma 2.6). *For any* $\sigma \in G_{\boldsymbol{Q}}$,

$$\sigma(\overrightarrow{P_3P_5}) = f_\sigma(x_{45}, x_{34})_{P_3}^{-1}\overrightarrow{P_3P_5}.$$

The fundamental idea of the proof is to embed a three-point-punctured projective line $\mathbf{P}^1 \setminus \{0, 1, \infty\}$ into X_5 "infinitesimally near" the line $\lambda = 0$, which is a divisor deleted from X_5^*. If we use the notion of tangential morphisms [**Ma**], the desired

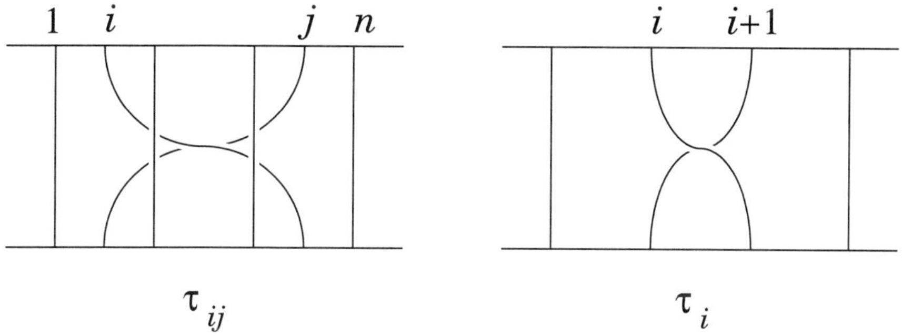

FIGURE 5. Generators of B_5

"embedding" is justified as the tangential morphism $Z \rightharpoondown X_5$ determined by the datum $(X_5, X_5^*, U, \lambda, Z)$, where $U = \mathrm{Spec}(\boldsymbol{Q}[\lambda, t, t^{-1}, (1-t)^{-1}, (1-\lambda)^{-1}, (1-\lambda t)^{-1}])$ and $Z = (\lambda = 0 \text{ in } U) \simeq \mathbf{P}_t^1 \setminus \{0, 1, \infty\}$.

2.7. To describe our "new-type" equations for the image of $G_{\boldsymbol{Q}}$ in $\widehat{\mathrm{GT}}$, we introduce *full* braid groups. Denote the full Artin braid group of 5 strings by B_5. The elements τ_{ij}'s $(i, j = 1, \ldots, 5, i \neq j)$ in Figure 5 generate B_5. We also put $\tau_i = \tau_{i,i+1}$ for $i = 1, \ldots, 4$ and usually use only $\tau_1, \tau_2, \tau_3, \tau_4$ and τ_{51}, The group B_5 is an extension of P_5 by $\mathfrak{S}_5$ and the center of B_5 coincides with that of P_5. Denote by B_5^* the quotient group of B_5 by the kernel of $P_5 \to P_5^*$. An element $s \in S(\mathcal{B})$ also induces an automorphism of B_5 (P_5, B_5^*, P_5^*) by $s(\tau_{ij}) = \tau_{s(i)s(j)}$ for any adjacent pair (i, j) of indices ($|i - j| = 1$ or $\{i, j\} = \{1, 5\}$) because they preserve adjacency of indices. (In general, if i, j are not adjacent, then $s(\tau_{ij})$ is not necessarily $\tau_{s(i)s(j)}$ but its conjugate.)

2.8. Now we introduce the tangential base points corresponding to the middle points of the edges of the pentagon $\mathcal{B}$. Let M_i $(i = 1, \ldots, 5)$ be the middle points of the edges of the pentagon $\mathcal{B}$, and for each of them let $\mathcal{B}_{M_i}$ be a simply connected open subset in $\mathcal{B}$ as in Figure 4. We regard them as tangential base points in the following way. Let L_{M_4} be the maximal extension of the function field $\overline{\boldsymbol{Q}}(\lambda, t)$ of X_5 unramified over X_5 in $\overline{\boldsymbol{Q}}((t-1/2))\{\{\lambda\}\}$. A tangential base point M_4 is defined to be attached to the coefficientwise action of $\sigma \in G_{\boldsymbol{Q}}$ on L_{M_4} extending that on $\overline{\boldsymbol{Q}}(\lambda, t)$. We define M_i for $i = 1, \ldots, 5$ to be $s(M_i) = M_{s(i)}$ for $s \in S(\mathcal{B})$.

Thus we have a set of tangential base points $V := \{P_i, M_i | i = 1, \ldots, 5\}$. For any $Q, Q' \in V$, we define $\overrightarrow{QQ'} \in \pi_1(\bar{X}_5; Q, Q')$ by analytic continuation backward along the path from Q to Q' inside $\mathcal{B}$. The stabilizer $S(\mathcal{B})$ of $\mathcal{B}$ induces an automorphism of the fundamental groupoid $\pi_1(X_5; V) = \{\pi_1(X_5; Q, Q')\}_{Q,Q' \in V}$ such that $s(\pi_1(X_5; Q, Q')) = \pi_1(X_5; s(Q), s(Q')), s(\overrightarrow{QQ'}) = \overrightarrow{s(Q)s(Q')}$ and $s(x_Q) = s(x)_{s(Q)}$ for any $Q, Q' \in V$ and $x_Q \in \pi_1(X_5; P)$.

3. Another proof of (III′)

In this section, we give another proof of the equation (III′) for $G_{\boldsymbol{Q}}$. Let $\overrightarrow{P_5M_5}$ be the path described in Figure 6.

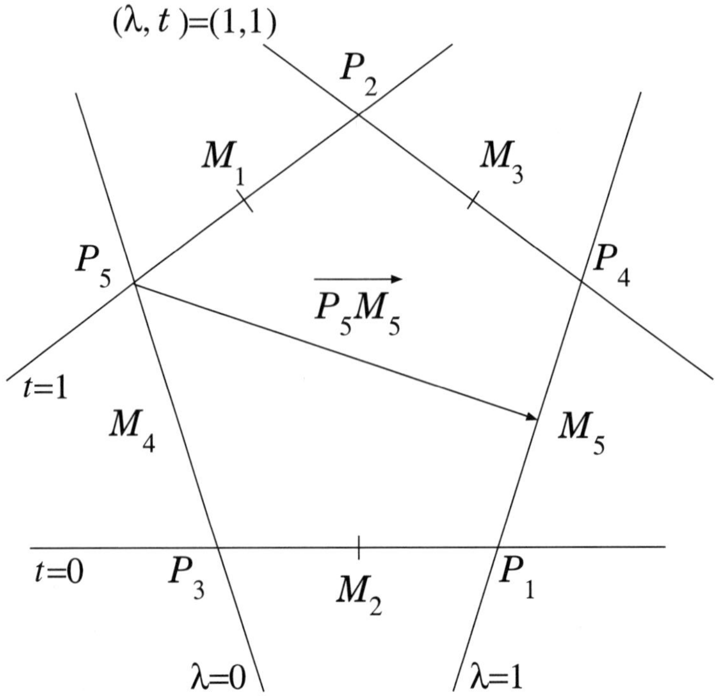

FIGURE 6. The path $\overrightarrow{P_5M_5}$

PROPOSITION 3.1. *For each* $\sigma \in G_{\mathbf{Q}}$, *we have*

$$\sigma(\overrightarrow{P_5M_5}) = f_\sigma(\tau_1\tau_3, x_{23})^{-1}\overrightarrow{P_5M_5}. \tag{3.1.1}$$

PROOF . We prove the proposition by embedding $\mathbf{P}^1 \smallsetminus \{0, 1, \infty\}$ into $X_5/\mathfrak{S}_5$ in a suitable way. For this, we embed $C_w = \mathbf{P}^1_w \smallsetminus \{0, \pm 1, \infty\}$ with coordinate w into X_5 by

$$\begin{aligned} \varphi : C_w &\longrightarrow X_5 \\ w &\longmapsto [0, w, 1, 1 + w, \infty], \end{aligned} \tag{3.1.2}$$

or, in terms of (λ, t)-coordinate, $\lambda = w, t = (1 + w)^{-1}$. Since $[0, -w, 1, 1 - w, \infty] = [w, 0, 1+w, 1, \infty]$ by $\xi \mapsto \xi + w$, there exists a morphism $\bar{\varphi}$ fitting with the following commutative diagram:

$$\begin{array}{ccc} C_w & \xrightarrow{\varphi} & X_5 \\ \beta\downarrow & & \downarrow \\ C_z & \xrightarrow[\bar{\varphi}]{} & X_5/\langle(1\ 2)(3\ 4)\rangle, \end{array} \tag{3.1.3}$$

where the morphism $\beta : C_w \to C_z = \mathbf{P}^1 \smallsetminus \{0, 1, \infty\}$ is defined by $z = \beta(w) = w^2$. Determine the image of the tangential base points $\overrightarrow{01}, \overrightarrow{10}$ on C_z under $\bar{\varphi}$. In the following, we denote the images of the tangential base points belonging to V in $X_5/\langle(1\ 2)(3\ 4)\rangle$ by the same symbols, and $\bar{\varphi}$ simply by φ. The image of $\overrightarrow{01}$ under

φ is "geometrically" equal to P_5, but we must consider the possible change of the "scaling" of the local parameter to respect $G_{\boldsymbol{Q}}$-action. The $G_{\boldsymbol{Q}}$-action on L corresponding to P_5 is determined via $L \simeq L_{P_5} \subset \overline{\boldsymbol{Q}}\{\{\lambda_5, t_5\}\}$, where

$$\lambda_5 = \frac{1-t}{1-\lambda t}, \quad t_5 = \lambda. \tag{3.1.4}$$

We calculate the pull-back of λ_5, t_5 by φ:

$$\begin{aligned} \varphi^*(\lambda_5) &= \frac{1-(1+w)^{-1}}{1-w(1+w)^{-1}} = \frac{(1+w)-1}{(1+w)-w} = w = z^{\frac{1}{2}}, \\ \varphi^*(t_5) &= w = z^{\frac{1}{2}}. \end{aligned} \tag{3.1.5}$$

Since $\varphi^*(\lambda_5^{1/N}) = \varphi^*(t_5^{1/N}) = z^{1/(2N)}$, $\varphi^* : \overline{\boldsymbol{Q}}\{\{\lambda_5, t_5\}\} \to \overline{\boldsymbol{Q}}\{\{z\}\}$ is $G_{\boldsymbol{Q}}$-equivariant. Hence we have $\varphi(\overrightarrow{01}) = P_5$.

Next we determine the image of $\overrightarrow{10}$, which is geometrically equal to M_5. The $G_{\boldsymbol{Q}}$-action on L corresponding to M_5 is determined via $L \simeq L_{M_5} \subset \overline{\boldsymbol{Q}}((s_5))\{\{\mu_5\}\}$, where

$$\mu_5 = \frac{1-\lambda}{1-\lambda t}, \quad s_5 = t - \frac{1}{2}. \tag{3.1.6}$$

The pull-back of μ_5 by φ is calculated as follows:

$$\varphi^*(\mu_5) = \frac{1-w}{1-w(1+w)^{-1}} = \frac{(1+w)(1-w)}{(1+w)-w} = 1-w^2 = 1-z. \tag{3.1.7}$$

Since $\varphi^*(\mu_5^{1/N}) = (1-z)^{1/N}$, $\varphi^* : \overline{\boldsymbol{Q}}((s_5))\{\{\mu_5\}\} \to \overline{\boldsymbol{Q}}\{\{1-z\}\}$ is $G_{\boldsymbol{Q}}$-equivariant. Hence we have $\varphi(\overrightarrow{10}) = M_5$. (We want to emphasize that it is just lucky that we need not change the scaling in this case.)

Thus we have $\varphi(p) = \overrightarrow{P_5 M_5}$. Let us determine the images of x and y. Parametrize the path on C_z corresponding to x by $z(\theta) = \varepsilon e^{2\pi i\theta}$ $(0 \leq \theta \leq 1)$. Since the image of the path under φ is represented by the braid in Figure 7, we have $\varphi(x_z) = \tau_1\tau_3, \varphi(y_z) = x_{23}$. The formula (3.1.1) follows from this.

□

Now we obtain the equation (III′) by applying the following general lemma for $\overrightarrow{P_5 M_5} = \overrightarrow{P_5 P_3}\,\overrightarrow{P_3 P_1}\,\overrightarrow{P_1 M_5}$.

LEMMA 3.2. *Let $Q_i \in V$ $(i = 0, 1, \ldots, n)$. For $\sigma \in G_{\boldsymbol{Q}}$, let $f_i(\sigma) \in \pi_1(X_5; \mathcal{B}) \simeq \hat{P}_5^*$ be the unique element satisfying $\sigma(\overrightarrow{Q_i Q_{i+1}}) = f_i(\sigma)_{Q_i}^{-1} \overrightarrow{Q_i Q_{i+1}}$ for $i = 0, 1, \ldots, n-1$. Then we have*

$$\sigma(\overrightarrow{Q_0 Q_n}) = (f_{n-1}(\sigma) \cdots f_0(\sigma))_{Q_0}^{-1} \overrightarrow{Q_0 Q_n}.$$

In particular, if $Q_0 = Q_n$ then $f_{n-1}(\sigma) \cdots f_0(\sigma) = 1$.

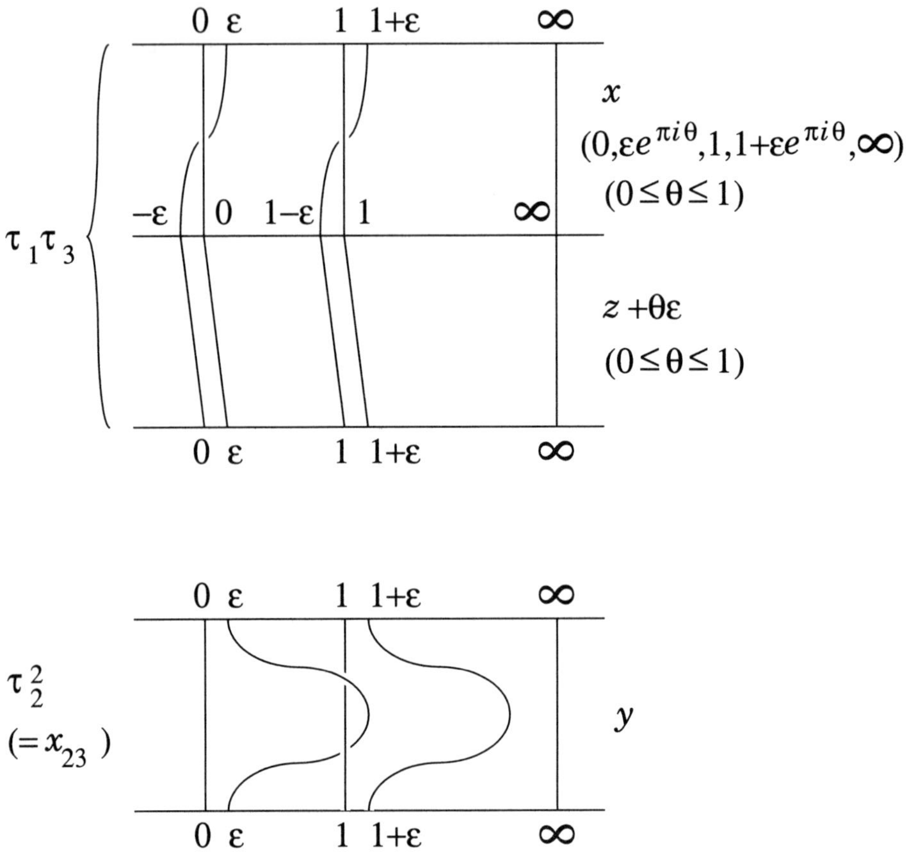

FIGURE 7. The braids corresponging to $\varphi(x)$ and $\varphi(y)$

PROOF . It suffices to show the case $n = 2$. Since $\overrightarrow{Q_0Q_2} = \overrightarrow{Q_0Q_1}\overrightarrow{Q_1Q_2}$,

$$\begin{aligned}
\sigma(\overrightarrow{Q_0Q_2}) &= \sigma(\overrightarrow{Q_0Q_1}\overrightarrow{Q_1Q_2}) \\
&= \sigma(\overrightarrow{Q_0Q_1})\sigma(\overrightarrow{Q_1Q_2}) \\
&= f_0(\sigma)_{Q_0}^{-1}\overrightarrow{Q_0Q_1} f_1(\sigma)_{Q_1}^{-1}\overrightarrow{Q_1Q_2} \\
&= f_0(\sigma)_{Q_0}^{-1} f_1(\sigma)_{Q_0}^{-1}\overrightarrow{Q_0Q_1}\overrightarrow{Q_1Q_2} \\
&= (f_1(\sigma)f_0(\sigma))_{Q_0}^{-1}\overrightarrow{Q_0Q_2}. \qquad \square
\end{aligned}$$

4. Some other embeddings

4.1. The key point of the proof in the last section is the diagram (3.1.3). So, to obtain more equations, we consider an embedding $\varphi : \mathbf{P}^1 \setminus S \to \mathcal{M}_{0,5}$ of a punctured projective line $\mathbf{P}^1 \setminus S$ into $\mathcal{M}_{0,5}$ making the following diagram commute:

$$\begin{array}{ccc} \mathbf{P}^1 \setminus S & \xrightarrow{\varphi} & \mathcal{M}_{0,5} \\ {\scriptstyle\beta}\big\downarrow & & \big\downarrow \\ \mathbf{P}^1 \setminus S_0 & \xrightarrow{\bar{\varphi}} & \mathcal{M}_{0,5}/G, \end{array} \tag{4.1.1}$$

where G is a subgroup of $\mathfrak{S}_5$, $\beta : \mathbf{P}^1 \to \mathbf{P}^1$ is a branched Galois covering with Galois group G unramified outside $S_0 = \{0, 1, \infty\}$, $S = \varphi^{-1}(S_0)$, and φ is equivariant with respect to G-actions on both sides.

Two diagrams

$$\begin{array}{ccc} \mathbf{P}^1 \setminus S & \xrightarrow{\varphi} & \mathcal{M}_{0,5} \\ {\scriptstyle\beta}\big\downarrow & & \big\downarrow \\ \mathbf{P}^1 \setminus S_0 & \xrightarrow{\bar{\varphi}} & \mathcal{M}_{0,5}/G, \end{array} \quad \text{and} \quad \begin{array}{ccc} \mathbf{P}^1 \setminus S' & \xrightarrow{\varphi'} & \mathcal{M}_{0,5} \\ {\scriptstyle\beta'}\big\downarrow & & \big\downarrow \\ \mathbf{P}^1 \setminus S_0 & \xrightarrow{\bar{\varphi}'} & \mathcal{M}_{0,5}/G', \end{array} \tag{4.1.2}$$

are said to be *equivalent* if there exist $\sigma \in \mathfrak{S}_5, \alpha : \mathbf{P}^1 \setminus S \to \mathbf{P}^1 \setminus S'$ and $\bar{\alpha} : \mathbf{P}^1 \setminus S_0 \to \mathbf{P}^1 \setminus S_0$ such that $G' = \sigma^{-1} G \sigma$ and they make everything commute.

4.2. We want the complete list of these diagrams. But there seem too many cases to determine completely. Then, here we restrict ourselves to the cases that

(A1) φ is defined over $\boldsymbol{Q}$.

Since the action of $\mathfrak{S}_5$ on $\mathcal{M}_{0,5}$ is defined over $\boldsymbol{Q}$, the action of G on $\mathbf{P}^1 \setminus S$ as the Galois group of the covering β is also defined over $\boldsymbol{Q}$ under this restriction. Hence G is included both in $\mathfrak{S}_5$ and in $\mathrm{PGL}(2, \boldsymbol{Q})$, that is, G is one of $C_2, C_2 \times C_2, D_3 \simeq \mathfrak{S}_3, D_4, D_6 \simeq \mathfrak{S}_3 \times C_2$.

REMARK 4.3. The assumption (A1) is purely technical. Indeed, when we deduced "equianharmonic equation" (II′) in [**NT1**], a morphism defined over $\boldsymbol{Q}(\zeta_3)$ (not over $\boldsymbol{Q}$) was used. In [**LNS2**], a morphism defined over $\boldsymbol{Q}(\zeta_5)$ plays a main role.

PROPOSITION 4.4. *Under the assumption (A1), an embedding $\varphi : \mathbf{P}^1 \setminus S \to \mathcal{M}_{0,5}$ fitting with the diagram (4.1.1) is equivalent to one of the following.*

(1) $G \simeq C_2, S = \{0, \pm 1, \infty\}, \beta(w) = w^2$:
  (a) $G = \langle (1\ 4) \rangle$:
    (i) $\varphi(w) = [0, 1 - w, 1, 2, \infty]$
    (ii) $\varphi(w) = [0, \dfrac{1 - w}{1 + w}, 1, \dfrac{2}{1 + w}, \infty]$
  (b) $G = \langle (1\ 2)(3\ 4) \rangle$:
    (i) $\varphi(w) = [0, w, 1, 1 + w, \infty]$

(2) $G \simeq C_2 \times C_2, S = \{0, \pm 1, \pm\sqrt{-1}, \infty\}, \beta(w) = \dfrac{4w^2}{(w^2 + 1)^2}$:
  (a) $G = \langle (1\ 3), (2\ 4) \rangle$:
    (i) $\varphi(w) = [0, \dfrac{w}{1 + w}, 1, \dfrac{1}{1 - w}, \infty]$

(3) $G \simeq \mathfrak{S}_3, S = \{0, \pm 1, 1/2, 2, (1 \pm \sqrt{-3})/2, \infty\}, \beta(w) = \dfrac{27w^2(w - 1)^2}{4(w^2 - w + 1)^3}$:
  (a) $G = \langle (1\ 3), (1\ 3\ 5) \rangle$:

(i) $\varphi(w) = [0, \frac{w-1}{w}, 1, \frac{(w-1)(w+1)}{w(w-2)}, \infty]$
(ii) $\varphi(w) = [0, \frac{w-1}{w}, 1, \frac{(w-1)^3(w+1)}{w^3(w-2)}, \infty]$
(iii) $\varphi(w) = [0, \frac{w-1}{w}, 1, -\frac{w+1}{w-2}, \infty]$
(iv) $\varphi(w) = [0, \frac{w-1}{w}, 1, \frac{(w-1)(w+1)^3}{w(w-2)^3}, \infty]$
(v) $\varphi(w) = [0, \frac{(w-1)^3(w+1)}{w^3(w-2)}, 1, \frac{(w-1)(w+1)^3}{w(w-2)^3}, \infty]$
(vi) $\varphi(w) = [0, \frac{(w-1)(w+1)}{w(w-2)}, 1, \frac{(w-1)^3(w+1)}{w^3(w-2)}, \infty]$

(b) $G = \langle (1\ 3)(2\ 4), (1\ 3\ 5)\rangle$:
(i) $\varphi(w) = [0, w, 1, \frac{1}{1-w}, \infty]$

(4) $G \simeq D_4, S = \{0, \pm 1, \pm 1 \pm \sqrt{2}, \pm\sqrt{-1}, \infty\}, \beta(w) = \frac{16w^2(w-1)^2(w+1)^2}{(w^2+1)^4}$:

(a) $G = \langle (1\ 2), (1\ 4)(2\ 3)\rangle$:
(i) $\varphi(w) = [0, \frac{1-2w-w^2}{(1-w)^2}, 1, \frac{2}{(1-w)^2}, \infty]$
(ii) $\varphi(w) = [0, \frac{(w-1)(w^2+2w-1)}{2(w^2+1)}, 1, \frac{(w-1)(w^2-2w-1)}{2w(w^2+1)}, \infty]$
(iii) $\varphi(w) =$
$[0, \frac{(w-1)^2(w^2+2w-1)}{8w}, 1, -\frac{(w-1)^2(w^2-2w-1)}{8w^3}, \infty]$

(5) $G \simeq D_6$ *does not happen.*

PROOF . We describe only the outline because the proof needs a terrible amount of case-by-case argument.

As mentioned above, by the assumption (A1) that everything is defined over $\boldsymbol{Q}$, the possible G is one of $C_2, C_2 \times C_2, D_3 \simeq \mathfrak{S}_3, D_4, D_6 \simeq \mathfrak{S}_3 \times C_2$. The determination of β for these cases is a classical result. The first crutial point to limit the cases is to consider that the image under f of a ramification point $w \in S$ of β must be a fixed point of the inertia subgroup at w in G. So, first we have to determine the fixed points by the elements of G.

LEMMA 4.5. *For the following representatives of the conjugacy classes of* $\mathfrak{S}_5$*, the fixed points of them in* X_5^* *are as in the table:*

$(1\ 2\ 3\ 4\ 5)$:	$[0, \frac{-1\pm\sqrt{5}}{2}, 1, \frac{1\pm\sqrt{5}}{2}, \infty]$,
$(1\ 3\ 2\ 4)$:	$[0, 1\pm\sqrt{-1}, 1, \pm\sqrt{-1}, \infty], [0, 0, 1, 1, \infty]$,
$(1\ 3\ 5)(2\ 4)$:	$[0, \frac{1\pm\sqrt{-3}}{2}, 1, \frac{1\pm\sqrt{-3}}{2}, \infty]$,
$(1\ 3\ 5)$:	$[0, \frac{1\pm\sqrt{-3}}{2}, 1, \frac{1\pm\sqrt{-3}}{2}, \infty], [0, \frac{1\pm\sqrt{-3}}{2}, 1, \frac{1\mp\sqrt{-3}}{2}, \infty]$,
$(1\ 2)(3\ 4)$:	$[0, \lambda, 1, 1+\lambda, \infty] (\lambda \in \mathbf{P}^1), [0, 0, 1, -1, \infty]$,
$(1\ 4)$:	$[0, \lambda, 1, 0, \infty] (\lambda \in \mathbf{P}^1), [0, 1, 1, 2, \infty]$.

Put $\varphi(w) = [0, \lambda(w), 1, t(w)^{-1}, \infty]$. Our next task is to bound the degree of the rational functions λ and t.

LEMMA 4.6. *Let C be a curve of genus g. If a morphism $f : C \to \mathbf{P}^1$ is of degree d with $\#f^{-1}(\{0,1,\infty\}) = n$, then $d \leq 2g + n - 2$. In particular, if $g = 0$ we have $d \leq n - 2$.*

Since $\lambda^{-1}(\{0,1,\infty\})$ and $t^{-1}(\{0,1,\infty\})$ must be included in the punctured point set S, this lemma gives an upper bound for the degrees of λ and t.

The actual determination of the list will have done with these consideration and long case-by-case arguments. We omit the detail here. □

By carrying out the similar calculation to the preceding section for the cases listed above, we obtain the formulae on the Galois action on paths. For a technical reason we again restrict ourselves to the cases that

> (A2) The inverse images of $\overrightarrow{01}_z$ and $\overrightarrow{10}_z$ by β are pointwise $\boldsymbol{Q}$-rational and their images under φ are geometrically vertices or middle points of pentagons (i.e. belong to the $\mathfrak{S}_5$-orbit of V up to scaling factors).

The cases satisfying this condition are (1-a-i), (1-a-ii), (1-b-i), (2-a-i), (3-a-ii), (3-a-iii), (3-a-v) and (3-b-i). Considering these cases we obtain the following list.

REMARK 4.7. By definition of f_σ (or, Proposition 2.6), we have

$$\sigma(\overrightarrow{P_3P_1}) = f_\sigma(x_{12}, x_{23})^{-1}\overrightarrow{P_3P_1} \tag{4.7.1}$$

We have already obtained in [**NT1**]

$$\sigma(\overrightarrow{P_3M_2}) = g_\sigma(x_{12}, x_{23})^{-1}\overrightarrow{P_3M_2} = \left(f_\sigma(x_{12}, \tau_1\tau_2\tau_1)x_{12}^{2\rho_2(\sigma)}\right)^{-1}\overrightarrow{P_3M_2}. \tag{4.7.2}$$

Adding these, this list covers, with $S(\mathcal{B})$-action, all the $\overrightarrow{QQ'}$ $(Q, Q' \in V)$.

PROPOSITION 4.8. *For $\sigma \in G_{\mathbf{Q}}$, we have the following equalities:*

(4.8.1)
$$\sigma(\overrightarrow{M_5M_4}) = \left(x_{12}^{-\rho_2(\sigma)} f_\sigma(\tau_4\tau_{51}\tau_4, x_{12}) x_{23}^{-\rho_2(\sigma)}\right)^{-1} \overrightarrow{M_5M_4} \qquad \textit{(from 1-a-i)},$$

(4.8.2)
$$\sigma(\overrightarrow{M_5P_5}) = \left(x_{12}^{-2\rho_2(\sigma)} x_{34}^{-2\rho_2(\sigma)} f_\sigma(\tau_2, x_{12}x_{34}) x_{23}^{-2\rho_2(\sigma)}\right)^{-1} \overrightarrow{M_5P_5} \qquad \textit{(from 1-a-ii)},$$

(4.8.3)
$$\sigma(\overrightarrow{P_5M_5}) = f_\sigma(\tau_1\tau_3, x_{23})^{-1} \overrightarrow{P_5M_5} \qquad (=(3.1.1), \textit{from 1-b-i}),$$

(4.8.4)
$$\sigma(\overrightarrow{P_5M_2}) = \left(f_\sigma(\tau_2^{-1}\tau_1^{-1}\tau_3^{-1}\tau_1^{-1}, \tau_1\tau_2\tau_1) x_{12}^{\rho_2(\sigma)} x_{34}^{\rho_2(\sigma)}\right)^{-1} \overrightarrow{P_5M_2} \qquad \textit{(from 2-a-i)},$$

(4.8.5)
$$\sigma(\overrightarrow{P_5M_5}) = \Big((\tau_4\tau_{51}\tau_4)^{6\rho_2(\sigma)-3\rho_3(\sigma)} f_\sigma(\tau_1^2\tau_3, \tau_4\tau_{51}\tau_4)$$

(4.8.6)
$$\tau_1^{-6\rho_2(\sigma)+6\rho_3(\sigma)} \tau_3^{-2\rho_2(\sigma)+3\rho_3(\sigma)}\Big)^{-1} \overrightarrow{P_5M_5}$$

(from 3-a-ii),

(4.8.7)
$$\sigma(\overrightarrow{M_4M_2}) = \left((\tau_1\tau_2\tau_1)^{2\rho_2(\sigma)-3\rho_3(\sigma)} f_\sigma(\tau_3\tau_4\tau_3, \tau_1\tau_2\tau_1)(\tau_3\tau_4\tau_3)^{-2\rho_2(\sigma)+3\rho_3(\sigma)}\right)^{-1} \overrightarrow{M_4M_2}$$

(from 3-a-iii),

(4.8.8)
$$\sigma(\overrightarrow{P_5P_1}) = \left(\tau_2^{8\rho_2(\sigma)-3\rho_3(\sigma)} \tau_4^{-6\rho_3(\sigma)} f_\sigma(\tau_1^2\tau_3, \tau_2\tau_4^2) \tau_1^{6\rho_3(\sigma)} \tau_3^{-8\rho_2(\sigma)+3\rho_3(\sigma)}\right)^{-1} \overrightarrow{P_5P_1}$$

(from 3-a-v),

(4.8.9)
$$\sigma(\overrightarrow{P_5P_1}) = \Big((\tau_3^{-1}\tau_2^{-1}\tau_4^{-1}\tau_3^{-1})^{2\rho_2(\sigma)-3\rho_3(\sigma)} f_\sigma(\tau_2\tau_3\tau_2\tau_{51}^{-1}, \tau_3^{-1}\tau_2^{-1}\tau_4^{-1}\tau_3^{-1})$$
$$f_\sigma(\tau_2^{-1}\tau_1^{-1}\tau_3^{-1}\tau_2^{-1}, \tau_2\tau_3\tau_2\tau_{51}^{-1})(\tau_2^{-1}\tau_1^{-1}\tau_3^{-1}\tau_2^{-1})^{-2\rho_2(\sigma)+3\rho_3(\sigma)}\Big)^{-1} \overrightarrow{P_5P_1}$$

(from 3-b-i).

PROOF . Here we describe the proof of (4.8.1) as the example appearing Kummer 1-cocycles coming from scaling factors. One can obtain the formulae for the remaining cases by similar calculation.

Put $C_w = \mathbf{P}^1_w \setminus \{0, \pm 1, \infty\}$ and $C_z = \mathbf{P}^1_z \setminus \{0, 1, \infty\}$. The morphism $\varphi : C_w \to X_5$ is, in terms of (λ, t)-coordinate, $\lambda = 1 - w, t = \frac{1}{2}$. Since $[0, 1 + w, 1, 2, \infty] = [2, 1 - w, 1, 0, \infty]$ by $\xi \mapsto 2 - \xi$, we have an morphism $\bar{\varphi} : C_z \to X_5/\langle(1\ 4)\rangle$ making the following diagram commute:

(4.8.10)
$$\begin{array}{ccc} C_w & \xrightarrow{\ \varphi\ } & X_5 \\ {\scriptstyle\beta}\downarrow & & \downarrow \\ C_z & \xrightarrow{\ \bar{\varphi}\ } & X_5/\langle(1\ 4)\rangle. \end{array}$$

Determine the image of the tangential base points $\overrightarrow{01}, \overrightarrow{10}$ on C_z under $\bar{\varphi}$. As before, we denote the images of M_4, M_5 in $X_5/\langle(14)\rangle$ by the same symbols, and $\bar{\varphi}$ simply by φ. The image of $\overrightarrow{01}$ under φ is "geometrically" equal to M_5. The $G_{\boldsymbol{Q}}$-action on L corresponding to M_5 is determined via $L \simeq L_{M_5} \subset \overline{\boldsymbol{Q}}((s_5))\{\{\mu_5\}\}$, where

$$\mu_5 = \frac{1-\lambda}{1-\lambda t}, \quad s_5 = t - \frac{1}{2}. \tag{4.8.11}$$

The pull-back of μ_5 by φ is calculated as follows:

$$\varphi^*(\mu_5) = \frac{1-(1-w)}{1-(1-w)/2} = \frac{2w}{1+w} = 2z^{1/2}(1-z^{1/2})^{-1}. \tag{4.8.12}$$

Hence $\varphi^* : \overline{\boldsymbol{Q}}((s_5))\{\{\frac{1}{2}\mu_5\}\} \to \overline{\boldsymbol{Q}}\{\{z\}\}$ is $G_{\boldsymbol{Q}}$-equivariant. Denote the tangential base point on X_5 corresponding to $\overline{\boldsymbol{Q}}((s_5))\{\{\frac{1}{2}\mu_5\}\}$ by M_5'. Then we have $\varphi(\overrightarrow{10}) = M_5'$.

Next we determine the image of $\overrightarrow{10}_z$, which is geometrically equal to M_4. Denote the tangential base point on C_w corresponding to $\overline{\boldsymbol{Q}}\{\{w_1\}\}$ $(w_1 := 1-w)$ by $\overrightarrow{10}_w$. Then, since $1-z = 1-w^2 = 2w_1(1-\frac{1}{2}w_1)$, $\beta(\overrightarrow{10}_w) = \frac{1}{2}\overrightarrow{10}_z$. Since the following calculation is done in terms of w, it is more convenient to consider the image of $\frac{1}{2}\overrightarrow{10}_z$. The $G_{\boldsymbol{Q}}$-action on L corresponding to M_4 is determined via $L \simeq L_{M_4} \subset \overline{\boldsymbol{Q}}((s_4))\{\{\mu_4\}\}$, where

$$\mu_4 = \lambda, \quad s_4 = t - \frac{1}{2}. \tag{4.8.13}$$

Since the pull-back of μ_4 by φ is $\varphi^*(\mu_4) = 1-w = w_1$, $\varphi^* : \overline{\boldsymbol{Q}}((s_5))\{\{\mu_5\}\} \to \overline{\boldsymbol{Q}}\{\{\frac{1}{2}(1-z)\}\}$ is $G_{\boldsymbol{Q}}$-equivariant. Hence we have $\varphi(\frac{1}{2}\overrightarrow{10}_z) = M_4$.

Let $\delta \in \pi_1(\bar{X}_5; M_5', M_5) \simeq \mathrm{Isom}_{\overline{\boldsymbol{Q}}(t)}(L_{M_5}, L_{M_5'})$ and $\delta' \in \pi_1(\bar{C}_z; \frac{1}{2}\overrightarrow{01}_z, \overrightarrow{01}_z) \simeq \mathrm{Isom}_{\overline{\boldsymbol{Q}}(t)}(L_{\overrightarrow{01}_z}, L_{\frac{1}{2}\overrightarrow{01}_z})$ be the canonical elements along the real axis. Compiling the above calculation, we obtain $\overrightarrow{M_5M_4} = \delta^{-1}\varphi(p_z\delta'^{-1})$, Since $\mu_5 = 0$ means $x_2 = x_3$, we have $\sigma(\delta) = \delta x_{23}^{-\rho_2(\sigma)}$ for any $\sigma \in G_{\boldsymbol{Q}}$. We also have $\sigma(p_z) = f_\sigma(x,y)^{-1}p_z$ and $\sigma(\delta') = \delta' y^{-\rho_2(\sigma)}$. Hence

$$\sigma(\overrightarrow{M_5M_4}) = \left(\varphi(y^{-\rho_2(\sigma)}f_\sigma(x,y))x_{23}^{-\rho_2(\sigma)}\right)^{-1}\overrightarrow{M_5M_4}. \tag{4.8.14}$$

Let us determine the images of x and y. The image of the path under φ is represented by the braid in Figure 8. (Notice that $\varphi(x) \neq \tau_2$ because $[0, 1-\varepsilon, 1, 2, \infty] \neq [0, 1, 1+\varepsilon, 2, \infty]$ in $X_5/\langle(1\ 4)\rangle$. To obtain the correct answer, we need the braid of the lower half in the figure, which is a trivial path in X_5.) We have $\varphi(x) = \tau_1^{-1}\tau_2^{-1}\tau_3^{-1}\tau_2^{-1}\tau_1^{-1} = \tau_4\tau_{15}\tau_4$. and $\varphi(y) = x_{12}$. Thus the formula (4.8.1) follows. □

From these formulae together with their images under $S(\mathcal{B})$-action, we can obtain a number of new-type equations which $\sigma \in G_{\boldsymbol{Q}}$ satisfies.

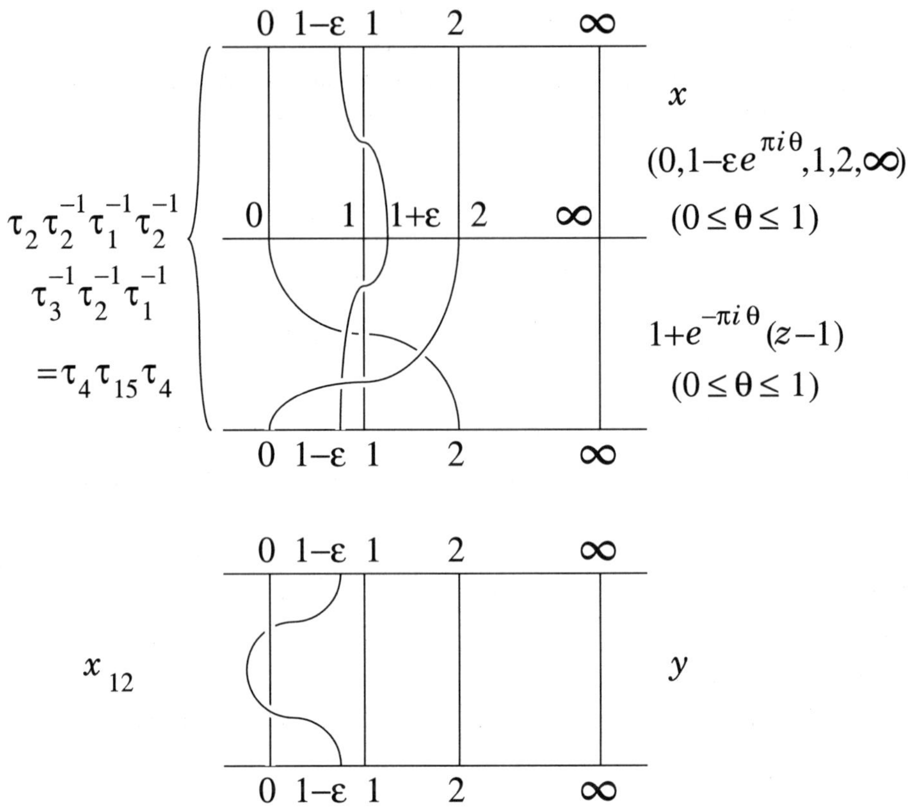

FIGURE 8. The braids corresponging to $\varphi(x)$ and $\varphi(y)$

4.9. By applying (4.8.1) to $\overrightarrow{M_5M_4}\theta(\overrightarrow{M_5M_4}) = 1$, we obtain

(4.9.1) $$x_{12}^{-2\rho_2} f(\tau_4\tau_{51}\tau_4, x_{12}) x_{23}^{-2\rho_2} f(\tau_4\tau_3\tau_4, x_{23}) = 1.$$

Similarly by (4.8.2), (4.8.3) and $\overrightarrow{M_5P_5}\overrightarrow{P_5M_5} = 1$, we have

(4.9.2) $$f(\tau_1\tau_3, x_{23}) x_{12}^{-2\rho_2} x_{34}^{-2\rho_2} f(\tau_2, x_{12}x_{34}) x_{23}^{-2\rho_2} = 1.$$

Since the paths in a equation lie in the same line in these cases,, these equations essentially follows from the equations arising from some double coverings of $\mathbf{P}^1 \setminus \{0, 1, \infty\}$ studied in [**NS, NT1, F**].

4.10. For $\overrightarrow{P_5M_5}$ and $\overrightarrow{P_5P_1}$, we have two formulae in the list coming from different (but homotopic) paths. Comparing them, we obtain the following:

(4.10.1) $$f(\tau_1\tau_3, x_{23}) = (\tau_4\tau_{51}\tau_4)^{6\rho_2-3\rho_3} f(\tau_1^2\tau_3, \tau_4\tau_{51}\tau_4)\tau_1^{-6\rho_2+6\rho_3}\tau_3^{-2\rho_2+3\rho_3},$$

(4.10.2) $$\begin{aligned}&\tau_2^{8\rho_2-3\rho_3}\tau_4^{-6\rho_3} f(\tau_1^2\tau_3, \tau_2\tau_4^2)\tau_1^{6\rho_3}\tau_3^{-8\rho_2+3\rho_3}\\ &= (\tau_3^{-1}\tau_2^{-1}\tau_4^{-1}\tau_3^{-1})^{2\rho_2-3\rho_3} f(\tau_2\tau_3\tau_2\tau_{51}^{-1}, \tau_3^{-1}\tau_2^{-1}\tau_4^{-1}\tau_3^{-1})\\ &\qquad f(\tau_2^{-1}\tau_1^{-1}\tau_3^{-1}\tau_2^{-1}, \tau_2\tau_3\tau_2\tau_{51}^{-1})(\tau_2^{-1}\tau_1^{-1}\tau_3^{-1}\tau_2^{-1})^{-2\rho_2+3\rho_3}.\end{aligned}$$

4.11. For any three points $Q_1, Q_2, Q_3 \in V$, the equality $\overrightarrow{Q_1Q_2}\overrightarrow{Q_2Q_3} = \overrightarrow{Q_1Q_3}$ gives a "triangle" equation. But, if we intend mechaninal computation, it seems more convenient to compare $\overrightarrow{QQ'}$ with the composite of paths along edges of $\mathcal{B}$ for any $Q, Q' \in V$;

(4.11.1)
$$x_{12}^{-\rho_2} f(\tau_4\tau_{51}\tau_4, x_{12}) x_{23}^{-\rho_2} = x_{45}^{-\rho_2} f(\tau_4\tau_{51}\tau_4, x_{45}) f(x_{23}, x_{12}) f(x_{45}, \tau_4\tau_3\tau_4) x_{45}^{\rho_2},$$

(4.11.2)
$$f(x_{45}, \tau_4\tau_{51}\tau_4) x_{45}^{\rho_2} f(x_{12}, x_{23}) f(x_{34}, x_{45}) = f(\tau_1\tau_3, x_{23}) \qquad (= (\mathrm{III}')),$$

(4.11.3)
$$f(x_{12}, \tau_1\tau_2\tau_1) f(x_{34}, x_{45}) = f(\tau_2^{-1}\tau_1^{-1}\tau_3^{-1}\tau_1^{-1}, \tau_1\tau_2\tau_1) x_{12}^{-\rho_2} x_{34}^{\rho_2},$$

(4.11.4)
$$f(\tau_4\tau_3\tau_4, x_{45}) f(x_{12}, \tau_1\tau_2\tau_1) = (\tau_1\tau_2\tau_1)^{4\rho_2-3\rho_3} f(\tau_3\tau_4\tau_3, \tau_1\tau_2\tau_1)(\tau_3\tau_4\tau_3)^{-6\rho_2+3\rho_3},$$

(4.11.5)
$$f(x_{12}, x_{23}) f(x_{34}, x_{45}) = \tau_2^{8\rho_2-3\rho_3} \tau_4^{-6\rho_3} f(\tau_1^2\tau_3, \tau_2\tau_4^2) \tau_1^{6\rho_3} \tau_3^{-8\rho_2+3\rho_3}.$$

4.12. It may be interesting to point out that one can find some equations of "5-cyclic" (or "pentagonal") type, though they are deduced from the above equations;

(4.12.1)
$$x_{12}^{-2\rho_2} f(\tau_4\tau_{51}\tau_4, x_{12}) x_{23}^{-2\rho_2} f(\tau_{51}\tau_1\tau_{51}, x_{23}) x_{34}^{-2\rho_2} \\ f(\tau_1\tau_2\tau_1, x_{34}) x_{45}^{-2\rho_2} f(\tau_2\tau_3\tau_2, x_{45}) x_{51}^{-2\rho_2} f(\tau_3\tau_4\tau_3, x_{51}) = 1,$$

(4.12.2)
$$x_{45}^{2\rho_2-3\rho_3} f(\tau_3\tau_4\tau_3, \tau_1\tau_2\tau_1) x_{12}^{2\rho_2-3\rho_3} f(\tau_{51}\tau_1\tau_{51}, \tau_3\tau_4\tau_3) x_{34}^{2\rho_2-3\rho_3} \\ f(\tau_2\tau_3\tau_2, \tau_{51}\tau_1\tau_{51}) x_{51}^{2\rho_2-3\rho_3} f(\tau_4\tau_{51}\tau_4, \tau_2\tau_3\tau_2) x_{23}^{2\rho_2-3\rho_3} f(\tau_1\tau_2\tau_1, \tau_4\tau_{51}\tau_4) = 1,$$

(4.12.3)
$$x_{45}^{-3\rho_3} f(\tau_1^2\tau_3, \tau_2\tau_4^2) x_{12}^{3\rho_3} x_{51}^{-3\rho_3} f(\tau_2^2\tau_4, \tau_3\tau_{51}^2) x_{23}^{3\rho_3} x_{12}^{-3\rho_3} \\ f(\tau_3^2\tau_{51}, \tau_4\tau_1^2) x_{34}^{3\rho_3} x_{23}^{-3\rho_3} f(\tau_4^2\tau_1, \tau_{51}\tau_2^2) x_{45}^{3\rho_3} x_{34}^{-3\rho_3} f(\tau_{51}^2\tau_2, \tau_1\tau_3^2) x_{51}^{3\rho_3} = 1.$$

5. Another geometric interpretation

5.1. Recently Schneps [**S1, S2**] introduced a new viewpoint into Grothendieck-Teichmüller theory. In addition to various (tangential) morphisms among moduli spaces considered in preceding works — coming from point-forgetting, point-duplicating or degeneration —, she proposed to discuss about morphisms into *special loci*, which are subvarieties of the moduli of curves consisting of the fixed points of an automorphism of finite order acting on the moduli, in other words, the moduli of curves with special automorphisms.

5.2. One of the most typical example of such *special morphisms* is $\mathbf{P}^1 \setminus \{0, 1, \infty\} = \mathcal{M}_{0,4} \to \mathcal{M}_{0,5}/\langle(1\,2)(3\,4)\rangle$ defined via $w \mapsto [0, w, 1, 1+w, \infty]$ considered in the proof of Proposition 3.1. Indeed, each point in the image is fixed by $(1\,4)(2\,3)$ because $[1+w, 1, w, 0, \infty] = [0, w, 1, 1+w, \infty]$ by $\xi \mapsto 1 + w - \xi$. From Lemma 4.5 one knows that this is the unique one-dimensional special locus in $\mathcal{M}_{0,5}$ up to the $\mathfrak{S}_5$-action.

5.3. One may want to understand the other examples in Proposition 4.4 from this point of view, though their images are not special loci in $\mathcal{M}_{0,5}$. In some cases the image of $\mathbf{P}^1 \setminus S \to \mathcal{M}_{0,5}$ is the image of a special locus in $\mathcal{M}_{0,n}$ $(n > 5)$ under a point-forgetting morphism. For example, the image of $\mathbf{P}^1 \setminus \{0, \pm 1, \infty\} \to \mathcal{M}_{0,5}; w \mapsto [0, 1-w, 1, 2, \infty]$, considered in the proof of (4.8.1), is the image of the special locus $\{[0, 1-w, 1, 1+w, 2, \infty]\}$ in $\mathcal{M}_{0,6}$, which is fixed by $(1\ 5)(2\ 4) \in \mathfrak{S}_6$. It might be happy if all the embeddings listed in Proposition 4.4 have natural geometric interpretation in the way like this, but the author does not know whether, for example, (3-a-v), which yields (4.8.8), permits such a natural interpretation.

5.4. In [**S2**], Schneps introduced a subgroup $\widehat{\mathrm{GS}}$ of $\widehat{\mathrm{GT}}$ as a new version of the Grothendieck-Teichmüller group respecting the two special embeddings in the first two level under a certain technical restriction on λ and some Kummer cocycles. Moreover, she showed that many known equations studied in preceding works can be induced from the two defining equation (*) and (**) of $\widehat{\mathrm{GS}}$. Here we give their "complete" forms removed any restriction on λ and Kummer cocycles:

$$(*)\quad \tau_3^{-1}\tau_2^{-1}x_{12}^{-2\rho_2}x_{34}^{-2\rho_2}\tau_2\tau_3 f(\tau_3^{-1}\tau_2\tau_3, \tau_3^{-1}\tau_2^{-1}x_{12}x_{34}\tau_2\tau_3)$$
$$\tau_3^{-1}x_{23}^{\frac{\lambda-1}{2}}\tau_3 f(x_{12}x_{34}, \tau_2)x_{12}^{2\rho_2}x_{34}^{2\rho_2}$$
$$= f(x_{51}, \tau_4 x_{51}\tau_4^{-1})x_{23}^{-\frac{\lambda-1}{2}} f(x_{34}, x_{23})x_{34}^{-\frac{\lambda-1}{2}}x_{51}^{\frac{\lambda-1}{2}} f(x_{12}, x_{51}),$$

$$(**)\quad (\tau_1\tau_3\tau_5)^{-3\rho_3-\frac{\lambda-1}{2}} f(\tau_1^{-1}\tau_3^{-1}\tau_4\tau_3\tau_2\tau_1, x_{12}x_{34}x_{56})(\tau_1^{-1}\tau_3^{-1}\tau_4\tau_3\tau_2\tau_1)^{1-\lambda}$$
$$f(x_{23}x_{45}x_{61}, \tau_1^{-1}\tau_3^{-1}\tau_4\tau_3\tau_2\tau_1)(\tau_2\tau_4\tau_{61})^{3\rho_3-\frac{\lambda-1}{2}}(\tau_5\tau_4\tau_3\tau_2\tau_1)^{\lambda-1}$$
$$= f(x_{12,3}, x_{34})f(x_{23}, x_{12})f(x_{23,4}, x_{1,23})f(x_{234,5}, x_{1,234})f(x_{45}, x_{23,4}).$$

To determine the effects of λ and Kummer cocycles, we need to write down special embeddings in terms of coordinate functions of $\mathcal{M}_{0,5}$ and $\mathcal{M}_{0,6}$ actually. Of course, these equations should be obtained by carring out the calculation in [**S2**] without any restriction on λ and Kummer cocycles. Here we do not intend it and just show that any element $\sigma \in G_{\boldsymbol{Q}}$ satisfies the above equations using already known equations freely if needed. One can easily obtain (*) from a combination of (I), (II), (III′) and (4.9.2). It may be convenient to use a "central extention version" $(\widetilde{\mathrm{II}})$ of (II):

$$(\widetilde{\mathrm{II}})\quad f(z,x)z^{\frac{\lambda-1}{2}} f(y,z)y^{\frac{\lambda-1}{2}} f(x,y)x^{\frac{\lambda-1}{2}} = (xyz)^{\frac{\lambda-1}{2}}$$

if xyz commutes with x, y and z.

To show (**), we use $X_6 = \mathcal{M}_{0,6}$, the moduli space of projective lines with ordered six marked points. By uniformizing an element of X_6 in the form $[\frac{u-1}{u}, 0, t, 1, \frac{1}{1-s}, \infty]$, we identify X_6 with an open subvariety of $\mathrm{Spec}(\boldsymbol{Q}(t,s,u))$. Similarly to the previous sections, we consider an embeddding

$$(5.4.1)\qquad \varphi : \mathbf{P}^1 \setminus \{0, 1, \infty, \rho, \bar{\rho}\} \longrightarrow X_6 \qquad (\rho = \frac{1+\sqrt{-3}}{2})$$
$$w \mapsto [\frac{w-1}{w}, 0, w, 1, \frac{1}{1-w}, \infty]$$

fitting with the commutative diagram

$$\begin{array}{ccc} \mathbf{P}^1 \smallsetminus \{0,1,\infty,\rho,\bar\rho\} & \xrightarrow{\varphi} & X_6 \\ {\scriptstyle\beta}\downarrow & & \downarrow \\ \mathbf{P}^1 \smallsetminus \{0,1,\infty\} & \xrightarrow{\bar\varphi} & X_6/\langle(1\ 3\ 5)\rangle, \end{array} \tag{5.4.2}$$

where $\beta(w) = (\sqrt{-3})^3 \dfrac{w(1-w)}{(1-\rho w)^3}$. The embedding φ is nothing but the special embedding into X_6 considered in [**S2**] Section 4.1, while the scaling factor $(\sqrt{-3})^3$ was not discussed there.

Let P (resp. Q) be the tangential base point on X_6 corresponding to $\boldsymbol{Q}\{\{t,s,u\}\}$ (resp. $\boldsymbol{Q}\{\{1-t,1-s,1-u\}\}$) and $\overrightarrow{PQ}$ the path from P to Q along the real axis. The equation (**) is obtained from the following:

PROPOSITION 5.5.

$$\begin{aligned} \sigma(\overrightarrow{PQ}) = \Big(&(\tau_1\tau_3\tau_5)^{-3\rho_3-\frac{\lambda-1}{2}} f(\tau_1^{-1}\tau_3^{-1}\tau_4\tau_3\tau_2\tau_1, x_{12}x_{34}x_{56}) \\ &(\tau_1^{-1}\tau_3^{-1}\tau_4\tau_3\tau_2\tau_1)^{1-\lambda} f(x_{23}x_{45}x_{61}, \tau_1^{-1}\tau_3^{-1}\tau_4\tau_3\tau_2\tau_1) \\ &(\tau_2\tau_4\tau_{61})^{3\rho_3-\frac{\lambda-1}{2}}(\tau_5\tau_4\tau_3\tau_2\tau_1)^{\lambda-1}\Big)^{-1}\overrightarrow{PQ} \qquad (\sigma\in G_{\boldsymbol{Q}}). \end{aligned} \tag{5.5.1}$$

Since β is not defined over $\boldsymbol{Q}$ but over $\boldsymbol{Q}(\rho)$, we are required to make a careful argument according as $\sigma \in G_{\boldsymbol{Q}(\rho)}$ or not, similarly to that appeared in [**NT1**] Section 5. We omit the detail here.

References

[A] G.Anderson, *The hyperadelic gamma function*, Invent. Math. **95** (1989), 63–131.

[AI] G.Anderson, Y.Ihara, *Pro-l branched coverings of* $\mathbf{P}^1$ *and higher circular l-units*, Part 1, Ann. of Math. **128** (1988), 271–293; Part 2, Intern. J. Math. **1** (1990), 119–148.

[Be] G.V.Belyĭ, *On Galois extensions of a maximal cyclotomic field*, Izv. Akad. Nauk USSR **43** (1979), 267–276.

[De] P.Deligne, *Le groupe fondamental de la droite projective moins trois points*, in "Galois Groups over $\mathbb{Q}$", Publ. MSRI 16 (1989), 79–298.

[Dr] V.G.Drinfel'd, *On quasitriangular quasi-Hopf algebras and a group closely related with* Gal$(\bar{\boldsymbol{Q}}/\boldsymbol{Q})$, Leningrad J. **2** (1991), 829–860.

[F] H.Furusho, *Geometric and arithmetic subgroups of the Grothendieck-Teichmüller group*, Math. Res. Letters **10** (2003), 97–108.

[I1] Y.Ihara, *Braids, Galois Groups, and Some Arithmetic Functions*, Proceedings of the International Congress of Mathematicians, Kyoto 1990.

[I2] Y.Ihara, *On the embedding of* Gal$(\bar{\boldsymbol{Q}}/\boldsymbol{Q})$ *into* $\widehat{GT}$, in "The Grothendieck theory of Dessins d'Enfants", London Math. Soc. Lect. Note Ser. 200, Cambridge University Press, 1994, 173–200.

[I3] Y.Ihara, *On beta and gamma functions associated with the Grothendieck-Teichmüller modular group*, in "Aspects of Galois Theory" (H. Voelklein et.al (eds.)), London Math. Soc. Lect. Note Ser. 256, Cambridge University Press, 1999, 144–179.

[I4] Ihara,Y., *On beta and gamma functions associated with the Grothendieck-Teichmüller modular group II*, J. reine und angew. Math. **527** (2000), 1–11.

[LNS1] P.Lochak, H.Nakamura, L.Schneps, *On a new version of the Grothendieck-Teichmüller group*, C. R. Acad. Sci. Paris **325** Ser. I (1997), 11–16.

[LNS2] P.Lochak, H.Nakamura, L.Schneps, *Eigenloci of* 5 *point configurations on the Riemann sphere and the Grothendieck-Teichmüller group*, Math. J. Okayama Univ. **46** (2004), 39–75.

[LS] P.Lochak, L.Schneps, *A cohomological interpretation of the Grothendieck-Teichmüllar group*, Invent. Math. **127** (1997), 571–600.

[Ma] M.Matsumoto, *Galois group* $G_{\mathbf{Q}}$*, Singularity* E_7*, and Moduli* $\mathcal{M}_3$, in "Geometric Galois Actions" 2. The Inverse Galois Problem, Moduli Spaces and Mapping Class groups, London Math. Soc. Lecture Note Ser. 243, Cambridge University Press, 1997, 179–218.

[N1] H.Nakamura, *Limits of Galois representations in fundamental groups along maximal degeneration of marked curves I*, Amer. J. Math. **121** (1999), 315–358.

[N2] H.Nakamura, *Galois representations in the profinite Teichmüller modular groups*, in "Geometric Galois Actions" 1. Around Grothendieck's Esquisse d'un Programme, London Math. Soc. Lecture Note Ser. 242, Cambridge University Press, 1997, 159–173.

[NS] H.Nakamura, L.Schneps, *On a subgroup of the Grothendieck-Teichmüller group acting on the tower of profinite Teichmüller modular groups*, Invent. Math. **141** (2000), 503–560.

[NT1] H.Nakamura, H.Tsunogai, *Harmonic and equianharmonic equations in the Grothendieck-Teichmüller group*, Forum. Math. **15**(6) (2003), 877–892.

[NT1*] H.Nakamura, H.Tsunogai, *Harmonic equations in the Grothendieck-Teichmüller group*, in "Communications in Arithmetic Fundamental Groups" (H.Nakamura ed.), RIMS proceedings **1267** (2002), 197–208.

[NT2] H.Nakamura, H.Tsunogai, *Harmonic and equianharmonic equations in the Grothendieck-Teichmüller group II*, this volume.

[S1] L.Schneps, *Special loci in moduli spaces of curves*, in "Galois Groups and Fundamental Groups", Publ. MSRI 41 (2003), 217–275.

[S2] L.Schneps, *Automorphism of curves and their role in Grothendieck-Teichmüller theory*, preprint (2004).

[T] H.Tsunogai, *New-type equations which the Galois images satisfy in the Grothendieck-Teichmüller group* (in Japanese), Waseda University ARISE proceedings IX, Number Theory (1998), 49–58.

Department of Mathematics
Sophia University
7-1, Kioi-cho Chiyoda-ku, Tokyo, 102-8554, JAPAN
E-mail address: tsuno@mm.sophia.ac.jp

Titles in This Series

416 **Toshitake Kohno and Masanori Morishita, Editors,** Primes and knots, 2006

415 **Gregory Berkolaiko, Robert Carlson, Stephen A. Fulling, and Peter Kuchment, Editors,** Quantum graphs and their applications, 2006

414 **Deguang Han, Palle E. T. Jorgensen, and David Royal Larson, Editors,** Operator theory, operator algebras, and applications, 2006

413 **Georgia M. Benkart, Jens C. Jantzen, Zongzhu Lin, Daniel K. Nakano, and Brian J. Parshall, Editors,** Representations of algebraic groups, quantum groups and Lie algebras, 2006

412 **Nikolai Chernov, Yulia Karpeshina, Ian W. Knowles, Roger T. Lewis, and Rudi Weikard, Editors,** Recent advances in differential equations and mathematical physics, 2006

411 **J. Marshall Ash and Roger L. Jones, Editors,** Harmonic analysis: Calderón-Zygmund and beyond, 2006

410 **Abba Gumel, Carlos Castillo-Chavez, Ronald E. Mickens, and Dominic P. Clemence, Editors,** Mathematical studies on human disease dynamics: Emerging paradigms and challenges, 2006

409 **Juan Luis Vázquez, Xavier Cabré, and José Antonio Carrillo, Editors,** Recent trends in partial differential equations, 2006

408 **Habib Ammari and Hyeonbae Kang, Editors,** Inverse problems, multi-scale analysis and effective medium theory, 2006

407 **Alejandro Ádem, Jesús González, and Guillermo Pastor, Editors,** Recent developments in algebraic topology, 2006

406 **José A. de la Peña and Raymundo Bautista, Editors,** Trends in representation theory of algebras and related topics, 2006

405 **Andrew Markoe and Eric Todd Quinto, Editors,** Integral geometry and tomography, 2006

404 **Alexander Borichev, Håkan Hedenmalm, and Kehe Zhu, Editors,** Bergman spaces and related topics in complex analysis, 2006

403 **Tyler J. Jarvis, Takashi Kimura, and Arkady Vaintrob, Editors,** Gromov-Witten theory of spin curves and orbifolds, 2006

402 **Zvi Arad, Mariagrazia Bianchi, Wolfgang Herfort, Patrizia Longobardi, Mercede Maj, and Carlo Scoppola, Editors,** Ischia group theory 2004, 2006

401 **Katrin Becker, Melanie Becker, Aaron Bertram, Paul S. Green, and Benjamin McKay, Editors,** Snowbird lectures on string geometry, 2006

400 **Shiferaw Berhanu, Hua Chen, Jorge Hounie, Xiaojun Huang, Sheng-Li Tan, and Stephen S.-T. Yau, Editors,** Recent progress on some problems in several complex variables and partial differential equations, 2006

399 **Dominique Arlettaz and Kathryn Hess, Editors,** An Alpine anthology of homotopy theory, 2006

398 **Jay Jorgenson and Lynne Walling, Editors,** The ubiquitous heat kernel, 2006

397 **José M. Muñoz Porras, Sorin Popescu, and Rubí E. Rodríguez, Editors,** The geometry of Riemann surfaces and Abelian varieties, 2006

396 **Robert L. Devaney and Linda Keen, Editors,** Complex dynamics: Twenty-five years after the appearance of the Mandelbrot set, 2006

395 **Gary R. Jensen and Steven G. Krantz, Editors,** 150 Years of Mathematics at Washington University in St. Louis, 2006

394 **Rostislav Grigorchuk, Michael Mihalik, Mark Sapir, and Zoran Šunić, Editors,** Topological and asymptotic aspects of group theory, 2006

393 **Alec L. Matheson, Michael I. Stessin, and Richard M. Timoney, Editors,** Recent advances in operator-related function theory, 2006

For a complete list of titles in this series, visit the
AMS Bookstore at **www.ams.org/bookstore/**.